Coping with Challenge

Welfare in Animals including Humans

Goal of this Dahlem Workshop:

To promote an interchange of ideas among scientists studying human psychological disorders, neuropsychology, or the behavior, physiology and immunology of other animals, about the attempts of individuals to adapt to or cope with their environment.

Report of the 87th Dahlem Workshop on
Coping with Challenge: Welfare in Animals including Humans
Berlin, November 12–17, 2000

Held and published on behalf of the
President, Freie Universität Berlin: P. Gaehtgens

Sponsored by: Deutsche Forschungsgemeinschaft

Coping with Challenge

Welfare in Animals including Humans

Edited by

D.M. Broom

Program Advisory Committee:
D.M. Broom, Chairperson
C.L. Coe, M.F. Dallman, R. Dantzer, D. Fraser,
K. Gärtner, D.H. Hellhammer, and N. Sachser

dahlem university press

Berlin

Customer Service: dahlem university press
Kaiserswerther Str. 16–18
14195 Berlin, Germany
Tel: +49 (30) 838 55053
Fax: +49 (30) 841 09103
email: dahlemup@zedat.fu-berlin.de

Visit our Home Page at
http://www.dahlem@fu-berlin.de

Die Deutsche Bibliothek - CIP - Einheitsaufnahme
Ein Titeldatensatz für diese Publikation ist bei
Der Deutschen Bibliothek erhältlich

Cover Credits: Figure from "L'Atmosphère: Météorologie Populaire," Camille Flammarion, Paris: Librairie Hachette et C., 1888

Cover design by Rössiger & Swoboda, Berlin, Germany

ISBN 3-934504-09-4

36 Figures, 12 Tables, 4 Photographs

Layout and typeset by Dahlem Konferenzen

Printed and bound by Druckhaus Köthen GmbH, Köthen, Germany

Contents

The Dahlem Konferenzen ix

List of Participants xi

Preface xv

1 **Coping, Stress, and Welfare** 1
D.M. Broom

2 **Is There a Neurobiology of Good Welfare?** 11
C.S. Carter

3 **What Is Important to Achieve Good Welfare in Animals?** 31
N. Sachser

4 **Life, Liberty, and the Pursuit of Happiness: Good Welfare in Humans** 49
S.K. Lutgendorf

5 **How Can We Recognize and Assess Good Welfare?** 63
M.S. Dawkins

6 **Group Report: Good Welfare: Improving Quality of Life** 79
U. Knierim, Rapporteur
C.S. Carter, D. Fraser, K. Gärtner, S.K. Lutgendorf, S. Mineka, J. Panksepp, and N. Sachser

7 **Can We Understand the Brain and Coping without Considering the Immune System?** 101
R. Dantzer

8 **Is There a Major Stress System in the Brain?** 111
R.J. Nelson

9 **Motivation and Coping** **123**
P. Jensen

10 **Is There a Major Stress System at the Periphery Other than the Adrenals?** **135**
S.W. Porges

11 **Group Report: Key Elements of Coping** **151**
J.M.H.M. Reul, Rapporteur
C.L. Coe, R. Dantzer, P. Jensen, S.L. Lightman, S.W. Porges, J. Rushen, V. Stefanski, and A.J. Zanella

12 **Intrinsic and Extrinsic Factors Modulating Physiological Coping Systems during Development** **169**
P.M. Plotsky, M.M. Sanchez, and S. Levine

13 **How and Why Coping Systems Vary among Individuals** **197**
J.M. Koolhaas, S.F. de Boer, B. Buwalda, B.J. van der Vegt, C. Carere, and A.G.G. Groothuis

14 **How Do Animals Cope with Social Problems?** **211**
M. Mendl

15 **Protective and Damaging Effects of Stress Mediators: Lessons Learned from the Immune System and Brain** **229**
B.S. McEwen

16 **Group Report: Key Sources of Variability in Coping** **249**
B. Forkman, Rapporteur
H.J. Blokhuis, D.M. Broom, S. Kaiser, J.M. Koolhaas, S. Levine, M. Mendl, P.M. Plotsky, and M. Schedlowski

17 **How Are Stress and Depression Interrelated?** **271**
M. Irwin

18 **What Are the Neurobiological Consequences of Stress?** **289**
E. Fuchs, M. Kramer†, and G. Flügge

19 **Stress and Sickness Decrease Food Intake and Body Weight: How Does This Happen? When Does This Adaptive Response Progress to Pain and Suffering?** **301**
M.F. Dallman

20 **Social Stress in Wild Mammals in Their Natural Habitat** **317**
D. von Holst

21 **Group Report: Pathological Consequences of Stress** **337**
P. Mormède, Rapporteur
M.F. Dallman, E. Fuchs, J.J. Heijnen, C. Heim, D.H. Hellhammer, M. Irwin, E.S. Paykel, D. von Holst, and S. von Hörsten

Author Index **356**

Subject Index **357**

The Dahlem Konferenzen

In 1974, the Stifterverband für die Deutsche Wissenschaft[1] in cooperation with the Deutsche Forschungsgemeinschaft[2] founded the *Dahlem Konferenzen*. It was created to promote an interdisciplinary exchange of scientific ideas as well as to stimulate cooperation in research among international scientists. Dahlem Konferenzen proved itself to be an invaluable tool for communication in science, and so, to secure its long-term future, it was integrated into the Freie Universität Berlin in January, 1990.

As has been evident over recent years, scientific research has become highly interdisciplinary. Now, before real progress can be made in any one field, the concepts, methods, and strategies of related fields must be understood and able to be applied. Coordinated research efforts, scientific cooperation, and basic communication between the disciplines and the scientists themselves must be promoted in order for science to advance.

To meet these demands, Dahlem Konferenzen created a special type of forum for communication, now internationally recognized as the *Dahlem Workshop Model*. These workshops are the framework in which coherent discussions between the disciplines take place and are focused around a topic of high priority interest to the disciplines concerned. At a Dahlem Workshop, scientists are able to pose questions and solicit alternative opinions on contentious issues from colleagues from related fields. The overall goal of a workshop is not necessarily to reach a consensus but rather to identify gaps in knowledge, to find new ways of approaching controversial issues, and to define priorities for future research. This philosophy is implemented at every stage of a workshop: from the selection of the theme to its breakdown in the discussion groups, from the writing of the background papers to the formulation of the group reports.

Workshop topics are proposed by leading scientists and are approved by a scientific board, which is advised by qualified referees. Once a topic has been approved, a Program Advisory Committee of scientists meets approximately one year before the workshop to delineate the scientific parameters of the meeting, select participants, and assign them their tasks. Participants are invited on the basis of their scientific standing alone.

Each workshop is organized around four key questions, which are addressed by four discussion groups of approximately ten participants. Lectures or formal presentations are taboo at Dahlem. Instead, concentrated discussion — within a group and between groups — is the

---

1 The Donors' Association for the Promotion of Sciences and Humanities, a foundation created in 1921 in Berlin and supported by German trade and industry to fund basic research in the sciences.

2 German Science Foundation.

means by which maximum communication is achieved. To facilitate this discussion, participants prepare the workshop theme prior to the meeting through the "background papers," the themes and authors of which are chosen by the Program Advisory Committee. These papers specifically review a particular aspect of the group's discussion topic as well as function as a springboard to the group discussion, by introducing controversies or unresolved problem areas.

During the workshop week, each group sets its own agenda to cover the discussion topic. Cross-fertilization between groups is both stressed and encouraged. By the end of the week, in a collective effort, each group has prepared a report reflecting the ideas, opinions, and contentious issues of the group as well as identifying directions for future research and problem areas still in need of resolution.

A Dahlem Workshop initiates and facilitates discussion between a certain number — necessarily restricted — of scientists. Because it is imperative that the discussion and communication should continue after a workshop, we present the results to the scientific community at large in the form of this published volume. In it you will find the revised background papers and group reports, as well as an introduction to the workshop theme itself.

The difference between proceedings of many conventional meetings and this workshop report will be easily discernable. Here, the background papers have not only been reviewed by formal referees, they have been revised according to the many comments and suggestions made by *all* participants. In this sense, they are reviewed more thoroughly than scientific articles in most archival journals. In addition, an extensive editorial procedure ensures a coherent volume. I am sure that you, too, will appreciate the tireless efforts of the many reviewers, authors, and editors.

On their behalf, I sincerely hope that the spirit of this workshop as well as the ideas and controversies raised will stimulate you in your work and future endeavors.

Wedigo de Vivanco
Dean of International Affairs and
Director of Dahlem Konferenzen
Freien Universität Berlin
Thielallee 66, 14195 Berlin, Germany

List of Participants with Fields of Research

HARRY J. BLOKHUIS Institute of Animal Science and Health, ID-Lelystad, P.O. Box 65, 8200 AB Lelystad, The Netherlands

Farm animal welfare, behavior, poultry

DONALD M. BROOM Dept. of Clinical Veterinary Medicine, University of Cambridge, Madingley Road, Cambridge CB3 0ES, U.K.

Scientific assessment of animal welfare

C. SUE CARTER Dept. of Psychiatry, 1601 W. Taylor Street, University of Illinois at Chicago, Chicago, IL 60612, U.S.A.

Behavioral endocrinology

CHRISTOPHER L. COE Dept. of Psychology, University of Wisconsin, 1202 West Johnson Street, Madison, WI 53706, U.S.A.

Psychoneuroimmunology, health psychology development, stress

MARY DALLMAN Dept. of Physiology, University of California at San Francisco, San Francisco, CA 94143–0444, U.S.A.

Regulation of autonomic, behavioral, and neurodocrine responses to chronic stressors

ROBERT DANTZER Integrative Neurobiology, INSERM U394, Rue Camille Saint-Säens, 33077 Bordeaux Cedex, France

Psychoneuroimmunology, effects of cytokines on behavior, stress, and welfare

BJØRN FORKMAN Dept. of Animal Science and Animal Health, The Royal Veterinary and Agricultural University, Groennegaardsvej 8, 1870 Frederiksberg C, Denmark

The effects of incentives on motivation; the cognitive abilities of chickens; personality and temperament in dogs

DAVID FRASER Faculty of Agricultural Sciences, and Centre for Applied Ethics, University of British Columbia, 2357 Main Mall, Vancouver V6T 1Z4, Canada

Animal welfare and the ethics of animal use

EBERHARD FUCHS Division of Neurobiology, Deutsches Primatenzentrum, Kellnerweg 4, 37077 Göttingen, Germany

Neurobiology of stress

KLAUS GÄRTNER Medizinische Hochschule Hannover, Zentrales Tierlabor, Postfach 610180, Carl Neuhaus Str. 1, 30623 Hannover, Germany

Animal physiology; comparative pathophysiology; laboratory animal science

J.J. (COBI) HEIJNEN Dept. of Immunology, Children University Hospital, Room KC03.068.0, P.O. Box 85090, 3508 AB Utrecht, The Netherlands

Psychoneuroimmunology: brain immune interactions with a special emphasis on autoimmunity; brain immune interactions in fatigue; perinatal manipulations — long-term influences on immune system development; molecular mechanisms of receptor regulation for neuroendocrine mediators

CHRISTINE HEIM Center for Psychobiological and Psychosomatic Research, Universität Trier, Friedrich-Wilhelm-Str. 23, 54290 Trier, Germany

Psychoneuroendocrinology

DIRK H. HELLHAMMER Center for Psychobiological and Psychosomatic Research, Universität Trier, Friedrich-Wilhelm-Str. 23, 54290 Trier, Germany

Psychobiology of stress

MICHAEL IRWIN Cousins Center for Psychoneuroimmunology, Mail Code 707624, Dept. of Psychiatry and Biobehavioral Sciences, University of California, Los Angeles, 300 UCLA Medical Plaza, Suite 3109, Los Angeles, CA 90095-7057, U.S.A.

Psychoneuroimmunology: health consequences and behavioral mechanisms

PER JENSEN Dept. of Animal Environment and Health, Swedish University of Agricultural Sciences, P.O. Box 234, 53223 Skara, Sweden

Domestication and side effects of selection on behavior; genetic dissection of behavioral strategies for coping in fowl

SYLVIA KAISER Netherlands Institute for Brain Research, Meibergdreef 33, 1105 AZ Amsterdam Z0, The Netherlands

Behavioral biology; prenatal stress; social stress

UTE KNIERIM Institut für Tierhygiene und Tierschutz, Tierärztliche Hochschule Hannover, Bünteweg 17p, 30559 Hannover, Germany

Animal behavior and welfare in poultry, environmental enrichment, animal welfare assessment

JAAP M. KOOLHAAS Dept. of Animal Physiology, University of Groningen, P.O. Box 14, 9750 AA Haren, The Netherlands

Physiological causes and consequences of aggressive behavior; individual differentiation in coping styles

SEYMOUR LEVINE Dept. of Psychiatry, Center for Neuroscience, University of California at Davis, Davis, CA 95616, U.S.A.

Developmental neuroendocrinology, stress

STAFFORD L. LIGHTMAN Bristol Royal Infirmary, University of Bristol, Marlborough Street, Bristol BS2 8HW, U.K.

Neuroendocrinology, stress, plasticity of brain response to patterns of stress

SUSAN K. LUTGENDORF Dept. of Psychology, University of Iowa, E11 Seashore Hall, Iowa City, IA 52242, U.S.A.

Psychoneuroimmunology in older adults and cancer patients; quality of life

MICHAEL MENDL Centre for Behavioural Biology, Dept. of Clinical Veterinary Science, University of Bristol, Langford House, Langford BS40 5DU, U.K.

Spatial and social memory in animals, and their susceptibility to disturbance by stressors; individual differences in behavioral characteristics and coping strategies

SUSAN MINEKA Dept. of Psychology, Northwestern University, 102 Swift Hall, Evanston, IL 60208, U.S.A.

Cognitive and behavioral (conditioning) approaches to understanding the etiology, maintenance, prevention, and treatment of anxiety disorders and depression; animal models of anxiety disorders; comorbidity of anxiety and depression

PIERRE MORMÈDE INSERM-INRA, Institut François Magendie, 33077 Bordeaux Cedex, France

Stress psychobiology; animal welfare; neurogenetics and stress

JAAK PANKSEPP Dept. of Psychology, Bowling Green State University, Bowling Green, OH 43403, U.S.A.

Behavioral neuroscience (neurology of animal emotions and motivations); affective neuroscience (seeking, rage, fear, separation distress, and play)

EUGENE S. PAYKEL Dept. of Psychiatry, University of Cambridge, Box 189, Addenbrooke's Hospital, Cambridge CB2 2QQ, U.K.

Depression: life stress, psychopharmacology, treatment and other aspects

PAUL M. PLOTSKY Stress Neurobiology Laboratory, Dept. of Psychiatry and Behavioral Sciences, Emory University School of Medicine, 1639 Pierce Drive, Suite 4105, Atlanta, GA 30322, U.S.A.

Neurobiology of stress, neuroendocrinology, neurobehavioral analysis, effects of perinatal environment on brain development and gene expression profiles

STEPHEN W. PORGES Dept. of Psychiatry, University of Illinois at Chicago, Chicago, IL 60612–7327, U.S.A.

Human neurobiology and development; neural regulation of the autonomic nervous system; developmental psychophysiology

HANS J.M.H.M. REUL Max Planck Institute of Psychiatry, Kraepelinstrasse 2, 80804 München, Germany

General: How does the brain adapt to stress? Specific research fields: HPA axis, corticosteroid reception regulation and function (RH, transcription factors (e.g., CREB), neurogenesis, neurotransmission (e.g., serotonin), behavior, neuroimmunology

JEFFREY RUSHEN Dairy and Swine Research and Development Centre, Agriculture and Agri-Food Canada, P.O. Box 90, Lennoxville, Quebec J1M 1Z3, Canada

Behavior and stress responses of dairy cattle, farm animal welfare

NORBERT SACHSER Abt. für Verhaltensbiologie, Institut für Neuro- und Verhaltensbiologie, Westfälische Wilhelms-Universität Münster, Badestr. 9, 48149 Münster, Germany

Stress and welfare, animal behavior, sociophysiology, mammalian social systems

MANFRED SCHEDLOWSKI Institute of Medical Psychology, Medical Faculty, Universität Essen, Hufelandstr. 55, 45122 Essen, Germany

Psychoneuroimmunology: stress effects on cellular immune functions; behavioral conditioned immunosuppression. Psychoneuroendocrinology: stress effects on neuro-endocrine functions

VOLKER STEFANSKI Lehrstuhl für Tierphysiologie, Universität Bayreuth, Universitätsstr. 30, 95440 Bayreuth, Germany

Psychoneuroimmunological studies in laboratory rats: psychosocial stress, behavior, endocrine status, immunological consequences and disease

DIETRICH VON HOLST Lehrstuhl für Tierphysiologie, Universität Bayreuth, Universitätsstr. 30, 95440 Bayreuth, Germany

Social behavior of animals, chemical communication, social stress, population biology

STEPHAN VON HÖRSTEN Abt. für Funktionelle und Angewandte Anatomie OE4120, Medizinische Hochschule Hannover, Carl-Neuberg-Str. 1, 30625 Hannover, Germany

Consequences of postnatal stress on behavioral, neuroendrocrine, and immunological adaptation in adult rats; role of neuropeptides in stress with focus on neuropeptide Y

ADROALDO J. ZANELLA Dept. of Animal Science, Michigan State University, 1230 Anthony Hall, East Lansing, MI 48824, U.S.A.

The impact of early environment on behavioral development and brain organization in domestic animals

Preface

Investigations of how easy or difficult it is for the individual to cope with the environment and of how great is the impact of positive or negative aspects of the environment on the individual are investigations of welfare. Welfare includes health as well as feelings and the state of various physiological and behavioral coping mechanisms, hence it varies over a range from very good to very poor and can be assessed scientifically. The welfare of individuals is a central concern for those carrying out quantitative studies in psychiatry and psychology. There are many parallels in the methodology with studies of the welfare of farm, companion, laboratory, and wild animals but too little exchange of expertise.

The opportunity to organize a workshop and produce this book was seen as a unique occasion to promote interchange of ideas among scientists studying welfare. Some are physiologists and immunologists investigating stress and its consequences. Others are concerned with the causation of abnormal behavior and associated brain and body physiology changes in farm animals, laboratory animals, or humans. Yet another group focuses on the causation and treatment of depression and other mental illnesses with multifactorial causes.

One year prior to the workshop the Program Advisory Committee (D.M. Broom, C.L. Coe, M.F. Dallman, R. Dantzer, D. Fraser, K. Gärtner, D.H. Hellhammer, and N. Sachser) carefully selected the participants, the topics and authors for the background papers, and defined the four topic areas which form the bulk of this book.

Good Welfare: Improving the Quality of Life

Much of the evidence used in welfare assessment indicates the extent of the problems of individuals, but it is also important to recognize and assess good welfare, that is, happiness, contentment, control of interactions with the environment, and possibilities to exploit abilities. It is desirable to assess the specific functioning of the brain when welfare is good in humans and other animals; the methods of recognizing when welfare is, or is likely to be good; and the factors which contribute to good welfare in humans and other species.

Key Elements of Coping

One definition of coping is "having control of mental and bodily stability." Whether coping is directed towards situational problems or inner emotions, it involves key bodily processes in the brain and at the periphery. It is important to understand brain, physiological, immunological, and behavioral functions which are involved in the motivational and emotional aspects of coping responses.

Sources of Variability in Coping

Individuals vary in how they cope with challenges. Some of this variation is a consequence of the use of different strategies in apparently similar circumstances. Genetic variation between individuals can affect such strategies as can variation in the environment during development. There are problems associated with coping with the social situation in nonhuman species and variation in the physiological measures which indicate that the coping system is failing.

Pathological Consequences of Stress

Stress has been defined as an environmental effect on an individual which overtaxes its control systems and results in adverse consequences, eventually reduced fitness, although not all scientists use the term in this way. Integrative studies are needed of the consequences of stress and, in particular, depression and the metabolic, cardiovascular, immunosuppressive, disease incidence, and general fitness reduction consequences. Evidence can be obtained from experimental and observational studies of human subjects, captive animals, and wild animals.

We have tried to define the gaps in our body of knowledge, identify the controversial areas and issues, and translate them into research directions and, at times, critical research projects and experiments and hope that the results will be valuable to the scientific community at large. The ideas and data are presented as papers modified in the light of discussion and comment from participants and the results of the group discussions.

The concepts that are central to this book are discussed in Chapter 1, but some editorial points should also be made. The terms "welfare" and "well-being" are both used in the text and have essentially the same meaning. However, in most of the scientific literature "welfare" has been used in a more precise way and is now widely used as a scientific and legal term. Hence it is used in most papers. Some of the hormones referred to in this book include the term "trophic." This word means "pertaining to or promoting growth" whereas "tropic" means "pertaining to or determining direction of movement." Since adrenocorticotrophic hormone is carried in the blood to the adrenal cortex and does lead to activity and growth in cells there, but does not predictably affect any direction of movement, it should not be called "adrenocorticotropic." The term "coping style" has been used in the literature, but style is defined in the Oxford English Dictionary as a "manner of writing, speaking, or doing" or the "collective characteristics of the writing or diction or artistic expression or way of presenting things or decorative methods proper to a person or school or period or subject." Hence, since the word is linked to art in the general sense or to a manner which is considered high quality or upper class, its usage is not appropriate for biological description. Coping strategy is used here, rather than coping style.

Acknowledgments

I was delighted to be asked to serve as the chairperson of this workshop. My task was greatly facilitated by the competence and enthusiasm of Klaus Roth, Dahlem's acting director, and the Dahlem team. On behalf of all participants, I thank them for their excellent support. I also thank the Program Advisory Committee for their guidance and the moderators and rapporteurs, who devoted considerable time and energy to ensuring that the workshop proceeded in a resolute but smooth and pleasant atmosphere. Finally, I thank the *Freie Universität Berlin* and the *Deutsche Forschungsgemeinschaft* for their financial support for this Dahlem Workshop. They ensured that the leading scientists in the field were able to exchange views and create a platform for better liaison in future, as well as producing some original ideas.

— D.M. Broom, Chairperson

1

Coping, Stress, and Welfare

D.M. BROOM

Department of Clinical Veterinary Medicine, University of Cambridge,
Cambridge CB3 0ES, U.K.

Keywords: adaptation, adrenal, awareness, challenge, control, coping, coping strategies, emotion, evolution, feelings, fitness, happiness, health, HPA, stress, suffering, welfare

CHALLENGE, COPING, AND ADAPTING

Every human and every other animal is subject to a variety of environmental challenges, which are more or less demanding upon its response systems. For an individual, the term "environmental" means from outside that individual but for any particular response system, "environmental" means from outside that system. Substantial challenges to animal functioning include those resulting from: pathogens, tissue damage, attack or threat of attack by a conspecific or predator, other social competition, complexity of information processing in a situation where an individual receives excessive stimulation, lack of key stimuli such as a teat for a young mammal or those associated with social contact for a social animal, lack of overall stimulation, and inability to control interactions with the environment. Hence, potentially damaging challenges may come from outside the body (e.g., many pathogens or causes of tissue damage) or from within it (e.g., anxiety, boredom, or frustration which come from the environment of a control system).

I refer to the systems that respond to such challenges or prepare for challenges as coping systems and coping means having control of mental and bodily stability (Broom and Johnson 1993). Coping attempts may be unsuccessful, in that such control is not achieved, but as soon as there is control, the individual is coping. Similar definitions, referring to use of a wide range of strategies, are used by White (1974), Monat and Lazarus (1985), Wechsler (1995), and Koolhaas et al. (1999). For some people, however, the term "coping" has a narrower meaning than this and refers only to dealing with challenges by the use of sophisticated

---

Coping with Challenge: Welfare in Animals including Humans
Edited by D.M. Broom

mental processes. In some cases, this may be because these people are considering a narrow range of challenges (e.g., Bramson 1981). A coping system, process, or strategy is a mechanism that helps the individual to cope and is in operation for some time before coping is achieved. The coping system of an individual may involve the use of one or more of several different strategies, employed to attempt to cope with a particular challenge. The system may operate successfully so that coping is achieved or may be unsuccessful in that the individual is harmed. For example, the dominant male house mouse, which must frequently act so as to control its territory, is using the best coping system that it can in a difficult situation but may live for a shorter time than an individual in a less stressful environment.

It is clearer to refer to coping rather than to adaptation because of the various biological meanings of adaptation. One sense of adaptation is the evolutionary process involving natural selection and leading to the formation of a successful trait. Another sense is the course of an effective behavioral response when learning occurs in some changed situation. The third, largely physiological, sense is the waning of a response after repeated stimulation. When coping systems function, adaptation is occurring but, as defined above, coping implies having control. Adaptation can occur to various degrees which do not go as far as control so, while coping attempts may result in only a certain degree of adaptation, coping is equivalent to successful adaptation.

Systems for attempting to cope with challenge may respond to short- or long-term problems, or sometimes to both. The responses to challenge may involve activity in parts of the brain, various endocrine, immunological, or other physiological responses as well as behavior. However, the more we learn about these responses, the clearer it becomes that these various types of response are interdependent. For example, not only do brain changes regulate bodily coping responses but adrenal changes have several consequences for brain function, lymphocytes have opioid receptors and a potential for altering brain activity, and heart-rate changes can be used to regulate mental state and hence further responses.

Regulation and control within an individual are achieved in two principal ways: negative feedback control and feedforward control (Broom and Johnson 1993). Negative feedback control involves detecting a displacement of body state from within a tolerable range and responding so as to counteract that displacement and return to within the range. An example is to eat when the receptors of blood glucose levels indicate that there is a carbohydrate deficiency within the body and to continue eating until the glucose levels are restored to a higher level. In feedforward control, a future deviation from a tolerable range of body state or from a desirable position is predicted and action is taken to keep within the tolerable range or at the desirable position. An example is when an ad lib fed cow (Metz 1975) predicts an overnight period when feeding will be difficult and eats enough to provide for her needs during that period. This means that meal size is proportional to interval to next feeding, rather than to interval since last feeding as would be the case if the control were by negative feedback. A further example on a different time scale is the migrating bird which eats and stores enough food, during a period of several days, to provide for its energy needs when it migrates over a great distance (Welty and Baptista 1988). The more that is understood about how vertebrate animals of various kinds regulate their lives, the clearer it becomes that feedforward control is widespread and normal in many control systems. A higher level of brain functioning, often involving considerable cognitive ability, is required for the kind of predictions that will allow effective feedforward control than for some reactive control systems.

FEELINGS AND EMOTIONS AS PART OF COPING SYSTEMS

Emotions are physiologically describable conditions in individuals characterized by electrical and neurochemical activity in particular regions of the brain, autonomic nervous system activity, hormone release, and peripheral consequences, including behavior. These emotions may result in feelings (e.g., sexual desire or anxiety), although emotion may exist without any accompanying feeling (e.g., as active regions of the amygdala with no cortical activity) (Guyton 1982). A "feeling" is a brain construct, involving at least perceptual awareness, which is associated with a life-regulating system, is recognizable by the individual when it recurs, and may change behavior or act as a reinforcer in learning. Awareness is taken here to mean "a state in which complex brain analysis is used to process sensory stimuli or constructs based on memory." In many circumstances, feelings involve mental images. For further explanation and examples, see Broom (1998) as well as Sommerville and Broom (1998). The idea that feelings increased fitness and hence evolved by natural selection was pointed out by Darwin (1872) and explained in detail by Cannon (1929). Cabanac (1979) argued, "We experience feelings of hunger because that is part of our mechanism for rectifying a food deficit and ... fear and pain because they are part of our body's way of removing us from situations which are life threatening." Wiepkema (1985) asserted that feelings are involved in monitoring the effectiveness of regulatory actions, being positive when the regulation is successful, and negative when it is not. The evolutionary advantage of having feelings has been emphasized by Dawkins (1977, 1990), Broom and Johnson (1993), and at length with emphasis on the role of feelings as reinforcers in learning by Broom (1998) and Rolls (1999).

Some coping systems include feelings as a part of their functioning, but other high or low level brain processes and other aspects of body functioning are also a part of attempts to cope with challenge. To understand coping systems in humans and other species, it is necessary to study a wide range of mechanisms, including complex brain functioning as well as simpler systems. Both fundamental work on basic mechanisms and the application of such knowledge to human or veterinary medicine and to animal management are worthwhile.

WELFARE

Investigations of how easy or difficult it is for the individual to cope with the environment and of how great is the impact of positive or negative aspects of the environment on the individual are investigations of welfare. Scientists who write about welfare and its assessment are generally agreed that (a) welfare refers to animals, including humans, but not to other organisms or inanimate objects, (b) the study of the welfare of individuals is a scientific discipline in which various measurements can be used as indicators, and (c) welfare varies over a range (Curtis 1986; Duncan 1987; Broom 1988), so welfare can be poor as well as good. There is also wide-ranging agreement that when assessing welfare, efforts should be made to assess degrees of suffering or happiness and the extent of any pathology and its consequences. Some authors accentuate feelings largely or exclusively (Duncan 1993, 1996) when referring to welfare, while others concentrate primarily on health aspects. Dawkins (1993) and Fraser et al. (1997) emphasize that both must be included. My definition of welfare — "the state of the individual as regards its attempts to cope with its environment" — includes feelings and health (Broom 1986, 1996, 1998). Welfare is a characteristic of an individual at a certain time, and the state of the individual can be assessed so welfare will vary on a range from very good to very poor. Welfare concerns how well the individual fares, or goes through life. Equivalent words in other languages include *bien-être*, *bienestar*, *bem estar*, *benessere*, *Wohlergehen*, *welzijn*, *velfærd*, and *dobrostan*.

Health, like welfare, can be qualified as good or poor and varies over a range. It refers to body systems, including those in the brain, which combat pathogens, tissue damage, or physiological disorder. All of this is encompassed within the broader term "welfare," so health is a part of welfare. Hence, it is not correct to say "consider the health and welfare" but rather "consider the welfare including the health." Using this definition of welfare, it is not logical to say "ensuring the welfare" but rather "ensuring the good welfare." Since welfare is a state, it is tautologous to refer to the welfare status or state of an individual.

The assessment of welfare (Broom and Johnson 1993) should be carried out in an objective way, taking no account of any ethical questions about the systems, practices, or conditions for individuals that are being compared. Once the scientific evidence about welfare has been obtained, ethical decisions can be taken.

Much of the evidence used in welfare assessment indicates the extent of the problems of individuals but it is also important to recognize and assess good welfare, i.e., happiness, contentment, control of interactions with the environment, and possibilities to exploit abilities. We should try to assess the specific functioning of the brain when welfare is good in humans and other animals; the methods for recognizing when welfare is, or is likely to be, good; and the factors that contribute to good welfare in humans and other species.

Good welfare, in general, and a positive status in each of the various coping systems should have effects that are a part of a positive reinforcement system, just as poor welfare is associated with various negative reinforcers. Good welfare should have various recognizable effects on individuals. We need to identify these so that the assessment of welfare is as effective at the good end of the range as at the bad end.

Feelings, which have a substantial role in many coping systems, are an important part of welfare. Many methods of assessing welfare give information about feelings. However, while the description of welfare is much concerned with the extent of any suffering or happiness, it does not concern only feelings. Measures of the extent of the effects of disease, or of injury, or of physiological and behavioral responses to adversity, all provide information about welfare whether or not feelings are involved. Evidence of absence of problems for an individual and of ability to carry out all normal activities indicate good welfare, even if there is no direct evidence of pleasure. Particular problems with a definition of welfare referring only to feelings are exemplified by the following examples, where most people would say that the welfare of the individual was poor but a welfare definition based only on feelings would not: an individual with a broken leg but asleep; an addict who has just taken heroin; an individual greatly affected by disease but unaware of it; or an injured individual whose pain system does not function.

Each assessment of welfare will pertain to a single individual and to a particular time range. In the overall assessment of the impact of a condition or treatment on an individual, a very brief period of a certain degree of good or poor welfare is not the same as a prolonged period. However, a simple multiplicative function of maximum degree and duration is often not sufficient. If there is a net extent of poor welfare reached by a certain time (Figure 1.1), the best overall assessment of welfare is the area under the curve thus produced.

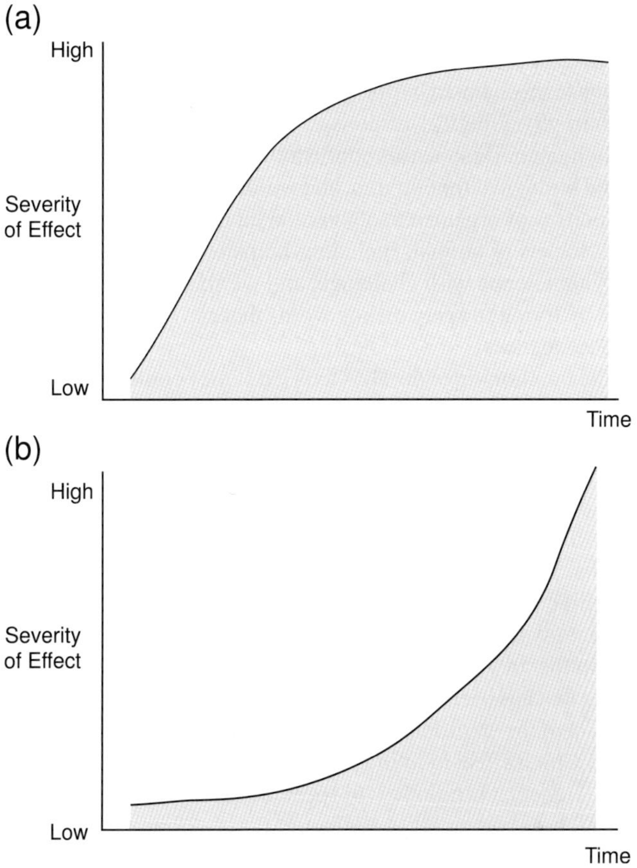

Figure 1.1 Overall effect on welfare up to a certain time is the area under the curve when severity of effect is plotted against time. This is greater in (a) than in (b). Modified after Broom (1999).

The policy that should be adopted in relation to the number of individuals affected is ethical rather than scientific. When many subjects are used in a study of the effects of a condition or treatment on welfare, a larger number of individuals with poor welfare overall indicates a greater problem than a smaller number. Hence, if a million broiler chickens have a problem, this is more important overall than one thousand chickens or one thousand cows or dogs with the same degree of problem. However, most people would consider that any individual whose welfare is very poor merits consideration, so decisions about policy are not just taken on the basis of the function of the severity of the problem and the numbers of individuals concerned.

STRESS

Most people who speak of stress refer to a situation in which an individual is subjected to a potentially or actually damaging effect of its environment. However, the usage of the term by many people, including Selye (1950, 1976), has been confusing. Stress has been referred to as an environmental change which affects an organism, as the process of affecting the organism, or as the consequences of effects on the organism. It has been limited by some to one kind of physiological response mechanism hypothalamic-pituitary-adrenal (HPA) cortex activity or to mental rather than physiological responses; it has also been regarded as a much more wide-ranging phenomenon. Despite this confusion, the concept of stress is very important if we are to understand biological functioning, and we should refine and use the term rather than discard it. The responses of organisms to potential adversity are among the most important functional systems for every animal, including humans. We know a lot about the various mechanisms for trying to cope with challenge, and we should facilitate the further development of the subject by trying to agree on key words that can be used to describe such mechanisms and their consequences.

Knowledge of the functioning of the HPA axis led some people to assume that stress can be equated with activity in this system. It was assumed that HPA activity occurred during all challenges to effective functioning and control within the individual; however, this was demonstrated to be untrue by, for example, Mason (1971). It is also well known that HPA activity is temporarily increased during courtship, mating, active prey catching, and active social interaction, none of which would be considered to be stressful by the majority of the general public or of scientists. To equate stress with HPA axis activity renders the word redundant because we can refer directly to HPA activity in a general or a specific way. Hence, usage of the term "stress" to refer only to HPA activity is considered unscientific and unnecessary by most scientists working in the area.

Another meaning that has been ascribed to stress makes it largely synonymous with stimulation. If every impact of the environment on an organism is called stress, then the term has no value. Many stimuli which affect individuals in beneficial ways would never be called stressors by most people. Hence, it is pointless to speak of stress as the process of response to any environmental event, however small. An effect of the environment on an individual that involves some response and neural development that is beneficial in the long-term is termed stimulation rather than stress. For most people, stress implies the effects of some challenge to the individual that disrupts homeostasis rather than just activating simple, energetically cheap control mechanisms.

A further area of general agreement among scientists studying the attempts of individuals to cope with challenge is that there are many coping systems. Hence it is incorrect to speak of "the stress response" if this means that there is only one. There are many different responses used by individuals in challenging and potentially adverse situations.

If stress implies some degree of adversity for the individual, the key question is: how much adversity? The definition of stress which I find most useful is: "stress is an environmental effect on an individual which overtaxes its control systems and results in adverse consequences, eventually reduced fitness" (modified after Broom 1983; Broom and Johnson 1993). The ultimate measure of fitness is the number of offspring reaching future generations, and there are many different ways in which challenges overtax control systems and have such effects. The environmental variable that has the effect on the individual can be called a stressor. Stress can have a variety of short- and long-term effects, including metabolic changes, cardiovascular malfunction, immunosuppression, increased incidence of parasitic bacterial and viral diseases, as well as psychological disorders, such as interludes of panic, anxiety, or depression.

Using this definition, stress may or may not involve the activation of the HPA axis, but stress is never good for an individual. If the eventual effect is good, it is stimulation or challenge but not stress. Consider, as an example, an individual, such as a human or zebra, that is detecting and responding to a predator, such as an approaching lion. The fitness of the prey individual is reduced if it is (a) killed by the predator, or if it is (b) chased and escapes but is (b1) injured, (b2) rendered less able to obtain food, or (b3) less able to combat disease, in such a way that survival or reproductive ability is impaired. In these cases the prey individual is stressed. However, if the prey individual sees the predator and easily hides or moves away with only slight elevation of heart rate and effort, it is not stressed. Stress is defined by the likely consequence and not by the mechanism of response, for there are very many possible responses.

When the term stress is limited to effects on individuals that are adverse, while natural selection will lead to the spread in the population of genes whose effects are to improve coping ability, genes which minimize the occurrence of stress will also spread. Individuals will not choose an option likely to result in stress unless there is some other functional system involved, i.e., unless any risk is likely to be counterbalanced by a significant benefit, such as increased chances of mating or avoiding starvation. Individuals do, of course, prepare for effective future responses by gaining useful experience. An action or experience which could be difficult and temporarily affect homeostasis might be advantageous overall because of the learning involved. Stimulation, some of which may be initially unpleasant, is necessary for the development of many aspects of systems for coping with challenge. Individuals that are prevented from having adequately varied experience may prove unable to cope with certain problems. Stimulatory effects that are ultimately beneficial are not stress.

Many other biological definitions of stress have been presented (e.g., Selye 1936; Banks 1982; Moberg 1987; Levine and Ursin 1991; Sapolsky 1992; and Chrousos 1998). Some of these refer to any disturbance of homeostasis, but this would mean that almost all functioning of body maintenance systems involves stress. Some refer to any nonspecific response or to any activation of the HPA axis, but again a high proportion of functional responses, some of which are clearly beneficial, would be included as stressful. Many of these definitions make reference to control systems in some way, but it is impossible to be precise if both good and bad consequences are included, and a more precise criterion is needed for the threshold beyond which adversity is sufficient for the effect to be called stress. At present, the likelihood of eventual fitness reduction seems to be the criterion that is most usable.

WELFARE OF HUMANS AND NONHUMANS

It is clear that there are many similarities between humans and nonhuman animals in the various kinds of coping mechanisms. Indeed, the question to pose is, "Are there any differences?" rather than the reverse. Studies of awareness and its various levels (Broom 1998; Sommerville and Broom 1998) make it clear that while some degrees of cognitive processing are widespread in animals, at least in vertebrates, others occur in humans but are rare or lacking in other species. There is a range in such complexity among animal taxa. Additional complex cognitive processing should make it more likely that there will be effective control of interactions with the environment. However, there are also drawbacks associated with having a high degree of understanding. Are these drawbacks as great as the advantages? I should expect that they are not, because the capabilities would probably not have evolved if there were no net benefit to the individuals bearing the genes which promoted them.

Some data collection on the extent of coping difficulties, or on the extent of failure to cope, in humans involves clinical observation or is part of research in which behavioral or physiological abnormalities are quantified. Other data are obtained from questionnaires. In any questionnaire study, there is the possibility that the respondent modifies answers to please the questioner, or to confuse the questioner, or in some other attempt to protect themselves. If there are any of these modifications of answers, the result is partially or wholly invalid, and this necessitates the inclusion of appropriate checking methods in the study as well as care in interpreting the results.

DEVELOPMENTAL AND GENETIC ASPECTS OF COPING SYSTEMS

Individuals vary in how they cope with challenges. Some of this variation is a result of the use of different strategies in apparently similar circumstances. Genetic variation between individuals can affect such strategies, as can variation in the environment during development. There are problems associated with coping with the social situation in nonhuman species and a variety of physiological measures can indicate that the coping system is failing.

When the coping strategies used by individuals and their effectiveness are investigated, it is clear that there are some genetic predispositions and that the functioning of all individuals is affected to some extent by early experience and by genetic factors. There is still much to discover about how each kind of source of individual variation has an effect on vulnerability to psychiatric disorders and on the methods of coping that are used. Interchange of ideas between those studying humans and other species is likely to be particularly valuable here. Indeed, such exchange of ideas about all aspects of the study of coping systems and welfare should be facilitated in future.

REFERENCES

Bramson, R.M. 1981. Coping with Difficult People. New York: Doubleday.

Broom, D.M. 1986. Indicators of poor welfare. *Br. Vet. J.* **142**:524–526.

Broom, D.M. 1988. The scientific assessment of animal welfare. *Appl. Anim. Behav. Sci.* **20**:5–19.

Broom, D.M. 1996. Animal welfare defined in terms of attempts to cope with the environment. Acta *Agric. Scand. Sect. A. Anim. Sci. Suppl.* **27**:22–28.

Broom, D.M. 1998. Welfare, stress and the evolution of feelings. *Adv. Study Behav.* **27**:371–403.

Broom, D.M. 1999. The welfare of vertebrate pests in relation to their management. In: Advances in Vertebrate Pest Management, ed. P.D. Cowern and C.J. Feare, pp. 309–329. Fürth: Filander Verlag.

Broom, D.M., and K.G. Johnson 1993. Stress and Animal Welfare. Dordrecht: Kluwer.

Cabanac, M. 1979. Sensory pleasure. *Q. Rev. Biol.* **54**:1–29.

Cannon, W.B. 1929. Bodily Changes in Pain, Hunger, Fear, and Rage. Boston: Branford.

Chrousos, G. 1998. Stressors, stress and neuroendocrine integration of the adaptive response. *Ann. N.Y. Acad. Sci.* **85**:311–335.

Curtis, S.E. 1986. Perception of thermal comfort by farm animals. In: Farm Animal Housing and Welfare, ed. S.H. Baxter, M.R. Baxter, and J.A.C. MacCormack. Curr. Top. Vet. Med. Anim. Sci. pp. 59–66 The Hague: Martinus Nijhoff.

Dantzer, R., and P. Mormède. 1983. Stress in farm animals: A need for re-evaluation. *J. Anim. Sci.* **57**:6–18.

Darwin, C. 1872. The Expression of the Emotions in Man and Animals. London: Murray.

Dawkins, M. 1977. Do hens suffer in battery cages? Environmental preferences and welfare. *Animal Behav.* **25**:1034–1046.

Dawkins, M. 1990. From an animal's point of view: Motivation, fitness and animal welfare. Behav. *Brain. Sci.* **13**:1–61.

Dawkins, M. 1993. Through Our Eyes Only. Oxford: W.H. Freeman.

Duncan, I.J.H. 1987. The welfare of farm animals: An ethological approach. *Sci. Prog.* **71**:317.

Duncan, I.J.H. 1993. Welfare is to do with what animals feel. *J. Agric. Environ. Ethics.* **6(2)**:8–14.

Duncan, I.J.H. 1996. Animal welfare definition in terms of feelings. *Acta agric. Scand. Sect. A. Anim. Sci. Suppl.* **27**:29–35.

Fraser, D., D.M. Weary, E.A. Pajor, and, B.N. Milligan. 1997. A scientific conception of animal welfare that reflects ethical concerns. *Animal Welfare* **6**:174–186.

Guyton, A.C. 1982. Human Physiology and Mechanisms of Disease, Philadelphia: Saunders.

Koolhaas, J.M., S.M. Korte, S.F. de Boer, B.J. van der Vegt, C.G. van Reenen, H. Hopster, I.C. de Jonge, M.A.W. Ruis, and H.J. Blokhuis. 1999. Coping styles in animals: Current status in behavior and stress-physiology. *Neurosci. Biobehav. Rev.* **23**:925–935.

Levine, S., and Ursin, H. 1991. What is stress? In: Stress: Neurobiology and Neuroendocrinology, ed. M.R. Brown, G. Koob, and C. Rivier, pp. 3–21. New York: Marcel Dekker.

Mason, J.W. 1971. A re-evaluation of the concept of "non-specificity" in stress theory. *J. Psychiat. Res.* **8**:323–33.

Metz, J.H.M. 1975. Time patterns of feeding and rumination in domestic cattle. *Meded. Landbhoogesch. Wageningen* **75**–12, 1–66.

Monat, A., and R.S. Lazarus. 1985. Stress and coping — Some current issues and controversies. In: Stress and Coping, ed. A. Monat and R.S. Lazarus, 1–12. New York: Columbia Univ. Press.

Rolls, E.T. 1999. The Brain and Emotion. Oxford: Oxford Univ. Press.

Sapolsky, R.M. 1992. Stress, the Ageing Brain and the Mechanisms of Neuron Death. Cambridge MA: MIT Press.

Selye, H. 1936. A syndrome produced by diverse nocuous agents. *Nature* **138**:32.

Selye, H. 1950. The Physiology and Pathology of Exposure to Stress. Montreal: Acta.

Selye, H. 1976. The Stress of Life, 2 ed. New York: McGraw-Hill.

Sommerville, B.A., and D.M. Broom. 1998. Olfactory awareness. *Appl. Anim. Behav. Sci.* **57**:269–286.

Wechsler, B. 1995. Coping and coping strategies: A behavioral view. *Appl. Anim. Behav. Sci.* **43**:123–134.

Welty, J.C., and L. Baptista. 1988. The Life of Birds, 4 ed., p. 515. New York: Saunders.

White, R.W. 1974. Strategies of adaptation: an attempt at systematic description. In: Coping and Adaptation, ed. G.V. Coelho, D.A. Hamburg, and J.E. Adams. New York: Basic Books.

Wiepkema, P.R. 1985. Abnormal behavior in farm animals: Ethological implications. *Neth. J. Zool.* **35**:279–289.

2

Is There a Neurobiology of Good Welfare?

C.S. CARTER

Deptartment of Psychiatry, University of Illinois at Chicago,
Chicago, IL 60612, U.S.A.

Keywords: social behavior, evolution, neurobiology, oxytocin, vasopressin, rats, voles

ABSTRACT

Good welfare is more than the absence of stress. It is a physiological reality, embedded in the concept of homeostasis, that active behavioral and other biological systems exist, and that these systems are capable of coping with stress and promoting physical and psychological "welfare."

This review will focus on neurobiological mechanisms for social behavior and social bonds, and mechanisms through which positive social experiences and social bonds can affect neurobiology and behavior and hence influence welfare. Specific neural systems, including those that incorporate the neuropeptides oxytocin, arginine vasopressin (AVP), opioids, catecholamines, and other hormones, coordinate some of the causes and effects of positive social experiences. Through an understanding of these systems, it is possible to begin to describe an integrative neurobiology of "good welfare." This review is also grounded on the premise that physiology and behavior are the products of evolution, which is manifested in the phylogenetic and ontogenetic processes that yield species-typical patterns of social behavior. However, inter- and intraspecific variation, as well as the tendency of physiologists to study rats, has slowed our understanding of this system.

INTRODUCTION

In the context of this Dahlem Workshop, I was asked by the organizers to address the following question: "Is there a neurobiology of good welfare?" As a starting point for a dialogue on the neurobiology of welfare, I propose here that welfare relies on active physiological and psychological states and the experience of those states. In addition, at the heart of welfare in mammals and birds are positive social experiences and social bonds.

Coping with Challenge: Welfare in Animals including Humans
Edited by D.M. Broom

Definitions of welfare arising from different points of view have been provided by other members of the workshop and in the group reports that arose from the conference. Research on welfare has often not involved studies of the nervous system. However, to understand the body's ability to cope with the challenges of daily life or even serious disease, it is also important to identify neural and neuroendocrine processes underlying good health and welfare in general. An emerging literature examines the neurobiology of positive social behaviors and experiences (reviewed by Carter et al. 1997; Panksepp 1998). From that research it is possible to identify neuroendocrine systems that are at the core of good welfare and which also impact on the capacity of animals including humans to cope with stressful experiences or "challenge." Other reward systems can affect good welfare. However, these are described in detail elsewhere (Panksepp 1998) and will not be the focus of this review.

Stress, Homeostasis, and Welfare

Much has been written about the problems associated with the use of the word "stress" (see, for example, Chrousos et al. 1995). This term is used to describe a variety of different states, responses, and physiological patterns. It is now well accepted that the nervous system is at the heart of the management of stress and indeed all challenges to homeostasis. However, it has been convenient to focus on hormones of the hypothalamic-pituitary-adrenal (HPA) axis, corticotrophin-releasing hormone (CRH), adrenocorticotrophic hormone (ACTH), glucocorticoids, and AVP. Thus, stress has been indicated either by levels of hormones or changes in hormones. Stress has become a field of research and the subject of a large body of scientific research. In contrast, much less is known regarding the mechanisms responsible for good welfare and positive behaviors, in part because of the problems listed in Table 2.1 (see also Figure 2.1).

The paucity of research on positive process is paralleled by a poverty of language, forcing scientists to define good welfare in negative terms, such as "systems that counteract the stress responses" or "anti-stress" (Uvnas-Moberg 1998). Concepts such as coping are directed toward positive processes but are of limited usefulness to neurobiologists, in part because they imply cognitive strategies and abilities that are not easily examined in the context of physiology. However, as described below, there is increasing evidence that good welfare, broadly defined, relies on active biological systems that are more than the absence of stress. Failure to attend to positive processes has produced a very one-sided view of stress, homeostasis, and associated physiology.

Processes that regulate welfare, of course, occur within the individual, in the form of behavior and physiology. However, by definition, social behaviors involve two or more individuals, including the tendency of organisms to come together or remain together. Positive social interactions are described by terms such as affiliation, social bonds, parental behavior, coalitions and alliances, symbiosis, mutualism, and altruism (Carter 1998). There are behavioral and physiological systems that support beneficial or health-promoting social interactions, conducted for mutual benefit or social support. It is likely that biological systems responsible for many types of positive interactions are based on shared biological processes. As described here, evidence for the existence of active systems responsible for certain positive components of homeostatic physiology can be derived from knowledge of neural systems containing peptide hormones, including oxytocin, AVP, and opioids, as well as catecholamines.

Table 2.1 Why have the processes responsible for good welfare and wellness been overlooked?

IMPORTANCE OF GOOD WELFARE
- Recognized primarily by absence
- Obscured by emphasis on stress and distress

THE CNS SYSTEMS THAT REGULATE GOOD WELFARE
- Are only now being identified because difficult to find in CNS
- Are ancient processes that may predate modern cognition
- Are regulated by peptides, steroid hormones, and other neurotransmitters

INTRA- AND INTERSPECIFIC VARIATION IS EXPECTED
- Species-typical patterns of behavior and physiological regulation
- Domestication targets this system
- Gender-typical variation, influenced by early hormones
- Individual differences that are modified by experience

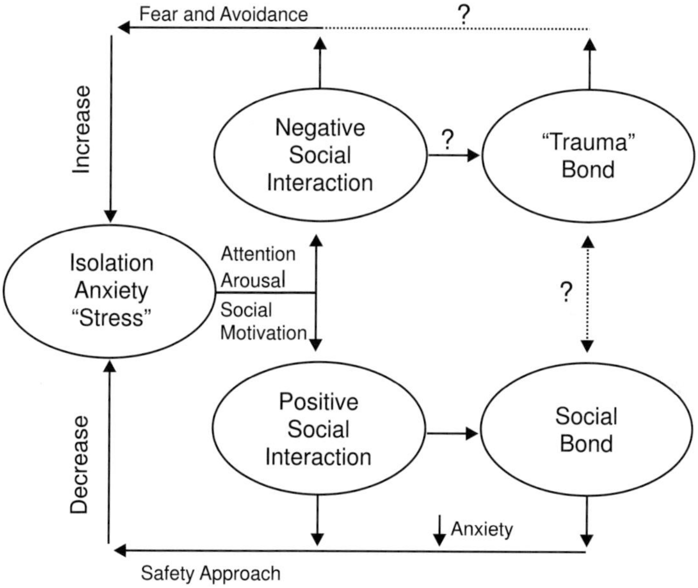

Figure 2.1 Behavioral and emotional correlates of social attachment.

SOURCES OF VARIATION IN SOCIAL BEHAVIOR, SOCIAL ORGANIZATION, AND LIFESTYLE

The behavioral patterns that constitute social behavior, in spite of their prevalence in the repertoire of animals, are often overlooked. In part this is because they are so common, and in part because they seem highly variable. However, patterns can be detected in positive social behaviors and in the neural functions underlying sociality.

Interspecific Variation

Patterns of social behavior, as well as the neural correlates of social behavior, vary across species. The neurobiology of good welfare is most easily understood in the context of evolution, phylogeny, and species-typical patterns of social organization. Interspecific variation initially slowed our understanding of social behavior, and physiologists have tended to focus on a small number of domestic species. However, as data have slowly accumulated from both domestic and nondomestic species, it has become apparent that rather than being a problem, instead interspecific variation provides insight into the mechanisms that regulate social behavior. For example, the neuroendocrine mechanisms regulating positive social behaviors and selective social bonds are particularly apparent in monogamous mammals, such as prairie voles. As described below, the use of molecular procedures for altering the gene-regulating specific peptides or their receptors support the hypothesis that species-typical patterns in behavior are related to oxytocin and AVP (Insel 1997; Young et al. 1998).

Intraspecific and Individual Variation

Another source of information regarding the neurobiology of welfare comes from within-species differences. For example, studies of sex differences (De Vries and Villalba 1997) or reproductive status (Carter 1998; Uvnas-Moberg 1998) are helpful to an understanding of the neurobiology of social behavior. Many of these differences involve the actions of gonadal and adrenal hormones and are beyond the scope of this review. Very recent studies also suggest that variation in AVP receptor binding can be correlated to both within- and between-species variations in social behavior (L. Young, pers. comm.).

Species Variation and Domestication

It is important, even for those whose primary interest is medicine or domestic animals, to study other animal models and to incorporate an evolutionary perspective into models of welfare. We rely here upon the fact that it is possible to extract from comparative data relationships between certain neurobiological processes and behavior.

Traditionally, physiological studies have focused on rats as subjects. Rats are polygynous but also very social, and both wild and domestic rats share with humans their omnivory of diet and ability to breed under a wide variety of environmental and social conditions. Wild rats show similarities with human populations and may share with humans a number of traits, including certain aspects of physiology. Under domestication, rats are constantly being selected for the ability to breed easily, for low levels of aggression, and probably for a comparatively inactive HPA axis. The result is a highly social creature, with behaviors and a variety of neurochemical processes altered significantly by human selection.

Of particular importance to our understanding of the neurobiology of welfare is the likelihood that domestication has had profound effects on endocrine systems. For example, it is likely that the HPA axis of wild rats is more reactive to stressors than that of domestic rats. In addition, rats breed easily and grow rapidly. This may have maximized the effects of reproductive hormones and reduced the effects of HPA axis hormones. In contrast, prairie voles, which have been subjected to less artificial selection (most animals studied are F1–F3 generation), have a very active HPA axis with levels of corticosterone and ACTH, even in the absence of stress, that are five to ten times those typical of domestic rats (DeVries et al. 1996). High levels of HPA axis activity are also found in other species, including highly social New World primates (Mason and Mendoza 1998) and rodents (Sachser et al. 1998). In addition to species differences in the production of hormones, there is also ample evidence of species variation in receptor sensitivity and distributions (reviewed by Taymans et al. 1997). The interpretation of endocrine data requires a broad understanding of physiology as well as an awareness of social systems.

Androgens and Species Differences

In general, the presence or absence of androgens can influence the neurobiology of welfare. Prairie voles have comparatively low levels of sex steroids, and especially androgens. In addition, in comparison with rats, the processes responsible for sexual differentiation in voles may not be as dependent on testosterone. For example, male prairie voles that are castrated on the first day of life show normal levels of male sexual behavior in response to adult androgens. In addition, exposure to extra testosterone during the first week of life does not increase the tendency of female prairie voles to show masculine sexual behavior, although perinatal treatment with corticosterone does facilitate mounting in adulthood (Carter and Roberts 1997). The downplaying of the developmental effects of androgens may have the consequence of increasing certain kinds of positive social behaviors later in life. In prairie voles this may account for some of the characteristics of social monogamy (Table 2.2).

NEUROBIOLOGICAL THEORIES OF SOCIAL BEHAVIOR

Fundamental to good welfare are positive social experiences, and especially social bonds. The physiological systems responsible for positive social experiences are also capable of modulating the body's response to challenge or stressors. Recent research on the neurobiology of mammalian social behavior has focused on oxytocin and AVP, as well as

Table 2.2 Characteristics of social monogamy.

- Pair bonding — long-lasting selective preferences
- Mate guarding — intrasexual aggression
- Biparental care — male helps with young
- Alloparenting — siblings help with young
- Reproductive and growth suppression of most young
- Incest avoidance
- Absence of sexual dimorphism in body size and appearance

Some of these traits may be inhibited directly or indirectly by high levels of androgens, especially during development.

hormones from the HPA axis. These systems are influenced by a variety of other hormones and neurotransmitters, including sex steroids, opioids, acetylcholine, serotonin, nitric oxide, glutamate, and gamma amino butyric acid (GABA). There is evidence, largely beyond the scope of this review, that each of these compounds plays a role in social behavior, both regulating these behaviors and contributing to the individual variation that typifies social behavior. These compounds also play a role in the functions of oxytocin and AVP.

The Physiology of Good Welfare

In mammals, ancient neuropeptide hormonal systems, including oxytocin, are at the core of integrated neural systems which regulate the expression and effects of positive processes (Carter 1998; Uvnas-Moberg 1998). Oxytocin is associated with positive social behaviors, social bonding, maternal behavior, and reduced reactivity of the HPA axis and sympathetic nervous system (Table 2.3; Figure 2.2).

Many components of behavior, such as motor patterns, are regulated, at least in part, by cognitive processes. However, the decision to engage in a given behavior and various mood

Table 2.3 Comparison of oxytocin and vasopressin.

	Oxytocin	Vasopressin
Synthesis (*released in CNS)	SON (PVN*)	SON (PVN*) (mAMY*) (bNST*)
Receptor #	1 known	3 known: V1a, V1b, V2
Production and receptors (species specific) regulated by	Estrogen Progesterone Testosterone	Testosterone Estrogen
Effect on HPA/corticosterone *acute* *chronic*	Increase in rats Decrease in voles and humans Decrease in rats	Increase in rats (no change in voles?) Increase
Fear of novelty	Decrease	Decrease
Social behavior	Increase	Increase with familiar partners (decrease with strangers)
Pair bonding	Increase	Increase — especially males
Parental behavior	Increase	Increase — especially males
Aggression and territoriality	Decrease	Increase — especially males

SON: supraoptic nucleus
PVN: paraventricular nucleus of the hypothalamus
mAMY: medial nucleus of the amygdala
bNST: bed nucleus of the stria terminalis

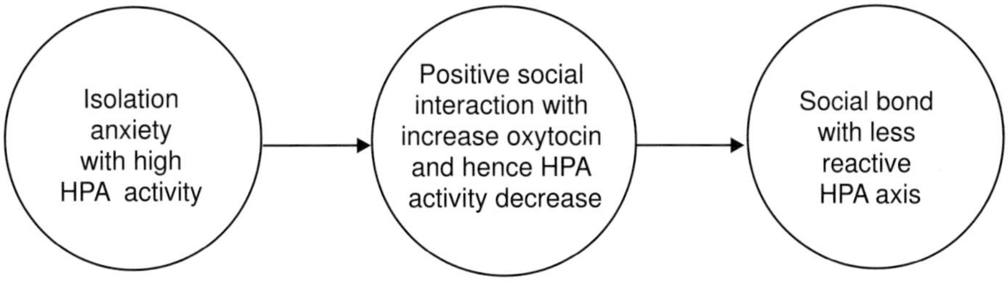

Figure 2.2 Behavioral, emotional, and neuroendocrine correlates of social attachment. HPA: hypothalamic-pituitary-adrenal axis.

states are strongly determined by visceral processes or autonomic states, which may in turn motivate the occurrence of specific behaviors (Porges 1998). Peptides, including oxytocin and vasopressin, regulate these visceral states or emotional feelings. Steroids, opioids, oxytocin, and/or vasopressin may induce a physiological process or "social motivation" that increases the probability of social interactions and the formation of social bonds. In addition, oxytocinergic processes, possibly due to their unique central position in the nervous system, may help to modulate autonomic and behavioral reactivity. Thus, by understanding the neurobiology of peptides, including oxytocin, we may gain a better understanding of the processes through which social bonds promote physical and emotional health.

Many of the actions of oxytocin are influenced by a related neuropeptide, AVP. AVP is often described as a stress hormone, since in rats it is capable of releasing ACTH (Chrousos et al. 1995). AVP's actions are associated with processes that defend the individual from threats of both a physical and emotional nature. An awareness of the novel properties of the neural systems that utilize oxytocin and AVP provides a useful perspective on the physiological basis of positive social experiences and welfare.

Oxytocin and Vasopressin

Both oxytocin and AVP are generally described as mammalian hormones. Related peptides, such as mesotocin, isotocin, and vasotocin, are found in other vertebrates and invertebrates. Oxytocin and AVP are synthesized primarily in the paraventricular nucleus of the hypothalamus (PVN) and supraoptic nucleus (SON). Most of this oxytocin and AVP is manufactured in magnocellular neurons and transported to the posterior pituitary gland, where it is secreted into the systemic circulation. Parvocellular neurons in the PVN also produce oxytocin and AVP, which are released and active within the CNS. In addition, AVP is produced within the medial nucleus of the amygdala (mAMY), and cell bodies containing both oxytocin and AVP are found in the bed nucleus of the stria terminalis (bNST) (De Vries and Villalba 1997). The central and peripheral release of the neuropeptides may be coordinated but can also occur independently.

Androgens play an important role in the production of AVP, particularly in the mAMY and bNST. The androgen-dependency of AVP synthesis may be of functional significance in explaining a number of sex differences in behavior and physiology (De Vries and Villalba 1997). For example, male–female differences in reactivity to environmental stressors may be influenced by sex differences in AVP.

Oxytocin and Arginine Vasopressin Receptors

At present only one receptor for oxytocin has been identified. Oxytocin receptors are found in several areas of the CNS, but in rats are concentrated in the anterior olfactory nucleus (AON), ventral medial hypothalamus (VMH), amygdala, dorsal motor nucleus of the vagus (DMX), locus coeruleus (LC), and raphe nucleus, as well as the dorsal horn of the spinal cord and the intermediolateral column of the spinal cord (Insel 1997).

Three subtypes of AVP receptors have been identified (V1a, V1b and V2) (De Vries and Villalba 1997). V2 receptors are most abundant in the kidneys and are implicated in water retention. V1b receptors are most common in the pituitary gland and found at low levels throughout the CNS, often in conjunction with oxytocin receptors. V1a receptors are found in the olfactory-limbic-hypothalamic system and brain stem, and are associated with blood vessels throughout the body. V1a receptor function has been implicated in cardiovascular function and especially blood pressure (Roozendaal, et al. 1993). In addition, AVP may affect a variety of social and aggressive behaviors (De Vries and Villalba 1997; Insel 1997; Carter 1998).

Centrally active oxytocin is synthesized in the PVN, and oxytocin has only one known receptor type. The central location of the oxytocin system, in an area which controls many other functions including the HPA axis, circadian rhythms and autonomic functions (Roozendaal, et al. 1993; Porges 1998), and the fact that the oxytocin system has a single type of receptor may allow this system to serve a unique integrative function. In contrast, centrally released AVP comes from more than one brain area (PVN, mAMY and bNST) and has multiple receptors; thus AVP may have separable functions and sites of action (De Vries and Villalba 1997).

Oxytocin and Arginine Vasopressin Systems Are Regulated Differently and Serve Different, but Interactive Functions

Oxytocin and AVP receptors seem to be unusually variable, at least when compared with receptors for other peptides such as the opioids (Insel 1997). The distributions of oxytocin and AVP receptors can differ across species and are also, in some cases, influenced by gender, reproductive status, and age. Receptors for both oxytocin and AVP are steroid sensitive, although these effects also vary across species. For example, in rats, oxytocin receptors throughout the CNS, and especially in the VMH, are upregulated by estrogen and progesterone; in contrast, in prairie voles, receptors in the olfactory system (AON), but not hypothalamus, were upregulated by estrogen (Witt et al. 1991). This finding probably has functional significance because rats, but not voles, use progesterone to trigger female sexual behavior.

In addition, peptide experiences, such as exposure to exogenous AVP, during development can have permanent effects on behavior (Stribley and Carter 1999) and the distribution of both AVP and oxytocin receptors, at least as measured by gene expression (Ostrowski 1998).

Hormones of the HPA Axis

Hormones of the adrenal gland, including the glucocorticoids and specifically cortisol or corticosterone, change in response to various kinds of social experiences (Mason and Mendoza 1998; Sachser et al. 1998). Corticosterone is abundant and easily measured, even in peripheral fluids such as saliva. The adrenal cortex is regulated primarily by ACTH from the anterior pituitary, which is under the control of CRH and AVP, synthesized in the hypothalamus. In addition, catecholamines, including dopamine (DA), norepinephrine (NE), and epinephrine (EPI), are produced in both the CNS and in the adrenal medulla.

Living conditions (isolation versus social housing), hierarchical reorganization, and other kinds of stressful experiences can alter the production of these hormones (Sachser et al. 1998). Glucocorticoids and CRH receptors in the CNS have been implicated in social behavior (DeVries et al. 1996). Corticosterone can both respond to and regulate physiology and behavior. However, it is more common to measure these hormones as a function of behavioral experiences than it is to study the causal effects of hormones of this axis. As explained below, modulatory roles have been identified for both oxytocin and AVP in the HPA axis. In addition, hormones of the HPA axis, and especially glucocorticoids, can affect every tissue of the body and have profound effects on the synthesis and actions of neuropeptides, including oxytocin and AVP (Liberzon and Young 1997).

Interactions between the HPA axis and neuropeptides, including oxytocin and AVP, form the basis for a neuroendocrine system capable of adjusting social behaviors to environmental and social demands, as well as the homeostatic needs of the individual organism.

Species Differences in Oxytocin

Of particular relevance to the initial interpretation of the effects of oxytocin on the HPA axis and wellness, is the fact that the response of rats to oxytocin or conditions that release oxytocin varies from that of humans and prairie voles. Both humans and prairie voles respond to oxytocin, or stimuli that release oxytocin, with reductions in corticosterone (reviewed in Carter 1998; Carter and Altemus 1997). In contrast, rats respond to oxytocin with an initial increase in corticosterone and blood pressure, although chronic exposure to oxytocin, even over a matter of hours, decreases corticosterone and blood pressure (Petersson et al. 1999). Rats also show a biphasic response to stress; acute stress releases oxytocin and chronic stress decreases oxytocin. In contrast, in humans, even acute stress can inhibit the release of oxytocin, at least as indexed by milk ejection (Carter and Altemus 1997).

Inter- and Intraspecific Variation in the Hypothalamic-Pituitary-Adrenal Axis

Between and within species, differences in the effects of social experiences on the HPA axis are commonly observed (Sachser et al. 1998; Mason and Mendoza 1998; Carter 1998). Such differences can only be understood when considered in light of phylogeny, ontogeny, and especially social organization. For example, both the types of stimuli, such as novelty or separation, which regulate distress calls and the release of HPA axis hormones, are highly species specific. Infant prairie voles, but not montane voles, show an immediate increase in corticosterone when separated from their mother (Insel 1997). Even the direction of an individual animal's reaction to novelty can vary as a function of its own history; for example, socially naive female prairie voles show a decrease in corticosterone levels in the presence of a stranger, while pair-bonded females show an increase in corticosterone under identical conditions (DeVries et al. 1996).

SOCIAL MONOGAMY AS A MODEL FOR THE ANALYSIS OF THE NEUROBIOLOGY OF WELFARE

Much physiological research has been done in rats. The selection of these polygynous mammals, bred to breed but not to bite, has colored the understanding of the physiology of social behavior. Thus, physiological studies in domestic rats, especially those aimed at understanding social behavior, may yield different conclusions than comparable studies in animals that have not undergone domestication. Other animals that have been selected for domestication must also share the rat's willingness to mate and live in the presence of humans, but may have been bred for behavioral traits that are different from those of rats. For example, domestic dogs were derived from a socially monogamous wolf ancestor. The capacity of dogs to bond with and live in the presence of humans probably takes advantage of this retained feature of the ancestral social system. The willingness of both canids and humans to form pair bonds creates a symbiosis that promotes good welfare in both species.

Differences in social behavior are the rule, not the exception. Even within a species, animals that are adapted to different habitats or that have experienced domestication may exhibit different patterns of social behavior. An attempt to develop a more general understanding of the concept of welfare requires an awareness of social organization and inter- and intraspecies variation, as well as an appreciation of those processes that are especially sensitive to both natural selection and domestication.

Mating Systems

One common method for categorizing species is based on mating systems, characterized by the presumed choice or number of sexual partners, and often focused on male behavior. The most common mating system in mammals is polygamy (many mates), specifically polygyny (many wives). Monogamy (one mate) is estimated to occur in about 3% of mammalian species (Kleiman 1977; Table 2.2), while polyandry is even less common. In birds, monogamy is widespread.

In mammals, female reproduction is constrained by gestation, birth, and lactation. In contrast, males have the option to invest less than females in their offspring, with the minimal investment being a contribution of sperm and genetic material. Mating systems have sometimes been described in terms of male fitness or reproductive success. However, defining mating systems in terms of the male tends to ignore female fitness, female–female interactions, and the importance of the larger family unit. In addition, within the last two decades DNA fingerprinting has revealed that at least some of the offspring raised by seemingly monogamous partners of most species are in fact due to extra-pair copulations. These create a problem for theories that focus on monogamy as a mating system and have encouraged scientists to shift their attention to social systems.

Social Systems

A vast array of social systems have been described, many of which are based directly on mating systems (such as polygamy or monogamy), while others may be defined by the living arrangements or genetic relationship among the members of a group. In comparison with mating systems, definitions of social systems are more likely to take into account the role of females and offspring. For example, many mammalian social systems can be described as matrilinear, and a variety of mammalian species live in groups in which young animals may remain in the natal family. The tendency to form extended families, often (but not always) based around a monogamous pair, has been described as communal or cooperative breeding (Carter and Roberts 1997).

Sexual versus Social Monogamy

Most of what has been written regarding monogamy focuses on sexual monogamy. The recent awareness that, in many species, sexual exclusivity is rare has created a challenge for the concept of sexual monogamy. Nonetheless, as described below, the traits of social monogamy are real and can play an important role in daily life and the welfare of both individuals and groups.

It is also interesting to note that the behavioral and physiological traits of social monogamy are very similar in diverse taxa, including for example, prairie voles, canids, mongooses, and several New World primates. Thus, it is likely that social monogamy has resulted from a number of independent evolutionary events. For this reason, the consistency of the features of social monogamy and its hormonal control is all the more remarkable. It is likely that the full range of traits of social monogamy cannot emerge in the presence of high levels of androgens (Carter and Roberts 1997; Table 2.2). By taking advantage of the developmental timing of exposure to androgens it may be possible to allow certain masculine features, such as sperm production or male sexual behavior, to be expressed, while other effects of androgens on the CNS and behavior are down-regulated. In addition, some species may rely more directly on neuropeptides, including AVP, to determine masculine patterns of physiology and behavior (e.g., territorial aggression) (Stribley and Carter 1999).

Peptides and Social Bonding

Monogamous rodents, including voles, can be studied both in the laboratory and in nature (Carter and Roberts 1997; Carter 1998), allowing us to examine the physiological mechanisms underlying social behaviors and in particular social bonds in the context of natural history and social organization. These studies reveal that the traits of mammalian monogamy typically occur together, forming what has been called a "syndrome" (Kleiman 1977; Table 2.2). The cardinal characteristic of monogamy is a male–female pair bond, defined by selective partner preferences and social interactions. Monogamous mammals, including males of these species, are social within their families, although they can show lethal aggression toward strangers. Pair bonding, mate guarding, and the other characteristics of monogamy are probably encouraged by down-regulating the effects of androgens. Examples of the role of oxytocin and AVP in these behaviors are described below.

High Sociality

Most socially monogamous mammals show high levels of social contact. Treatment by injection into the cerebrospinal fluid (ICV) with either oxytocin or AVP approximately doubles social contact in prairie voles (Cho et al. 1999). Based on the behavioral effects of AVP and species differences in the AVP receptor, Young and his associates (1998) created transgenic mice carrying the gene for the AVP receptor. These transgenic animals expressed the AVP receptor in a pattern that was different from the wildtype mouse and similar in some brain regions to prairie voles. The AVP receptor transgenic animals also showed high levels of social contact resembling that seen in normal prairie voles. However, this social behavior was not selective and thus could not be interpreted as pair bonding.

Additional information regarding the endocrine basis of social behavior comes from studies of "knock-out" mice, created with selective genetic lesions. Deletions of the gene that encodes the oxytocin peptide in mice produces an animal capable of all aspects of reproduction except milk ejection (Nishimori et al. 1996; Russell and Leng 1998). However, oxytocin knock-out mice do not show normal social behavior and tend to be more aggressive than normal mice (Winslow et al. 2000).

Pair Bonds

Placing a reproductively naïve prairie vole with a stranger of the opposite sex begins the process that will lead to a selective partner preference. In rats and mice the presence of a novel conspecific leads to an increase in the activity of the HPA axis and elevated corticosterone concentrations. This is not the case in prairie voles of either sex (DeVries et al. 1996). Corticosterone concentrations in unstressed reproductively naïve prairie voles are very high (600–800 ng/ml), and upon pairing levels fall, within less than one hour, to approximately half of their basal levels. Treatment with oxytocin, even in the absence of a partner, has the same effect. When prairie voles are pretreated with an oxytocin antagonist, corticosterone levels no longer decline upon pairing. Social contact and selective social behaviors also increase in prairie voles of both sexes treated with oxytocin, while pretreatment with an oxytocin antagonist interferes with partner preference formation following either oxytocin treatments or prolonged cohabitation. Thus, under these conditions, both the development of partner preferences and the social regulation of the HPA axis are affected by oxytocin.

The development of partner preferences can occur after nonsexual cohabitation but is hastened by mating. Mating or vaginal cervical stimulation is known to release oxytocin, and it was this fact that originally suggested the notion that oxytocin might play a role in pair bonding (Carter 1998).

Exogenous AVP also facilitates the formation of new partner preferences in both male and female prairie voles. In addition, partner preferences in response to AVP are not formed when animals are pretreated with an AVP antagonist. However, the endogenous production of AVP is androgen dependent, especially in the mAMY and bNST. Testosterone treatments increase AVP synthesis and mating experience can increase AVP levels in males (De Vries and Villalba 1997). Endogenous AVP probably plays a major role in pair bonding in male prairie voles and may encourage sex differences in social bonding. When oxytocin or AVP are given ICV, either can stimulate social contact in prairie voles; however, access to both the oxytocin and AVP receptors is necessary for either peptide to induce pair bonding. Injection of either oxytocin or AVP antagonists can block the formation of pair bonds (Cho et al. 1999).

One of the earliest theories of social attachment was based on the notion that the physiological and emotional responses to separation were similar to the responses to opiate withdrawal (Panksepp et al. 1994). This theory was based on the premise that shared neural mechanisms are involved in the experience of pain and the formation of place preferences. Support for a role for endogenous opioids in maternal bonding comes from research in sheep (Keverne and Kendrick 1991). However, in adult rats, attempts by Panksepp and his colleagues (reviewed in Nelson and Panksepp 1998) to block the rewarding effects of social behavior with opioid receptor antagonists have not supported this hypothesis. Although, opioids may not be essential for social reward, at least in adult rats, these peptides are known to play a role in the regulation of the endogenous release of oxytocin (Russell and Leng 1998), and thus may have an indirect regulatory role in behavioral and endocrine processes that are influenced by oxytocin.

Sex Differences in the Effects of Stress

Stressful experiences (3 minutes of swimming) or corticosterone treatments immediately prior to pairing can facilitate the onset of heterosexual partner preferences in male prairie voles (DeVries et al. 1996). In addition, pretreatment with exogenous corticosterone facilitates the effects of low doses of AVP (DeVries and Carter, unpublished). However, exposure to a stressor or corticosterone inhibits the development of a preference for a new male partner in females (DeVries et al. 1996). In addition, pretreatment with corticosterone inhibits the effects of exogenous oxytocin on pair bonding. It is interesting that a stressful experience does facilitate the development of preferences for other females (DeVries and Carter, unpublished data). Males who remain within the natal family have no opportunity to mate and must leave the family to reproduce. Therefore, it is critical for males of this species to be able to form new bonds, even under stressful conditions. For female prairie voles, reproduction can still occur within the security of the family, although mechanisms are in place to prevent inbreeding. In female prairie voles, stress could discourage the formation of a pair bonding and encourage females to return to the security of other females. Sex differences in the response to stress are probably not limited to prairie voles and must be taken into account in assessments of welfare.

It is interesting to note that new partner preferences can develop in prairie voles during times when gonadal hormones are naturally low or even following gonadectomy. However, developmental exposure to steroid hormones can influence the expression of social bonds in this species. For example, animals that are exposed to androgens during development are especially likely to seek social contact with strangers, rather than siblings (Carter and Roberts 1997).

Separation Distress and Social Buffering

Social separation in both male and female prairie voles is followed by an increase in corticosterone concentrations (Williams and Carter, unpublished data). When reunited with a partner, corticosterone concentrations dropped to below baseline (about 50%) in previously paired males and female. However, if previously paired animals and separated animals were placed with an unfamiliar animal of the opposite sex, corticosterone concentrations increased (DeVries et al. 1996). Separation from the mother also induces increased corticosterone concentrations in infant prairie voles, but not nonmonogamous montane voles (Insel 1997). Such findings suggest that adult separation responses may be related to those seen in infants.

Experiments on separation distress have tested the capacity of peptides, including opioids, to prevent behavioral changes during separation (reviewed Panksepp 1998; Nelson and Panksepp 1998). The intense calling behavior of isolated domestic chicks declines following various treatments, including injections of oxytocin. In rat pups, separation cries were inhibited by central treatments with oxytocin (Insel 1997).

A Working Model for the Role of Peptides and Adrenal Steroids in Social Bonding and Separation

Stressful experiences (such as pregnancy and parturition), anxiety, neophobia, and isolation often precede the formation of social attachments (Figure 2.1). These circumstances may increase social motivation and subsequent social interactions. Positive social interactions, in turn, could be rewarding, and in species or individuals that possess the capacity to form attachments, positive social bonds would follow. Both positive social interactions and social bonds could function to provide a sense of safety and reduce anxiety or the negative feelings associated with stressful experiences. The stress-reducing effects of social interactions may be especially important in females. In fact, males may seek stressful experiences and, at least in the case of voles, may respond to such experiences in a manner that is different from that of females (DeVries et al. 1996).

As reviewed here, oxytocin has been implicated in positive social behaviors (Table 2.3; Figure 2.2). In addition, oxytocin is capable of regulating the HPA axis (DeVries et al. 1996; Uvnas-Moberg 1998). Reductions or fluctuations in oxytocin activity might also account for some of the symptoms of social separation. Oxytocin is well positioned to influence both the behavioral (Carter 1998) and autonomic (Porges 1998) symptoms that follow the loss of an attachment object.

Natural conditions associated with high concentrations of oxytocin may facilitate both social contact and selective social interactions associated with social attachment, pair bonding, and reduced HPA reactivity (Table 2.2; Figure 2.2). AVP, in contrast, is associated with behaviors that might be broadly classified as "defensive," including enhanced arousal, attention or vigilance, increased aggressive behavior, and a general increase in HPA reactivity. AVP synthesis is androgen dependent; AVP can facilitate alertness and territorial aggression, and even pair bonding, yet does not down-regulate the HPA axis. Differential effects of oxytocin and AVP may contribute to male–female differences in the vulnerability to stress-related illnesses, including those that are influenced by the social environment and that may affect females and males differently.

Neural Substrates of Social Behavior and Responsivity to Challenge

There are many levels at which steroids and peptides can interact. A description of the neuroanatomical substrates of social behavior is beyond the scope of this review. Possible mechanisms through which peptides, steroids, and other neurochemicals may influence social behavior and/or the response to stress are listed in Table 2.4 and detailed elsewhere (Carter 1998; Carter and Keverne 2002). In general, those areas of the nervous system that have been implicated in social behaviors, both positive and negative, are also areas containing receptors

Table 2.4 Mechanisms through which steroids and peptides may influence social behavior and the response to stress.

DEVELOPMENTAL REGULATION OF:
- Species-typical traits
- Sexual dimorphism in nervous system
- Alterations in peptide sensitivity or reactivity by altering hormones or neurotransmitters or their receptors

IN ADULTHOOD – EFFECTS ON:
- Peptide synthesis or release
- Peptide receptors or binding
- Altering behaviors that release peptides
- Peptide–peptide interactions
- Steroid–steroid interactions
- Steroid–peptide–neurotransmitter interactions

PRIMARY NEURAL TARGETS (not mutually exclusive):
- Sensory processes (including especially olfactory and somatosensory systems)
- Integrative areas of the limbic system and related brain areas (including olfactory system, amygdala–hippocampus, nucleus accumbens, prelimbic cortex, and selected areas of the hypothalamus including mPOA, VMH, PVN)
- The HPA axis (PVN) and autonomic nervous system (DMX)
- Motor systems, including spinal cord

mPOA: medial preoptic area VMH: ventromedial hypothalamus
PVN: paraventricular nucleus DMX: dorsal motor nucleus of the vagus, i.e., 10th cranial nerve

for oxytocin, AVP, and/or opioids. In rodents and sheep there is evidence that intact olfactory and vomeronasal systems and their connections to more central neural structures are important for social behavior and especially pair bonding. In addition, lesions of the extended amygdala, including the corticomedial and central nucleus, interfere with social behavior. Oxytocin, AVP, and opioids may also act within the amygdala to regulate autonomic functions, which allows these peptides to influence both behavior and emotional states.

One possible mechanism through which oxytocin affects pair bonding might involve social learning and reward. Studies of the cellular basis of pair bonding in prairie voles, but not in nonmonogamous voles, have revealed high densities of oxytocin receptors in the nucleus accumbens (NAcc) and prelimbic cortex (PLC) (Insel 1997). These areas also contain dopamine receptors and have been implicated in reward and drug abuse. Pair bond formation was blocked when oxytocin antagonist was injected directly into the NAcc or PLC, but not other brain regions (L. Young, pers. comm.). In addition, mating both facilitates pair bonding and releases dopamine in the NAcc. It is likely that neuropeptides and catecholamines have a major role in the rewarding consequences of social interactions. Species differences in these systems, including differences in the neural interactions between compounds such as oxytocin and dopamine, may help to explain species differences in the types of experiences that contribute to good welfare.

PEPTIDES AND HUMAN BEHAVIOR

Very little is known regarding the effects in humans of centrally administered oxytocin. Peptide hormones, including oxytocin and AVP, do not readily cross the blood–brain barrier and are usually administered centrally (ICV) to reach the brain. Nasal sprays have been used to promote milk letdown and have been used in a few behavioral studies; however, the extent to which such compounds reach the brain is not known. The data that do exist are primarily correlational and must be interpreted accordingly.

Naturally Occurring Changes in Peptides

Two aspects of mammalian life, birth and breast-feeding, are clearly associated with the release of oxytocin (Russell and Leng 1998). The naturally occurring changes in peptides associated with birth and lactation offer opportunities to correlate behavioral and hormonal events. Among the neuroendocrine adaptations that accompany both birth and subsequent lactation are hormonal changes that may promote selective social interactions, including high levels of physical contact between parent and child and maternal–infant bonding.

Birth and Breast-feeding

Birth is a hormonally complex event that is difficult to study. In addition, very strong social attachments between adults and children can occur in the absence of birth. Although, the experiences associated with birth may play a role in maternal bonding in humans, studies of the physiology of mother–infant interactions are at present primarily correlational (Uvnas-Moberg 1998).

Lactation is the defining property of mammalian existence and until modern times was essential to mammalian reproduction. Oxytocin plays an essential role in milk letdown (Russell and Leng 1998). The study of human lactation, in conjunction with animal research, provides an opportunity to develop hypotheses regarding the behavioral effects of oxytocin. However, it is important to realize that other hormones, including prolactin, the endogenous opioids, and various steroids, regulate lactation.

It has been found that lactating women interact more positively with their babies, directing more touching and smiling toward their infants than do bottle-feeding mothers. It also is reported that nursing, versus bottle-feeding, mothers are more likely to describe positive mood states and be less reactive to stressful experiences (reviewed in Carter and Altemus 1997; Uvnas-Moberg 1998).

Lactation Inhibits Reactivity to Stressful Experiences

In general, neuroendocrine reactivity of the adrenal axis is reduced during lactation. Successful lactation involves the nervous system and neuroendocrine adaptations of the HPA and gonadal axes. Oxytocin in particular plays a pivotal role in the integration of the behavioral and physiological processes unique to female physiology. Based primarily on research with nonhuman animals, one can speculate that oxytocin and AVP, and interactions between these hormones and steroid hormones might regulate dynamic behavioral states, including the capacity of an individual to respond to both social and physical challenges. It has been known since the early 1970s that lactating female rodents showed reduced adrenal reactivity, often indexed by reduced corticosterone secretion, following exposure to stressors such as ether, surgical trauma, and electric shock (Thoman et al. 1970). In rats, injections of hypertonic saline normally are considered stressful and are expected to release corticosterone. However, during lactation there is a selective inhibition of normal hypothalamic stress responses. In women, both ACTH and cortisol concentrations fall during a bout of breast-feeding. There is also evidence that peripheral injections of oxytocin can inhibit ACTH and cortisol release in both men and women. In addition, oxytocin injections can inhibit the release of ACTH and cortisol, which normally follow treatments with CRH, AVP plus CRH, or exercise (Carter and Altemus 1997).

Altemus and associates (1995) have examined the effects of physical stress in lactating versus recently delivered, bottle-feeding women. In that study, women were given treadmill exercise to 90% of their VO2 max. The two groups were matched in age and weeks postpartum. The peak blood lactate level, a measure of exercise intensity, was similar in both groups, and lactating and nonlactating subjects had similar basal concentrations of ACTH and cortisol. ACTH, cortisol, and AVP increased following exercise in bottle-feeding women, as would be expected in normal controls. However, the magnitude of the increase in ACTH, cortisol, and AVP in response to exercise stress was blunted in the lactating women. Thus, lactating women show a marked inhibition of stress hormone secretion in response to exercise, which was not seen in postpartum women who bottle fed their infants. Taken together, these studies suggest for humans and other mammals that lactation and oxytocin can reduce physiological reactivity to various stressors.

Lactation also influences the activity of other neural systems that have been implicated in the management of psychological stress. For example, catecholamine responses to stress are reduced in lactating rats. Suckling also increases central production of GABA in rats and sheep. GABA is an inhibitory neurotransmitter, known to play an important role in the regulation of anxiety and behavioral reactivity. Lactating females do not show the expected neuronal activation in cortical neurons following exposure to an excitatory amino acid, suggesting that the functional modifications associated with lactation extend beyond the hypothalamus to include cortical functions (Abbud et al. 1993).

Since oxytocin and lactation can alter adrenal function and glucocorticoid secretion, and because oxytocin is found in the thymus, it is to be expected that this might have immunological consequences. In general, recent research on the effects of lactation on immunological parameters is uncommon in humans or other animals. Among the few studies available is work indicating that lactation is associated with enhanced inflammatory reactions to endotoxin and ozone in rats. We have recently observed that the responses of lymphocytes to mitogen stimulation is altered (generally increased) in lactating rats and breast-feeding women. In addition, lymphocyte proliferation, which normally declines during stress, did not do so in lactating women (Redwine et al. 2001). Measures of cardiac vagal tone, which may be indicative of good welfare (Porges 1998), were also higher in lactating versus bottle-feeding mothers (Altemus et al. 2001).

Clinical research also indicates that biological changes associated with pregnancy and lactation may protect some women from mental disorders. In women with a history of panic disorder, panic symptoms tend to decline in pregnancy and remain low during the lactation period (reviewed in Carter and Altemus 1997). These results suggest that patterns of infant feeding may influence a mother's mental health and thus her ability to deal with the demands of child rearing. This, in turn, has consequences for both parents and their offspring.

SUMMARY

The neurobiology of positive social behavior and social experiences is becoming apparent. Positive social behaviors and the systems that allow these behaviors to emerge have both direct and indirect consequences for welfare. However, understanding the nature of this system, including its neurobiology, requires an integrative perspective, which acknowledges the fact that both social behaviors and the neurochemical systems that affect social behavior are inherently species, gender, age, and context dependent.

ACKNOWLEDGMENTS

Much of this manuscript was written during a residency at the Bellagio Conference Center. I am very grateful for support from the Rockefeller Foundation and Gianna Celli, which made the completion of the document possible. I am especially thankful to Kerstin Uvnas-Moberg and Stephen Porges for providing intellectual stimulation and helping me to understand the physiology of positive social behavior. Support from the U.S. Department of Defense, National Science Foundation (BNS 7925713, 8506727, 8719748), and the National Institutes of Health, including the Institute of Child Health and Human Development (HD 16679) and the National Institute of Mental Health (MH 45836), was essential to the completion of the laboratory research described here.

REFERENCES

Abbud, R., G.E. Hoffman, and M.S. Smith. 1993. Cortical refractoriness to N-methyl-D,L-aspartic acid (NMDA) stimulation in the lactating rat: Recovery after pup removal and blockade of progesterone receptors. *Brain Res.* **604**:16–23

Altemus, M., P.A. Deuster, E. Galliven, C.S. Carter, and P.W. Gold. 1995. Suppression of hypothalamic-pituitary-adrenal axis responses to stress in lactating women. *J. Clin. Endocrinol. Metabol.* **80**:2954–2959.

Altemus, M., L.S. Redwine, Y.-M. Leong, C.A. Frye, S.W. Porges, and C.S. Carter. 2001. Responses to laboratory psychosocial stress in postpartum women. *Psychosom. Med.* 2001. **63**:814–821.

Carter, C.S. 1998. Neuroendocrine perspectives on social attachment and love. *Psychoneuroendocrinology* **23**:779–818.

Carter, C.S., and M. Altemus. 1997. Integrative functions of lactational hormones in social behavior and stress management. *Ann. NY Acad. Sci.* **807**:164–174.

Carter, C.S., and E.B. Keverne. 2002. The neurobiology of social affiliation and pair bonding. In: Hormones, Brain, and Behavior, ed. D. Pfaff et al., New York: Academic, in press.

Carter, C.S., I.I. Lederhendler, and B. Kirkpatrick, eds. 1997. The Integrative Neurobiology of Affiliation. *Ann. NY Acad. Sci.* **807x.**

Carter, C.S., and R.L. Roberts. 1997. The psychobiological basis of cooperative breeding. In: Cooperative Breeding in Mammals, ed. N.G. Solomon and J.A. French, pp. 231–266. New York: Cambridge Press.

Cho, M.M., A.C. DeVries, J.R. Williams, and C.S. Carter. 1999. The effects of oxytocin and vasopressin on partner preferences in male and female prairie voles (*Microtus ochrogaster*). *Behav. Neurosci.* **113**:1071–1080.

Chrousos, G.P., R. McCarty, K. Pacak, G. Cizza, E. Sternberg, P.W. Gold, and R. Kvetnansky, eds. 1995. Stress: Basic Mechanisms and Clinical Implications. *Ann. NY Acad. Sci.* **771.**

DeVries, A.C., M.B. DeVries, S.E. Taymans, and C.S. Carter. 1996. Stress has sexually dimorphic effects on pair bonding in prairie voles. *Proc. Natl. Acad. Sci. USA* **93**:11,980–11,984.

De Vries, G.J., and C. Villalba. 1997. Brain sexual dimorphism and sex differences in parental and other social behaviors. *Ann. NY Acad. Sci.* **807**:273–286.

Keverne, E.B., and K.M. Kendrick. 1991. Morphine and corticotropin releasing factor potentiates maternal acceptance in multiparous ewes after vaginocervical stimulation. *Brain Res.* **540**:55–62.

Kleiman, D. 1977. Monogamy in mammals. *Q. Rev. Biol.* **52**:39–69.

Insel, T.R. 1997. A neurobiological basis of social attachment. *Am. J. Psychiatry* **154**:726–735.

Liberzon, I., and E.A. Young. 1997. Effects of stress and glucocorticoids on CNS oxytocin receptor binding. *Psychoneuroendocrinology* **22**:411–422.

Mason, W.A., and S.P. Mendoza. 1998. Generic aspects of primate attachments: Parents, offspring, and mates. *Psychoneuroendocrinology* **23**:765–778.

Nelson, E.E., and J. Panksepp. 1998. Brain substrates of infant-mother attachment: Contributions of opioids, oxytocin, and norepinephrine. *Neurosci. Biobehav. Rev.* **22**:437–452.

Nishimori, K., L.J. Young, Q. Guo, Z. Wang, T.R. Insel, and M.M. Matzuk. 1996. Oxytocin is required for nursing but is not essential for parturition or reproductive behavior. *Proc. Natl. Acad. Sci. USA* **93**:11,699–11,704.

Ostrowski, N.L. 1998. Oxytocin receptor mRNA expression in rat brain: Implications for behavioral integration and reproductive success. *Psychoneuroendocrinology* **23**:989–1004.

Panksepp, J. 1998. Affective Neuroscience. New York: Oxford Univ. Press.

Panksepp, J., E. Nelson, and S. Siviy. 1994. Brain opioids and mother-infant social motivation. *Acta Paediatr.* **Suppl. 397**:40–46.

Petersson, M., T. Lundeberg, and K. Uvnas-Moberg. 1999. Short-term increase and long-term decrease of blood pressure in response to oxytocin-potentiating effect of female steroid hormones. *J. Cardiovasc. Pharmacol.* **33**:102–108.

Porges, S.W. 1998. Love: An emergent property of the mammalian autonomic nervous system. *Psychoneuroendocrinology* **23**:837–861.

Redwine, L.S., M. Altemus, Y.-M. Leong, and C.S. Carter. 2001. Lymphocyte responses to stress in postpartum women: Relationship to vagal tone. *Psychoneuroendocrinology* **26**:241–251

Roozendaal, B., G.H.M. Schoorlemmer, J.M. Koolhaas, and B. Bohus. 1993. Cardiac, neuroendocrine, and behavioral effects of central amygdaloid vasopressinergic and oxytocinergic mechanisms under stress-free conditions in rats. *Brain Res. Bull.* **32**:573–579.

Russell, J.A., and G. Leng. 1998. Sex, parturition and motherhood without oxytocin? *J. Endocrinol.* **157**:343–359.

Sachser, N., M. Durschlag, and D. Hirzel. 1998. Social relationships and the management of stress. *Psychoneuroendocrinology* **23**:891–904.

Stribley, J.M., and C.S. Carter. 1999. Developmental exposure to vasopressin increases aggression in adult prairie voles. *Proc. Natl. Acad. Sci. USA* **96**:12,601–12,604.

Taymans, S.E., A.C. DeVries, M.B. DeVries, R.J. Nelson, T.C. Friedman, M. Castro, S. Detera-Wadleigh, C.S. Carter, and G.P. Chrousos. 1997. The hypothalamic-pituitary-adrenal axis of prairie voles (*Microtus ochrogaster*): Evidence for target tissue glucocorticoid resistance. *Gen. Comp. Endocrinol.* **106**:48–61.

Thoman, E.B., R.L. Conner, R.L., and S. Levine. 1970. Lactation suppresses adrenal corticosterone activity and aggressiveness in rats. *J. Comp. Physiol. Psychol.* **70**:364–369.

Uvnas-Moberg, K. 1998. Oxytocin may mediate the benefits of positive social interaction and emotions. *Psychoneuroendocrinology* **3**:819–835.

Winslow, J.T., E.F. Hearn, J. Ferguson, L.J.Young, M.M. Matzuk, and T.R. Insel. 2000. Infant vocalization, adult aggression, and fear behavior of an oxytocin null mutant mouse. *Horm. Behav.* **37**:145–155.

Witt, D.M., C.S. Carter, and T.R. Insel. 1991. Oxytocin receptor binding in female prairie voles: Endogenous and exogenous oestradiol stimulation. *J. Neuroendocrinol.* **3**:155–161.

Young, L.J., Z. Wang, and T.R. Insel. 1998. Neuroendocrine bases of monogamy. *Trends Neurosci.* **21**:71–75.

3

What Is Important to Achieve Good Welfare in Animals?

N. SACHSER

Department of Behavioral Biology, Universität Münster,
48149 Münster, Germany

Key words: affective state, age, animal welfare, biorhythm, coping, domestication, emotion, endocrine stress responses, environment, environmental stimulation, evolution, expectation, genes, good life, individuality, mental state, natural selection, ontogeny, physiological state, predictability, sex, social experience, social organization, social support, social stability, stress

ABSTRACT

The welfare of an animal is a state that can vary between very good and very bad. Good welfare is achieved when the individual's expectations match the perceptions of the internal and external environment or, should there be a mismatch, the organism is able to implement a program that will adequately remove the difference. Both expectations and physiological and behavioral coping programs have been brought out by natural selection and, thus, represent the requirements of the species-specific ecological niche. They have been changed significantly, however, by the process of domestication, and in higher vertebrates they can be influenced distinctly by experiences during ontogeny. As a consequence, great but predictable variation exists in what is important to achieve good welfare in animals of different species, in wild and domestic subjects of the same species, and in different individuals of the same genotype. Furthermore, the causation of good welfare is influenced by various internal and external factors including sex, age, and genetic predisposition, on the one hand, as well as the predictability, contorllability, and intensity of environmental stimuli, on the other. Living a "natural" life normally comprises prolonged states of poor welfare. Under human control it would be possible to protect animals from all challenge that possibly causes stress. However, a good life is not a life without challenge as long as the challenge does not overtax the individual's capacity to cope.

INTRODUCTION

All animals have a range of systems whose function is to attempt to cope with challenge or adversity (Broom and Johnson 1993). Traditionally, the behavioral and physiological responses

Coping with Challenge: Welfare in Animals including Humans
Edited by D.M. Broom

to such environmental stimuli have been the subject of stress research. Indeed, this discipline has contributed fundamentally to our understanding of endocrinological and immunological responses that ultimately trigger the etiology of disease (e.g., Henry and Stephens 1977; von Holst 1998). However, it was not until recently that the question was asked: what contributes to an animal's good welfare rather than to what constitutes a stress? Jumping to conclusions one might argue that the absence of stress is good welfare. But freedom from stress, for example, hunger, thirst, pain, injury, or disease, is nothing but an essential prerequisite. What then, in principle, is good welfare, and what basic mechanisms bring it out? Answers are given in the following section.

A squirrel monkey has different requirements to achieve good welfare than the closely related titi monkey (Mendoza and Mason 1986; Poole 1997). A domestic pig differs distinctly in this aspect from its ancestor, the wild boar. Even members of the same stock of guinea pigs may vary conspicuously in what brings out good welfare (Sachser et al. 1994). To explain such differences, the section on THE GENESIS OF GOOD WELFARE discusses their genesis. From a biologist's point of view, this involves the effects of evolution, domestication, and ontogeny (Sachser 1998).

The welfare of an animal is a state that can vary between very good and very bad (Broom and Johnson 1993; Broom 1998). It differs between individuals and it can change with the course of time. The section on FACTORS INFLUENCING THE CAUSATION OF GOOD WELFARE discusses the most important internal and external factors that influence the causation of good welfare. These involve sex, age, and genetic predisposition, on the one hand, as well as the predictability, controllability, and intensity of environmental stimuli, on the other. Finally, in the section on GOOD WELFARE AND GOOD LIFE, the question is posed as to whether the experience of good welfare during every moment of an animal's life really represents a good life.

Biologists discuss welfare issues on different levels. They may view not only the individual, but also the group, the population, the species, or the ecosystem (Fraser 1999). For example, they may believe that it is right to kill some elephants in order to maintain a healthy population, or they impose special restraints on certain animals in order to preserve a rare species. Many such actions may conflict with the good welfare of each individual animal (Fraser 1999). Thus, different theoretical frameworks are needed to analyze good welfare on different levels. This contribution, however, is exclusively about the good welfare of individual nonhuman animals.

A BASIC MODEL OF GOOD WELFARE

Good welfare of an animal is a state, and deviations from this state can be detected by the organism. If such deviations are perceived, the individual will try to activate the relevant physiological and behavioral systems to return to a state of good welfare. Thus in principle, the control of good welfare can be described in terms of a regulatory (homeostatic) model. Accordingly, good welfare is related to a set of expected values (*Sollwerte*/set values) ranging from optimal blood sugar concentrations to the availability of mating partners during the reproductive season. The expected values are compared with the various inputs (*Istwerte*/actual values). These include intrinsic stimuli, being somatic or psychological in nature, or extrinsic stimuli from the physical or psychosocial world. If there is a significant mismatch between any relevant expected value and environmental input, behavioral and physiological systems will be activated. If these responses result in a removal of the observed mismatch, good welfare will be maintained (e.g., Wiepkema 1985, 1987; Broom and Johnson 1993; Clark et al. 1997). Thus, in a broad sense, the welfare of an individual is its state as regards its attempts to cope with its environment (Broom 1986; Broom and Johnson 1993), and good welfare is achieved when the individual's expectations match the perceptions of the internal and external world (Figure 3.1a). In general, this is the case when the animal functions well in the sense of good health, growth, and normal working of behavioral and physiological systems, including reproduction. In addition, in those species in which emotions have evolved (e.g., in mammals), good welfare requires also that the individuals feel well, in the sense of experiencing positive affective states (Fraser et al. 1997; Broom 1998; Dawkins 1990; Knierim et al., this volume).

The total of all expected values relating to good welfare can be regarded as the individual's "expectation copy" (termed by analogy with Archer, cited in Sachser and Lick 1991). It includes standards (e.g., the optimal body temperature or the instinctive knowledge of key stimuli) that have been formed in the course of evolution and are strongly similar for individuals of the same species, sex, and age. However, the expectation copy also includes information acquired during ontogeny (e.g., social experiences) that may differ distinctly between individuals. The expectation copy is not fixed in space and time, but is of dynamic character, and its actual constitution is influenced by a variety of internal and external factors (Clark et al. 1997; Wiepkema 1985, 1987). The use of the term "expectation" does not mean that the animals are consciously aware of their expectations.

A mismatch of elements of the expectation copy and the environmental input can be of varying significance to the animal's good welfare. If a hungry dog is deprived of food and there is danger of starving, then its welfare is very poor. If the same dog expects liver but offal is offered, its good welfare obviously is not really impaired. Following Dawkins (1990), differences concerning environmental aspects for which the individual shows inelastic demands, the so-called necessities, represent much more serious threats to good welfare than luxuries for which the individual has an elastic demand.

A discrepancy between the individual's expectations and the environmental input can be caused in different ways. On one hand, the environmental input may represent a stressor that impinges on the organism (Figure 3.1b), for example, a lion attacking a zebra or a potentially harmful virus, for example, influenza, invading a cow. Whether or not good welfare can be maintained in such situations depends on how well or how badly the organism copes and how much difficulty it is having in coping. In general, attempts to cope, that is, attempts to maintain or to regain mental and bodily stability, include the functioning of body repair systems, immunological defenses, emergency physiological responses, and a variety of behavioral responses (Broom and Johnson 1993). The zebra will immediately activate its sympathetic-adrenomedullary system and try to escape from the predator. In case of success there is no longer a discrepant input, that is, the lion is no longer around, and good welfare can be regained. The virus will be detected by the cow's immune system and successive immune responses will attack and try to destroy this pathogen. As a consequence, the discrepant input is removed and good welfare can be maintained.

On the other hand, the individual might perceive that the environment lacks expected stimuli and resources. For example, a chaffinch might find a habitat unsuitable to establish a territory and a parrot kept singly as a pet might miss the company of a social partner. In such

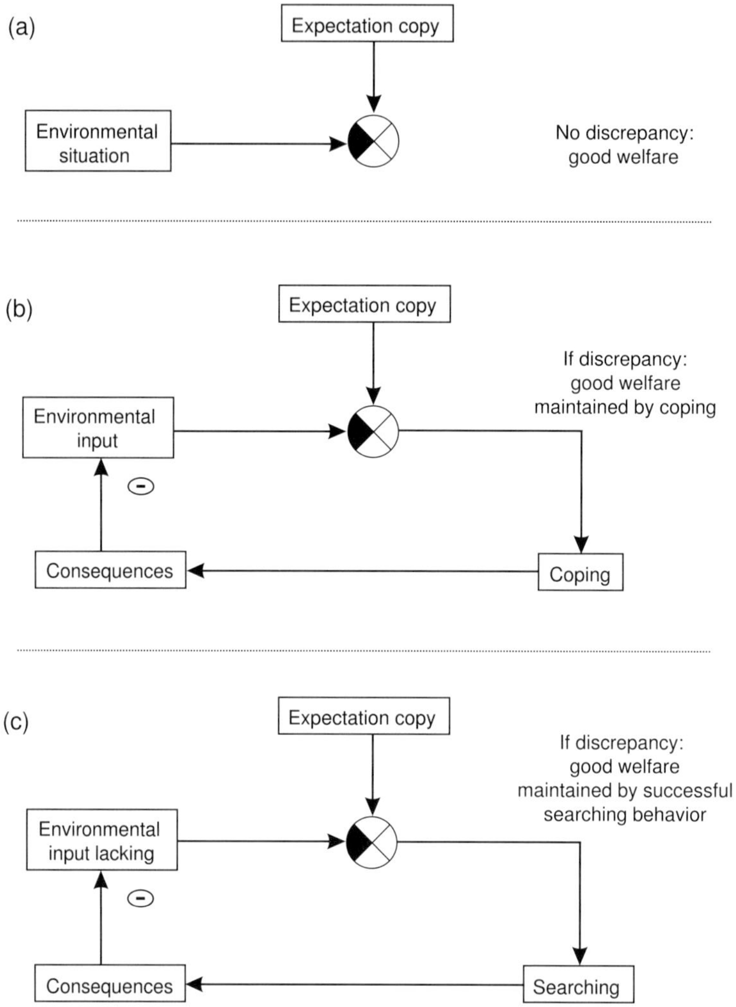

Figure 3.1 A homeostatic model of good welfare: (a) if the animal's expectation copy (the total set of expected values relating to good welfare) matches the perceptions of the environment, good welfare is achieved; (b) if a significant discrepancy is detected between any element of the expectation copy and the environmental input, e.g., if a stressor impinges on the organism, physiological and behavioral coping systems will be activated whose consequences are to remove the discrepant input or to remove the animal from it (feedforward mechanisms are also possible, see text); (c) if the environment lacks expected stimuli, searching behavior will be activated whose consequence is to find the required input.

situations the animals will show searching behavior in order to meet these requirements. Good welfare will be maintained in case of success (Figure 3.1c), that is, if the chaffinch finds an appropriate place to live and if the parrot meets a companion.

If a discrepancy exists between the expectation copy and the environmental situation that can neither be removed by successful coping nor by searching behavior, then the homeostatic model points to a third way in which good welfare could be maintained: the modification of the expectation copy (not shown in Figure 3.1). Unfortunately, data concerning this issue are largely lacking in animals.

Simple negative feedback models, as presented in Figure 3.1, have been criticized as somewhat oversimplistic. There are, for example, cases in which the performance of behavior itself does have significant consequences for welfare which are not necessarily related to functional requirements. Hens, for example, go through nest-building sequences during pre-laying behavior, even though the nest they created previously is still available (Hughes and Duncan 1988). In addition, animals do not just cope with existing challenges, they also anticipate threats to their welfare (Dawkins, this volume). In such cases, feedforward control frequently induces a change in the homeostatic control system before a disruption actually occurs (Broom and Johnson 1993). Therefore, the model presented in Figure 3.1 does not claim to cover all aspects of the control of welfare. Rather, it emphasizes the three key variables that have to be taken into account to elucidate what is important to achieve good welfare in animals: the individual's expectation copy, the environmental input, and the array of behavioral and physiological programs available to remove any significant difference between expectation and existing or anticipated input.

THE GENESIS OF GOOD WELFARE

Evolution and Welfare

During the process of evolution, organisms adapted to specific parts of their environment according to their species. The members of any given species are, therefore, equipped with physiological, motivational, and behavioral systems to cope with the demands of their specific environmental compartment and their expectations match the requirements of this specific ecological niche. However, living a "natural" life in a "natural" environment is by no means synonymous with the experience of good welfare (Broom and Johnson 1993; Fraser et al. 1997; Dawkins 1998). In the long term, for animal species, the individual number of a given population is more or less the same. In every generation, though, many more offspring are produced than would be needed to keep individual numbers constant. From this an intense competition arises between the animals of the same species for survival and reproduction that is regularly associated with high degrees of stress, injury, disease, and death (e.g., Darwin 1859; von Holst 1998). Those individuals that cope best with the environmental challenges and that are most successful in acquiring matings will transmit their genes into the next generation. To achieve this goal, infants are killed, rivals are injured to death, and females are raped. Animals of some species even engage in behaviors that lead to their own death in order to pass on their genes: *Antechinus stuartii*, a small marsupial, lives in Australia in small, well-defined territories. At the beginning of the mating season, which lasts two to three weeks, the territories break down and all the males become involved in extensive fighting and copulation. These interactions evoke extreme physiological stress responses that lead to the death of all males at the end of the reproductive period. The survival of the population is only guaranteed through the pregnant females (Bradley et al. 1980). These effects, which take place in the same way every year, lead without doubt to pain, suffering, injury, and early death. However, only those males that engage in these activities during the reproductive period have any chance of transmitting their genes to the next generation. If a male decided to enjoy a comfortable life, rather than engaging in deadly activities, this would not be favored by natural selection. Thus, ultimately organisms are designed to maximize their lifetime reproductive success rather than to experience all components of good welfare (Sachser and Beer 1995; Dawkins 1998). Indeed, selection favors feelings and various physiological and behavioral systems that are associated with fitness (Broom 1998; Dawkins, this volume).

Living a "natural" life for most animals includes prolonged states of stress. Thus, in wild animals living under human-made housing conditions, good welfare cannot be achieved by simply providing environments that copy the animals' natural way of living and allow them to perform their full "behavioral repertoires" (Fraser et al. 1997, Dawkins 1998). This does not mean, however, that the knowledge of an animal's natural life in its natural environment is insignificant. In contrast, there is increasing evidence that not copying, but orienting towards the species' "natural" life contributes greatly to the understanding of what is important to achieve good welfare (see also Knierim et al., this volume).

Natural selection, for example, has brought out different types of social organization that represent adaptations to the ecological conditions under which the species lives. Although most social systems can be characterized by a certain degree of flexibility (e.g., Sachser 1998), at least in mammals typical basic forms of social organization can be assigned to most species: many animals, like the tiger or the hamster, live a solitary life, that is, the adult individuals meet regularly only for mating and the offspring leave their mothers around puberty. Other species form stable social units: gibbons and beavers live in monogamous pairs, hamadryas baboons and plains zebras establish harems, chimpanzees and rats organize themselves in mixed groups of several females and several males. These different types of social organization obviously have a strong effect on which situations and processes will become stressful for an individual (Sachser and Beer 1995). For example, species that are territorial and solitary-living in their natural habitat respond quite differently to being held in isolation than do members of socially-living species: in captivity the solitary hamster has to be kept individually to achieve good welfare; by contrast, in the group-living guinea pig, housing singly is related to stress. Zoos sometimes advertise that tigers live in families, that is, father, mother, and offspring. From a human point of view, this may seem pleasant; for a solitary animal, it probably is not.

Mendoza and Mason (1986) supplied the most convincing data to support the hypothesis that the social organization which exists under natural conditions has a decisive influence on the physiological stress responses of an individual. Squirrel monkeys in the wild live in large, mixed-sex groups in which the males associate predominantly with other males, and females predominantly with other females. On the other hand, the closely related titi monkeys form family groups consisting of a monogamous pair and one or two offspring. If individuals of these South American primate species are held in the laboratory in pairs of one male and one female, they respond quite differently to experimental changes of their social situation. Those males and females that live monogamously in their natural habitat respond to the separation from their partner with an increase of the stress hormone cortisol, but those animals that normally live in large groups do not. Correspondingly, titi monkey males show a clear increase of cortisol if they notice a same-sex conspecific in the neighboring enclosure, while the hormone concentrations of squirrel monkeys do not change in the same situation (Figure 3.2).

	Titi monkeys	Squirrel monkeys
"Natural" social organization	monogamous	large groups of several males and females
Glucocorticoids after separation from partner	↑	⟷
Glucocorticoids after exposure to a same-sex conspecific	↑	⟷
Social constellation required for good welfare	E + G	GG+EEE

↑ increase ⟷ no change

Figure 3.2 "Natural" social organization and welfare: Comparison between titi monkeys (*Callicebus moloch*) and squirrel monkeys (*Saimiri sciureus*); for references and original data see Mendoza and Mason (1986); Poole (1997).

To summarize, wild animals kept in captivity will have good welfare under environmental conditions that match the species-specific expectations that have been brought out by natural selection, for example, one social partner in the titi monkey, several social partners in the squirrel monkey (Mendoza and Mason 1986; Poole 1997). However, challenges that might overtax an individual's behavioral and physiological coping systems (e.g., predators, adverse weather conditions, lack of food) are not required to achieve good welfare even when these events occur frequently in the natural habitat of the species.

Domestication and Welfare

The history of humankind has been attended by domestic animals for thousands of years, and most individuals that live under human control are domesticated. During their domestication, wild animals have to adapt, for a period of many generations, to their new conditions in captivity and human care, resulting in long-term genetic changes, and eventually become domestic animals. Domestication, therefore, involves an evolutionary process based on the shift in selective pressures: with removal of organisms from some natural selection pressures to new natural selection by artificial environments and artificial selection by humans. This process is always accompanied by distinct changes in morphology, physiology, and behavior (Künzl and Sachser 1999). As a consequence, wild and domestic animals require different conditions and resources to achieve good welfare although, in a biological sense, they still belong to the same species.

Guinea pigs, for example, were domesticated 3000–6000 years ago in the highlands of South America. Their wild ancestor, the cavy, is among the most common and widespread

		Wild	Domestic
Behavior	Aggressive	+	—
	Sociopositive	—	+
	Courtship	—	+
	Vocal	—	+
	Attentive	+	—
Endocrine stress response	Adrenocortical system (cortisol)	+	—
	Adrenomedullary system (epinephrine)	+	—

+ frequent/high – rare/low

Figure 3.3 Behavioral and physiological consequences of domestication: Comparison between wild (*Cavia aperea*) and domestic guinea pigs (*Cavia aperea* f. *porcellus*); for reference and original data see Künzl and Sachser (1999).

rodents of South America. Interestingly, the behavior patterns are similar in the domesticated and the wild animals; distinct differences, however, occur in behavioral frequencies and thresholds (Figure 3.3). Domestic guinea pigs are distinctly less aggressive than their wild ancestors and they display more sociopositive behaviors. In addition, overt courtship behavior is more frequently expressed, and a lower threshold for vocalization is found. Last, they are distinctly less attentive to their physical environment (Künzl and Sachser 1999). Owing to these differences, social interactions between members of the wild and domesticated forms proceed in a completely different way. When domestic guinea pigs are kept in breeding groups of one adult male and several adult females, the mature sons and daughters will integrate rather peacefully into the social system of the groups and all animals will cohabitate in a nonaggressive and nonstressful way (Sachser 1998). The resulting large mixed-sex colonies can even be regarded as the most appropriate way to keep domestic guinea pigs. When adult wild cavies are kept in breeding groups of one male and several females, a completely different picture emerges: the daughters integrate into the linear dominance hierarchy of the females. In contrast, the father and his sons become rather incompatible when the sons attain sexual maturity. Then, in most cases, they must be taken out of the groups because otherwise the father will injure and even kill them (Sachser 1998). Thus it is nearly impossible to keep adult males together in the presence of females. The social composition favoring good welfare of wild guinea pigs, therefore, is one adult male and one or several females. In contrast to the domestic form, colony-housing cannot be recommended.

On a physiological level, wild animals respond to stressors with a much stronger activation of the pituitary-adrenocortical and the sympathetic-adrenomedullary systems — the organism's two stress axes — than their domesticated counterparts. If, for example, wild and domestic guinea pigs are placed into unfamiliar enclosures, the wild animals will react with a significantly stronger increase of their plasma cortisol concentrations. If both forms are handled, plasma epinephrine and norepinephrine concentrations will be several times higher in the wild form (Figure 3.3). Significantly lower cortisol levels in response to ACTH application point to a generalized reduction in stress responsivity in the domestic guinea pigs (Künzl and Sachser, unpublished data). In general, this trait can be regarded as the physiological correlate of the reduced alertness, nervousness, and sensitivity of the domesticated animals compared to their wild counterparts. While the reduced stress responsivity obviously helps domestic animals to adapt to artificial housing conditions, it is counterselected in wild animals in their natural habitats by natural selection (Künzl and Sachser 1999). As a consequence, under human housing conditions, good welfare can be achieved much easier in domesticated animals than in their wild ancestors.

In summary, during the process of domestication organisms have adapted to artificial housing conditions. Thus, good welfare is not simply achieved by providing these animals with the requirements for which their wild ancestors had been selected. Many expectations of domesticated animals (e.g., permanent availability of food, controlled environmental conditions) refer to the artificial environments created by humans. Furthermore, their physiological and behavioral systems are designed to cope well with the specific challenges of human environments, such as handling or being housed at high individual numbers (see, for example, Sachser 1998). At the same time, these coping systems may be overtaxed when the animals are challenged by "natural" stressors. The reference system to recognize what might be important to achieve good welfare in domesticated animals, therefore, is a generous human housing system (e.g., a richly structured enclosure for laboratory mice) rather than the ecological niche of the wild ancestor (Broom and Johnson 1993). On the other hand, domesticated animals still bear the heritage of their wild counterparts. For example, studies by Stolba, Jensen, and Wood-Gush (cited in Broom and Johnson 1993) of modern pig breeds living in seminatural conditions reveal many similarities to wild boar. Thus, we should not expect that domestic animals are able to adjust to any artificial housing. Rather, they require at least some essentials that their wild ancestors had been selected for. Dogs, for example, were derived from group-living wolves, a species with strong social bondings among group members. As in the wild ancestor, to achieve good welfare for a dog still requires the availability of a social bonding partner, which can be a member of the same species or a human. Cats were derived from a solitary, territorial species and still in the domesticated form, stable spatial structures seem to be more important for good welfare than the permanent presence of interaction partners, irrespective of whether these are cats or humans.

Ontogeny and Welfare

Nearly 50 years ago, experiments with rhesus monkeys by Harlow and coworkers made it clear for the first time that experiences during ontogeny might be crucial for the welfare of animals (cited in Sachser and Beer 1995). Monkeys raised in isolation are fearful and depressed in new situations, while they are hyperaggressive in response to unfamiliar members of the same species and cannot build "normal" social relationships. These findings are not just restricted to primates. All mammals and the majority of birds probably require adequate socialization in order to be able to communicate successfully and interact with members of the same species and to achieve good welfare during their life.

Predominantly in mammals and birds, many expectations are determined by experiences during early sensitive phases. Hereby, characteristics of prospective social and mating partners are fixed in the same way as preferences for specific foods or habitats. From these imprinting-like processes, severe problems can arise with respect to welfare: a bird of an endangered species will not find the expected mating partner because it was imprinted sexually on the human being that raised the bird. An orangutan might suffer severely when it is released into its natural habitat and be forced to live the "natural" solitary life of its species, when in actuality it would prefer to remain in the companionship of humans that have raised it in captivity. For these animals, good welfare is achieved only when the expectations brought out by early experiences are met by the current environmental situation. It is noteworthy that in many reptiles, amphibians, fish, and nonvertebrates, the expectations related to welfare and the array of coping systems available are obviously less influenced by learning processes than in birds and mammals (cf. earlier section on Evolution and Welfare). As a consequence, captivity-reared individuals of the first-mentioned species can be released into nature without problems, given that the species-specific niche is available. In contrast, most attempts to release endangered bird and mammalian species fail (Arzdorf 1990). One of the major reasons for this disastrous result is obviously that the animals' expectations and coping patterns are strongly influenced by the experiences made during behavioral development in artificial housing conditions, which frequently do not meet the requirements of the natural habitat.

In many mammals, good welfare during the early phases of life requires close contact with members of the same species, predominantly the mother, to provide the young animals with a sense of security. Only if this secure base is available can they explore the environment and new objects in a nonanxious way (Gandelman 1992). For decades it was believed that such early experiences were primarily important for the development of behavioral strategies and good welfare in later life. Recent studies, however, point to the time around puberty. During this time, social relationships change dramatically. Conspecifics of the same sex become rivals and animals of the opposite sex become attractive as mates. Thus, it is not surprising that also the time around puberty is crucial for the acquisition of significant coping patterns needed during the animal's future life (Sachser et al. 1998).

For example, when two male domestic guinea pigs that have grown up in different large mixed-sex colonies are placed into an unfamiliar enclosure in the presence of an unfamiliar female, they quickly establish stable dominance relationships without displaying overt aggression. They just look at each other to assess who is the dominant and who has to retreat. No significant changes in serum cortisol (Figure 3.4) or norepinephrine concentrations are found either in the dominant or in the subdominant male. Thus, the opponents arrange in a nonstressful way. Interestingly, the males do not court the female earlier than several hours after the begin of the experiment (Sachser et al. 1994). A completely different picture emerges when two males that had grown up either singly or in pairs consisting of one male and one female are confronted in the same way. In such confrontations, high levels of aggressive behavior are displayed and escalated fighting is frequent. The males begin to court the female immediately after the beginning of the experiment. It normally takes several days until dominance relationships are established and many experiments have to be stopped to avoid irreversible injuries and even death of the losers. As can be seen from Figure 3.4, distinct increases in cortisol concentrations are found in both males during the early phase of the confrontations. In the dominant males these values decline to pre-experiment values and are no

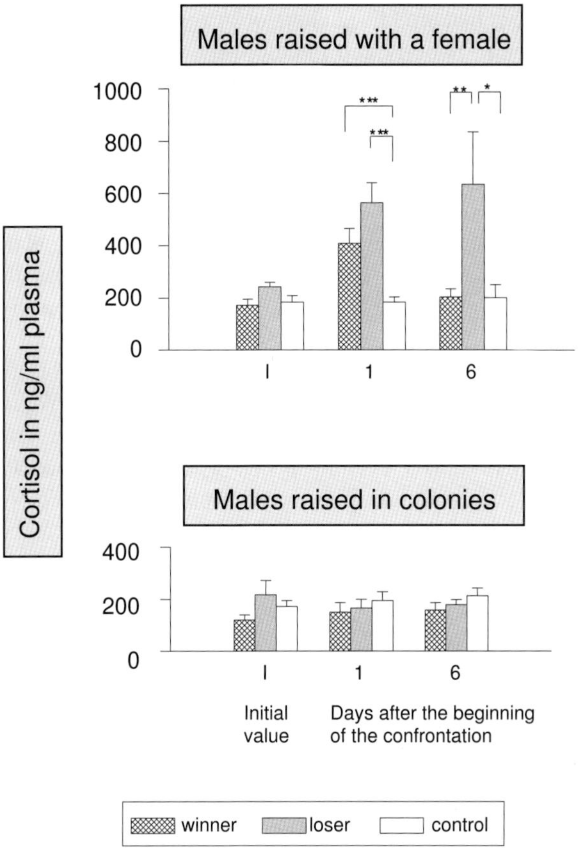

Figure 3.4 Consequences of rearing conditions for endocrine stress responses in male guinea pigs. Plasma-cortisol concentrations of controls ($N = 6$–8) and confronted animals (winners, $N = 5$–7; losers, $N = 3$–7) 20 h before (initial values) and at different times after the begin of the confrontations. The confronted males were either reared in pairs consisting of one male and one female, or in different large colonies. Values are means + SEM. *$p < 0.05$, **$p < 0.01$, ***$p < 0.001$. Reprinted from Sachser et al. (1998) by permission of Pergamon.

longer different from those of controls. In contrast, in losers high stress hormone concentrations persist (Sachser and Lick 1991; Sachser et al. 1994).

How do these differences evolve? In colonies, growing males learn the social "rules of the game" of their species around puberty through interactions with older dominant males (Sachser et al. 1994, 1998). Two things are learned above all: (a) submissive behavior if the other animal is dominant and (b) not to court every female. Thus, the expectation copy of a colony-reared male is: "if I do not court the female, I will not be attacked/if I am dominant my opponent will retreat/if he is dominant, I have to retreat and he will not attack me" (Sachser and Lick 1991). Thus, during the confrontation, these males' expectations will be met and they have the required behavioral coping patterns irrespective of whether they will become dominant or subdominant animals. As a consequence, good welfare is maintained in both the winner and the loser. Males that are raised alone or in a pair have no agonistic experiences with dominant males around puberty and therefore do not learn these rules. As a consequence, they respond to unfamiliar females and males in a different way: "if I encounter a female I immediately will court/if I encounter a courting male, I immediately will attack." The dominant's expectations will be met and his welfare is good. The subordinate, however, will experience a significant difference between his expectations and his current situation. As a consequence, severe degrees of stress will arise (Sachser and Lick 1991). In summary, these and further data show that a causal relationship exists between the social experiences made around puberty and the availability of those behavioral coping patterns that are needed to maintain good welfare during the establishment of dominance relationships (Sachser et al. 1994, 1998).

In group-living animals, the presence of members of the same species is an essential prerequisite to achieve good welfare (see earlier section on Evolution and Welfare). Groups, as a general phenomenon, are socially stratified and distinct differences in status have been described among the adult individuals, even in species where close cooperation, social tolerance, and amicable relations predominate. In stable social systems, established dominance relationships result in predictable behavior and, as a consequence, differences in social status do not lead to differences in stress and health. Thus, both high- and low-ranking individuals can live a good life (Sachser 1994). However, to build up a stable social system requires that the individuals are able to establish and to respect dominance relationships. Whether these social abilities are realized depends, at least in mammals, on social experiences made during behavioral development (see above). In these animals, therefore, the requirements to achieve good welfare in adulthood are (a) the experience of security during an early phase of life and (b) the learning of social rules around puberty.

FACTORS INFLUENCING THE CAUSATION OF GOOD WELFARE

Internal Factors

Welfare is a state that can change from moment to moment. Its current constitution is influenced by various external and internal factors. Among the most powerful internal factors are the experiences made during ontogeny and the evolutionary history of the species that is represented in each animal (see above). In addition, sex and age also significantly influence whether or not good welfare can be achieved under the given environmental circumstances.

In many species the life-styles of both sexes vary distinctly. Thus, it is not surprising that good welfare may require different environmental conditions for males and females. For example, in their natural habitats, male mice establish territories and live a solitary life; the females, in contrast, frequently organize themselves in groups of same-sex individuals. If, in the laboratory, groups of female mice are provided with spacious cages that are structured richly, all animals have excellent welfare (Kaiser et al. 1999; cf. section on External Factors). In contrast, in groups of males, the same environment activates a territorial program resulting in a distinct increase of aggressive behavior, extreme endocrine stress responses, injuries, and even death (Kaiser et al. 1999; Poole 1997; Sachse, unpublished data).

A newborn requires different things to achieve good welfare than a juvenile, and a juvenile differs strongly in this respect from a sexually mature animal. In wolves, for example, cubs need their mothers overall, juveniles love to play with their peers, and adult males and females appreciate filling the alpha positions of their packs. Interestingly, in their natural habitats, animals only rarely survive beyond their reproductive phase, that is, senescence is not favored by natural selection. Under human control, however, animals can reach considerably high ages. Every dog owner knows that the requirements for good welfare change with old age. Surprisingly, there are no systematic scientific studies on this topic and in the index of current textbooks on behavioral biology or welfare you will not find an entry for "age."

Additionally, four further internal factors have to be mentioned:

1. Individual genetic predisposition: for example, whether animals of the same species tackle problems in a more active or in a more passive way (proactive versus reactive coping styles) obviously has — at least in mice — a genetic background (Koolhaas et al., this volume).
2. Animal's physiological state: for example, a doe of a red deer in estrus requires a stag, the same animal during pregnancy seeks the companionship of related females and during delivery she prefers to stay by herself.
3 Mental state: for example, after a painful experimental procedure a rat might associate every white coat with pain; in contrast, a zebra fish or a *Drosophila* fly owing to their lesser cognitive abilities, might not.
4. Animal's biorhythms: for example, during spring and summer a group of migrating birds has good welfare in an aviary under constant laboratory conditions. However, during the time of migration their endogenous biorhythm signals the necessity to migrate. Good welfare, possibly, cannot be maintained if this behavior is prevented.

Finally, individual differences are a crucial factor in whether or not good welfare can be achieved (Wiepkema 1987). For example, two male mice of the same age and of the same inbred strain (i.e., more or less genetically identical individuals) that have been raised and tested under identical conditions may nevertheless prove to be consistently different with respect to anxiousness, curiosity, and behavioral strategies. Thus, individuality, (i.e., consistent individual differences) has to be considered in all reflections on the causation of welfare, as long as natural sciences cannot derive this factor from other parameters in a satisfactory way.

External Factors

As mentioned before, when two male guinea pigs that did not experience agonistic interactions around puberty are confronted in an unfamiliar enclosure in the presence of an unfamiliar female, high levels of aggression and stress occur (see section on Ontogeny and Welfare; Sachser et al. 1994). An even higher incompatibility between the opponents is found if the enclosure is familiar to both males. In contrast, in the absence of a female, the rivals arrange themselves in a low aggressive and nonstressful way (Sachser et al. 1994). Thus whether or not good welfare can be achieved in such a situation depends strongly on the specific features of the current environmental situation.

Owing to their various evolutionary histories, different species are in need of different external factors to achieve good welfare. In addition, in some species these requirements can underlie strong modification brought out by individual experiences during ontogeny (see above). Nevertheless, some general relationships between external factors and good welfare are evident.

1. Modern stress research shows that situations of uncertainty or unpredictability are a major source of severe stress responses. Thus, the predictability and controllability of external factors is a most important prerequisite to achieve good welfare (Wiepkema 1987; von Holst 1998). This accounts for both the animal's physical and social environment. In stable social systems, the predictability of social events is high. Therefore, in such systems neither a low social status nor a high number of individuals will necessarily lead to higher social stress. Thus, the integration into a stable social network in which every animal knows and accepts its social position, allows every individual to live a good life (Sachser 1994; Sachser et al. 1998). Even the stability of the social environment during pregnancy may be decisive for the offspring's' behavior and welfare. In the domestic guinea pig, daughters of mothers who lived in an unstable social environment during pregnancy are characterized by a distinct behavioral and endocrinological masculinization as well as by higher indices of stress than daughters of mothers who lived in a stable social environment during this period of life (Kaiser and Sachser 1998).

2. Appropriate environmental stimulation favors good welfare; in contrast, lack of adequate stimulation and overstimulation might overtax the individual's coping attempts (Broom and Johnson 1993). Many domestic, laboratory, and zoo animals are kept in impoverished environments that are characterized by a lack of novelty and few opportunities for exploration. Various behavioral abnormalities occur under such conditions. One way to improve the quality of life in these animals is to enrich their housing conditions environmentally. Female laboratory mice, for example, living in spacious terraria which are structured richly by a second floor, a plastic inset, stairs, ropes, scaffolding, etc. (designated as complex enrichment) show high frequencies of sociopositive and play behaviors while stereotypic behavior and aggression rarely occur (Figure 3.5). By contrast, when the animals are housed in spatially nonstructured small cages — the standard housing in biomedical laboratories all over the world — agonistic and stereotypic behaviors are frequent and play and sociopositive behaviors seldom occur. Interestingly, the spontaneous behavior of animals that live in standard cages with a simple enrichment, consisting of a plastic box with several openings and wooden scaffolding, does not differ distinctly from that of standard housed mice (Figure 3.5). Thus, these data not only point to the importance of an appropriate environmental enrichment to achieve good welfare, but also to shortcomings of a too simple enrichment (Kaiser et al. 1999). It cannot be derived from the present findings, however, which specific factors (more space and/or different furnishings) made the difference between the simple and the complex enrichment.

Whereas welfare problems related to lack of adequate stimulation have become a frequent topic of research, not much is known about the inputs that cause problems of overload. It is possible, however, for animals that those situations in which there is too much to decide (Broom and Johnson 1993) or in which the sensory systems are overstrained by environmental stimulation are not compatible with good welfare.

3. A major environmental factor that promotes good welfare is the presence of a noncompeting member of the same species. Its ability to buffer endocrine stress responses is called social support. Frequently, such support cannot be provided, however, by any conspecific, but the ability to give social support is restricted to bonding partners. In most mammalian species, mothers are important bonding partners for their infants. As a general rule, mothers

	Housing conditions		
	Standard	Simple enrichment	Complex enrichment
Agonistic behavior	+	+	–
Sociopositive behavior	–	–	+
Stereotypic behavior	+	0	–
Play behavior	0	–	+

+ frequent   0 medium   – rare

Figure 3.5 Environmental enrichment and good welfare in female laboratory mice. Spontaneous behavior of 36 NMRI outbred mice was recorded from 28–50 days of age. Animals living in groups of four females were housed in (a) standard laboratory cages (Makrolon type III, 37 × 21 × 15 cm; standard housing), (b) Makrolon type III cages, which were enriched with a plastic box with several openings and a wooden scaffolding (simple enrichment); (c) spacious terraria five times the size of a Makolon type III cage and structured richly by a second floor, stairs, ropes, scaffolding, etc. (complex enrichment). Differences between +, 0, and – are at least $p < 0.05$. For reference and original data, see Kaiser et al. (1999).

provide social support in a very effective way. In addition, in those species in which social bondings between adult individuals are part of the natural social organization, the respective bonding partner can also provide social support (Henry and Stephens 1977; Sachser et al. 1998; von Holst 1998). For example, in large colonies of guinea pigs, male–female bonding is very strong. As Figure 3.6 shows, the various types of social relationships differ remarkably. Serum cortisol concentrations increase significantly when a male is placed singly in an unfamiliar cage. The presence of a strange female or of one with whom he is merely acquainted has little effect. There is a sharp reduction of the stress response, however, when a female to whom he is bonded is placed in the cage with him (Sachser et al. 1998). In general, the stress buffering effect of a bonding partner is reflected by the reduced activities of the pituitary-adrenocortical and the sympathetic-adrenomedullary axes. Changes in the activation of the first system point to changes in an individual's degree of security, while changes in the activity of the latter system reflect changes in the degree of arousal (Henry and Stephens 1977; von Holst 1998). On a psychological level, the bonding partner, therefore, can be regarded as a "security-giving and arousal-reducing structure." This is true irrespective of whether the bonding partner is the mother, in the case of an infant, or a male or female, in the case of an adult individual (Sachser et al. 1998).

GOOD WELFARE AND GOOD LIFE

Using biological knowledge, welfare can be assessed in animals in a scientific way. Furthermore, it can be elucidated in which way good welfare can be achieved. But how is good

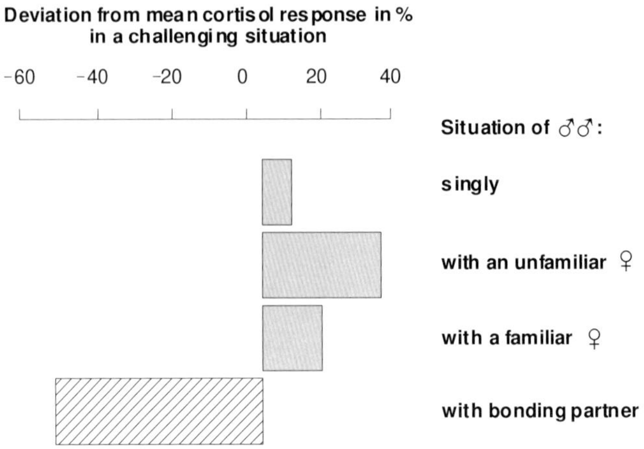

Figure 3.6 Social support from bonding partner. Male guinea pigs ($N = 10$) were tested in four situations. They were placed into an unfamiliar enclosure (= challenging situation) singly, together with an unfamiliar female, together with a familiar female with whom no bonding existed, or together with the female bonding partner. Plasma-cortisol concentrations increased significantly in all four situations (from this a mean cortisol response was calculated); however, significant variations were found between groups ($p < 0.002$): cortisol titers were lowest in 9 out of 10 males when they were tested together with their bonding partners. For reference and original data, see Sachser et al. (1998).

welfare linked to a good life? Does the experience of good welfare during every moment of an animal's life really represent a good life? For wild animals living in their natural habitats, the last question is rather hypothetical. The physical and social environment will permanently change, often in a random and nonpredictable way. Although behavioral and physiological machinery has evolved to enable the organism to cope with these challenges, states of severe stress are part of most animals' natural life (see section on Evolution and Welfare and von Holst, this volume).

Under human control, it is possible, in principle, to protect animals from all environmental challenge and stress. However, some stimulation, which is disadvantageous in the short term, may have beneficial effects in the long term (Broom and Johnson 1993). For example, handling of young rats during the nursing period constitutes a stressor just as cold, heat, and electrical shock. This early stress, however, promotes physical well-being in later life, for example, resistance to cancer is enhanced and recovery from stress occurs with greater rapidity (e.g., Gandelmann 1992). Furthermore, in male guinea pigs that have been raised with a female and have never been involved in aggressive interactions, the experience of about an hour of agonistic encounters with unfamiliar males around puberty is clearly stressful. This experience, however, enables the animals to associate with unfamiliar members of the same species in the future in a less aggressive and less stressful way (Sachser et al. 1994).

Thus, a good life obviously is not a life without challenge. Rather, it is important that challenge does not overtax the individual's capacity to cope. In addition, if we accept that affective states have evolved in some species (e.g., mammals) and that they play a critical role in shaping behavior programs (Wiepkema 1985, 1987; Dawkins 1990; Broom 1998; Fraser and Duncan 1998), it can be assumed that strong positive emotions will arise during and after the process of successful coping. Thus, to prevent these animals from overcoming any form of environmental challenge would also prevent them from the experience of strong positive feelings.

Finally, it has to be stated that any conception of what constitutes a good life for an animal inherently involves values because it pertains to what is better or worse for animals. Decisions about whether an animal should be challenged at all, and if yes, how severe the challenge should be in order to live a good life, must be based in part on ethics and cannot be derived exclusively from biological data (Broom and Johnson 1993; Fraser et al. 1997).

ACKNOWLEDGMENTS

I thank Trevor G. Cooper, Gerti Dücker, Hubert Hendrichs, Dietrich von Holst, and the participants of the Dahlem Workshop for valuable comments on the manuscript. The research was supported by the German Science Foundation (DFG) and the Volkswagen Foundation.

REFERENCES

Arzdorf, W. 1990. Stand, Möglichkeiten und Grenzen von Zucht und Auswilderung bedrohter Tierarten als Beitrag zum Artenschutz. Gefördertes Forschungs- und Entwicklungsvorhaben des Bundesministeriums für Umwelt, Naturschutz und Reaktorsicherheit. Cologne: Zoologischer Garten.

Bradley, A.J., I.R. McDonald, and A.K. Lee. 1980. Stress and mortality in a small marsupial (*Antechinus stuartii*, Mcleay). *Gen. Comp. Endocrinol.* **40**:188–200.

Broom, D.M. 1986. Indicators of poor welfare. *Br. J. Vet.* **142**:524–526.

Broom, D.M. 1998. Welfare, stress, and the evolution of feelings. *Adv. Study Behav.* **27**:371–403.

Broom, D.M., and K.G. Johnson. 1993. Stress and Animal Welfare. London: Chapman and Hall.

Clark, J.D., D.R. Rager, and J.P. Calpin. 1997. Animal well-being. I. General considerations. *Lab. Animal Sci.* **47**:564–570.

Darwin, C. 1859. The Origin of Species. London: Murray.

Dawkins, M.S. 1990. From an animal's point of view: Motivation, fitness, and animal welfare. *Behav. Brain Sci.* **13**:1–61.

Dawkins, M.S. 1998. Evolution and animal welfare. *Q. Rev. Biol.* **73**:305–328.

Fraser, D. 1999. Animal ethics and animal welfare science: Bridging the two cultures. *Appl. Animal Behav. Sci.* **65**:171–189.

Fraser, D., D.M. Weary, E.A. Pajor, and B.N. Miligan. 1997. A scientific conception of animal welfare that reflects ethical concerns. *Animal Welfare* **6**:174–186.

Fraser, D., and I.J.H. Duncan. 1998. "Pleasures," "pains," and animal welfare: Towards a natural history of affect. *Animal Welfare* **7**:383–396.

Gandelman, R. 1992. The Psychobiology of Behavioral Development. New York: Oxford Univ. Press.

Henry, J.P., and P.M. Stephens. 1977. Stress, Health, and the Social Environment. A Sociobiologic Approach to Medicine. New York: Springer.

Hughes, B.O., and I.J.H. Duncan. 1988. The notion of ethological "need", models of motivation and animal welfare. *Animal Behav.* **36**:1696–1707.

Kaiser, S., D. Classen, and N. Sachser. 1999. Auswirkungen unterschiedlicher struktureller Anreicherungen auf das Spontanverhalten weiblicher Labormäuse (Stamm NMRI). Aktuelle Arbeiten zur artgemäßen Tierhaltung 1998. *KTBL-Schrift* **382**:56–62.

Kaiser, S., and N. Sachser. 1998. The social environment during pregnancy and lactation affects the female offspring's endocrine status and behaviour in guinea pigs. *Physiol. Behav.* **63**:361–366.

Künzl, C., and N. Sachser. 1999. The behavioral endocrinology of domestication: A comparison between the domestic guinea pig (*Cavia aperea* f. *porcellus*) and its wild ancestor, the cavy (*Cavia aperea*). *Horm. Behav.* **35**:28–37.

Mendoza, S.P., and W.A. Mason. 1986. Contrasting responses to intruders and to involuntary separation by monogamous and polygynous new world monkeys. *Physiol. Behav.* **38**:795–801.

Poole, T. 1997. Happy animals make good science. *Lab. Animals* **31**:116–124.

Sachser, N., 1994. Social stratification and health in non-human mammals. In: Social Stratification and Socioeconomic Inequality, vol. 2, ed. L. Ellis, pp. 113–121. Westport: Praeger.

Sachser, N. 1998. Of domestic and wild guinea pigs. Studies in sociophysiology, domestication, and social evolution. *Naturwissenschaften* **85**:307–317.

Sachser, N., and R. Beer. 1995. Long-term influences of social situations and socialisation on adaptability in behaviour. In: Research and Captive Propagation, ed. U. Gansloßer, J.K. Hodges, and W. Kaumanns, pp. 207–214. Fürth: Filander.

Sachser, N., D. Hierzel, and M. Dürschlag. 1998. Social relationships and the management of stress. *Psychoneuroendocrinology* **23**:891–904.

Sachser, N., and C. Lick. 1991. Social experience, behavior, and stress in guinea pigs. *Physiol. Behav.* **50**:83–90.

Sachser, N., C. Lick, and K. Stanzel. 1994. The environment, hormones and aggressive behaviour — A five-year study in guinea pigs. *Psychoneuroendocrinology* **19**:697–707.

Von Holst, D. 1998. The concept of stress and its relevance for animal behavior. *Adv. Study Behav.* **27**:1–131.

Wiepkema, P.R. 1985. Abnormal behaviours in farm animals: Ethological implications. *Neth. J. Zool.* **35**:279–299.

Wiepkema, P.R. 1987. Behavioural aspects of stress. In: Biology of Stress in Farm Animals, ed. P.R. Wiepkema and P.W.M. van Adrichem, pp. 113–133. Dordrecht: Martinus Nijhoff.

4

Life, Liberty, and the Pursuit of Happiness

Good Welfare in Humans

S.K. LUTGENDORF
Deptartment of Psychology, University of Iowa, Iowa City, IA 52242, U.S.A.

Keywords: positive psychology, hypothalamic-pituitary-adrenal axis, immunity

> You can have your wealth, your power, your achievement and recognition, but when the bookkeeping is done, the best bottom line is health and love of family.
>
> —Mike Donner, age 93

ABSTRACT

Although positive factors are important in human psychology, until recently there has been little empirical examination of good welfare. This essay focuses on various models of welfare, including Maslow's hierarchy of needs and Ryff's multidimensional model of well-being. Internal influences such as temperament and external influences such as income, age, occupation, and social support that appear to influence welfare are also examined. Relationships of welfare with stress response systems, such as the sympathetic adrenomedullary and hypothalamic-pituitary-adrenocortical axes and immune correlates of welfare are discussed. Future research in this area should include examinations of physiological systems such as oxytocin and opioid systems, which may specifically underlie states of good welfare.

INTRODUCTION

Although the quest for happiness and life satisfaction seems to occupy a central place in human motivation, investigations of human unhappiness and suffering have received much more attention than investigations of good welfare in the psychological literature (Diener 1984; Jahoda 1958). In fact, an electronic search of *Psychological Abstracts* from 1887 until the present identified 136,728 articles on the topics of anger, anxiety and depression, whereas

Coping with Challenge: Welfare in Animals including Humans
Edited by D.M. Broom

only 851 abstracts referred to joy, 2958 happiness, and 5701 life satisfaction (Myers 2000). However, a shift in attention to more positive aspects of functioning and a science of positive psychology is highlighted by the dedication of the first issue in the year 2000 of the *American Psychologist* to a special issue that focused on "Happiness, Excellence, and Optimal Human Functioning." In this chapter, I will first cover conceptualizations of good welfare in humans and then will discuss their relationships with the immune response and possible mechanisms underlying such relationships.

MOTIVATIONAL THEORIES OF GOOD WELFARE

Historically, several figures have contributed to the conceptualization of good welfare. Freud (1961) proposed that "Liebe und Arbeit" or "love and work" were essential components to human welfare. Jung (1933) proposed the concept of individuation, suggesting that individuals need to pursue their own unique potential to be psychologically healthy. Other theorists have approached this issue from a life-cycle perspective. These include Erickson (1959), who considered the key to good welfare to be adequately meeting the appropriate challenges at different stages of the life cycle, and Neugarten (1973), who described factors underlying successful personality development in adulthood and old age.

Maslow (1968) outlined one of the most well-known theoretical formulations regarding what is necessary for good welfare. His conceptualization was based on the assumption of a fundamental motivational striving towards positive growth, which was considered to be latent in all humans. Maslow introduced the concept of a hierarchical model of human needs. At the lowest level was satisfaction of basic biological needs, such as food, clothing, shelter, and safety. Maslow proposed that only after these basic needs were fulfilled could individuals pursue needs at higher levels, such as those for social contact, creative expression, and fulfillment of their own unique potential. Although Maslow's model has enjoyed great popularity, it has also been criticized, particularly with respect to the hierarchical nature of need fulfillment, and his underlying assumptions have been difficult to substantiate with empirical research. For example, beyond the lowest level of the hierarchy, there is little evidence that human motives or needs are ordered in the exact sequence that Maslow proposed (Wahba and Bridewell 1976). In fact, other theorists, such as Bowlby (1988), who emphasized the key role of attachment in development, might have put factors such as love and social contact at the base of the hierarchy as no less essential to human survival than food, clothing, shelter, and other biological needs.

In contrast to theorists who emphasized more psychological aspects of welfare, Selye (1974) took a more biological approach to the concept of the optimal state of life challenge and positive motivation. He proposed the concept of "eustress," suggesting that the ideal life situation was not one without stress, but a life situation that presented enough stimuli to challenge, but not overstimulate the individual.

INDIVIDUAL DIFFERENCE CHARACTERISTICS AND WELFARE

Whereas Maslow and Selye described motivational models that would apply to all people, the question has been posed whether it is an individual's life circumstances or specific traits about them that determine whether or not they feel good about their lives. Antonovsky (1987), an Israeli psychologist, addressed this question by attempting to understand why, if stress is injurious to health, did Holocaust survivors, who had undergone some of the most traumatic stress known to humankind, not all fall prey to severe chronic illness. In examining these survivors, and who stayed well versus those who got sick, Antonovsky (1987) developed a concept which he called "sense of coherence." He described coherence as having three components: (a) confidence in the predictable, structured, and understandable nature of one's environment (comprehensibility), (b) belief in one's ability to meet environmental demands successfully (manageability), and (c) ability to interpret demands as meaningful challenges worthy of investment of energy (meaningfulness). Coherence is seen as a trait characteristic similar to resilience, which enables an individual to maximize appropriate coping resources in any specific situation (Antonovsky 1987). This factor has been found to be a strong predictor of physical and mental health outcomes and life satisfaction in adults of many ages throughout the world (e.g., Callahan and Pincus 1995; Sagy et al. 1990; Steiner et al. 1996).

Hardiness

About the same time, Kobasa (1979) investigated characteristics that differentiated high stress executives who remained healthy from those who got sick. Whereas she identified only a modest correlation between life stress and the incidence of illness, the characteristics she came to define as "hardiness" were able to differentiate these two groups of executives. She described a "hardy" personality as consisting of three facets that were quite similar to those comprising the sense of coherence: (a) seeing life events as *challenges* rather than as threats, (b) having a sense of *commitment* to one's activities, and (c) having a sense of *control* regarding one's life and activities. Although there has been debate about the key elements of hardiness, and the nature of its effects on health have been challenged (Funk and Houston 1987), Kobasa's work has been instrumental in stimulating research on how personality affects health and tolerance of stress.

Optimism

The trait of dispositional optimism is another individual difference characteristic that has been associated with good welfare. Optimism is defined as a general tendency to hold positive expectancies for one's future (Scheier and Carver 1993). Although the cognitive components of optimism have been emphasized by some theorists, this construct also involves motivation and emotion. Optimism has been associated with qualities such as positive mood, perseverance and effective problem solving, occupational success, good health, long life, and freedom from trauma (Peterson 2000). Early in the 20th century, theorists and philosophers argued that optimism prolongs suffering and that it is therefore better in the long run to face the harsh truths of reality. For example, Freud argued that although it was a common emotional state, optimism was illusory. Freud's ideas led to the conception that mental health involves accurate perceptions of reality. Views of optimism began to change in the 1960s and 1970s, as cognitive psychology identified positive biases contained in language, memory, and thought patterns (Peterson 2000). Lazarus (1983) described a notion he called "positive denial," which was associated with good welfare during adversity. Taylor (1989) reported that a general positive bias or the maintenance of positive illusions, however unfounded, is a component of good welfare. Taking this argument one step further, Tiger (1979) argued that optimism is one of the most defining and adaptive characteristics of the human species. In his book, *Optimism: The Biology of Hope* (1979), Tiger proposed that in the course of evolution, optimism results from positive selection because it conferred an adaptive advantage by counteracting the fear and paralysis that thoughts of mortality might engender. He also argued that optimism developed along with our cognitive abilities and capacity for culture. From this perspective, optimism is not a derivative of other psychological characteristics, rather it is inherent in the human psyche (Buss 2000).

Scheier and Carver (1987, 1992) considered optimism a dispositional trait, or personality variable. In fact, a twin study indicated that the heritability of optimism is about 25% (Plomin 1994). Carver and Scheier found associations between optimism and relatively good physical health in a sample of college students and have also determined that optimists cope with stress differently than do pessimists (Scheier et al. 1986). Optimists are more likely to engage in action-oriented, problem-focused coping and have better welfare under challenge than do pessimists. They also are more willing to seek social support and more likely to emphasize the positive in their appraisals of stressful events. In comparison, pessimists are more likely to deal with challenge by giving up or engaging in denial. Carver and colleagues (1993) also found that the effects of optimism on mood in breast cancer patients are mediated by active coping. Specifically, optimism in these patients was associated with greater use of coping strategies such as accepting reality, looking at the positive aspects of a situation, using humor, and active problem-solving. In fact, optimists more realistically accepted their situation, sought information, and made the best of it. Similarly, Taylor and colleagues (1992) found that among men who tested both positive and negative for HIV, optimists have more positive attitudes, less self-blame, less fatalism, and lower distress.

The more adaptive strategies utilized by optimists, hardy individuals, and individuals with a high sense of coherence have led some theorists to propose that positive mood is evolutionarily adaptive. For example, Watson (2000) argues that general euthymic (positive mood) states are essential to ensure that humans obtain the resources necessary for survival of the species (e.g., shelter, food, sexual partners). A positive selection bias would have been conferred on individuals with energy, confidence, and adaptive coping strategies, in contrast to individuals possessing qualities associated with negative mood. Thus, individuals with lower levels of positive mood would have been at an evolutionary disadvantage.

Can optimism be learned, or are we victims of our dispositions? Seligman (1991) states that optimism can be learned and has written a book on that thesis. He proposed that people have habitual patterns for explaining negative events, called explanatory styles, and that such styles are thought to influence the perception of helplessness following adversity. The particularly pernicious aspect of a negative explanatory style is that it is habitual. If individuals can be taught to identify their bias towards a negative explanatory style, they can learn to be more flexible in their use of explanatory style as circumstances warrant and can change their bias to a more optimistic explanatory style when appropriate. Seligman (1991) proposed that individuals can learn to utilize optimism with flexibility, when appropriate, as opposed to invoking a reflexive or habitual pessimism over which the individual has no control.

INTERNAL VERSUS EXTERNAL DETERMINANTS OF WELFARE

Numerous studies have addressed the question of what is needed to achieve good welfare. Perceived "life satisfaction," "subjective well-being," and "happiness" have tended to be the outcome variables of choice in many studies. Happiness has been variously defined as a transitory feeling of joy, a more general feeling of life satisfaction, and a balance between positive and negative affect (Bradburn 1969, Ryff 1989). In answer to the question, "what is the good life," Diener (2000) proposed that this is a question each individual must answer for themselves, and thus arrived at the construct "subjective well-being," which he equated with perceived happiness.

Happiness

What does it take for people to consider themselves as happy? When 7,204 college students in 42 countries were asked how important is "life satisfaction," "happiness," or "money," only 6% of the total sample rated money as more important than happiness, while 69% rated happiness highest on the importance scale (Diener 2000). A 1976 study estimated that ten resources, including income, number of friends, religious faith, intelligence, and education, contributed only 15% to reported "happiness," with the rest of the variance presumably derived from internal sources (Campbell et al. 1976). Similarly, physical attractiveness was found to be only minimally associated with reports of positive and negative affect and life satisfaction among college students (Diener et al. 1995). Even more surprisingly, several studies have indicated that physical health, even among the elderly, is minimally correlated with perception of welfare (Okun and George 1984). External factors such as age, gender, education, and type of occupation have been shown to have minimal associations with self-reports of happiness. An individual's reported happiness appears to be stable over the life span, suggesting that this may be a stable personality trait and not just influenced by transitory states. Even employment does not contribute greatly to happiness unless one is involuntarily unemployed. In contrast, numerous studies indicate that people who have strong social supports as well as those who have strong spiritual or religious belief systems report that they are happier (Myers 2000).

Social Support

Social support, frequently defined as the degree of one's perceived satisfaction with social relationships, is an important psychological resource that contributes to good welfare (Cohen and Wills 1985). Emotional support, including receipt of understanding, concern, and reassurance from others, is thought to be important because it can help to restore self-esteem, decrease feelings of isolation, and provide for the important feelings of connection with others. Instrumental support, such as receipt of help, information, and advice from other people, can increase feelings of control by providing information and practical assistance that may help individuals cope more effectively with potential stressors (House and Kahn 1985; Helgeson and Cohen 1996). There is a substantial body of research relating higher levels of social support to lower morbidity and mortality from a variety of diseases as well as to better welfare (e.g., House et al. 1988; Uchino et al. 1996).

Income appears to be related to happiness only insofar as it is necessary for affording life's basic necessities. Specifically, a relation between income and happiness exists only in countries where the gross national product (GNP) is less than US$ 8000. When the GNP rises above US$ 8000, worldwide, the relationship between per capita income and welfare is no longer detectable. In the United States, happiness is reported to be lower among the very poor. However, the very rich are only slightly happier than the average American (Myers 2000). Additionally, studies of people whose income increased or decreased in the last decade showed no difference in rates of happiness (Diener et al. 1993). Thus, other than the importance of a religious or spiritual framework and social supports, there is a lack of evidence supporting a contribution of external factors to reported happiness.

The concept of the "hedonic treadmill" (Brickman and Campbell 1971) has been proposed, suggesting that as people rise in their accomplishments and possessions, their expectations also rise, thus leading to continued craving and dissatisfaction. People habituate to goals once they are achieved, and these new acquisitions no longer make them happy, and thus they strive for more. Interestingly, this concept is similar to a fundamental tenet underlying many eastern spiritual traditions, i.e., that the basic nature of humans is to strive for what they do not have; therefore, happiness is forever fleeting unless the individual is able to rise above the craving that is an essential part of human nature (e.g., Bodhi 1994; Meher Baba 1987). Although there is evidence supporting the existence of such a "treadmill," alternate evidence suggests that individuals adapt to their circumstances, both positive and negative, over time and return to their original level of happiness (or set-point). The evidence supporting the existence of a happiness set-point suggests that happiness or life satisfaction is strongly influenced by temperament (Diener 2000). This role of temperament in perception of happiness is consistent with findings of high concordance of happiness among twins separated by adoption, suggesting that a large percentage of the variance in perceived welfare is due to heritability (Tellegen et al. 1988). Additionally, as discussed below, an apparent bias towards reported "happiness" has emerged in several large-scale investigations of life satisfaction, suggesting that there may be a general affective set-point located in the positive range of emotions.

For example, repeated studies by the National Opinion Research Council in the U.S. indicate that generally three in ten Americans say they are "very happy," six out of ten say they are "pretty happy," and only one in ten says they are "not too happy" (Myers 2000). An aggregated study from 916 surveys of 1.1 million people in 45 nations (Myers and Diener 1996) indicated that the average rating of perceived welfare on a scale from 0 (very unhappy, completely dissatisfied with life) to 10 (the high extreme) was approximately 6.75. In a survey conducted in the Detroit area, residents were asked to identify, on a Faces Scale, "Which face comes closest to expressing how you feel about your life as a whole?" More than 90% of the people picked one of the three happy faces, suggesting a general perception of good welfare from an independent type of assessment (Andrews and Withey 1976).

In summarizing the existing knowledge on happiness, Watson (2000) concluded that happiness is primarily a subjective phenomenon that is much more highly influenced by subjective internal processes such as beliefs, temperament, and set-point than by external factors such as one's income and occupation. He also argued that even though there may be genetic and biological underpinnings to perceived welfare, these may function to constrain the upper limits of welfare for a particular individual. However, individual actions to improve one's lot still make a difference in perceived welfareas most people have not reached the upper bound of their biological "reaction range."

Core Characteristics of Good Welfare

Although there have been many investigations of "well-being," "satisfaction," and "happiness," defining the essential characteristics of good welfare has been relatively neglected. Ryff (1989; Ryff and Keyes 1995) analyzed many earlier perspectives of well-being and life satisfaction, and integrated them into a more parsimonious summary of what constitutes good welfare. Points of convergence in prior theories were what Ryff described as the core dimensions of well-being:

1. self-acceptance (positive evaluation of oneself),
2. positive interpersonal relationships,
3. autonomy (sense of self-determination),
4. environmental mastery (capacity to effectively manage one's life and environment),
5. belief that one's life is purposeful and meaningful,
6. personal growth (sense of continued growth and development as a person).

A factor analysis of these findings on a national sample of 1,108 individuals upheld this six-factor model. However, one superordinate factor was found. This suggests that the six factors may be a function of this overarching factor of good welfare. As in previous studies, demographic factors and gender were not significant predictors of welfare. However, the relative presence of each of these factors varies with age. For example, environmental mastery and autonomy increased linearly with age, but purpose in life and personal growth decreased following midlife. There was no change in self-acceptance and positive relations with others over the life span.

In a recent review of the literature, Watson (2000) elaborated on the meanings of purpose and growth, arguing that not only attaining goals, but the very process of striving after goals is critical for happiness. He proposed that happiness rests on the perceptions that there are goals worth pursuing and that one is making reasonable progress towards these goals.

QUALITY OF LIFE AMONG MEDICAL POPULATIONS

Many of the above conceptualizations have referred to healthy individuals. An alternate perspective in viewing the issue of welfare involves how medical patients view their life satisfaction (Spitzer 1987). The importance of "quality" and not just "quantity" of life to medical patients was highlighted by Karnofsky and colleagues (1948) when they developed a measure of functional abilities for cancer patients as an adjunct to the more traditional statistic "length of survival" time as an indicator of treatment success. About the same time, the World Health Organization (WHO) adopted a definition of health in its constitution stating, "Health is not only the absence of infirmity and disease, but also a state of physical, mental, and social well-being" (WHO 1947, p. 29). The "individual's ability to function and derive satisfaction from a variety of roles" is seen as a core attribute of quality of life (Wenger et al. 1984, p. 908), which is composed of (a) functional capacity, or the ability to perform activities of daily life in social, intellectual, emotional, and economic domains; (b) perceptions of welfare and satisfaction with life; and (c) physical impairment, including objective disease symptoms and disability.

A more phenomenological approach to quality of life has emphasized the primacy of subjective perception of current ability to function as compared with a patient's internalized standards of what is (a) possible or ideal (Cella and Tulsky 1990; Ziller 1974) or, alternatively, (b) with their ability to achieve their life ambitions (Cohen 1982). The ability to alter expectations, and thereby increase perceived satisfaction, has been called "response shift" (Schwartz and Sprangers 2000). The presence of response shift in medical patients is highlighted by an interesting study in which healthy subjects, postmyocardial infarction (MI) patients, and bronchial cancer patients were asked to rate their general satisfaction with the quality of their life. A much larger proportion of MI and cancer patients than healthy adults rated themselves as very satisfied, presumably due to their alterations in their expectations of what might be possible for them (Muthny et al. 1990). Such findings highlight the importance of understanding the degree to which individuals can adapt to adverse circumstances and still consider themselves to have high life satisfaction.

WELFARE AND THE IMMUNE RESPONSE

Given this attention to positive psychology, it is important to address how factors such as optimism, sense of purpose and meaning, happiness, and social support may affect physiology, in particular, the immune response. There has been a paucity of research examining physiological mechanisms by which positive psychological factors might influence immunity. One of the important questions concerns whether positive states have direct positive effects on physiology (i.e., are these positive factors helpful whether or not the person has an illness or life challenge) or whether the effects of positive factors on physiology are buffering effects (i.e., they are only operative in the face of life stress or challenge). A second question concerns whether these states operate by attenuation of what are traditionally thought as stress response mechanisms (e.g., the sympathetic-adrenomedullary axis, and the hypothalamic-pituitary-adrenal (HPA) axis or whether there are different physiological underpinnings to positive states.

Direct Protective Role of Positive States

Positive life events and positive psychological states have been shown to have both direct and buffering effects on immune function. For example, in a sample of fifty law students in their first semester of law school, situational optimism was associated with better mood and a more adequate immunological response as indicated by higher numbers of CD4+ (helper) T cells and greater natural killer (NK) cell cytotoxicity (Segerstrom et al. 1998). Direct effects of positive daily events on immune function have been reported in a community study of adults, among whom more vigorous salivary antibody (sIgA) responses to oral challenges with a novel antigen were demonstrated by individuals reporting greater incidence of positive daily events (Stone et al. 1994). In an epidemiological sample of over 300 older adults, Lutgendorf and colleagues have recently found that both social support and church attendance predict lower levels of IL-6 both concurrently and six years prospectively (unpublished data).

A reliable positive relationship between social support and functional immune measures has been supported by a meta-analysis of 19 studies whose researchers examined the association between social support and measures of functional immunity (Uchino et al. 1996). Several models have been proposed by which social support may affect immune function. A direct model proposes that even in the absence of stress, social support is beneficial for health-related outcomes (Cohen and Wills 1985). These direct effects may be derived via enhancement of positive behavioral health practices and attitudes, or by the comfort of social relationships (Cohen and Wills 1985). For example, a recent investigation of men participating in a prostate-specific antigen (PSA) screening program found that high perceived stress was associated with elevated levels of PSA. In contrast, satisfaction with social support was associated with lower levels of PSA. There was no interaction effect, suggesting that social support did not buffer the relationship between stress and PSA.

Buffering Effects of Positive States

Support for a buffering hypothesis comes from several studies reporting effects of positive states on the immune response, but only in the presence of negative mood, negative conditions, or life stress. For example, one report indicated that greater daily positive mood was related to higher levels of NK cell activity in a sample of healthy nonelderly women, but only in those women who experienced some negative mood as well (Valdimarsdottir and Bovbjerg 1997). Vitaliano and colleagues (1998) found that uplifts in the presence of challenges explained a significant portion of NK cytotoxic activity in older adult caregivers with cancer histories. Lutgendorf and colleagues (1999) found that older adults facing anticipated voluntary housing relocation who had a low sense of coherence showed very low levels of NK cell activity, while NK cell activity of movers with high sense of coherence showed less compromise, and was commensurate with levels of healthy controls (Lutgendorf et al. 1999). This suggests that during the course of a life transition among older adults, a sense of coherence may be protective and may buffer the effects of challenge upon the immune response. A similar finding was reported by Bower and colleagues (1998), when they studied 40 HIV-seropositive men who had recently experienced the loss of a close friend or partner to AIDS. Only the men who reported finding meaning in the experience maintained their levels of CD4+ helper T cells over the follow-up period. Those bereaved individuals who did not think through the events related to the bereavement, or who did think through the events but failed to find meaning in their loss, showed a decline in CD4+ helper cells over a 2–3 year follow-up period. These effects remained significant even controlling for initial CD4+ cell counts, health habits, and number of HIV-related symptoms. In addition, emergence of a sense of meaning following bereavement was associated with a significantly lower rate of AIDS-related mortality over a 4–9 year follow-up period. These findings highlight the importance of positive factors, such as development of a sense of meaning, as potential buffers to the negative effects of the stress of bereavement on the immune response.

Another investigation examined effects of a ten-week randomized cognitive behavioral stress management intervention among 18 early stage breast cancer patients and ten wait-list controls following their initial treatment. The intervention was designed to increase positive appraisals and sense of control, decrease negative mood, and teach relaxation skills in the face of the life stressor of breast cancer treatment and adjusting to life with breast cancer.

Women participating in the intervention reported greater positive growth from the experience of breast cancer and also showed a greater increase in T-cell lymphoproliferative response to stimulation by CD-3 than did wait-list controls. In addition, greater increases in self-report of positive growth were significantly associated with the immune increases, suggesting a relationship between this positive factor and the immune enhancement noted in the study. These findings suggest that helping patients with chronic and life-threatening illnesses find ways to appraise their life situation, to allow greater positive understanding, may have positive physiological effects, including effects on the immune response (McGregor et al., under review). Although several studies have suggested relationships between aspects of the immune response and breast cancer prognosis, it is not known whether these findings will have clinical significance.

A variety of studies have shown the ability of social support to moderate the negative effects of life stress on immune functioning (Uchino et al. 1996). For example, social support has been shown to moderate decrements in immune responses to acute laboratory stressors (Kiecolt-Glaser et al. 1994) and to the chronic stress of Alzheimer's caregiving (Esterling et al. 1994). Specifically, caregivers reporting greater emotional and tangible social supports and greater closeness in their relationships showed less decrements in cytokine-enhanced NK cell activity and less physician visits for infectious illnesses than caregivers reporting lower levels of social support (Esterling et al. 1994). Greater perceived social support has been related to higher NK cell activity and greater lymphoproliferative responses to mitogens in an older sample of spouses of urologic cancer patients (Baron et al. 1990). Among Stage I and II breast cancer patients following surgery, Levy et al. (1991) found that greater NK cell cytotoxicity was related to patients' perceptions of emotional support from their spouse, greater perceived support from their physician, and seeking social support as a coping strategy.

Psychological Mechanisms Underlying These Relationships

What might be the psychological mechanisms underlying these types of findings? It has been noted that optimists tend to engage in better health practices. For example optimistic men being tested for AIDS had fewer anonymous sex partners (Taylor et al. 1992), and optimistic patients in a cardiac rehabilitation program lowered body fat and coronary risk and increased exercise more than nonoptimistic patients (Shepperd et al. 1996). Additionally, optimists are more likely to cope more adaptively, to try to accept reality, to grow personally from bad experience, to show planful coping, and are less likely to engage in avoidance coping (Carver et al. 1993). Individuals with greater social support are thought to be equipped with better information, to engage in positive health practices and preventive health behaviors, and to have earlier identification of illnesses, as well as the stress-reducing properties of social support (Cohen and Wills 1985).

According to the buffering model, positive factors such as sense of coherence, optimism, or social support would influence the appraisal of stressful situations. These factors would likely (a) increase the probability that a situation will be perceived as nonthreatening, thus decreasing distress or arousal accompanying a potentially stressful situation, (b) increase the probability that difficulties would be seen as challenges that could be successfully managed, thus increasing coping abilities and decreasing perceived stress, and (c) decrease potential physiological reactivity to the situation (Antonovsky 1987).

Specifically, a general tendency to appraise stressor-associated stimuli as benign or challenging rather than as threatening would likely result in attenuated SNS or HPA reactivity in response to stressful life situations, because they would be perceived as less threatening. Additionally, lower overall levels of distress and arousal would result in lower tonic levels of sympathetic arousal and HPA hormones accompanying challenging life situations. Taken together, these factors may create a situation of less strain, or allostatic load (McEwen 1998), in individuals with high levels of positive factors. Consistent with this argument, lower physiological reactivity in response to a laboratory stressor has been observed in individuals with high sense of coherence (McSherry and Holm 1994).

In the MacArthur studies on successful aging, Seeman and colleagues (1994; Seeman and McEwen 1996) argued that social support appears to buffer against neuroendocrine overactivation in the elderly and demonstrated that such protection may contribute to better health, cognitive function, and longevity among older adults. It is possible that similar pathways underlie buffering relationships of other positive psychological factors, such as sense of coherence or optimism.

In addition to attenuation of stress pathways, it is likely that there are other physiological pathways underlying positive states. These include neuropeptide and hormonal systems such as opioids or the oxytocin system which have been proposed to be associated with positive social relationships (see Carter, this volume). Pathways underlying relationships of positive factors with physiology are largely still to be determined and may represent an important direction for future research. Examination of interventions to increase positive factors is also an important frontier for future research.

SUMMARY

Good welfare is an important concept in human motivation, yet until recently empirical examination of this area has been limited. This chapter has focused on various conceptualizations of welfare and examined influences that appear to underlie it. Internal and external contributions to good welfare were examined. These various dimensions may shed light on what aspects of an individual's welfareare potentially changeable and which are resistant to change. Relationships between well-being and immune function were explored. The critical importance of welfare, both for both mental and physical health, suggests that an important task of modern societies is to find ways to increase perceptions of good welfare as a way of buffering the deleterious physiological effects of modern social and technological life. Understanding the physiological pathways that underlie states of good welfare is an important research direction for the future.

ACKNOWLEDGMENTS

I would like to gratefully acknowledge Nan Rothrock, Erin Costanzo, Phillip Ullrich, Erica Johnsen, Anna Hoffman, James Bunde, Heena Maiseri, Megan Septer, Vanessa Parks, Colleen Clarke, Margaret Henderson, and Tracy Bronemann for their helpful assistance in developing this paper.

REFERENCES

Andrews, F.M., and S.B. Withey. 1976. Social Indicators of Well-being: America's Perception of Life Quality. New York: Plenum.

Antonovsky, A. 1987. Unraveling the Mystery of Health: How People Manage Stress and Stay Well. San Francisco: Jossey-Bass.

Baron, R., C. Cutrona, D. Hicklin, D. Russel, and D. Lubaroff. 1990. Social support and immune function among spouses of cancer patients. *J. Personality & Soc. Psychol.* **59**:344–352.

Bodhi, B. 1994. The Noble Eightfold Path: The Way to the End of Suffering. Kandy, Sri Lanka: Buddhist Publ. Soc.

Bower, J.E., M.E. Kemeny, E.S.D. Taylor, and J.L. Fahey. 1998. Cognitive processing, discovery of meaning, CD4 decline, and AIDS-related mortality among bereaved HIV-seropositive men. *J. Consult. & Clin. Psychol.* **42**:861–865.

Bowlby, J. 1988. Attachment. New York: Basic Books.

Bradburn, N.M. 1969. The Structure of Psychological Well-being. Chicago: Aldine.

Brickman, P., and D.T. Campbell. 1971. Hedonic relativism and planning the new society. In: Adaptation-level Theory, ed. M.H. Appley, pp. 287–305. New York: Academic.

Buss, D.M. 2000. The evolution of happiness. *Am. Psychol.* **55**:15–23.

Callahan, L.F., and T. Pincus. 1995. The sense of coherence scale in patients with rheumatoid arthritis. *Arthritis Care & Res.* **8**:28–33.

Campbell, A., P.E. Converse, and W.L. Rodgers. 1976. The Quality of American Life. New York: Russell Sage Foundation.

Carver, C., C. Pozo, S.D. Jarris, V. Noriega, M.F. Scheier, D.S. Robinson, A.S. Ketcham, F.L. Moffat, and K.C. Clark. 1993. How coping mediates the effect of optimism on distress: A study of women with early stage breast cancer. *J. Personality & Soc. Psychol.* **65**:375–390.

Cella, D., and D.S. Tulsky. 1990. Measuring quality of life today: Methodological aspects. *Oncology* **4**:29–38.

Cohen, C. 1982. On the quality of life: Some philosophical reflections. *Circulation* **66(3)**:29–33.

Cohen, S., and T.A. Wills. 1985. Stress, social support, and the buffering hypothesis. *Psychol. Bull.* **98**:310–357.

Diener, E. 1984. Subjective well-being. Psychological well-being. *Psychol. Bull.* **95**:542–575.

Diener, E. 2000. Subjective well-being: The science of happiness and a proposal for a national index. *Am. Psychol.* **55**:34–43.

Diener, E., E. Sandvik, L. Seidlitz, and M. Diener. 1993. The relationship between income and subjective well-being: Relative or absolute? *Soc. Indic. Res.* **28**:195–223.

Diener, E., B. Wolsic, and F. Fujita. 1995. Physical attractiveness and subjective well-being. *J. Personality & Soc. Psychol.* **5469**:120–129.

Erickson, E. 1959. Identity and the life-cycle. *Psychol. Iss.* **1**:18–164.

Esterling, B., J. Kiecolt-Glaser, J. Bodnar, and R. Glaser. 1994. Chronic stress, social support, and persistent alterations in the natural killer cell response to cytokines in older adults. *Health Psychol.* **13**:291–298.

Freud, S. 1961. Civilization and Its Discontents. New York: Norton.

Funk, S.C., and B.K. Houston. 1987. A critical examination of the hardiness scale's validity and utility. *J. Personality & Soc. Psychol.* **53**:572–578.

Helgeson, V.S., and S. Cohen. 1996. Social support and adjustment to cancer: Reconciling descriptive, correlational, and intervention research. *Health Psychol.* **15**:135–148.

House, J.S., and R.L. Kahn. 1985. Measures and concepts of social support. In: Social Support and Health, ed. S. Cohen and L. Syme, pp. 83–108. Orlando: Academic.

House, J.S., K.R. Landis, and D. Umberson. 1988. Social relationships and health. *Science* **241**: 540–545.

Jahoda, M. 1958. Current concepts of positive mental health. New York: Basic Books.

Jung, C.G. 1933. Modern Man in Search of a Soul. Trans. W.S. Dell and C.F. Baynes. New York: Harcourt Brace and World.

Karnofsky, D.A., W.H. Ablemann, L.F. Craver, and J.H. Burchenal. 1948. The use of the nitrogen mustards in the palliative treatment of carcinoma. *Cancer* **1**:634–652.

Kiecolt-Glaser, J., W. Malarkey, J. Cacioppo, and R. Glaser. 1994. Stressful personal relationships: Endocriine and immune function. In: Handbook of Human Stress and Immunity, ed. R. Glaser and J. Kiecolt-Glaser, pp. 321–339. San Diego: Academic.

Kobasa, S.C.O. 1979. Stressful life events, personality and health: An enquiry into hardiness. *J. Personality & Soc. Psychol.* **37**:1–11.

Lazarus, A. 1983. The costs and benefits of denial. In: Denial of Stress, ed. S. Benitz, pp. 1–30. New York: Intl. Univ. Press.

Levy, S., R. Herberman, M. Lippman, T. D'Angelo, and J. Lee. 1991. Immunological and psychosocial predictors of disease recurrence in patients with early-stage breast cancer. *Behav. Med.* **17**:67–75.

Lutgendorf, S.K., P.P. Vitaliano, T. Tripp Reimer, J. Harvey, and D. Lubaroff. 1999. Sense of coherence buffers effects of life stress on NK activity in healthy older adults. *Psychol. & Aging* **14**:552–563.

Maslow, A. 1968. Toward a Psychology of Being, 2nd edition. Princeton: Van Nostrand Reinhold.

McEwen B.S. 1998. Stress, adaptation, and disease. Allostasis and allostatic load. *Ann. NY Acad. Sci.* **840**:33–44.

McSherry, W.C., and J.E. Holm. 1994. Sense of coherence: Its effects on psychological and physiological processes prior to, during, and after a stressful situation. *J. Clin. Psychol.* **50**:476–487.

Meher Baba. 1987. Discourses. Myrtle Beach, SC: Sheriar Press.

Muthny, F.A., U. Koch, and S. Stump. 1990. Quality of life in oncology patients. *Psychother. & Psychosom.* **54**:145–160.

Myers, D.G. 2000. The funds, friends, and faith of happy people. *Am. Psychol.* **55**:56–67.

Myers, D.G., and E. Diener. 1996. The pursuit of happiness. *Sci. Am.* **274**:54–56.

Neugarten, B.L. 1973. Personality change in late life: A developmental perspective. In: The Psychologicy of Adult Development and Aging, ed. C. Eisdorfer and M.P. Lawton, pp. 311–335. Washington, D.C.: Am. Psychol. Assoc.

Okun, M.A., and L.K. George. 1984. Physician and self ratings of health, neurotocisim, and subjective well-being among men and women. *Personality & Indiv. Diff.* **5**:533–539.

Peterson, C. 2000. The future of optimism. *Am. Psychol.* **55**:44–45.

Plomin, R., M.F. Scheier, C.S. Bergeman, N.L. Pedersen, J.R. Nesselroade, and G.E. McClearn. 1994. Optimism, pessimism, and mental health: A twin/adoption analysis. *Personality & Indiv. Diff.* **13**:921–930.

Ryff, C.D. 1989. Happiness is everything, or is it? Explorations on the meaning of psychological well-being. *J. Personality & Soc. Psychol.* **57**:1069–1081.

Ryff, C.D., and C.L.M. Keyes. 1995. The structure of psychological well-being revisited. *J. Personality & Soc. Psychol.* **69**:719–727.

Sagy, S., A. Antonovsky, and I. Adler. 1990. Explaining life satisfaction in later life: The sense of coherence model and activity theory. *Behav. Health & Aging* **1**:11–25.

Scheier, M.F., and C.S. Carver. 1987. Dispositional optimism and physical well-being: The influence of generalized outcome expectancies on health. *J. Personality* **55**:169–210.

Scheier, M.F., and C.S. Carver. 1992. Effects of optimism on psychological and physical well-being: Theoretical overview and empirical update. *Cog. Ther. & Res.* **16**:201–228.

Scheier, M.F., and C.S. Carver. 1993. On the power of positive thinking: The benefits of being optimistic. *Curr. Dir. Psychol. Sci.* **2**:26–30.

Scheier, M.F., J.K. Weintraub, and C.S. Carver. 1986. Coping with stress: Divergent strategies of optimists and pessimists. *J. Personality & Soc. Psychol.* **51**:1257–1264.

Schwartz, C.E., and M.A.G. Sprangers, eds. 2000. Adaptation to changing health: Response shift in quality-of-life research. Washington, D.C.: Am. Psychol. Assoc..

Seeman, T.E., L.F. Berkman, D. Blazer, and J.W. Rowe. 1994. Social ties and support and neuroendocrine function: The MacArthur studies of successful aging. *Ann. Behav. Med.* **16**:95–106.

Seeman, T.E., and B.S. McEwen. 1996. Impact of social environment characteristics on neuroendocrine regulation. *Psychosom. Med.* **58**:459–471.

Segerstrom S.C., S.E. Taylor, M.F. Kemeny, and J.L. Fahey. 1998. Optimism is associated with mood, coping, and immune change in response to stress. *J. Personality & Soc. Psychol.* **74**:1646–1655.

Seligman, M.E.P. 1991. Learned Optimism. New York: Knopf.

Selye, H. 1974. Stress Without Distress. Philadelphia: Lippincott.

Shepperd, J.A., J.J. Maroto, and L.A. Pbert. 1996. Dispositional optimism as a predictor of health changes among cardiac patients. *J. Res. Personality* **30**:517–534.

Spitzer, W.O. 1987. State of Science 1986: Quality of life and functional status as target variables for research. *Chronic Dis.* **40**:465–471.

Steiner, A., K. Raube, A.E. Stuck, H.U. Aronow, D. Draper, L.Z. Rubenstein, and J.C. Beck. 1996. Measuring psychosocial aspects of well-being in older community residents: Performance of four short scales. *Gerontologist* **36**:54–62.

Stone, A.A., J.M. Neale, D.S. Cox, A. Napoli, H. Valdimarsdottir, and E. Kennedy-Moore. 1994. Daily events are associated with a secretory immune response to an oral antigen in men. *Health Psychol.* **13**:440–446.

Taylor, S.E. 1989. Positive Illusions. New York: Basic Books.

Taylor, S.E., M.E. Kemeny, L.G. Aspinwall, S.C. Schneider, R. Rodriguez, and M. Herbert. 1992. Optimism, coping, psychological distress, and high-risk sexual behavior among men at risk for AIDS. *J. Personality & Soc. Psychol.* **63**:470–473.

Tellegen, A., D.T. Lykken, T.J. Bouchard, Z.K.J. Wilcox, N.L. Segal, and S. Rich. 1988. Personality similarity in twins reared apart and together. *J. Personality & Soc. Psychol.* **54**:1031–1039.

Tiger, L. 1979. Optimism: The Biology of Hope. New York: Simon and Schuster.

Uchino, B.N., J.T. Cacioppo, and J.K. Kiecolt-Glaser. 1996. The relationship between social support and physiological processes: A review with emphasis on underlying mechanisms and implications for health. *Psychol. Bull.* **119**:488–531.

Valdimarsdottir, H.B., and D.H. Bovbjerg. 1997. Positive and negative mood: Association with natural killer cell activity. *Psychol. & Health* **12**:319–327.

Vitaliano, P.P, J.M. Scanlan, H.D. Ochs, K. Syrjala, I.C. Siegler, and E.A. Snyder. 1998. Psychosocial stress moderates the relationship of cancer history with natural killer cell activity. *Ann. Behav. Med.* **20**:98–108.

Wahba, N., and L. Bridewell. 1976. Maslow reconsidered: A review of research on the need hierarchy theory. *Org. Behav. & Human Perf.* **15**:212–240.

Watson, D. 2000. Mood and Temperament. New York: Guilford.

Wenger, N.K., M.E., Mattson, C.D. Furberg, and J. Elinson. 1984. Assessment of quality of life in clinical trials of cardiovascular therapies. *Am. J. Cardiol.* **54**:908–913.

World Health Organization (WHO). 1947. Constitution of the World Health Organization. *WHO Chronicles* **I**:29.

Ziller, R.C. 1974. Self-other orientation and quality of life. *Soc. Indic. Res.* **1**:301–327.

5

How Can We Recognize and Assess Good Welfare?

M.S. DAWKINS

Department of Zoology, University of Oxford, Oxford OXl 3PS, U.K.

Keywords: animal welfare, emotion, feelings, health, psychological state

ABSTRACT

Animal welfare is a broad term which includes two components: physical health and psychological health. Physical health is fundamental and poses no particular problems since disease, injury, and deformities are universally and uncontentiously recognized as signs of poor welfare. Psychological health, however, is more problematical since animals do not just cope with existing challenges, they also anticipate threats to their welfare and behave as if it has already been diminished even before any physical challenge has occurred. Deciding whether animals are in a good psychological state or are suffering depends on establishing whether the animal is in a positive or negative emotional state which, in turn, can be established through what they find positively or negatively reinforcing. Other welfare indicators, such as activity patterns or hormone levels, should be empirically investigated to see whether or not they correlate with reinforcement value rather than simply collected together in checklists.

INTRODUCTION

One of the key concepts of this workshop was the idea of welfare, human or animal, in relation to coping with challenge. "Coping" implies that an organism is responding physiologically, or behaviorally, to something in its environment in such a way that it nullifies the potentially damaging effects of some challenge, such as a change in temperature. Fraser and Broom (1990) define coping as "having control of mental and bodily stability." Thus an animal "copes" with a drop in the temperature of its environment by various physiological and behavioral changes that result in its core body temperature remaining constant despite the vagaries of the outside world. Of course, if the drop in temperature is too large and the animal is unable to respond adequately, then core body temperature will drop and its welfare will be

Coping with Challenge: Welfare in Animals including Humans
Edited by D.M. Broom © 2001 Dahlem University Press

impaired even to the point where it may die. Hence the idea that animals may attempt, but fail, to cope with challenges (Broom and Johnson 1993).

I have no difficulty in seeing how the use of the word "coping" to define welfare applies to such obvious disturbances to health and other aspects of welfare as food shortages, water shortages, disease, predation, and other guises in which death stalks the world of wild animals. Natural selection favored animals that were able to respond to such challenges, i.e., that coped, and it was very harsh on the ones that made no response or attempted to cope and failed. However, some people consider that all that animals do is to attempt to react to things that are done to them. The dictionary definition of coping, "to contend or to deal with successfully," can be taken to imply that coping involves challenge first and reaction afterwards. If we ask, "How is she coping with the death of her mother?," we imply that there was a major event to which someone may have reacted in any of a number of ways. If we ask someone how they are and they reply, "I'm coping," we take this as a pretty low rating of their welfare as it suggests they are just about able to deal with each sling and arrow of outrageous fortune, managing to stagger back onto their feet before the next one comes along.

If all there were to welfare were reacting, our job would be a great deal simpler than I believe it is. We could call in the vet or the doctor, and he or she would look at the patient and declare that, yes, the wound was healing and that the individual was coping with the injury or that, yes, the temperature had dropped but that the animal was coping behaviorally by building a bigger nest.

What I believe is missing from this reactive view of welfare is that both humans and nonhumans also have highly developed mechanisms for anticipating challenges to their welfare, including health, that are quite distinct from the ones they use to cope with challenges that have already occurred. These anticipatory mechanisms range from the very simple, (even some seeds "anticipate" spring by germinating only after a cold snap, a preparation for climatic change which does not require cognitive processing) to the highly sophisticated, for example, you filling in your health insurance forms years before you need medical treatment. However, as I shall argue, it takes the recognition of good welfare into far more difficult territory than simply responding to an existing challenge.

The distinction between reaction and anticipation and its relevance to assessing good welfare can best be illustrated by stepping back for a moment from welfare itself and taking an evolutionary view of the animals we are addressing. Even though most of these animals are now domesticated or kept in environments very different from the ones in which they originally evolved, they still carry with them the physiological and behavioral legacy inherited from their wild ancestors. It is this legacy, encompassing a whole repertoire of defenses against dangers in the present and the possible ones that might come along in the future, that makes the assessment of good welfare so problematic. Thereafter, having learned some lessons from an evolutionary approach, we can then move onto how we might use these to help in the assessment of welfare.

EVOLUTION, FITNESS, AND WELFARE

For wild animals, life is a constant battle against different sources of danger, such as not getting enough to eat, falling victim to a predator, being overheated, frozen, and so on. At any given time, an animal may be under threat from several of these potential disasters and, to make matters worse, doing something about one of them may actually increase the risk from others. For example, going to a waterhole to drink may reduce the danger from dehydration but substantially increase the risk from predation. Increasing reproductive hormone levels may enhance the probability of finding a mate but reduce the availability of energy available for the immune system to combat disease.

Animals still living in the environments in which their ancestors evolved attempt to pick a path through life that minimizes these various dangers. They eat long before starvation becomes a serious threat. They hide in trees so that their predators rarely even see them. Such animals are "on course" or going well towards high surviving and reproducing. Now the literal meaning of "welfare" is "going well" (it has the same components as "farewell"), so good welfare can be interpreted as the animal being on course towards high reproductive success. It follows that if we want "indicators" of welfare, we have to look for indicators that an animal's fitness (survival and reproduction) has been threatened in some way. Since there are many different causes of death and reproductive failure, and animals have been selected to respond to them in different ways, it follows that there will be no single indicator that the animal's fitness is under threat. An animal whose fitness (and welfare) is threatened by a predator will have a completely different set of indicators from an animal whose fitness (and welfare) is threatened by dehydration. Similarly, the way an animal responds to a ground predator may be different from the way it responds to an aerial predator and different again from the way it responds to an aggressive conspecific. The situation is further complicated by the fact that how the animal responds will depend not only on what the danger is but also on what other dangers are threatening it at the same time. An animal that is short of food may still not eat if the search for food would expose it to high risks of predation or threaten its offspring.

The first lesson, from an evolutionary approach, is thus that there are no universal indicators of welfare. How the animal responds will depend on where the threat or threats to its welfare originates. We have no a priori reason to expect that the responses to different dangers will have much, if anything, in common. Animals do not have simple or universal indicators of welfare because they do not have simple or universal ways of responding to threats to their fitness. They have a whole range of different behaviors in their repertoire. This is disappointing to those people who look for comprehensive lists of indicators of good and poor welfare, but we now know enough about how animals respond to different challenges that we should have learned this lesson (Rushen 1986b; Fraser 1995; Dawkins 1998).

Although it is inconvenient of animals not to have provided us with such lists of good and bad welfare, we should not really be surprised. Animals do not have indicators of welfare for our benefit. What they have are physiological and behavioral responses to a wide variety of situations, and it is up to us to read the signs. That may mean careful and often time-consuming study of the way an animal responds in a wide variety of different situations. Even then, we will only be able to draw conclusions about that species in those situations and may have to revise our ideas completely when we turn to a new species and new challenges. We have only to think of the wide variety of ways we respond to threats to our own welfare, being hungry, bereaved, exhausted, bored, and so on, to see how dangerous oversimplification can be. Not only do we have different physiological and behavioral responses to these different situations, even more importantly, we have totally different conscious experiences and completely different ideas about what we would want done about them. Welfare, its loss and restoration, is no simple matter, because the world in which we and other animals evolved is no simple matter either.

The second lesson we can draw from an evolutionary approach comes, as I have already indicated, from the fact that animals have evolved two types of mechanisms to meet these dangers, some of which involve reacting to danger and coping while others involve anticipating danger that has not yet happened. Animals have, of course, a wide variety of ways of responding to damage that has already occurred, including wound-healing, immune responses, and behavioral ones such as remaining immobile, biting an attacker to prevent further attack, and so on. Animals, however, are not mere passive recipients of fate, like pebbles being rolled along by a river. They can be active participants with ways of anticipating and getting out of harm's way long before a threat even appears above the horizon. They make use of regularity or predictability in the environment to predict the likely occurrence of one event following another. For example, migrating birds use the fact that shortening day length regularly precedes the advent of cold weather to fly to a warmer climate before their fitness is seriously affected by cold weather itself. If they waited until they could no longer find enough food before moving, or were incapacitated by cold, they would be more likely to die than by responding to a stimulus like shortening day length that occurs well before the real threat to their fitness.

Such anticipatory mechanisms, while giving organisms the edge in the struggle for existence, also make the assessment of welfare more difficult. Not only do we have to see whether damage has already been done to the organism, we also have to see whether the organism is responding as if it were anticipating danger that might be done to it in the future. A migratory bird that was prevented from migrating by being kept in an aviary, when other members of its species were flying away, might be undamaged and healthy, and in this sense could be said to have good welfare; however, its frantic attempts to escape, which can be interpreted as behaving as if it were in danger, should leave considerable doubt about its true welfare. It is, therefore, important to distinguish three categories of responses that are, as it were, the raw materials for the assessment of welfare: (a) those that show that damage has already been done, (b) mechanisms for repairing existing damage, and (c) those for anticipating future damage. Examples of these are shown in columns II–IV of Table 5.1.

Column I shows some of the major causes of death and reproductive failure in animals. Column II shows some of the indications that death or reproductive failure are imminent, at least if no steps are taken. Sometimes these changes are brought about by the animal's own defense and repair mechanisms (Column III) and sometimes by the intervention of a veterinary surgeon or doctor, so that the animal does not actually die. The items listed in Columns I and II and, to a certain extent, in III are therefore the physical measures of animal welfare. They are the primary measures of welfare (Dawkins 1998) in that they indicate that the animal's fitness has already been reduced and the animal is either on a decline towards death or is taking desperate steps to avoid it.

In Column IV, however, are some of the methods animals use to avoid having to bring their repair or damage limitation mechanisms into action at all, by making sure they do not get injured or dehydrated in the first place. For example, many prey animals reduce their chances of being caught by flocking or herding together with members of their own species (Krebs and Davies 1993). This does not need any conscious planning ahead. It is simply that, for some species, individuals that have a rule of thumb to respond to conspecifics by keeping close to

Table 5.1 Ways in which animals respond to threats to their fitness and welfare.

I. Threat to fitness (cause of death or reproductive failure)	II. Indicator that fitness is already reduced (I. more likely)	III. Repair mechanisms (counteracting I. and II.)	IV. Means of avoiding reduction in fitness (avoiding I., II., and III.)
Major injury (e.g., predation)	Minor injury	Wound healing	Escape, hiding, social grouping, exercising, etc.
Disease	Debilitating ill health	Immune defenses, resting	Avoidance, cleaning behavior, grooming
Starvation	Loss of weight	Eating, conserving energy, immobility	Searching for food, hoarding
Dehydration	Lack of water to cells	Drinking, water conservation	Seeking water, staying in cool place

them survive better than those that move away from them. Gregariousness thus becomes part of good welfare for such animals, not just in the obvious sense that they are less likely to be injured and killed by a predator if they stay close to other individuals, but in the further sense that if they are separated from their flock or herd they become distressed and behave as if their welfare had already been threatened by making desperate attempts to find them again. Because in the wild being alone is a good predictor of being caught and eaten, being alone *itself* becomes a source of poor welfare, something that the animal will take steps to remedy if it can. In fact, much animal behavior, such as migration, hoarding food, grooming, and seeking cover, can be seen as anticipating future threats to fitness rather than coping with present dangers. This means that preventing an animal from doing a behavior may not lead to immediate death; however, we should still regard it as a potential welfare problem, since the animal may have been selected to behave as if death were imminent because of the contingent connection in the wild between not doing the behavior and dying.

We can now see why the distinction between repair mechanisms and anticipation mechanisms for combating threats to fitness has such important implications for animal welfare. Minor reductions in fitness that continue (Column II) or repair mechanisms that are unable to operate (Column III) will often lead to death and thus can be uncontroversially used as indicators of poor welfare. However, many of the anticipation mechanisms (Column IV) have no such dire consequences. Healthy animals can behave as if their fitness were threatened when in fact it is not. This is particularly likely to occur in zoos or farms where humans provide protection and food, but the animals may be kept without some key feature of their natural environment. A member of a social species kept on its own or a migratory bird kept in a small cage might both be in apparently good health but still be behaving as if they were in great danger. They are behaving as if they were anticipating threats to fitness that never actually materialized. This leads to what might be called the Paradox of Animal Welfare. Mechanisms that were originally evolved to aid the animal's survival, such as fear, can be seen as causes of poor welfare in their own right, particularly in the unnatural environments in which we humans so often keep their descendants.

There is also a third important lesson that we can learn from an evolutionary approach. While some reactive and anticipatory mechanisms may lead to reduced welfare if they are activated, we should be cautious about assuming that all of them necessarily do. "Welfare" for most people is a word extended only to organisms that can consciously experience pain or suffering. We may feel there are ethical issues in the preservation of works of art or giant redwoods but the extra ethical issues raised by what we do to animals comes from a widespread, if often implicit, conviction that vertebrates and possibly some invertebrates are capable of consciously feeling pain and experiencing suffering. What animals feel is of major importance in the concept of welfare. Ian Duncan (1993) said that it is the only issue, and even people who are skeptical about whether other species consciously experience anything are now bound in many countries of the world by legislation that assumes that they are capable of experiencing pain and suffering and that regulates the treatment of animals in experiments, agriculture, and other areas where they come into contact with humans.

It is therefore salutary to take a look at organisms that are clearly incapable of conscious experience and to make sure that what we have identified as the signs of good and bad welfare would not include them. If it would, we may have either to modify what we mean by "welfare" or be a great deal more discriminating in our choice of ways of assessing welfare if we want "welfare" to include conscious experience. There is one group of organisms that provides a particularly powerful challenge to how we assess welfare.

Plants can show signs of damage (Column II), they can repair damage with wound healing and the equivalent of an immune system (Column III), and they can also show "anticipation" of the future. I have already mentioned the example of seeds that will not germinate unless they have been subjected to low temperatures followed by warming. They make use of the contingent fact that spring follows winter to "anticipate" when conditions will be favorable for germination (Column IV). Deciduous trees drop their leaves in autumn, again anticipating or predicting environmental conditions that would be unfavorable to them.

These plant examples show that "anticipatory mechanisms" do not necessarily imply conscious experiences. Thwarted anticipatory mechanisms in Column IV do not, in other words, necessarily result in the conscious experience of suffering. In the case of seeds, chemical reactions that are temperature-dependent automatically trigger germination. Natural selection simply favors organisms that responded to certain proximate cues in their environments because those cues were contingently and reliably related to other events that were either damaging or favorable to the organism. The seed and tree examples make it clear that there is a wide diversity of anticipatory mechanisms, so that before we can convincingly incorporate them into our assessment of welfare, we need to look in more detail at exactly how these anticipatory mechanisms work.

Rolls (1999) has recently introduced an important distinction between different kinds of anticipatory mechanisms that has very important implications for animal welfare, because it is a theory of the evolution of emotions. Emotions, both the positive and the negative ones, are of particular importance in the assessment of welfare. Rolls makes a distinction between simple tropisms and taxes, on the one hand, and learning to obtain reinforcers, on the other.

Tropisms and taxes are innate, fixed responses that lead to approach or avoidance behavior, such as responses to light in plants or humidity gradients in woodlice. These fixed responses enable the organism to "choose" environments that are favorable to them and avoid those that are not and, while the mechanism may be simple ("hard-wired" in animals, difficult to call it even this in plants), the results can be surprisingly sophisticated. For example, Kelly (1992) has described the behavior of the parasitic plant dodder (*Cuscuta europaea*) which grows over bushes, such as hawthorns, and feeds from them by inserting small pegs into their stems. These plants do not insert their pegs into bushes in a poor state of nutrition and keep growing until they find more nutritious hosts to feed on. Their growth has much in common with the behavior of optimally foraging small birds, which move on from one patch to another when their rate of food intake drops below a certain level, but it is all accomplished by a series of relatively simple chemical reactions. Rolls' point is that where choice mechanisms are "hard-wired," or whatever the equivalent is in plants, there is no need for an intervening emotional state. It does not help the dodder plant or the woodlouse to have a positive emotion of pleasure whenever an attractive stimulus is encountered, and it does not add anything functionally to its ability to remove itself from potentially dangerous or harmful situations to have negative emotions.

Rolls then goes on to argue, however, that one of the major steps in evolution occurred when organisms evolved the ability to learn arbitrary connections between stimuli and reinforcers or learn to do completely arbitrary actions to achieve certain goals. If an organism can learn to associate almost any stimulus with, say, a food reward, then natural selection can no longer "build in" the rules for action. If it can learn to go away from a red light because it signals danger and towards a green one because it signals food but can equally well make the opposite associations, then simple innate rules of thumb such as "move towards red" are no longer adequate. Also, if it can then balance the conflicting demands of avoiding a predator and gaining food by weighing up the relative values of the two colored lights or their natural equivalent, it is going to need a common currency for evaluating the priority it should give to one or the other.

That common currency is emotion. Negative emotions, such as fear, act as negative reinforcers in the sense that the animal learns to avoid or not to repeat one set of arbitrary actions if fear is the consequence of those actions. Positive emotions, such as pleasure, act as positive reinforcers that lead animals to repeat other arbitrary actions (see also Broom 1998). Rolls argues that only organisms with the capacity for such arbitrary learning to obtain reinforcers have emotions. In this, he differs from Cabanac (1992) who sees a pleasure-suffering continuum as common to any organisms exhibiting choice. His argument is essentially an evolutionary one. Only organisms with the capacity for building up associations using rewards and punishments needed emotions. Before that, there was no need for emotions. All behavior could be hard wired.

Although this was not a symposium about consciousness, it is, however, important to point out in passing that postulating an emotion does not necessarily imply conscious experience of that emotion (Rolls 1999). All that is meant by "emotion" is a brain state common to several different situations, produced by reinforced or anticipated responses. A useful property of an emotional state is that it is a continuing state. Consciousness may indeed be associated with the capacity for using rewards and punishments to build up arbitrary associations, but it may equally be associated more narrowly with the evolution of higher cognitive functions, such as language (Rosenthal 1993; Dennett 1996; Rolls 1999). Rolls' theory provides a justification for distinguishing different types of choice mechanism, not a guide to where consciousness arose. Where we individually decide to believe that consciousness comes in is a different question (Dawkins 2001). Nevertheless, by focusing our attention on the evolution of emotions, and by using what animals find positively and negatively reinforcing, we have an essential tool for assessing emotional state. I will return to this point below but first I want to clarify just why the assessment of good and bad welfare has been, and continues to be, such a problem.

PROBLEMS WITH THE RECOGNITION OF GOOD AND POOR WELFARE

The search for reliable indicators of welfare has been the goal of much research over the last twenty years (Appleby and Hughes 1997) and that goal still eludes us (Fraser 1995). Although it is widely accepted that there are no single indicators of good and bad welfare (Dawkins 1980; Broom and Johnson 1993; Mason and Mendl 1993) and that we must take a broad view with health, biochemical, physiological, and behavioral measures, there is still no agreement on how we should go about this and, in particular, what we should do with the various measurements when we have collected them. Should we make checklists of all of these measures of, say, animals kept under two different husbandry systems and decide one is better for welfare than the other because one gives us a score of 75, whereas the other only gives a score of 60? One obvious difficulty with such a procedure is that the answer we get will depend critically on the weighting that is given to the different components and whether they are truly independent of one another. A second is that the different components may actually give conflicting answers. For example, physiological measures, such as heart rate, may not correlate very well with outward behavioral signs (Duncan and Hughes 1988). Yet another is that many of the so-called indicators of poor welfare, such as raised levels of corticosteroids, can only be understood in the context of the animal's normal adaptive responses to changes in the environment (Sapolsky 1994). This means that it is completely erroneous to think of any rise in such an "indicator" as showing poorer welfare, and we are left with the problem of how much of a change indicates poor welfare.

Much of the research on welfare indicators is carried out on the assumption that there are indicators of welfare, if only we could find them, and that it is only our own stupidity or lack of effort that has prevented us from finding them so far or putting them together into the right package. I have argued that there are no single indicators or simple checklists. The task ahead of us is more complex than we have yet admitted. In taking an evolutionary approach, I have argued that we should stop assuming that there are general indicators of welfare because animals have different ways of responding to different sources of danger. Rather, we should expect them to respond differently in different situations and, consequently it will be more fruitful to set about the task of documenting the responses they make, for example, to specific situations such as to the appearance of predators (Jones 1996) or overheating than to continue the search for general indicators of welfare. General indicators of welfare, i.e., responses common to different sorts of threat, may exist but they should be demonstrated, not assumed.

We should also stop assuming that there are indicators of good welfare that are distinct from those of poor welfare, in other words that each simple measure allows us to recognize and assess good welfare and distinguish it from poor welfare. A good example of how this assumption has led people astray is the use that has been made of measurements of glucocorticoids, sometimes called "stress" hormones. Because increased concentrations of glucocorticoids are often associated with situations that humans find undesirable, it is easy to assume that all situations where glucocorticoid concentrations are raised are indicative of "poor welfare." However, what is often not taken into account is that increased concentrations of corticosteroids are also associated with experiences that humans find pleasurable, such as sex and the anticipation of food (Toates 1995).

Changes in levels of glucocorticoids are thus not necessarily an indication of poor welfare but rather of a preparation for action that may be either aversive or pleasurable for the animal. This should not come as a surprise. Hormonal changes occur as a natural part of the lives of animals as these are an integral part of how their behavioral machinery works. Glucocorticoids, for example, are part of the means animals have of mobilizing stores of glucose and this is something that occurs whether the animal is preparing to run away from a predator (it is fearful) or to run after its prey. Both the prey animal that flees and the predator that runs after it need to mobilize stores of energy for the chase and would thus show an elevation of so-called "stress" hormones; however, the psychological state of the two might be quite different. One is fearful, a state we might associate with reduced welfare; the other is excited in anticipation of a good meal, its welfare probably enhanced.

The problem here is that the emotional state of the animal, that is, whether it is experiencing positive or negative emotions, is being judged purely on one autonomic measure, namely, hormonal state, which may give extremely variable results (Freeman 1985; Barnett and Hemsworth 1990; Rushen 1991). Even taking many different physiological measurements here will not necessarily distinguish the fearful prey from the happy predator. Both will have increased heart rates, both will have raised body temperatures and so on, because running fast, either away from danger or towards something desirable, both require the body to mobilize energy and get going as fast as possible. Thus the problem is not that we need to take into account more physiological measurements that are all under the control of the autonomic nervous system. The problem is that autonomic responses themselves, which tend to be what applied ethologists mean when they talk about physiological measures of welfare, are often more indicative of general excitement or arousal than whether the animal is finding the source of that excitement aversive or pleasurable.

This point can be further illustrated by reference to what is now known about human emotions, where it has been recognized that autonomic responses, such as changes in heart rate, temperature, and hormone levels, are a very unreliable guide to what a person may be actually feeling (Wagner 1989; Cacioppo et al. 1993). Oatley and Jenkins (1996) distinguish three systems underlying emotions in humans:

1. cognitive/verbal: the verbal reports that people give on what they are feeling at any one time,
2. autonomic: changes in temperature, heart rate, hormone levels, and so on, that humans have in common with other species,
3. behavior/expressive: the behavior and facial expressions that people show with different emotions.

Although studying each of these three gives a "measure" of emotion, they do not necessarily correlate with each other. Sometimes, for example, strong subjective emotions occur with no obvious autonomic changes, as when someone experiences a rapid switch from excitement to fear on a roller coaster. This does not mean that the change in emotional experience has no physiological basis. It just means that it is probably due to a subtle change in brain state rather than the obvious autonomic changes that are what are usually referred to as physiological measures of welfare. At other times, the emotion we experience and report corresponds to several different kinds of autonomic change or one kind of autonomic change, such as heart rate, can be shown to accompany very different emotions (Frijda 1986; Wagner 1989; Cacioppo et al. 1993).

The significance of using studies on humans in our quest for measurements of psychological health in animals is that humans are the one species where we have a window into what another organism is experiencing because we can ask them and, at the same time, humans have many of the same autonomic responses as do other species (Panksepp 1998). But it is clear from the study of human emotions that if all we had was a knowledge of the autonomic responses that were occurring in someone — what their heart rate was, what their breathing rate was, and some measure of their hormones — then we would not be able to tell whether that person was angry, fearful, or just plain excited. This is not because there are no physiological measures of positive and negative emotions, but that these are more probably a function of subtle changes in the central nervous system, not in the autonomic nervous system which serves both the positive and the negative emotions. One day, perhaps soon, we shall have the technology to monitor the brain states of freely moving animals, and we shall be able to see which parts of the brain are active when they are in a variety of situations. We may even be able to designate certain situations as pleasurable or aversive depending on which parts of the brain are active. At the moment, however, we do not yet have technology to tell us what the brain of an actively moving animal is doing and the only physiological measurements we currently have at our disposal are autonomic ones which, as I have argued, do not reliably enable us to distinguish different emotional states.

The underlying problem with the search for methods of assessing good welfare is that we have underestimated the amount of knowledge we need. All we have are the responses that animals make to, or in anticipation of, different kinds of danger, and these can be variable, context-dependent, and overlap with the responses they make to things that are fitness enhancing. Many of them have much in common with the responses made by organisms that no one believes consciously experience anything, namely plants. Except in the rather special case of offspring signaling their "need" to their parent (Weary and Fraser 1995), the animal is not providing us with "indicators." It could be responding to danger, being stimulated, being thwarted in its attempts to escape from danger, searching for something it does not have, and so on. Thus an animal may show reduced activity because it has everything it needs and is not seeking anything further (its welfare is good). On the other hand, reduced activity may be a sign of boredom and that the animal's health will deteriorate through lack of exercise. We cannot simply correlate the responses an animal makes with different conditions and make assumptions that we have hit on welfare indicators. We need an independent validation that what we are using bears some relation to suffering and poor welfare, on the one hand, or to good welfare, on the other. That, in turn, may mean much more detailed, species-specific and context-specific studies than we have yet undertaken.

THE ASSESSMENT OF GOOD WELFARE: WHERE WE MIGHT GO IN THE FUTURE

I can summarize so far by saying that good welfare has two components: physical and psychological. Injury, disease, deformity, and other physical signs of poor health are widely agreed to mean reduced welfare (Broom and Johnson 1993; Fraser 1995) and are signs that damage has already been done, even though the animal may now be coping and on the way to repairing that damage. However, absence of obvious physical damage is no guarantee of good welfare because of the possibility that poor welfare may also result from such things as confinement, restriction, or understimulation. Even when no obvious veterinary symptoms are apparent, the animal may be in a poor psychological health, which essentially means that it is experiencing negative emotions, such as fear, boredom or hunger. Good welfare is therefore recognized by a combination of good physical health and positive feelings, which we group together under the heading of "pleasure" or "contentment."

Emotional state is far more difficult to assess than physical health. Emotions are often difficult to judge correctly from their outward manifestations, particularly the autonomic ones so often used by applied ethologists as the physiological measures of welfare, and partly because emotions are often anticipating damage that might happen unless some action is taken rather than manifestations of actual damage. They are not so much coping with immediate challenge as taking avoiding action so that the challenge does not actually occur. To make life even more complicated, anticipation mechanisms can be mimicked by plants which, unless we are all very much mistaken, do not have emotions at all and to which the term psychological health could not be attributed. Fortunately, a possible way forward comes out of confronting what appears to be the most difficult problem of all, namely, that the states we refer to as poor welfare are so diverse that at first sight it seems impossible to say what they have in common in us, let alone to infer anything about them in other species. There is, however, a common thread. We talk about suffering from fear, suffering from exhaustion, suffering from grief, suffering from hunger, and so on. Although our behavior, when we experience these different states, is quite variable as are our physiological responses and, even more importantly, the subjective experiences we have, there must be something about them that enables us to use the one word "suffering" and expect to be understood or treated in appropriate ways. What these experiences have in common is that they are all unpleasant and that we take steps to avoid or alleviate them wherever we can. They are also states we would rather not be in if we had the choice. Clearly, we cannot access the pleasantness or otherwise of another person's experience directly, but we can look for evidence that they are taking behavioral or physiological steps to alleviate their situation or avoid similar ones in the future. Unpleasant emotions can thus be accessed through choice behavior but, as already argued, not all choice mechanisms stem from unpleasant emotions (Rolls 1999). A major task in assessing welfare in both humans and nonhuman animals is, therefore, ascertaining what they find positively and negatively reinforcing. This may mean actively investigating very specific situations. As an example, I would like to cite work done some years ago by Jeff Rushen, because I think it shows very clearly how it is possible to give answers to what appears at first sight to be impossible questions by doing relatively simple experiments.

Rushen (1986a) was concerned with the welfare aspects of the handling of sheep when they were restrained for shearing purposes. By all obvious behavioral criteria, sheep find having their wool sheared a cause of at least temporarily poorer welfare. They struggle violently and run away if they can. Rushen used an aversion-learning technique to show what parts of the shearing process were most aversive to the sheep. Sheep were placed individually in a race (corridor) down which they ran. At the far end of the race, they were treated in one of three ways: they were allowed to run unhindered back to the flock, they were restrained for a few minutes in a sheep-handling machine, or they were put in the machine and subjected to simulated shearing (clippers were moved backwards and forwards but no wool was removed). The sheep were then subjected to these procedures for a total of seven separate trials and on each trial their reluctance or willingness to run down the race was recorded. Sheep that were not handled continued to run down the race without hesitation and without needing to be pushed. Sheep that were restrained, however, showed great reluctance, and by the fourth trial it was clear that the sheep that experienced both restraint and shearing had to be pushed continuously to make them go down the race. Rushen concluded that sheep find both restraint and shearing aversive but they find shearing more aversive than restraint alone.

Rushen next wanted to know whether the welfare of sheep could be improved by electro-immobilization, because the manufacturers of commercial electro-immobilization devices claimed that these were an effective and humane way of restraining animals and preventing distress. Using the same technique of measuring reluctance to run down a race after experience of physical or electro-immobilization, he found that electro-immobilization was significantly more aversive than physical restraint. It appeared that the electro-immobilization did not reduce the sheep's perception of pain but acted more like a neuromuscular blocking agent. Rushen further argued that this aversion-learning technique gave a more reliable and direct insight into the sheep's view of the shearing process than commonly available physiological measures, such as corticosteroid or β-endorphin levels. The reduction in the sheep's willingness to run down the race was directly proportional to the amount of current applied (voltage levels and duration), whereas these physiological measures showed no difference in response to shearing, physical restraint, and electro-immobilization.

Rushen's work shows that it is possible to assess reductions in psychological health "from the animal's point of view." By using aversion-learning techniques, in which the results depend on the experience the animal has had of particular situation, he is tapping into the emotional state of the animal, its reluctance to run down a race indicating its anticipation of events that it wants to avoid. What we need now are many more studies that use this kind of approach and then relate it to other measures of welfare. For example, it might turn out that, unlike Rushen's sheep example, a hormone measure or the occurrence of a specific behavior might correlate very closely with what the animals found positively or negatively reinforcing. High levels of hormone X could be present whenever the animal was in the presence of what the animal indicated were aversive stimuli and low levels when they were in the presence of what they found rewarding. In this case, the level of hormone X could be used to assess welfare. On the other hand, it might turn out that, as with Rushen's sheep, hormone levels correlated very poorly with what the animals found positively or negatively reinforcing as did a behavior such as "immobility." The point is that it is an empirical matter whether such correlations exist or not. We cannot assess welfare by collecting together a series of measurements, however sophisticated, and making assumptions about what they mean. We have to establish empirically their relationship to positive and negative reinforcers. This would necessitate some painstaking research: How does this species, or even this sex or this age-class, respond under certain conditions? What does this animal find positively and negatively reinforcing? How does it rank various commodities in terms of their reinforcing properties? Does this ranking vary with circumstances such as what alternatives are available, whether the animal is alone or with others and so on?

Since such a program would necessitate the collection of a large number of facts, many of which would, individually, be quite limited in their application, a fruitful way forward might be to focus first on the solution of certain pressing practical problems, such as Rushen did in asking whether electro-immobilization was less or more humane than mechanical restraint for sheep when being sheared. Other obvious targets might be the testing of various proposed "alternatives" for commonly used experimental procedures used in research or specific environmental enrichments that have been recommended for zoo and farm animals.

We can thus conclude:

1. Physical health is the basis of all good welfare.
2. Psychological health, which is the second element of good welfare, is more complex than can possibly be encompassed within a small number of general indicators of good and poor welfare or even a checklist of many.
3. Emotional states, at least to the extent that they are positive or negative to the organism, can be assessed through what is empirically demonstrated to be positively or negatively reinforcing.
4. Positive or negative reinforcement is the ultimate arbiter of emotion and other measures should be empirically related to it.

REFERENCES

Appleby, M.C., and B.O. Hughes. 1997. Animal Welfare. Wallingford: CAB International.

Barnett, J.L., and P.H. Hemsworth. 1990. The validity of physiological and behavioral measures of animal welfare. *Appl. Animal Behav. Sci.* **25**:177–187.

Broom, D.M. 1998. Welfare, stress and the evolution of feelings. *Adv. Study Behav.* **27**:371–403.

Broom, D.M., and K.G. Johnson. 1993. Stress and Animal Welfare. London: Chapman and Hall.

Cabanac, M. 1992. Pleasure: The common currency. *J. Theor. Biol.* **155**:173–200.

Cacioppo, J.T., D.J. Klein, G.C. Berntson, and E. Hatfield. 1993. The psychophysiology of emotion. In: Handbook of Emotions, ed. M. Lewis and J.M. Hatfield, pp. 119–142. New York: Guilford.

Dawkins, M.S. 1980. Animal Suffering: The Science of Animal Welfare. London: Chapman and Hall.

Dawkins, M.S. 1998. Evolution and animal welfare. *Q. Rev. Biol.* **73**:305–328.

Dawkins, M.S. 1999. The role of behavior in the assessment of poultry welfare. *World's Poultry Sci. J.* **55**:295–303.

Dawkins, M.S. 2001. Who needs consciousness? *Animal Welfare* **10**:319–329.

Dennett, D.C. 1996. Kinds of Minds: Towards an Understanding of Consciousness. London: Weidenfeld and Nicolson.

Duncan, I.J.H. 1993. Welfare is to do with what animals feel. *J. Agric. Env. Ethics* **6 (2)**:8–14.

Duncan, I.J.H., and B.O. Hughes. 1988. Can the welfare of poultry be measured? In: Science and the Poultry Industry, ed. J. Hardcastle, pp. 24–25. London: AFRC.

Fraser, D. 1995. Science, values and animal welfare: Exploring the "inextricable connection." *Animal Welfare* **4**:103–117.

Fraser, A.F., and D.M. Broom. 1990. Farm Animal Behaviour and Welfare. Wallingford: CAB International.

Freeman, B.M. 1985. Stress and the domestic fowl: Physiological fact or fantasy? *World's Poultry Sci. J.* **141**:45–51.

Frijda, N.H. 1986. The Emotions. Cambridge: Cambridge Univ. Press.

Jones, R.B. 1996. Fear and adaptability in poultry: Insights, implications and imperatives. *World's Poultry Sci. J.* **152**:131–174.

Kelly, C.K. 1992. Resource choice in *Cuscuta europaea.Proc. Natl. Acad. Sci. USA* **89**:12.
Krebs, J.R., and N.B. Davies. 1993. An Introduction to Behavioural Ecology. Oxford: Blackwell Scientific.
Mason, G., and M. Mendl. 1993. Why is there no simple way of measuring animal welfare? *Animal Welfare* **2**:301–320.
Oatley, K., and J.M. Jenkins. 1996. Understanding Emotions. Oxford: Blackwell.
Panksepp, J. 1998. Affective Neuroscience: The Foundations of Human and Animal Emotions. Oxford: Oxford Univ. Press.
Rolls, E.T. 1999.The Brain and Emotion. Oxford: Oxford Univ. Press.
Rosenthal, D.M. 1993. Thinking that one thinks. In: Consciousness, ed. M. Davies and G.W. Humphreys. Oxford: Blackwell.
Rushen, J. 1986a. Aversion of sheep to electro-immobilization and mechanical restraint. *Appl. Animal Behav. Sci*. **15**:315–324.
Rushen, J. 1986b. The validity of behavioral measures of aversion: A review. *Appl. Animal Behav. Sci.* **16**:309–323.
Rushen, J. 1991. Problems associated with the interpretation of physiological data in the assessment of animal welfare. *Appl. Animal Behav. Sci.* **28**:381–386.
Sapolsky, R.M. 1994. Why Zebras Don't Get Ulcers: A Guide to Stress and Stress-related Diseases. New York: W.H. Freeman.
Toates, F. 1995. Stress: Conceptual and Biological Aspects. New York: Wiley.
Wagner, H. 1989. The peripheral physiology and differentiation of emotions. In: Handbook of Social Psychophysiology, ed. H. Wagner and A. Mainstead, pp. 78–98. New York: Wiley.
Weary, D., and D. Fraser. 1995. Calling by domestic piglets: Reliable signals of need? *Animal Behav.* **50**:1047–1055.

Standing left to right:
Norbert Sachser, Sue Mineka, David Fraser, Klaus Gärtner
Seated left to right:
Susan Lutgendorf, Sue Carter, Ute Knierim, Jaak Panksepp

6

Group Report: Good Welfare

Improving Quality of Life

U. KNIERIM, Rapporteur

C.S. CARTER, D. FRASER, K. GÄRTNER, S.K. LUTGENDORF, S. MINEKA, J. PANKSEPP, and N. SACHSER

Keywords: human welfare, animal welfare, affect, well-being, welfare assessment, behavior, evolution, neurobiology, neuroendocrinology, psychology, physiology

ABSTRACT

A wide range of fields — psychology, neuroscience, physiology, ethology, and veterinary medicine — provide complementary insights into what constitutes a state of "good welfare" in humans and other species, and how good welfare can be assessed and achieved. Welfare should be viewed as involving multiple elements, including both the functioning of the organism and its affective states. Moreover, good welfare is not merely the absence of stress, illness, and negative affective states; rather, it includes the experience of positive affective states, and thus requires indicators beyond those used to identify welfare problems.

Human welfare, or well-being, can be described using core dimensions such as self-acceptance, positive relations with others, autonomy, environmental mastery, purpose in life, and personal growth. Some of these features, such as positive social relations and control over the environment, have apparent parallels in nonhuman species, while others do not. Humans and many other species have behavioral indicators of positive affect, although these need to be interpreted carefully. Moreover, certain behavioral activities that promote fitness appear to be accompanied by positive affective states, whereas behavior that responds to immediate threats to fitness may be motivated more purely by negative states such as fear and pain. Physiological processes can be identified that appear to reflect positive affect, but these are not readily monitored using peripheral measures or in noninvasive ways.

The environmental prerequisites for good welfare vary from species to species, depending on its evolutionary history, perhaps as modified by domestication and selective breeding, together with individual differences caused by epigenetic factors such as early experience. A key element is appropriate stimulation including appropriate complexity, intensity, social interaction, and the presence of certain key features that will vary according to the species. In domesticated species, certain types of

Coping with Challenge: Welfare in Animals including Humans
Edited by D.M. Broom © 2001 Dahlem University Press

genetic selection can limit the animals' capacity for good welfare. In humans, good welfare can also be promoted by appropriate cognitive interventions.

At present, behavioral measures, including verbal reports in humans, are the primary means of assessing the positive affective states that are important for good welfare. In the future, enhanced understanding and techniques in neurobiology, especially at the subcortical level, and in neuroendocrinology, especially neuropeptide systems incorporating oxytocin, vasopressin, endogenous opioids, and corticotrophin-releasing hormone, may provide further physiological indicators. Future research should also refine existing behavioral and psychological methods, and should identify measures or interventions and their underlying mechanisms that improve quality of life.

INTRODUCTION

Scientists have directed a great deal of effort toward identifying and addressing those threats to organisms, both human and nonhuman, that jeopardize their welfare. Such efforts have been a major preoccupation of many fields of science including psychiatry, psychology, pathology, veterinary medicine, stress physiology, and animal welfare science. In contrast, the positive state of "good welfare," despite its obvious importance for both humans and nonhumans, has received far less explicit scientific attention. Moreover, there has been relatively little cross-fertilization between scientific research on human welfare and that on animal welfare.

This chapter attempts to bring "good welfare" more explicitly into scientific discourse by providing initial answers to the questions: what constitutes good welfare, what are its prerequisites, and how can it be recognized or assessed? It is a collaboration involving a wide range of fields focused on humans and nonhumans in order to highlight insights from studies of human welfare that are useful in the study of animal welfare and vice versa.

What Is "Good Welfare"?

Scientific research on the "welfare" of animals and humans arose because of public concerns about the quality of life of animals and humans. If research is to help address these concerns, then the concept of welfare used by scientists must adequately reflect them.

Stafleu et al. (1996) describe the distortion that can arise when scientists try to operationalize a socially relevant concept such as animal welfare. As a first stage, scientists may try to redefine welfare in terms of a scientific theory, such as the theory of evolution (equating good welfare with fitness[1]) or the theory of stress biology (equating welfare with lack of stress). Next, they may try to operationalize welfare in terms of a measure related to the chosen theory, such as the number of offspring produced as a measure of fitness or the level of cortisol in the blood as a measure of stress. At each of these steps, objective scientific study becomes more feasible, but the research becomes more and more removed from the original concept. As a result, research often fails to answer the very questions that caused it to be undertaken.

---

[1] We are using "fitness" in the way the term is used in evolutionary biology, namely to refer to the reproductive success of an organism.

To avoid this situation, we need to define welfare in a way that adequately reflects concerns that people have about quality of life. In the case of animal welfare, these concerns include two key issues: (a) that the animal should *function* well in the sense of good health, normal growth, capacity for reproduction, and normal working of behavioral and physiological systems; and (b) that the animal should *feel* well, in the sense of experiencing positive affective states (comfort, contentment) and avoiding or minimizing suffering (Fraser et al. 1997; Dawkins, this volume). The former issue, regarding biological functioning, lends itself to conventional biological research. However, even at the low end of the welfare scale, where there is obvious evidence of biological malfunctioning such as disease and injury, public concerns tend to be focused especially on the unpleasant affective states that the malfunctioning is assumed to denote. Affect is arguably even more defining of welfare at the positive end of the scale where states such as comfort, contentment, and pleasure have few obvious, unambiguous indicators in biological functioning.

Similarly, good quality of life for humans has been defined as being physically, functionally, socially, and emotionally well. In addition to the commonalities noted here between human and animal welfare, there are some elements necessary for human welfare, such as the need to find a sense of purpose or to express creativity, which presumably do not apply to nonhuman species.

Affect and Welfare

The need to include affective states in welfare research, especially the affective states of nonhuman animals, raises an important issue in the philosophy of science. Before the early twentieth century, researchers in both biology and comparative psychology often drew conclusions about affective states in animals (e.g., Darwin 1872). Under the influence of Positivism, however, many twentieth-century scientists declared that because affective states cannot be observed directly, they fall outside the realm of scientific enquiry. Thus, many scientists in both ethology and comparative psychology avoided any attempt to understand affective states, especially of animals, and attempted to explain behavior without reference to states such as fear, frustration, pleasure, or even pain and hunger.

By the late twentieth century, these limitations were clearly beginning to break down (e.g., Bekoff 2000). Nonetheless, when scientists addressing questions of animal welfare attempt to understand the affective states of animals, they have remarkably little in the way of accepted theory or methodology to draw on, and in some cases they meet continued resistance from a scientific establishment that is not yet ready to accept such attempts as a part of science.

The central role of affective states in good welfare also raises questions about the range of species to which the concept of good welfare can properly be applied. If it could be demonstrated that no animals other than humans are capable of experiencing affective states, then many concerns about animal welfare could be dismissed. Such a position has often been attributed to René Descartes (e.g., Singer 1990) and was certainly held by at least some of Descartes' followers (Preece 1999). A similar position has been argued by some modern philosophers (e.g., Carruthers 1992) and by scientists, such as Bermond (1997), who claim that the experience of affective states is confined to human beings and perhaps a very few other species.

These claims have not been widely accepted by the public, nor indeed by the majority of the scientific community. Rather, it appears to be assumed by a wide range of philosophers (e.g., Midgley 1983) and scientists (e.g., Duncan 1996; Broom 1998; Dawkins and Sachser, both this volume) that affective states do occur in mammals and likely in a wide range of vertebrate species. For present purposes, much of our discussion will deal with mammalian species and to some extent with birds; however, we have no wish to foreclose the discussion of which taxonomic groups are capable of experiencing affective states.

What Is *Good* Welfare?

Research on the welfare of both humans and animals has tended to focus on the negative end of the scale rather than the positive. As Lutgendorf (this volume) notes, "investigations of human unhappiness and suffering have received much more attention than investigations of good welfare in the psychological literature." Similarly, research on animal welfare has tended to focus on welfare problems or challenges such as cold, injury, disease, over-crowding, or aggression whereas states of "good welfare" have attracted relatively little research. Likewise, studies of affective states in animals have dealt mainly with fear, pain, separation distress, and other presumably aversive states rather than positive states such as pleasure, comfort, and contentment. Hence, the study of *good* welfare presents a largely new challenge.

A good deal of theory regarding both welfare and affective states has been underlain by the relatively simplistic, if often unstated, assumption that welfare and affect can be represented along a single continuum running from negative to positive. For example, Young (1959) postulated a "hedonic continuum" ranging from strongly negative to strongly positive affect, and Cabanac (1992) saw the balance of pleasure and displeasure as a scale that serves as the "common motivational currency" for behavioral decisions. Similarly, Broom and Johnson (1993) and many others speak of a scale running from poor to good welfare.

While this simple, unidimensional metaphor for welfare or affect may be useful for certain descriptive purposes, it may also misguide us when we attempt to analyze or operationalize the concepts. At least certain constituents or indicators of good welfare may not simply be the opposite of constituents or indicators of poor welfare. If poor animal welfare can be demonstrated by illness, slow growth, suppressed immunity, and high levels of stress hormones, it does not necessarily follow that the reverse of these measures is definitive of good welfare; rather a different or expanded set of indicators may be needed. Likewise, human studies show that the two primary dimensions of emotional experience, positive affect and negative affect, are largely independent of one another. For example, a person with very high levels of negative affect can be quite capable of also experiencing moderate levels of positive affect, as tends to occur in clinical anxiety, or, alternatively, very low levels of positive affect, including anhedonia or the inability to experience pleasure, as tends to occur in clinical depression.

Multidimensionality of Welfare

Unidimensional metaphors for welfare or affect may also mislead us into seeking a single integrative function for what is inherently a multidimensional concept (Fraser 1995; Dawkins, this volume). Attempts to operationalize human welfare have identified a number of distinct factors (see Lutgendorf, this volume). For example, Ryff and Keyes (1995; Ryff and Singer

1998) have integrated many of the previously existing perspectives regarding well-being and life satisfaction into a multidimensional model of human well-being. They described core dimensions of human well-being as: self-acceptance, positive relations with others, autonomy, environmental mastery, purpose in life, and personal growth. The independence of these factors has been supported empirically in large-scale community samples.

Animal welfare also involves a large number of factors (Broom and Johnson 1993) for which there is no logically or empirically correct combining formula (Fraser 1995). In fact, some of the dimensions of human welfare may be highly applicable to animals. As with Ryff's dimension of environmental mastery, the welfare of animals is enhanced by their having control over changes in their environment and being able to predict events in the social and physical world (Mineka et al. 1986; Wiepkema 1987). Positive social relations are also applicable to animals; in several mammalian species, it has been shown that the presence of a bonding partner inhibits potentially harmful neuroendocrine and immunological responses to aversive events. This holds true irrespective of the bonding partner being the mother, in the case of an infant, or another adult in the case of adult animals (Sachser et al. 1998).

In summary, therefore, we must begin an analysis of good welfare, including positive affect, by searching for factors that (a) may be independent, to a degree, of elements of bad welfare and negative affect, and (b) involve multiple elements which we may not be able to reduce to simple integrative indices.

CHARACTERIZING GOOD WELFARE

With that background, and having defined good welfare as meaning both functioning well and feeling well, we summarize scientific approaches from different disciplines that are relevant to the issue of good welfare.

Insights from Human Psychology

There have been many conceptualizations of human welfare. In medical contexts, welfare is often referred to as "quality of life," which is defined as the individual's ability to function and derive satisfaction from a variety of roles (Wenger et al. 1984).

One of the best known theoretical formulations was proposed by Maslow (1968), who introduced the concept of a hierarchical scale of human needs. At the lowest level is satisfaction of basic biological and survival needs. Once these needs are fulfilled, individuals can address needs such as those for social contact, for creative expression, and for fulfillment of their own unique potential.

Although much research on positive life factors has used unidimensional concepts such as "happiness" and "life satisfaction," approaches such as that proposed by Ryff (1989, Ryff and Keyes 1995) provide multidimensional models of welfare (see earlier section on Multidimensionality of Welfare).

Insights from Clinical Veterinary Medicine

In veterinary medicine, clinical diagnosis includes the assessment of the welfare of the patient using the appearance and behavior of the animal as indicators. Clinical examination typically begins with an evaluation of the general state of health by visual appraisal based on posture, including carriage of head and tail or lying postures, facial expression, vocalizations, general reactivity and activity, i.e., demeanor and temperament, patterns of movement, and general bodily condition (e.g., Hornbuckle 1992; Speirs and Wrigley 1997; Baumgartner and Ketz Riley 1999). Good welfare is assumed if the animal's activity and vocalizations are appropriate to its age, gender, breed or type, and physiological condition, i.e., if it has clear bright eyes, good posture, clean healthy skin, good fur or plumage, adequate nutritional condition, and an appropriate response when disturbed (e.g., Council of Europe Standing Committee 1995). In terms of activity, special attention is paid to locomotion, getting up and lying down, feeding and drinking behavior, rumination, exploration, grooming or preening, play, and social behavior within a group. Sleeping rhythm, postures, and location are also of diagnostic value if these can be ascertained. The assessment of deviations from the normal presupposes a good knowledge of the species under examination and of the typical range of individual differences. Although this kind of welfare assessment is a long-standing component of clinical veterinary medicine, and considerable clinical experience has developed, very little effort has been made to discuss, refine, and validate the methods formally. This may be due to the fact that in this context, the diagnosis of good or impaired welfare is not the primary goal, but one diagnostic tool among others to detect illness or ascertain good health.

Insights from Evolutionary Biology

Since the time of Darwin, and arguably long before, it has been assumed that affective states are adaptations that motivate animals to perform behaviors beneficial for their survival and reproductive success (i.e., fitness). Table 6.1 lists examples of negative affective states commonly experienced by humans and corresponding threats to fitness that appear to give rise to these states. Judging from human experience, it seems plausible to postulate that these states motivate behaviors that tend to correct or prevent the corresponding threat to fitness. This is not to say that the relevant behaviors occur only when such states are experienced; eating, for example, may become sufficiently habitual that actual hunger rarely occurs.

Humans and nonhuman animals, however, often spend a large fraction of their time in behavior that does not appear designed to cope with any immediate threat to fitness. Examples include play, exploration, renewing territorial markings, grooming others, caring for offspring of others, storing food, and eating at times when nutritional needs are met. Such activities may well make a positive contribution to fitness on average, as long as they are performed at times when the cost of performance is not unduly high. For example, through play, individuals may develop motor or social skills, and through exploration they may detect resources that will be useful in the future. Drawing again on human experience, it appears that activities of this type are largely motivated, not by negative affective states analogous to pain or hunger, but by the pleasure derived from performing them.

This suggests that there are two complementary types of motivation: one, based on negative affective states, which deals largely with more immediate threats to fitness, and a second, based on positive affective states, which causes individuals to perform behavior which conveys fitness benefits on average, at times when more immediate threats have been resolved (Fraser and Duncan 1998). For a given species, therefore, we should be able to identify activities that form a prominent part of behavior, which can plausibly be seen as contributing on

Table 6.1 Examples of negative affective states and corresponding threats to fitness.

Negative Affective State	Threats to Fitness
Hunger	Negative energy balance
Thirst	Dehydration
Pain	Injury
Fear	Danger
Separation distress in young and parent	Danger of losing protection of parent, inability to protect young

average to fitness, but which do not appear to meet any pressing threat to fitness. According to this view, such activities should be pleasant to perform, or be positive incentives, in traditional behavioral terminology. Thus, to promote good welfare, as opposed to merely avoiding welfare problems associated with negative affective states, we need to allow members of the species to perform these types of behavior and thus experience the corresponding positive affect. Obviously the species' natural life-style, taking account of factors such as gender and age, can help to identify such activities and the environmental conditions necessary for their performance (see section on PREREQUISITES FOR GOOD WELFARE and Sachser, this volume).

Insights from Stress Physiology

Traditional stress physiology has centered on two systems: the hypothalamic-pituitary-adrenocortical (HPA) system originally studied by Hans Selye (e.g., 1932), and the sympathetic-adrenomedullary (SAM) system made famous by Walter Cannon (e.g., 1914). In their monograph *Stress, Health, and the Social Environment*, Henry and Stephens (1977) noted that these two stress systems are frequently triggered by central nervous processes ("emotions"), which are especially likely to occur in situations characterized by uncertainty or unpredictability, as well as in situations where the individual's control over the situation is threatened or impossible. If the stressful situation persists, different physiological stress reactions are induced, depending on the appraisal of the situation and the resulting behavioral strategy. In this traditional conception, active attempts at obtaining control over a situation (e.g., fight and flight) are characterized by a predominant activation of the SAM system and may, if chronically activated, lead to pathological processes such as cardiovascular disease. Passive perception of defeat or loss of control is characterized by HPA activation and, in the long term, by negative effects on almost all body functions as well as anxious and depressed emotional states. Although speculative in terms of the actual feelings involved, these stress responses have a bipolar aspect: the opposite pole to the SAM anger–fear response appears to be a state of serenity and relaxation. The opposite pole to the HPA response that accompanies phenomena, such as depression and loss-of-attachment, may be a subjective feeling of positive affect which may, for instance, arise from clear, stable dominance relationships or strong bonds to partners (von Holst, this volume; for an extension of Henry's ideas, see von Holst 1998; for another point of view, see Porges, this volume). Thus, under the viewpoint of traditional stress physiology, good welfare relates to the repeated experience of a low activation of the HPA and SAM systems, which are presumably associated with feelings of relaxation and positive affect.

Insights from Neurobiology

Discussions about relatively nonspecific peripheral stress response systems are now being extended by neurobiological research into more specific emotional states.

In mammals, and probably other vertebrates, behavior indicative of emotional processes is controlled by evolutionarily ancient neural systems situated in subcortical regions of the brain. To characterize the emotional systems that exist, we need to consider empirical evidence of (a) categories of emotional behaviors organized by the brain, (b) types of basic affective processes that humans can agree upon from an analysis of their subjective experiences, and (c) the types of neural networks that exist in the brain related closely to these processes. If all three lines of evidence match up, we can have some confidence in the existence of basic emotional systems that need to be considered in any systematic discussion of welfare issues.

So far, the study of emotional behavior patterns that can be evoked by localized electrical and chemical stimulation of the brain has provided the best evidence for characterizing types of emotional states. Various distinct types of emotional behaviors can be evoked from specific brain areas, and to all appearances these behavioral displays are accompanied by positive and negative behavioral tendencies, which may be indicative of affective states (Panksepp 1998).

On the basis of these lines of evidence, one provisional scheme identifies seven executive networks involving emotional processes:

1. a SEEKING[2] system of exploring and investigating resources in the environment
2. a RAGE system that causes invigoration when resource-holding capacity is compromised,
3. FEAR system(s) that are aroused when survival is rapidly threatened,
4. a variety of sexual LUST systems that are substantially different in males and females,
5. a CARE promoting system that instigates or facilitates maternal nurturance,
6. a SADNESS or PANIC system that mediates separation distress vocalizations and that promotes affective states of mind that accompany acute feelings of social isolation, and probably
7. a PLAY system that provokes the vigorous social play commonly seen in the young of many species (Panksepp 1998).

Presumably these executive systems have widespread brain representations that help coordinate behavioral, physiological, and psychological processes.

These systems have evolved around specific neuropeptide neuromodulators and general brain amino-acid transmitter networks, all of which are modulated and assisted by nonspecific arousal/attention control systems, such as norepinephrine and serotonin. It is a working assumption that activation of these brain systems is important in the occurrence of emotions;

[2] Capitalization is used, in line with the convention of Panksepp (1998), to emphasize that the referents are specific neuropsychological systems which are incompletely understood.

for example, that the activation of the executive systems for separation distress, in some presently unknown way, helps generate the unpleasant feelings of sadness in humans.

This schema is not meant to be comprehensive, as there are also various consummatory pleasures and displeasures from interacting with specific environmental objects and events. It is presented simply as one empirically based way to proceed in generating a working theoretical model for the types of emotional states that we need to consider when we discuss welfare issues. Presumably each of these systems could be monitored by appropriate and careful analysis of the relevant behavior patterns of various species of animals and of verbal reports in humans. Some of these networks (e.g., FEAR, RAGE, and SADNESS systems) may help create negative (unpleasant) affective states that promote escape and avoidance. Others (e.g., SEEKING, LUST, CARE, and PLAY) may create positive affective states that generate appetitive and approach behaviors. Presumably the affective states promoted by these systems provide intrinsic signals for fitness-promoting activities in specific environments. Presumably also, the impact of these states in the nervous system can be amplified and attenuated by general arousal processes, and they can be conditioned and molded by experience.

Insights from Neuroendocrinology

The nervous system, including the central (CNS) and autonomic nervous system and certain neuroendocrine systems, plays a pivotal role in welfare. Particularly important are specific neural and neuropeptide systems, including those that incorporate oxytocin, vasopressin, and corticotrophin-releasing hormone (CRH), which regulate social behavior as well as certain other neuroendocrine and autonomic processes.

Oxytocin and vasopressin are primarily mammalian peptides and will be used as examples here. However, related and functionally similar peptides, including mesotocin and vasotocin, are found in other vertebrates, including birds. Thus, at least some of the concepts described here for mammals may have functional parallels in other vertebrates. In addition, there is a vast number of additional neuropeptides, such as opioids, luteinizing hormone-releasing hormone (LH-RH), cholecystokinin (CCK), and prolactin, that figure heavily in social affective processes (Panksepp 1998).

Oxytocin and vasopressin are centrally produced, primarily in the hypothalamus, and partly secreted into the periphery at the level of the posterior pituitary. Receptors for these peptides occur in a variety of tissues and may be regulated by various other hormones, including gonadal and adrenal steroids. Thus, it is possible for these peptides to have a wide range of actions. The classical and peripheral effects of oxytocin include uterine contractions and milk ejection in females, and vas deferens contractions in males. Within the CNS, oxytocin may bind to receptors in brain areas that influence the HPA axis, fear, and other aspects of autonomic activity (Porges 1998, this volume) and behavior. The net effects of oxytocin include reductions in SAM–HPA axis activity and an increased tendency to show positive social behavior. In general, and especially when produced or administered chronically, oxytocin exerts anti-stress effects, in part perhaps by antagonizing the actions of vasopressin and CRH (Carter 1998). Vasopressin is known for its effects on blood pressure and other cardiovascular functions and as an anti-diuretic hormone acting at the level of the kidneys. Many of the effects of vasopressin and CRH are in opposition to those of oxytocin, and vice versa.

Neuropeptides, and especially interactions among oxytocin, vasopressin, and CRH or related hormones, are ideally positioned to coordinate the central effects of social and nonsocial experiences. Centrally produced neuropeptides, aside from those that are secreted at the posterior pituitary, presumably act primarily within the nervous system. Moreover, the magnocellular neurons that release oxytocin and vasopressin at the posterior pituitary are separate from those releasing the peptides within the CNS. Given their location within the CNS and the fact that peripheral secretion is not always coordinated with central secretion, both oxytocin and vasopressin are difficult to study or directly manipulate.

Nonetheless, knowledge of the physiological functions and interactions among these systems, and especially the effects of oxytocin and endogenous opioids, provide a new perspective on mechanisms through which social behavior may modulate the response of an organism to a stressful experience. Because these hormones are released in a social context and also affected by social experience, they are capable of mediating and responding to certain aspects of good or bad welfare.

Because the peptide hormones can have long-lasting effects, they also provide mechanisms through which experience, especially during the perinatal period, can lead to enduring changes in behavior or physiology. For example, nurturance, other positive experiences, or exposure to oxytocin or opioids in early life may produce changes in peptide hormones and their receptors, which in turn allow individuals to be more resilient to stressors such as those experienced during later life. Conversely, exposure to a harsh environment or vasopressin in early life may be associated with higher levels of aggression or other forms of defensive reactivity in adulthood (Stribley and Carter 1999).

Animal models, and especially highly social mammals, offer the opportunity to understand the mechanisms through which neuropeptides may modify and encode the effects of social and nonsocial stimuli (Panksepp et al. 1980, 1985; Carter, this volume). Because the oxytocin system is especially identified with positive experiences as well as the health consequences of such experiences (Carter 1998; Uvnas-Moberg 1998), knowledge of this system can be useful in predicting the physiological and behavioral consequences of manipulations of the social environment, and in interpreting studies in which peripheral hormones, such as the glucocorticoids or adrenocorticotrophic hormone (ACTH), or autonomic functions are measured.

KEY SOURCES OF VARIATION IN REQUIREMENTS FOR GOOD WELFARE

Natural Selection

During the process of evolution, organisms became adapted to specific parts of the environment. Thus, the members of any given species are equipped with physiological, motivational, and behavioral systems to cope with the demands of their specific environmental niche, and we can expect the requirements for good welfare to be related to the features of this niche. Since no two species occupy the same ecological niche, it has been argued that the prerequisites for good welfare will be unique for every species (Dawkins, this volume). However, species can be grouped according to similar life-styles that were brought about by natural selection. The knowledge of the life-style thus allows predictions about what will favor good welfare (Carter, and Sachser, both this volume). For example, species may require a particular kind of substratum for locomotion or specific environmental features for body-care activities. Social requirements can also be predicted from the species' natural social organization. For example, the roughly 4,250 species of mammals can be classified according to a few basic patterns of social organization. Associated with certain types of social systems are consistent patterns of behavior and anatomy. Those species that are territorial and solitary in their natural habitat respond quite differently to being held in isolation than do members of socially living species. In captivity, the solitary European hamster has to be kept individually to achieve good welfare; by contrast, in the group-living guinea pig, housing the animals singly leads to poorer welfare (Sachser 1994). Furthermore, all socially monogamous animals require a social partner, and separation of a bonded pair results in stress (e.g., Mendoza and Mason 1986).

Domestication

Domestication can exert important effects on the prerequisites for good welfare. During domestication, animals have had to adapt, over hundreds or thousands of generations, to living in captivity and under human protection. Many important natural selection pressures, such as predation, food shortage, or adverse weather, were therewith reduced or eliminated. At the same time, artificial selection favored tameness, high reproductive rate, and specific physical and behavioral traits. This has led to changes in morphology, physiology, and behavior (Herre and Röhrs 1990; Künzl and Sachser 1999; Carter, this volume).

The domestication of farm, laboratory, and certain companion species has resulted in a reduction of the relative weight of certain organs, especially the heart, brain, and a few sense organs. This reflects a functional reduction of those organs during domestication, an alteration which Röhrs (1991) interpreted as a type of regressive evolution.

On an intraspecies allometric scale, Herre and Röhrs (1990) and Röhrs (1991) calculated a reduction in heart weight by around 35% in many domesticated species (except the dog and cat) due predominantly to a reduction in the ventricles. Reduction of brain weight reaches 25–35% for species with a high degree of cephalization (sheep, pig, cat, dog), and 8–13% in species with lower cephalization (mouse, rat, rabbit). Among the different brain areas, the telencephalon and diencephalon, especially their areas for visual and olfactory functions, are most obviously reduced (e.g., neocortical part of telencephalon by 30–38% in the sheep, dog, pig). The limbic structures (septum, hippocampus, and parahippocampal cortex — schizocortex) are also remarkably reduced. Reduction of the olfactory and visual functions has also been recorded. Domesticated pigs have only 66% of the olfactory epithelia and a similar reduction in the number of olfactory cells, compared with wild pigs. Visual function is reduced by a reduction of the retina caused by a loss of eye volume (20–35%) in domestic animals (rabbit, pig, dog) and reduction (15–47%) of the optic tract, lateral geniculate body, and primary visual cortex.

Domestication has also caused a remarkable reduction in the weight of adrenal glands of many species. Lowered responsiveness of the HPA and SAM axes is described by Sachser (this volume). The reduction of the limbic structures correlates with a lowered motivation for performing spontaneous behavior in domestic animals (Price 1998). Laboratory rats and mice exhibit decreased flight and fight behaviors and changes in their learning ability

(Barnett 1969; Smith 1972, Blanchard et al. 1995). Domestic pigs and poultry use different strategies than their wild counterparts when searching for food (Jensen, this volume).

Guinea pigs and most other species show modified social and aggressive behavior, enabling them to live harmoniously in large mixed-sex groups, although this does not apply as strongly to adult males of most species.

This selection of animals for a generally lower responsiveness to environmental impacts presumably promotes good welfare under artificial housing conditions (see also Sachser, this volume). However, genetic adaptations to specific housing conditions are not likely to occur because housing systems and conditions tend to change rapidly, and parent stocks are often kept in different systems than the offspring that are raised for meat. In fact, most domestic animals closely resemble their wild ancestors in the main features of their behavioral repertoire (e.g., Kruijt 1964; Stolba and Wood-Gush 1989). This genetic heritage sets certain limits to their ability to cope successfully with environmental conditions that are very different from the ones in which their ancestors evolved.

Individual Differences

Differences between individuals are another important source of variation. Both physiology and behavior are affected by genetic differences and by a variety of factors, including social experiences throughout the life span, with particular vulnerability to changes in early life (Sachser, this volume). Important epigenetic factors include those associated with gender and those that are influenced by environmental challenge or stress.

Within a species, males and females often differ along various dimensions, and these differences reflect hormonal changes during development. In addition, social systems, including polygamy or monogamy, as well as the traits of individual animals may be modified by the presence of hormones at different times throughout the life span. Because of their powerful and long-lasting effects, steroid hormones may help to determine both sex differences and individual differences in anatomy, physiology, and behavior. For example, the relative presence or absence of androgens and glucocorticoids may influence genital dimorphism, the tendency to show masculine or feminine behaviors, parental behavior, and so forth. Underlying or in addition to the effects of steroids are the developmental actions of peptide hormones. Changes in either steroids or peptides can produce long-term (probably life-long) changes in anatomy and behavior. In general, exposure to stressful experiences or glucocorticoids tends to masculinize genetic females and feminize genetic males (Ward and Ward 1985; Pfaff 1999).

In addition to genetic and maturational variables, there are numerous experiential variables that result in individual differences. For example, early deprivation or separation experiences often have adverse effects on development, including increased susceptibility to negative effects of future stressors (Panksepp 2001). In addition, exposure to external, uncontrollable traumatic events can also lead to a learned helplessness syndrome and increased susceptibility to future stress. By contrast, early experiences with mastery and control can help to immunize the organism against some of the adverse effects of future stressors (Peterson et al. 1993; Mineka and Zinbarg 1996).

Individual differences in temperament and personality also significantly affect proclivities of both humans and nonhumans for positive and negative emotions, which are in turn related to certain health variables. Starting very early in life, certain temperamental or personality characteristics predispose some individuals to experience negative emotions more readily than individuals without those characteristics. For example, Kagan and colleagues (1994) have shown that infants and young children with high levels of behavioral inhibition (shyness, high levels of negative emotional reactivity to novel stimuli, social avoidance, and withdrawal) are especially vulnerable to multiple phobias and multiple anxiety disorders in later childhood and adolescence as well as panic attacks and major depression in early adulthood (e.g., Caspi et al. 1998; Schmidt and Schulkin 1999; Hayward et al. 2000; Panksepp 2001). Behavioral inhibition in human children is moderately stable (e.g., Emde et al. 1992; Plomin et al. 1993; Lemery et al. 1999). Behavioral inhibition has also been studied in rats, cats, dogs, and rhesus monkeys with similar results regarding negative emotional consequences (see Mineka and Zinbarg 1996). Related personality traits in adulthood, such as neuroticism or negative affectivity, have also been shown to be risk factors for the development of anxiety and unipolar depressive disorders (e.g., Clark et al. 1994, for a review; Hayward et al. 2000), as well as for certain negative health variables (Watson and Pennebaker 1989).

Another personality variable that has been shown to be a risk factor for depression and possibly social phobia is introversion — the introverted pole of the extraversion dimension. Several, but not all, longitudinal studies have found introversion predicted subsequent symptoms of depression and subsequent diagnosis of major depression (e.g., Hirschfeld et al. 1989; Clark et al. 1994; Krueger et al. 1996). Thus several different temperamental and personality characteristics may be considered as risk factors for negative emotions and negative welfare.

With respect to determinants of positive affectivity in humans, Watson (2000) makes a strong case that internal processes, such as beliefs and temperament, are much more influential than external factors, such as income and job. Genetic and other biological factors may thus place constraints on upper limits of well-being for a particular person, and this "reaction range" may provide a physiological substrate upon which environmental influences act.

Numerous studies have found supportive evidence for a moderate heritability of temperamental and personality dimensions. For example, in adults many studies converge on an estimate of approximately 50% heritability for extraversion and 40% for neuroticism (see Plomin et al. 1997). There is also moderate heritability of childhood temperamental dimensions (Plomin et al. 1997; Schmitz et al. 1999). Nonshared environmental factors (e.g. siblings' different experiences at school) appear to account for most of the rest of the variance not accounted for by genetic variables (Plomin et al. 1997).

Recently, a specific gene related to anxiety and neuroticism appears to have been identified; the gene affects the brain's ability to use serotonin through its effect on the serotonin transporter (Lesch et al. 1996). However, the size of the relationship between this transporter gene and neuroticism accounts for only about 4% of the variance in people's neuroticism levels (Goldman 1996).

The personality characteristics described thus far seem to constitute risk factors for emotional disorders or upper limits of welfare. In addition there are personality characteristics whose adaptiveness varies depending on the animal's environment. For example, in a number of species, males can be characterized as having proactive or reactive coping strategies in stressful situations (Koolhaas et al., this volume). The proactive strategy, characterized by high levels of aggression and active avoidance, seems to be most adaptive in stable, highly predictable environments, such as a stable colony. The reactive coping strategy, characterized by more passive avoidance and caution, seems to be more adaptive under variable and unpredictable conditions.

PREREQUISITES FOR GOOD WELFARE

Prenatal Influences

Certain prerequisites for good welfare need to be in place before birth. Genetic disorders in humans and other animals may, for example, lead to problems that cause pain or ill health, and limit normal life. In animals, selective breeding for extreme traits has led to anatomical problems (e.g., difficult breathing in dogs bred for flat faces) or high risk of disease (e.g., mastitis in dairy cattle bred for high milk yield). Breeding for behavioral traits can sometimes lead to welfare problems; for example, breeding for high food intake in broiler chickens in combination with more efficient food conversion has meant that the long-lived parent birds must be severely feed-restricted and appear to live with chronic hunger (Savory et al 1993; Savory and Lariviere 2000).

Additional prenatal influences can arise from the social environment in which the mother lived during pregnancy. These can have profound effects on the offspring's behavior and welfare in later life (Kaiser and Sachser 1998, 2001).

Postnatal Ontogeny

A second set of prerequisites arises during postnatal ontogenetic development. Young animals of many species are especially vulnerable to attack and predation; hence they require a "secure base" (Bowlby 1988) involving attachment to a parental figure or a secure environment. Development of normal functioning of stress response systems may also require specific types of care and contact with the parental figure; for example, Liu et al. (2000) have shown that licking and grooming in rat pups are important for shaping their stress responsiveness. Animals that have received little licking and grooming will have increased numbers of glucocorticoid receptors, increased levels of CRH in the hypothalamus and other brain regions, as well as increased levels of peripheral ACTH and cortisol. This will result in an overreactive HPA system, high circulating levels of cortisol, and high levels of fear and anxiety. Rat pups whose mothers display high levels of licking and grooming have normal levels of CRH and cortisol. Thus, in addition to a safe social environment, the individual may need the system-modulating effects of highly specific types of interaction to achieve good welfare.

To function well later in life, animals also need a set of social and other skills that are developed during ontogeny, sometimes during sensitive periods or through important types of stimulation such as juvenile play and social integration during puberty (Sachser et al. 1994; Panksepp 1998; Sachser, this volume).

Living Conditions throughout Life

Based on evolutionary history, perhaps as modified by domestication, we should be able to identify certain environmental conditions important for welfare (Sachser, this volume).

1. Appropriate social contact is a key prerequisite for good welfare in social species and a key means of buffering the effects of stressors. This will depend, however, on the type of social organization typical of the species. For socially monogamous animals, for example, the ability to select and remain with a bonded social partner is important, whereas the constant presence of a conspecific may actually be stressful for solitary territorial species.
2. Rather than an environment that involves minimal stimulation and minimal challenge, the level of stimulation and challenge should be within a range that is suitable for the species. For example, demands for vigorous activity may be much higher for active predatory animals than for sedentary herbivores. Moreover, the degree of predictability and control are important in determining an animal's response to a situation, and need to be within a range that is suitable for the species. For humans, and very likely also for other species, optimal levels of stimulation resemble a U-shaped function with limits that probably differ for different individuals. Both very high and very low levels of stimulation may be aversive.
3. In many respects the important features of the environment are not simply the level of complexity and stimulation, but the presence of key features that allow the animal to respond to challenges in ways appropriate for the species and to perform those types of behavior that are thought to give rise to positive affect. Examples such as play, exploration, and grooming are given above. Rest is another key element of behavior which serves as a means of alleviating many welfare problems. Species differ in their environmental requirements for rest, and although these requirements are poorly studied for many species, accommodating resting behavior is important for good welfare.

HOW DO POSITIVE STATES INTERACT WITH AND INFLUENCE NEGATIVE STATES?

There are several ways that positive states may interact with and influence negative states. For humans, a cognitive filter influences appraisal of an environmental event as well as the physiological, emotional, and behavioral response to it. According to the transactional model of stress (Lazarus and Folkman 1984), it is the individual's perception of environmental events as threatening or not (primary appraisal) and the individual's appraisal of their own resources to meet the challenge of this threat (secondary appraisal) that result in the experience of "stress." If individuals have a positive bias (optimism, positive affectivity, high sense of self-efficacy, etc.), they may appraise the event as being less threatening or perceive themselves as being more able to meet the challenge. These perceptions may thereby result in lower levels of stress reactivity, for example lower SAM or HPA activation, and thus lower ultimate negative endpoints.

A similar buffering of negative states may occur when the organism (human or nonhuman) is in the presence of social support or other condition indicative of safety. In this situation, oxytocin or brain opioid systems may be activated, which may modulate the stress response system to create a state of lowered stress and, thereby, less HPA activation (Sachser et al. 1998; von Holst 1998). It appears that the basic systems that generate positive and negative states interact abundantly in ways that have not yet been sufficiently characterized.

Human research on panic disorder shows how both social support and cognitive variables, such as perceptions of control or predictability, can have dramatic effects on the affective states experienced when panic and other emotional states are provoked. Individuals with panic disorder are much more likely to experience panic attacks in studies in which they are exposed to various biological challenge procedures than are normal people or other psychiatric controls. These biological challenges, such as taking drugs like yohimbine or inhaling air with altered amounts of carbon dioxide, have physiological and physical effects, such as altered levels of carbon dioxide in the blood, increased breathing and heart rate; these often culminate in a panic attack for individuals with panic disorder. If these panic attacks were induced exclusively through specific biological mechanisms one would not expect cognitive or social variables to influence their occurrence. However, an individual with panic disorder who is allowed to bring a trusted companion to the panic provocation laboratory is much less likely to panic in response to a biological challenge than is one without a companion (Carter et al. 1995). Likewise a person with panic disorder who is told they have some control over administration of the biological challenge agent is much less likely to panic than is someone not told they have control (Sanderson et al. 1989). Finally, simply being given detailed information about what to expect after a panic provocation agent has been administered (predictability) is sufficient to block panic as well (Clark 1997). In each case social or cognitive variables dramatically modulate and block the autonomic activation and the associated subjectively terrifying experience of a panic attack.

ASSESSING GOOD WELFARE

Since good welfare is not just the absence of stress, illness, and injury, assessing good welfare requires measures that go beyond indicators of such negative states. As noted by Dawkins (this volume) simple, universal indicators of good welfare do not exist. Nevertheless, it is possible to use the tools of science in assessing good welfare. Such assessment, however, must be based on the thorough knowledge of the species' biology and of variation related to domestication, breeding, sex, and age.

In the field of human psychology, one of the early approaches to assessing welfare was the Karnofsky scale (Karnofsky et al. 1948), developed in the late 1940s to assess qualitative functioning of cancer patients along with the more traditional medical indicator of survival time. This scale focuses specifically on what a patient is able to do, rather than how they feel about their functional abilities, and allows assessment of functional status along a range from completely functional to completely disabled. Since Karnofsky's time, a multitude of quality-of-life self-report scales have been developed, most of which assess the welfare of medical patients along several dimensions, including how well a patient can function physically, how well they can fulfill their social or occupational roles, their level of fulfillment, and their emotional well-being (e.g., Cella et al. 1993). In healthy populations, welfare has been assessed by a variety of self-report surveys. These include single item instruments such as the FACES scale (Andrews and Withey 1976), which shows a series of 7 faces and asks the respondent to identify "which face comes closest to expressing how you feel about your life as a whole?" Other single-item surveys have asked individuals how happy they are on a 10-point scale, from "not too happy" to "very happy" (Myers and Diener 1996). Assessment of single traits have been done using instruments such as the Philadelphia Geriatric Center Morale Scale (Lawton 1972), the Life Satisfaction Index (Neugarten et al. 1961), the Life Orientation test (a measure of optimism; Scheier and Carver 1985), and the Sense of Coherence Scale (Antonovsky 1987; see Lutgendorf, this volume). An alternate definition of happiness has been "the net of positive over negative affect," and this construct has been measured via scales of positive and negative affectivity such as the Positive and Negative Affect Scale (PANAS, Watson et al. 1988) and the Affect Balance Scale (Moriwaki 1974), the latter of which measures the balance between positive and negative emotionality in an individual's life. Ryff's (1989) Well-being Scale is one of the most comprehensive instruments for assessment of welfare . This is an 86-item multidimensional self-report instrument that assesses six core dimensions of welfare identified earlier in this paper.

Because humans can report their affective states directly, they can potentially serve as an experimental model that might help to develop methods such as brain-scanning to guide further studies on animals. In the animal field, the assessment of welfare has traditionally focused on pathology, stress physiology, and other indicators of welfare problems. Because the assessment of good welfare requires attention to states of positive affect, it presents new challenges, but there are several avenues that show promise.

Unfortunately, the neurophysiological events associated with positive affect do not lend themselves to straightforward measurement. For the relevant peptide hormones, such as oxytocin and endogenous opioids, the important variables are likely to be their activity in specific parts of the brain. This activity is not always associated with parallel changes in the systemic circulation. Brain-imaging may provide appropriate measures in the future, but basic knowledge and the accuracy of the technology are currently insufficient. At present there are no direct brain measures of affective states in humans or other animals. In the search for other physiological indicators, measures of variability may be more useful than absolute values.

Currently, behavioral measures are an important source of information for recognizing good welfare, but such measures need to be developed and tested. Traditional clinical measures of reactivity, demeanor, and appearance could be refined. Types of behavior indicative of positive affect can also be identified and monitored. For example, some species appear to have vocal signals that indicate contentment. Rats, in addition to a vocalization related to pain, have two general types of ultrasonic vocalizations: one exhibited when anticipating affectively negative events (22 kHz lower-frequency "complaints"), the other when anticipating positive events (50 kHz high-frequency "chirps," Burgdorf et al. 2000, 2001; Panksepp and Burgdorf 2000).

A related approach is to identify common types of behavior which, in the natural life of the species, appear to enhance fitness in ways that are not simply responses to immediate threats to fitness (Fraser and Duncan 1998). One example is play. Play tends to be suppressed at times when the animal is experiencing some threat to fitness, such as food deprivation or social stress. Moreover, the neural substrates underlying play appear to give rise to strong positive affect, and play is often accompanied by signs that appear to signal enjoyment (Panksepp 1993, 1998). Certain comfort behaviors (e.g., grooming) and positive social behaviors (e.g., social grooming) are potential further examples.

A more specifically experimental approach involves identifying conditions that the animal finds reinforcing in the sense of supporting instrumental conditioning. By itself, the fact that a resource is reinforcing does not allow us to determine whether it supports a positive affective state, relieves a negative affective state, or both. However, according to one hypothesis (Fraser and Duncan 1998) where motivation is based on a strong negative state such as pain or hunger, the demand for the reward is likely to be relatively "inelastic;" that is, the animal is likely to expend more and more effort when this is necessary to obtain the reward. In contrast, if a reward promotes a positive affective state but does not relieve a negative one, demand is likely to be more "elastic," with the animal accepting a lower level of reward when the cost of obtaining it is increased. Although instrumental conditioning experiments have methodological problems (e.g., creating a test that uses response patterns appropriate for the species and motivational state; Fraser and Matthews 1997), they may provide a valuable, if time-consuming, means of achieving insights into the affective states of animals.

FUTURE DIRECTIONS OF RESEARCH

Good welfare includes both functioning well and feeling well. To date, most research on health and welfare, in both humans and nonhumans, has tended to focus on the functioning of bodily systems (behavioral, physiological, immunological) and has tended to emphasize negative conditions (disease, stress, pathology) instead of positive ones. Affective states, especially the positive affective states indicative of good welfare, have received comparatively little research. As a result, our knowledge about indicators and prerequisites for positive affective states is limited. Thus, one of the major aims of future research is to develop valid methods for assessing affective states by such means as:

- validation of behavioral and other indicators of affective states, including measures based on instrumental conditioning,
- better understanding of the relationship between physiological measures and affect in humans, and of the use of humans as models for the study of affective states in other species,
- development of brain-based and peripheral neuroendocrine measures of positive affect.

The development of reliable assessment methods will permit research on the prerequisites and effects of good welfare by answering questions such as:

- What interventions improve welfare in humans and what are the mechanisms by which they operate?
- What rearing conditions and genetic selection criteria promote adaptability and capability to experience positive affect in nonhumans?
- How do positive affective states influence negative ones, and to what extent can improved welfare reduce the incidence or severity of disease?

The ultimate goal of such work is to produce a theory of welfare that will make it possible to predict the requirements for good welfare in the members of different species and for different individuals of the same species that vary according to sex, age, personality, etc. A central aspect of such a theory will concern the relationships between environmental conditions, affective states, and behavior. Understanding these relationships will enable us to improve the quality of life in animals including humans.

ACKNOWLEDGMENTS

We are grateful to all the participants of the Dahlem Workshop for their contributions in the discussion of topics presented in this chapter, especially to Dietrich von Holst and Erwin Lemche.

REFERENCES

Andrews, F.M., and S.B. Withey. 1976. Social Indicators of Well-being: Americans' Perceptions of Life Quality. New York: Plenum.

Antonovsky, A. 1987. Unraveling the Mystery of Health: How People Manage Stress and Stay Well. San Francisco: Jossey-Bass.

Barnett, S.A. 1969. The Rat, a Study in Behavior, pp. 93–94. Chicago: Aldine Publishing.

Baumgartner, W., and C.J. Ketz Riley. 1999. Klinische Propädeutik der inneren Krankeiten und Hautkrankheiten der Haus- und Heimtiere, 4th ed. Berlin: Verlag Paul Parey.

Bekoff, M., ed. 2000. The Smile of a Dolphin. New York: Discovery Books.

Bermond, B. 1997. The myth of animal suffering. In: Animal Consciousness and Animal Ethics, ed. M. Dol, S. Kasanmoentalib, S. Lijmbach, E. Rivas, and R. van den Bos, pp. 125–143. Assen: Van Gorcum.

Blanchard, R.J., S. Parmigiani, R. Angullana, S.M. Weiss, and D.C. Blanchard. 1995. Behaviors of Swiss-Webster and C57BL/6N Sin mice in a fear/defense test battery. *Aggr. Behav.* **21**:21–28.

Bowlby, J. 1988. A Secure Base: Clinical Applications of Attachment Theory. London: Routledge.

Broom, D.M. 1998. Welfare, stress, and the evolution of feelings. *Adv. Study Behav.* **27**:371–403.

Broom D.M., and K.G. Johnson. 1993. Stress and Animal Welfare. London: Chapman and Hall.

Burgdorf, J., B. Knutson, and J. Panksepp. 2000. Anticipation of rewarding brain stimulation evokes ultrasonic vocalizations in rats. *Behav. Neurosci.* **114**:1–8.

Burgdorf, J., B. Knutson, J. Panksepp, and T.S. Shippenberg. 2001. Evaluation of rat ultrasonic vocalizations as predictors of the conditioned aversive effects of drugs. *Psychopharmacology* **255**:35–42.

Cabanac, M. 1992. Pleasure: The common currency. *J. Theor. Biol.* **155**:173–200.

Cannon, W.B. 1914. The emergency function of the adrenal medulla in pain and the major emotions. *Am. J. Physiol.* **33**:356–372.

Caspi, A., T.E. Moffitt, D.L. Newman, and P.A. Silva. 1998. Behavioral observations at age 3 years predict adult psychiatric disorders: Longitudinal evidence from a birth cohort. In: Annual Progress in Child Psychiatry and Child Development: 1997, ed. M.E. Hertzig et al., pp. 319–331. Bristol PA: Brunner/Mazel Inc.

Carruthers, P. 1992. The Animals Issue: Moral Theory in Practice. Cambridge: Cambridge Univ. Press.

Carter, C.S. 1998. Neuroendocrine perspectives on social attachment and love. *Psychoneuroendocrinology* **23**:779–818.

Carter, M.M., S.D. Hollon, R. Carson, and R.C. Shelton. 1995. Effects of a safe person on induced distress following a biological challenge in panic disorder with agoraphobia. *J. Abn. Psychol.* **104**:156–163.

Cella, D.F., D.S. Tulsky, G. Gray, et al.1993. The functional assessment of cancer therapy scale: Development and validation of the general measure. *J. Clin. Oncol.* **11**:570–579.

Clark, D.M. 1997. Panic disorder and social phobia. In: Science and Practice of Cognitive Behaviour Therapy, ed. D.M. Clark and C.G. Fairburn, pp. 121–153. Oxford: Oxford Univ. Press.

Clark, L.A., D. Watson, and S. Mineka. 1994. Temperament and personality in mood and anxiety disorders. *J. Abn. Psychol.* **103**:103–116.

Council of Europe Standing Committee. 1995. Recommendation concerning domestic fowl (*Gallus gallus*), adopted by the Standing Committee of the European Convention for the protection of animals kept for farming purposes on 28 November 1995. Strasbourg: Council of Europe.

Darwin, C. 1872. The Expression of the Emotions in Man and Animals. London: John Murray.

Duncan, I.J.H. 1996. Animal welfare defined in terms of feelings. *Acta Agric. Scand. Sect. A, Animal Sci.*, **Suppl. 27**:29–35.

Emde, R.N., R. Plomin, J.L. Robinson, J.S. Reznick, J.J. Campos, R. Corley, J.C. DeFries, D.W. Fulker, J. Kagan, and C. Zahn-Waxler. 1992. Temperament, emotion, and cognition at 14 months: The MacArthur Longitudinal Twin Study. *Child Dev.* **63**:1437–1455.

Fraser, D. 1995. Science, values and animal welfare: exploring the "inextricable connection." *Animal Welfare* **4**:103–117.

Fraser, D., and I.J.H. Duncan. 1998. "Pleasures," "pains," and animal welfare: Toward a natural history of affect. *Animal Welfare* **7**:383–396.

Fraser, D., and L.R. Matthews. 1997. Preference and motivation testing. In: Animal Welfare, ed. M.C. Appleby and B.O. Hughes, pp. 159–173.Wallingford: CAB International.

Fraser, D., D.M. Weary, E.A. Pajor, and B.M. Milligan. 1997. A scientific conception of animal welfare that reflects ethical concerns. *Animal Welfare* **6**:187–205.

Goldman, D. 1996. High anxiety. *Science* **274**:1483–1484.

Hayward, C., J.D. Killen, H.C. Kraemer, and C.B. Taylor. 2000. Predictors of Panic Attacks in Adolescence. *J. Am. Acad. Child & Adolesc. Psychiatry* **39**:207–214.

Henry, J.P., and P.M. Stephens. 1977. Stress, Health, and the Social Environment. A Sociobiologic Approach to Medicine. New York: Springer.

Herre, W., und M. Röhrs. 1990. Haustiere, zoologisch gesehen. 2nd ed. Stuttgart: Fischer.

Hirschfeld, R.M., G.L. Klerman, P. Lavori, M.B. Keller, P. Griffith, and W. Coryell. 1989. Premorbid personality assessments of first onset of major depression. *Arch. Gen. Psychiatry* **46**:345–350.

Hornbuckle, W.E. 1992. General physical examination of the cat and dog. In: Handbook of Small Animal Practice, ed. R.V. Morgan, 2nd ed., p. 3. New York: Churchill Livingstone.

Kagan, J., N. Snidman, D. Archus, and J.S. Reznick. 1994. Galen's Prophecy: Temperament in Human Nature. New York: Basic Books.

Kaiser, S., and N. Sachser. 1998. The social environment during pregnancy and lactation affects the female offsprings' endocrine status and behaviour in guinea pigs. *Physiol. & Behav.* **63**:361–366.

Kaiser, S., and N. Sachser. 2001. Social stress during pregnancy and lactation affects in guinea pigs the male offsprings' endocrine status and infantilizes their behaviour. *Psychoneuroendocrinology* 26:503–519.

Karnofsky, D.A., W.H. Ablemann, L.F. Craver, and J.H. Burchenal. 1948. The use of the nitrogen mustards in the palliative treatment of carcinoma. *Cancer* **1**:634–652.

Krueger, R.F., A. Caspi, T.E. Moffitt, P.A. Silva, et al. 1996. Personality traits are differentially linked to mental disorders: A multitrait-multidiagnosis study of an adolescent birth cohort. *J. Abn. Psychol.* **105**:299–312.

Kruijt, J.P. 1964. Ontogeny of social behaviour in Burmese Red Junglefowl (*Gallus gallus spadiceus* Bonnaterre). *Behaviour* **Suppl. 12**:1–201.

Künzl, C., and N. Sachser. 1999. The behavioral endocrinology of domestication: A comparison between the domestic guinea pig (*Cavia aperea* f. *porcellus*) and its wild ancestor, the cavy (*Cavia aperea*). *Horm. Behav.* **35**:28–37.

Lawton, M.P. 1972. The dimensions of morale. In: Research, Planning, and Action for the Elderly, ed. D. Kent, R. Kastenbau, and S. Sherwood, pp.144–165. New York: Behavioral Publications.

Lazarus, R.S., and S. Folkman. 1984. Stress, Appraisal, and Coping. New York: Springer.

Lemery, K.S., H.H. Goldsmith, M.D. Klinnert, and D.A. Mrazek. 1999. Developmental models of infant and childhood temperament. *Dev. Psychol.* **35**:189–204.

Lesch, K.P., D. Bengel, A. Heils, S.Z. Sabol., B.D. Greenberg, S. Petri, B. Jonathan, C.R. Müller, D.H. Harmer, and D.L. Murphy. 1996. Association of anxiety-related traits with a polymorphism in the serotonin transporter gene regulatory region. *Science* **274**:1527–1531.

Liu, D., J. Diorio, J.C. Day, D.D. Francis, and M.J. Meaney. 2000. Maternal care, hippocampal synaptogenesis and cognitive development in rats. *Nature Neurosci.* **3**:799–806.

Maslow, A. 1968. Toward a Psychology of Being, 2nd ed. Princeton: Van Nostrand Reinhold.
Mendoza, S.P., and W.A. Mason. 1986. Contrasting responses to intruders and to involuntary separation by monogamous and polygynous new world monkeys. *Physiol. Behav.* **38**:795–801.
Midgley, M. 1983. Animals and Why They Matter. Athens GA: Univ. of Georgia Press.
Mineka, S., M. Gunnar, and M. Champoux. 1986. Control and early socioemotional development: Infant rhesus monkeys reared in controllable versus uncontrollable environments. *Child Dev.* **57**:1241–1256.
Mineka, S., and R. Zinbarg. 1996. Conditioning and ethological models of anxiety disorders: Stress-in-Dynamic-Context Anxiety models. In: Perspectives on Anxiety, Panic, and Fear, 43rd Ann. Nebraska Symp. on Motivation, ed. D. Hope, pp. 135–211. Lincoln NE: Univ. of Nebraska Press.
Moriwaki, S.Y. 1974. The Affect Balance Scale: A validity study with aged samples. *J. Gerontol.* **29:**73–78.
Myers, D.G., and E. Diener. 1996. The pursuit of happiness. *Sci. Am.* **274**:54–56.
Neugarten, B.L., R.J. Havighurst, and S. Tobin. 1961. The measurement of life satisfaction. *J. Gerontol.* **16**:134–143.
Panksepp, J. 1993. Rough-and-tumble play: A fundamental brain process. In: Parents and Children Playing, ed. K.B. MacDonald, pp. 147–184. Albany NY: SUNY Press.
Panksepp, J. 1998. Affective Neuroscience: The Foundations of Human and Animal Emotion. New York: Oxford Univ. Press.
Panksepp, J. 2001. The long-term psychobiological consequences of infant emotions: prescriptions for the twenty-first century. Infant Mental Health J. 22:132–173.
Panksepp, J., and J. Burgdorf. 2000. 50k-Hz chirping (laughter?) in response to conditioned and unconditioned tickle-induced reward in rats: Effects of social housing and genetic variables. *Behav. Brain Res.* **115**:25–38.
Panksepp, J., B. H. Herman, T. Villberg, P. Bishop, and F.G. DeEskinazi. 1980. Endogenous opioids and social behavior. *Neurosci. Biobehav. Rev.* **4**:473–487.
Panksepp, J., S. M. Siviy, and L.A. Normansell. 1985. Brain opioids and social emotions. In: The Psychobiology of Attachment and Separation, eds. M. Reite and T. Fields, pp. 3–49. New York: Academic.
Peterson, C., S.F. Maier, and M.E.P. Seligman. 1993. Learned Helplessness: A Theory for the Age of Personal Control. New York: Oxford Univ. Press.
Pfaff, D. 1999. Drive: Neurobiological and Molecular Mechanisms of Sexual Motivation. Cambridge MA: MIT Press.
Plomin, R., J.C. De Fries, G.E. McClearn, and M. Rutter. 1997. Behavior Genetics, 3rd ed. New York: W.H. Freeman.
Plomin, R., R.N. Emde, J.M. Braungart, J.J. Campos, R. Corley, D.W. Fulker, J. Kagan, J.S. Reznick, J.L. Robinson, C. Zahn-Waxler, and J.C. DeFries. 1993. Genetic change and continuity from fourteen to twenty months: The MacArthur Longitudinal Twin Study. *Child Dev.* **64**:1354–1373.
Porges, S.W. 1998. Love: An emergent property of the mammalian autonomic nervous system. *Psychoneuroendocrinology* **23**:837–861.
Preece, R. 1999. Animals and Nature: Cultural Myths, Cultural Realities. Vancouver: Univ. of British Columbia Press.
Price, E.O. 1998. Behavioural genetics and process of animal domestication. In: Genetics and Behavior of Domestic Animals, ed. T. Grandin, pp. 31–65. San Diego: Academic.
Röhrs, M. 1991. Einflüsse der Domestikation auf Versuchstiere. In: Qualitätskriterien der Versuchstierforschung, ed. K. Gärtner, pp. 27–45. Weinheim: VCH.
Ryff, C.D. 1989. Happiness is everything, or is it? Explorations on the meaning of psychological well-being. *J. Personality & Soc. Psychol.* **57**:1069–1081.
Ryff, C.D., and C.L.M. Keyes. 1995. The structure of psychological well-being revisited. *J. Personality & Soc. Psychol.* **69**:719–727.
Ryff, C.D., and B. Singer. 1998. The contours of positive human health. *Psychol. Inq.* **9**:1–28.

Sachser, N. 1994. Sozialphysiologische Untersuchungen an Hausmeerschweinchen. Gruppenstrukturen, soziale Situation und Endokrinium, Wohlergehen. Berlin: Verlag Paul Parey.

Sachser, N., D. Hierzel, and M. Dürschlag. 1998. Social relationships and the management of stress. *Psychoneuroendocrinology* **23**:891–904.

Sachser N., C. Lick, and K. Stanzel. 1994. The environment, hormones and aggressive behaviour: A five-year-study in guinea pigs. *Psychoneuroendocrinology* **19**:697–707.

Sanderson, W.C., R.M. Rapee, and D.H. Barlow. 1989. The influence of an illusion of control on panic attacks induced via inhalation of 5.5% carbon dioxide-enriched air. *Arch. Gen. Psychiatry* **46**:157–162.

Savory, C.J., and J.M. Lariviere. 2000. Effects of qualitative and quantitative food restriction treatments on feeding motivational state and general activity level of growing broiler breeders. *Appl. Animal Behav. Sci.* **69**:135–147.

Savory, C.J., K. Maros, and S.M. Rutter. 1993. Assessment of hunger in growing broiler breeders in relation to a commercial restricted feeding programme. *Animal Welfare*. **2**:131–152.

Scheier, M., and C. Carver. 1985. Optimism, coping, and health: Assessment and implications of generalized outcomes expectancies. *Health Psychol.* **4**:219–247.

Schmidt, L.A., and J. Schulkin, eds. 1999. Extreme Fear, Shyness and Social Phobia: Origins, Biological Mechanisms and Clinical Outcomes. New York: Oxford Univ. Press.

Schmitz, S., D.W. Fulker, R. Plomin, C. Zahn-Waxler, R.N. Emde, and J. DeFries. 1999. Temperament and problem behaviors during early childhood. *Intl. J. Behav. Dev.* **23**:333–335.

Selye, H. 1932. The general adaptation syndrome and the diseases of adaptation. *J. Clin. Endocrinol.* **6**:117–152.

Singer, P. 1990. Animal Liberation, 2nd ed. New York: Avon Books.

Smith, R.H. 1972. Wildness and domestication in *Mus musculus*, a behavioral analysis. *J. Comp. & Physiol. Psychol.* **79**:22–29.

Speirs, V.C., and R.H. Wrigley. 1997. Clinical Examination of Horses, pp. 20–21. Philadelphia: W.B. Saunders.

Stafleu, F.R., F.J. Grommers, and J. Vorstenbosch. 1996. Animal welfare: Evolution and erosion of a moral concept. *Animal Welfare* **5**:225–234.

Stolba, A., and D.G.M. Wood-Gush. 1989. The behaviour of pigs in a semi-natural environment. *Animal Prod.* **48**:419–425.

Stribley, J.M., and C.S. Carter. 1999. Developmental exposure to vasopressin increases aggression in adult prairie voles. *Proc. Natl. Acad. Sci. USA* **96**:12,601–12,604.

Uvnas-Moberg, K. 1998. Oxytocin may mediate the benefits of positive social interaction and emotions. *Psychoneuroendocrinology* **23**:819–835.

Von Holst, D. 1998. The concept of stress and its relevance for animal behavior. *Adv. Study Behav.* **27**:1–131.

Ward, I.L., and O.B. Ward. 1985. Sexual behavior differentiation: Effects of prenatal manipulations in rats. In: Handbook of Behavioral Neurobiology, vol. 7, ed. N. Adler, D. Pfaff, and R.A. Goy, pp. 77–98. New York: Plenum.

Watson, D. 2000. Mood and Temperament. New York: Guilford Press.

Watson, D., L.A. Clark, and A. Tellegen. 1988. Development and validation of brief measures of positive and negative affect: The PANAS scales. *J. Personality & Soc. Psychol.* **54**:1063–1070.

Watson, D., and J.W. Pennebaker. 1989. Health complaints, stress, and distress: Exploring the central role of negative affectivity. *Psychol. Rev.* **96**:234–254.

Wenger, N.K., M.E. Mattson, C.D. Furberg, and J. Elinson, eds. 1984. Assessment of Quality of Life in Clinical Trials of Cardiovascular Therapies, p. 908. New York: LeJaq.

Wiepkema, P.R. 1987. Behavioural aspects of stress. In: Biology of Stress in Farm Animals, ed. P.R. Wiepkema and P.W.M. van Adrichem, pp. 113–133. Dordrecht: Martinus Nijhoff.

Young, P.T. 1959. The role of affective processes in learning and motivation. *Psychol. Rev.* **66**:104–125.

7

Can We Understand the Brain and Coping without Considering the Immune System?

R. DANTZER

Integrative Neurobiology, INSERM U394, 33077 Bordeaux Cedex, France

Keywords: Stress, coping, emotion, depression, mitogen-induced lymphoproliferation, natural killer cell activity, inflammation, cytokine

ABSTRACT

Stress has long been known to alter the host resistance to pathogenic microorganisms. The development of psychoneuroimmunology as a scientific discipline has allowed us to describe more precisely the effects of stress on immune responses and to study the mechanisms that are involved in these effects. Coping has turned out to be a key factor in the effects of stressors on immunity, as illustrated by the results of numerous studies carried out in humans and other animals. The realization that peripheral immune events can alter brain functions has clarified the reasons why the immune system is so sensitive to stress influences. The brain needs to be recruited in conditions of immune challenge so as to allow the development of sickness behavior and downregulate activated immunocytes. This recruitment of brain functions is likely to alter coping processes both directly and in the long term. The existence of reciprocal relationships between the brain and the immune system provides important new perspectives for understanding the pathophysiology of stress and coping.

HISTORICAL PERSPECTIVE

The scientific discipline of immunology has developed independently of neurosciences and psychology. This is not unusual in science. Each scientific field tends to grow independently of the others, especially at the beginning; then, as the field matures, a few individuals at the margin of the field start looking for possible intersections with other fields. This gives birth to new branches that are usually looked down upon by those who stay in the main stream. These

Coping with Challenge: Welfare in Animals including Humans
Edited by D.M. Broom © 2001 Dahlem University Press

new branches can either prosper or regress and die, depending on the pertinence of the questions that are asked and the technical means that are available to answer them.

The study of the reciprocal interactions between the immune and the central nervous systems is certainly a good example of a success story, although it was far from evident at the start. Initially located at the margin of behavioral neuroscience, the discipline of psychoneuroimmunology grew up from an interest in the way stress affects resistance of the host to infection and other immune-mediated diseases, including (improperly) cancer. The hypothalamic-pituitary-adrenal (HPA) cortex axis response normally helps the organism to restore homeostasis. However, it can have deleterious effects if prolonged or excessive. This is particularly the case because of glucocorticoids, which have immunosuppressive properties. According to one, influential theory of stress, heightened emotional arousal is accompanied by elevated concentrations of stress hormones which, in turn, impact on physiology and metabolism, and favor the induction of diseases. Hans Selye referred to these diseases as diseases of adaptation, to mean that they represent failure of the adaptation process that is normally triggered by exposure to stressors (Selye 1973). The particular organ affected during the general adaptation syndrome is the one which is the weakest for genetic reasons or because of previous pathological problems.

To test the validity of the implications of the theory of stress, a number of experiments were carried out in the 1950s with the objective of assessing the influence of laboratory stressors on the course of immune-mediated pathologies and on immune functions. All of these experiments confirmed that the immune system is very sensitive to the effect of stressors. For instance, exposure to high intensity noise increased mortality in virus-infected mice and decreased production of interferon, an antiviral agent manufactured by leukocytes (Chang and Rasmussen 1965). In the same manner, physical restraint decreased resistance of mice to herpes-like viruses and led to a reduction in endotoxin- and virus-induced synthesis of interferon (Rasmussen et al. 1957). However, even at this early stage of investigation, it was already apparent that the effects of HPA activity, in terms of progression of the disease process, were not always negative and depended on the type of pathogen under investigation. For instance, the same physical restraint that decreased resistance to viruses in laboratory rodents attenuated somatic symptoms in a model of multiple sclerosis known as experimental allergic encephalomyelitis (Levine et al. 1962). Chickens housed in an environment with a high level of social interactions were more susceptible to viral infections but more resistant to bacterial and parasitic infections (Gross 1972, 1976).

The main problem with all of these studies was that the limited knowledge which was then available on mechanisms of immunity did not allow interpreters of the results to go beyond phenomenology. The tendency was to consider the observed effects as an artifact due to the immunosuppressive effects of glucocorticoids. This field of research, therefore, drew very little attention from other disciplines, including immunology, and quickly subsided. The second wave of research came only in the 1980s, when progress in cellular immunology combined with advances in neuroendocrinology allowed researchers to investigate the mechanisms underlying the influence of stressors on immunity in a more sophisticated way. This research actually developed in several directions at the same time, including the demonstration of the role of controllability in the effects of stress on both lymphocyte proliferation and tumor rejection and development (Laudenslager et al. 1983; Sklar and Anisman 1979; Visintainer et al. 1982); the discovery that immune cells are able to produce neuropeptides, such as pro-opiomelanocortin (Smith et al. 1982); and the observation of an association between the immune response and changes in neuroendocrine activity (Besedovsky et al. 1983). These new findings, combined with evidence of a functional autonomic innervation of the bone marrow and lymphoid organs, including the thymus and lymph nodes (Felten et al. 1984; Williams et al. 1981), initiated psychoneuroimmunology or neuroimmunomodulation, as some scientists preferred to call the discipline, in an effort to get away from the supposedly soft behavioral sciences and base the new science on less disputed cellular and molecular biology.

The net result of all these efforts is that nowadays most immunology textbooks contain a section on the interactions between the immune and the central nervous systems. During the last decade, neuroscientists whose interest was initially restricted to neurons and neurotransmitters have incorporated the fact that the brain also contains glial cells. Microglial and macroglial cells are not just for packaging neurons. They play an important role in brain homeostasis and, from the view point of psychoneuroimmunology, produce soluble mediators, including cytokines, that mediate the brain response to inflammation and to injury (Merrill and Benveniste 1996) and certainly intervene in development and brain plasticity (Patterson and Nawa 1993). Psychologists, working initially on the relationship between stress and disease, are now fully aware of the existence of a bidirectional communication between the brain and the immune system, so that not only mental events influence immune responses but also vice versa, immune events influence brain functions (Dantzer and Kelley 1989).

THE RELATION BETWEEN COPING AND IMMUNITY

Since different hormones have different effects on immune functions, the immune consequences of the physiological activation that occurs in response to environmental factors should depend on the patterning of the hormonal reaction. However, this patterning does not necessarily reflect the type of emotion that is experienced by the subject. Measurements of changes in circulating cortisol levels and activity of the autonomic nervous system, in contrast with intuitive thinking, do not demonstrate that sadness is different from joy and elation (Brown et al. 1993; Levenson 1992). Appraisal of the situation and possibilities of action are actually more important than affects for determining the pattern of hormonal response that is elicited by the emotional stimulus (Dantzer and Kelley 1989). This means that the impact of stressors on the immune system should be critically dependent on the ability of the individual to cope in an appropriate way with the situation. Although the methodology used for assessing the effects of stress on immunity has mainly relied on nonspecific immune responses, such as mitogen-induced lymphoproliferation and natural killer cytotoxicity, the results are remarkably consistent between studies. In general, the inability to cope with the stressor has detrimental effects on cellular immunity, whereas the ability to cope exerts a protective effect. The rate of tumor growth in mice exposed to a single session of inescapable electric shocks, 24 hours following tumor cell transplantation, was accelerated whereas mice exposed to escapable electric shock did not differ from nonshocked mice in time of tumor appearance and tumor size (Sklar and Anisman 1979). In rats, a single session of inescapable, but not of escapable, shock significantly reduced the incidence of rejection of transplanted tumor cells (Visintainer et al. 1982). The detrimental effects of uncontrollability are likely to be mediated by changes in immune responses since inescapable electric shock decreases cellular immune responses but escapable shock does not. Rats submitted to inescapable electric shock on the first day and re-exposed to a few "reminder" shocks on the second day displayed a lower proliferative response of blood lymphocytes to T-cell mitogens than rats submitted to escapable electric shocks or apparatus-control animals (Laudenslager et al. 1983). The ability to predict the occurrence of a stressor is also important since rats submitted to unpredictable electric shock had lower proliferative responses of splenocytes to T-cell mitogens than rats submitted to predictable shock or nonshocked animals (Mormède et al. 1988).

Many animal species are gregarious and form enduring social bonds. These social variables have profound influences on physiology. Familiar and positive social relationships exert buffering effects on the stress response, whereas the formation of new relations and the disruption of established ones elicit considerable physiological activation. Over the last fifteen years, there has been a growing body of primate research demonstrating these two facets of the effects of psychosocial factors on immune responses. Proliferation of lymphocytes in response to mitogens was reduced in young pigtail macaque monkeys following separation from their social companions (Reite et al. 1981). The same effect was observed in both mother and infants following separation of mother–infant pairs of bonnet macaques (Laudenslager et al. 1982). In squirrel monkeys injected with a benign virus to allow measurement of the primary antibody response, neutralizing antibody levels were reduced tenfold in separated infants compared with control infants who remained with their mothers. However, this effect occurred only in infants that were separated from their mothers by being moved to a novel cage, but not in infants who were permitted to remain in the home cage in which they had been reared. In the same manner, providing infants with familiar companions during the separation prevented the reduction in antibody responses (Coe 1993). These observations illustrate the concept that the immune consequences of a stressful event are very much dependent on the cognitive and affective processes that are involved in the appraisal of this event. Individual differences are important in this respect. For instance, when lymphocyte proliferation responses of separated infant pigtail macaques and bonnet macaques were studied for each individual following separation, it appeared that only some infants displayed the decreased reactivity to mitogens (Laudenslager et al. 1990). Immune changes occurred primarily in those infants that were the most reactive to the separation process, as assessed by agitation and distress vocalizations on the first day, and slouched posture on the subsequent days.

The findings of altered immune responses in emotionally distressed experimental animals have prompted a search for the possible occurrence of similar changes in human beings. These studies are by necessity carried out on circulating populations of leukocytes, of which the composition and functions do not reflect in a very accurate way the ability of the organism's immune system to protect against pathogenic processes.

Short-term effects of emotions on immune functions can be observed under appropriate experimental circumstances. For example, a 12-min mental arithmetic task induced rapid immune changes, in the form of a release of CD8 suppressor/cytotoxic T cells and NK cells in the circulation, and, in young subjects only, an increase in NK cell activity (Naliboff et al. 1991). Exposure to a stressor consisting of mild electric shock and loud white noise administered over a 30-min period in an unpredictable and intermittent manner resulted in a lowered lymphocyte proliferation response to mitogens in subjects who could control the stressor, whereas subjects who could not control the stressor were not affected, despite increased anger and frustration (Weisse et al. 1990).

A few studies have looked for possible association between temperament traits and immune responsiveness. In view of the proposed importance of emotional repression in the pathogenesis of cancer, eighty first-year undergraduates were asked to complete a personality questionnaire and write an essay about a stressful event that had happened in their life (Esterling et al. 1990). The degree of emotional disclosure was scored according to the ratio of emotional-to-total words used in the writing. Antibody titers against Epstein-Barr virus were measured in a blood sample collected immediately after writing. High levels of antibody titers were observed in repressors independently of their degree of emotional disclosure, but also in sensitizers who were low disclosers, which can be interpreted to suggest that individuals who are aware of their heightened emotional response to environmental events, but choose not to express it, are more at risk than individuals who are not inhibited.

In clinically oriented studies, the emphasis has been mainly on negative life events and negative affective states. In the first case, Janice Kiecolt-Glaser and her colleagues have repeatedly observed altered immune competence, in the form of reduced NK cell activity, decreased lymphoproliferative responses, and reduced cellular immune control over latent herpes viruses resulting in increased antibody titers to Epstein-Barr virus, in adults experiencing difficulties in their marital life, spouses of patients afflicted with Alzheimer's disease, and medical students at the time of academic examination (Glaser and Kiecolt-Glaser 1994). In the second case, the possible link between depression and susceptibility to disease has been examined using immune responsiveness as a mediating factor. Depression can occur in many different contexts, including the context of loss. Bereavement is associated with altered immune competence although the observed changes and the exact relationship between depressed mood and immunity are not always clear. In one of the first studies on this topic, decreased lymphocyte proliferation was reported in spouses of women who were terminally ill with breast cancer (Schleifer et al. 1983). Compared with immune responses tested before the death, lymphocyte responses were reduced 1 month after the death and tended to return towards baseline during the 4–14 month follow-up period. In another study, however, women who had been widowed for less than 6 months, women whose husbands were diagnosed with terminal lung cancer, and women whose husbands were healthy did not differ in their NK cell activity (Irwin et al. 1987). This variability between studies can be explained partly on methodological grounds, especially the large magnitude of interindividual differences in immune responses and the lack of preselection of specific cell subsets to be used in the immune assays.

There is a strong belief in the medical community for the existence of a strong association between depressed mood and decreased immune function in patients who suffer from depressive illness. However, a closer examination of the abundant literature on depression and immune competence reveals that reduced lymphocyte responsiveness is mainly observed among the clinically depressed, particularly older patients diagnosed with unipolar disorders severe enough to warrant hospitalization. Other comparisons of immune system function and depression do not provide a consistent picture. In addition, the real importance of dysphoric mood per se in the immune alterations observed in these studies is doubtful (Dantzer et al. 1999). An important question is whether the immune changes that occur in depressed patients are large enough to result in an altered sensitivity to pathogenic microorganisms and consequently a higher prevalence of clinical or subclinical infectious diseases. Although this point is dealt with to some extent by Irwin (this volume), it is worth pointing out that the occurrence of even a mild disease can affect coping, for example, by altering the way patients seek social support.

In view of the many pathways that link the brain to the immune system, it is not easy to dissect out the precise mechanisms that are responsible for stress-induced immune alterations. Two main approaches are used for this purpose: (a) correlation studies which aim at finding possible associations between changes in the blood levels of a specific hormone and the immune-dependent variable; and (b) intervention studies in which causality is explored by pharmacological tools opposing the action of a given mediator or mimicking its effect. Nowadays, more and more emphasis is put on the possibility that stress and coping actually alters immune competence by changing the balance between T helper 1 (Th1) and T helper 2 (Th2) lymphocytes. The Th1–Th2 response switch is considered to be a crucial event in several immunopathologies. Predominance of Th1 over Th2 responses is associated with inflammation and cellular immunity, whereas predominance of Th2 over Th1 responses is association with humoral immunity. Both glucocorticoids (Rook 1999) and β-endorphin produced by immunocytes in conditions of stress (Panerai and Sacerdote 1997), have been proposed as causes of a shift in the balance toward a Th2-like response.

THE INFLUENCE OF IMMUNE EVENTS ON BRAIN FUNCTIONS

Behavioral neuroscience is concerned with the way behavioral processes are represented in the brain. Neuroanatomy traces the neuronal circuits that process sensory stimuli and organize behavior in terms of structures and communication networks. The functional relevance of the neurotransmitters that are synthesized and released in this circuit is identified with the help of neuropharmacology, using relatively simple tools such as specific agonist and antagonist drugs or more sophisticated approaches, such as animals in which a gene is either deleted by homologous recombination or overexpressed. Overall, the brain is seen as a more or less complex wiring diagram with interneuronal communication mediated by specialized neurotransmitters. The immune system is given no role in this conception. The brain has always been considered as being separated from the rest of the body by virtue of the blood–brain barrier and endowed with a state of immunological privilege. This is also because the brain parenchyma is devoid of lymphocytes, and there is little or no expression of the molecules of the major histocompatability complex. New findings obtained over the last two decades have forced a revision of this restrictive view. Activated T cells have the ability to penetrate the brain parenchyma in search of the antigen, and microglial cells can function as antigen-presenting cells (Hickey and Kimura 1988). In addition, many of the soluble mediators that serve communication within the immune system and between immune cells and other cell types are expressed in the brain together with their receptors. Inflammatory reactions are generated in the intact brain in response to peripheral infection or to internal causes, such as seizure and stroke. Brain inflammation also participates in certain neurodegenerative diseases, such as Alzheimer's disease and HIV-induced dementia. This has enabled correlations to be made between inflammatory signs and mental illness, especially depression, and this has opened a new frontier of research.

The previous considerations are important in relation to welfare. It is commonly assumed that ill individuals are suffering; however, the reality of this phenomenon and its mechanisms have remained elusive. It is now evident that a peripheral inflammation induces the expression of cytokines in the brain and that these locally produced cytokines are responsible for sickness behavior. In terms of behavioral organization, sickness behavior is like fear, the expression of a central motivational state that is triggered by recognition of pathogenic microorganisms via cells of the immune system and organizes perceptions and actions of the sick individual (Hart 1988). This response is adaptive in the sense that it enables the organism to cope better with infection. However, like fear, sickness behavior can be maladaptive if its duration exceeds that of the infectious episode, if its intensity is exaggerated, or if its occurrence takes place independently of any immune event.

The possibility that sickness behavior can develop in response to nonimmune stressors has been given some consideration, using molecular and physiological approaches. In the first case, there are several instances of stress-induced elevations of cytokine expression in the brain of animals submitted to restraint or to inescapable electric shock, although the exact conditions that are responsible for such a response are still elusive. In the second case, there is evidence that stress can induce a true fever which, in contrast to hyperthermia, is marked by an increase in the thermoregulatory set point and associated with the production of prostaglandins, certainly via the activation of the inducible form of cyclo-oxygenase (COX-2). There is still a gap in the available literature, since there has been no attempt to relate those stress-induced functional changes that appear to be mediated by peripherally or centrally released cytokines to a stress-induced production of cytokines, and conversely, the functional consequences of most of the stress-induced changes in cytokine levels in the brain have not been investigated. In either case, the mechanisms responsible for the effects of stress on cytokine production are not known. An obvious possibility is the bacterial and endotoxin translocation and clearance that occurs when the gastrointestinal mucosal barrier is impaired, as a result of the redistribution of blood flow from visceral organs to the brain and muscles in stressed individuals. Since bacterial translocation is more likely when animals have been starved, the nutritional status of stressed animals is certainly important for the effects of stress on cytokine production.

There is some limited evidence that indicates activation of the brain cytokine system can alter coping mechanisms. In particular, the group of Tilders has shown that a previous episode of sickness sensitizes the response of the HPA axis to novelty (Tilders 1999). The underlying mechanisms of this cross-sensitization phenomenon are represented by a phenotypic shift in the neurons of the paraventricular nucleus of the hypothalamus that, under the effect of sickness, co-express vasopressin and corticotrophin-releasing hormone (CRH).

The development of sickness behavior critically depends on the transfer of the immune information from the periphery to the brain. The first studies carried out on the mechanisms by which peripheral immune stimuli signal the brain to induce fever, activation of the HPA axis, and sickness behavior emphasized the importance of fenestrated parts of the blood–brain barrier, known as circumventricular organs, for allowing blood-borne pro-inflammatory cytokines to act on brain functions. The discovery in the mid-1990s that subdiaphragmatic section of the vagal nerves attenuates the brain effects of systemic cytokines, together with the demonstration of an inducible brain cytokine compartment, shifted the attention from circumventricular organs to neural pathways in the transmission of the immune message to the brain. Since then, neuroanatomical studies have confirmed the existence of a fast route of communication from the immune system to the brain via the vagal nerves. This neural pathway is complemented by a humoral pathway that involves cytokines produced at the level of the circumventricular organs and the choroid plexus and at the origin of a second wave of cytokines produced in the brain parenchyma. Depending on their source, these locally produced cytokines can either activate neurons that project to specific brain areas or diffuse by volume transmission into the brain parenchyma to reach their targets. Activation of neurons by cytokines can be direct or indirect via prostaglandins. The exact way the neural pathway of transmission interacts with the humoral pathway remains to be elucidated.

CONCLUSION: PROBLEMS AND OPEN QUESTIONS

As a result of recent advances in the study of the reciprocal relationships between the immune system and the brain, it is easy to answer positively the question of the relevance of the immune system to the brain and its functions, including coping mechanisms. However, in relation to the issue of welfare, there are still a number of issues that need to be addressed for a proper consideration of immunity:

1. Are stress-induced changes in immune responses a better indicator of coping than traditional neuroendocrine markers?
2. Are stress-induced changes in immune responses relevant to health? In particular, does stress-induced immunosuppression favor the emergence of infectious microorganisms that contribute to disease?
3. Is the brain cytokine compartment involved in the response of the organism to nonimmune stimuli?
4. How specific is the transmission of the immune information to the brain via vagal nerves? In particular, is there a specific pattern of immune changes in the brain associated with previously identified vagally mediated brain states?
5. How is the processing of information organized in a relatively primitive brainstem structure such as the nucleus tractus solitarius?
6. Can phylogenetic and ontogenetic approaches shed some light on the organization of the brain–immune relationships?

REFERENCES

Besedovsky, H., A. del Rey, E. Sorkin, M. Da Prada, R. Burri, and C. Honegger. 1983. The immune response evokes changes in brain noradrenergic neurons. *Science* **221**:564–566.

Brown, W.A., A.D. Sirota, R. Niaura, and T.O. Engebretson. 1993. Endocrine correlates of sadness and elation. *Psychosom. Med.* **55**:458–467.

Chang, S.R., and A.F. Rasmussen, Jr. 1965. Stress-induced suppression of interferon production in virus-infected mice. *Nature* **205**:623–625.

Coe, C.L. 1993. Psychosocial factors and immunity in nonhuman primates: A review. *Psychosom. Med.* **55**:298–308.

Dantzer, R., and K.W. Kelley. 1989. Stress and immunity: An integrated view of relationships between the brain and the immune system. *Life Sci.* **44**:1995–2008.

Dantzer, R., E.E. Wollman, and R. Yirmiya, eds. 1999. Cytokines, Stress, and Depression. New York: Kluwer/Plenum.

Esterling, B.A., M.H. Antoni, M. Kumar, and N. Schneiderman. 1990. Emotional repression, stress disclosure responses, and Epstein-Barr viral capsid antigen titers. *Psychosom. Med.* **52**:397–410.

Felten, D.L., S. Livnat, S.Y. Felten, S.L Carlson, D.L Bellinger, and P. Yeh. 1984. Sympathetic innervation of lymph nodes in mice. *Brain Res. Bull.* **13**:693–699.

Glaser, R., and J. Kiecolt-Glaser, eds. 1994. Human Stress and Immunity. San Diego: Academic.

Gross, W.B. 1972. Effect of social stress on occurrence of Marek's disease in chickens. *Am. J. Vet. Res.* **33**:2275–2279.

Gross, W.B. 1976. Plasma steroid tendency, social environment and Eimeria necatrix infection. *Poultry Sci.* **554**:1508–1512.

Hart, B.L. 1988. Biological basis of the behavior of sick animals. *Neurosci. Biobehav. Rev.* **12**:123–137.

Hickey, W.F., and H. Kimura. 1988. Perivascular microglial cells of the CNS are bone marrow-derived and present antigen *in vivo*. *Science* **239**:290–292.

Irwin, M., M. Daniels, E.T. Bloom, T.L. Smith, and H. Weiner. 1987. Life events, depressive symptoms, and immune function. *Am. J. Psychiatry* **144**:437–441.

Laudenslager, M.L., P.E. Held, M.L. Boccia, M.L. Reite, and J.J. Cohen. 1990. Behavioral and immunological consequences of brief mother-infant separation: A species comparison. *Dev. Psychobiol.* **23**:247–264.

Laudenslager, M.L., M. Reite, and R.J. Harbeck. 1982. Suppressed immune response in infant monkeys associated with maternal separation. *Behav. Neural Biol.* **36**:40–48.

Laudenslager, M.L., S.M. Ryan, R.C. Drugan, R.L. Hyson, and S.F. Maier. 1983. Coping and immunosuppression: Inescapable but not escapable shock suppresses lymphocyte proliferation. *Science* **221**:568–70.

Levenson, R.W. 1992. Autonomic nervous system differences among emotions. *Psychol. Sci.* **3**:23–27.

Levine, S., R. Strebel, E.J. Wenk, and P.J. Harman. 1962. Suppression of experimental allergic encephalomyelitis by stress. *Proc. Soc. Exp. Biol. Med.* **109**:294–298.

Merrill, J.E., and E.N. Benveniste. 1996. Cytokines in inflammatory brain lesions: helpful and harmful. *Trends Neurosci.* **19**:331–338.

Mormède, P., R. Dantzer, B. Michaud, K.W. Kelley, and M. Le Moal. 1988. Influence of stressor predictability and behavioral control on lymphocyte reactivity, antibody responses and neuroendocrine activation in rats. *Physiol. Behav.* **43**:577–583.

Naliboff, B.D., D. Benton, G.F. Solomon, J.E. Morley, J.L.Fahey, E.T. Bloom, T. Makinodan, and S.L. Gilmore. 1991. Immunological changes in young and old adults during brief laboratory stress. *Psychosom. Med.* **53**:121–132.

Panerai, A.E., and P. Sacerdote. 1997. Beta-endorphin in the immune system: A role at last? [see comments]. *Immunol. Today* **18**:317–319.

Patterson, P.H., and H. Nawa. 1993. Neuronal differentiation factors/cytokines and synaptic plasticity. *Cell* **72(l)**:23–37.

Rasmussen, A.F., Jr., J.T. Marsh, and N.Q. Brill. 1957. Increased susceptibility to herpes simplex in mice subjected to avoidance-learning stress or restraint. *Proc. Soc. Exp. Biol. Med.* **96**:183–189.

Reite, M., R. Harbeck, and A. Hoffman. 1981. Altered cellular immune response following peer separation. *Life Sci.* **29**:1133–1136.

Rook, G.A. 1999. Glucocorticoids and immune function. *Baillieres Best Pract. Res. Clin. Endocrinol. Metab.* **13**:567–581.

Schleifer, S.J., S.E. Keller, M. Camerino, J.C. Thornton, and M. Stein. 1983. Suppression of lymphocyte stimulation following bereavement. *J. Am. Med. Assoc.* **250**:374–377.

Selye, H. 1973. The evolution of the stress concept. *Am. Sci.* **61**:692–699.

Sklar, L.S., and H. Anisman. 1979. Stress and coping factors influence tumor growth. *Science* **205**:513–515.

Smith, E.M., W.J. Meyer, and J.E. Blalock. 1982. Virus-induced corticosterone in hypophysectomized mice: A possible lymphoid adrenal axis. *Science* **218**:1311–1312.

Tilders, F.J.H., and E.D. Schmidt. 1999. Cross-sensitization between immune and non-immune stressors. In: Cytokines, Stress, and Depression, ed. R. Dantzer, E.E. Wollman, and R. Yirmiya, pp. 179–197. New York: Kluwer/Plenum.

Visintainer, M.A., J.R. Volpicelli, and M.E. Seligman. 1982. Tumor rejection in rats after inescapable or escapable shock. *Science* **216**:437–439.

Weisse, C.S., C.N. Pato, C.G. McAllister, R. Littman, A. Breier, S.M. Paul, and A. Baum. 1990. Differential effects of controllable and uncontrollable acute stress on lymphocyte proliferation and leukocyte percentages in humans. *Brain Behav. Immun.* **4**:339–351.

Williams, J.M., R.G. Peterson, P.A. Shea, J.F. Schmedtje, D.C. Bauer, and D.L. Felten. 1981. Sympathetic innervation of murine thymus and spleen: Evidence for a functional link between the nervous and immune systems. *Brain Res. Bull.* **6**:83–94.

8

Is There a Major Stress System in the Brain?

R.J. NELSON

Departments of Psychology and Neuroscience, Ohio State University, Columbus, OH 43210–1222, U.S.A.

Key words: CRH, brain, stress, receptors, GABA

ABSTRACT

Individuals have evolved many mechanisms to combat the Second Law of Thermodynamics (entropy). The nervous system has been shaped over eons by natural selection to perceive, mobilize, and cope with the forces of nature that act to hasten the demise of living creatures. These forces are commonly called stressors. Individuals require energy to attempt to cope with stressors, and the most common stress response in vertebrates includes mobilization of energy via the release of glucocorticoids and epinephrine from the adrenal glands. The brain stress system is defined as the neural circuitry necessary and sufficient to evoke stress responses. Although brain systems involved in sensation, perception, attention, learning, memory, motor output, and cortical processing have been considered part of the brain stress system in the recent past, the inclusion of most of the brain in the stress system may become unwieldy and potentially imprecise. A better approach may be to focus on specific environmental demands, specific physiological adjustments, and specific brain and behavioral responses in each instance. These numerous systems certainly contribute to the welfare of animals, including humans, and are generally linked to the brain stress system. However, to evaluate welfare in animals, it is very difficult to monitor brain neurotransmitters or cytokines. Thus, only the hypothalamic-pituitary-adrenal axis and its hormones will be considered.

INTRODUCTION

Life is challenging. The pressure of survival and reproduction takes its toll on every individual living on the planet; eventually and inevitably the "wear and tear" of life leads to death. Mechanisms have evolved to delay death presumably because, all other things being equal, conspecific animals that live longer tend to leave the most successful offspring. In the

Darwinian "game of life," individuals who leave the most successful offspring "win." Although some of the variation in longevity merely reflects good fortune, a significant part of the variation in longevity among individuals of the same species reflects differences in the ability to cope with the demands of living. All living creatures are dynamic vessels of equilibria, or homeostasis. Any perturbation to homeostasis requires energy to restore the original steady-state. An individual's total energy availability is partitioned among many competing needs, such as growth, cellular maintenance, immune function, reproduction, and thermogenesis (Figure 8.1). During environmental energy shortages, nonessential processes such as growth and reproduction are suppressed. If homeostatic perturbations require more energy than is readily available after nonessential systems have been inhibited, then survival may be compromised. All living organisms currently exist because of evolved adaptations that allow individuals to cope with energetically demanding conditions. These demanding conditions may range from a well-fed scientist facing a deadline to complete a paper to a reindeer searching for food on the tundra to a rodent displaying a fear-induced freezing posture when it detects the odors of a predator. Some neuroendocrine coping mechanisms are engaged in all of these cases, as well as in many other situations.

The goal of this chapter is to determine if there is a major stress system in the brain. Because description should always precede formal analyses in science (Tinbergen 1951), it is important to agree on what is meant by "stress system." Here, it is defined as the neural circuitry in the brain that is necessary and sufficient to induce the "stress response." In addition to describing the neuroanatomy of such a putative system, it would also be desirable to elucidate the mechanisms underlying this neural circuit. Integrative mechanisms would include identifying the chemical signals and modulators of stress (e.g., neurotransmitters, neurohormones, hormones, and cytokines) and their receptor subtypes. Then, the molecular mechanisms should be identified within this brain stress system. This idealized, rational approach to the study of a stress system in the brain has not been realized. In some areas, researchers are making remarkable progress at identifying the genetic and molecular mechanisms of stress with little regard for the integrative systems to which these molecular mechanisms contribute. In other areas, scientists are still struggling to parse out the interactive effects of behavioral or emotional factors such as fear and anxiety to stress responsiveness. Obviously, a holistic approach is necessary to understand the brain stress system, perhaps more importantly than for other neural systems.

A large proportion of the stressors which have been described act upon the body to increase energy consumption significantly above some resting, or basal, level (Dallman et al. 1995). Hence, a critically important stress system in the brain is the hypothalamic-pituitary-adrenal (HPA) axis, and this system should be monitored to evaluate welfare. However, when this is done, sampling artifacts must be taken into account. If blood samples are obtained, then the stress of the painful sampling techniques might complicate the findings. To monitor hormone concentrations, a noninvasive method should be used. For example, fecal or urinary glucocorticoid concentrations can be monitored long-term without disturbing the animal.

Sensation, perception, emotion, experience, and stress-coping skills modulate the HPA axis. Furthermore, motor responses (e.g., escape, avoidance) also are engaged as stress-coping mechanisms. One might include the entire brain as part of the stress system. This approach, albeit inclusive, is not very helpful in discerning mechanisms. Where do we

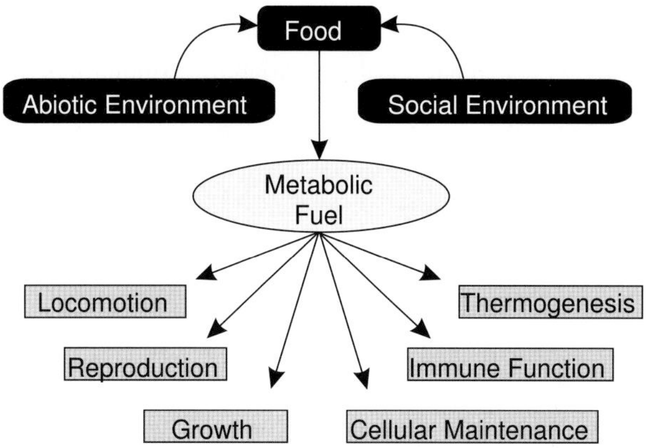

Figure 8.1 Competing physiological and behavioral demands require metabolic energy. When energetic demands are higher than the energy that is immediately available, a stress response often occurs.

draw the line? Should the brain regions mediating sensation, perception, emotion, learning, and memory be considered part of the stress system of the brain? They appear to be either secondary systems that interact with the primary stress system in the brain, or if we discard the notion of a primary stress system in the brain, then these systems may be considered specific stress systems that are engaged by different classes (e.g., psychological versus physical) or types (e.g., sodium or water shortages) of stressors.

Investigators of the various systems for coping with challenge may have been deflected by the assertion of nonspecificity of effects of stressors on the body and a unitary system underlying stress responsiveness (Selye 1936). Many now believe that we should discard the concept of nonspecificity and focus more attention on the specific environmental demands, the specific physiological adjustments, and the specific behavioral responses in each situation. With modern neurochemical and neuroendocrine techniques, it is possible to observe that different hormones and different neurotransmitters are recruited during exposure to different types of stressors.

An example of individual differences in response to stress is that of two food-restricted monkeys examined by Sapolsky (1992). Although both individuals were equally deprived of any nutrition, and both should have experienced equal energy deficits and equivalent elevated stress hormone concentrations, one monkey, who was fed a nonnutritive flavored placebo solution, did not display elevated glucocorticoid secretion. It seems that the critical factor is that the first monkey did not *perceive* things to be as stressful as the second one (Sapolsky 1992). Thus, brain systems that involve sensation, perception, various coping strategies, learning, and memory could be considered as part of a "stress system." If these brain systems are included in the brain "stress system," then little of the brain is now excluded from the neural circuitry.

Another important factor that modulates stress is emotion. In the literature, stress is often conflated with emotional states such as anxiety, fear, depression, and in humans, bereavement, sorrow, or even elation (Kopin 1995). Anxiety is not stress. It may serve as a stressor. Anxiety may increase cardiac tone, and the increased energetic fuel necessary to increase respiration, blood pressure, and pulse rate is preceded by glucocorticoid secretion.

Most scientists now accept the power of psychological variables to modulate stress physiology (Sapolsky 1992, 1998). What are these variables? Among individuals of vertebrate species, the main psychological factors that modulate stress responses and lead to poor welfare include: (a) lack of control, (b) lack of predictability, and (c) a lack of outlets for dealing with frustration.

PSYCHOLOGICAL VARIABLES

Lack of Control

Individuals suffer fewer stress-related pathologies if they can control the situation causing the stress (Gatchel et al. 1989). Controling a potentially stressful situation is a type of coping. In one early study (Weiss 1968), rats were subjected to intermittent electric shocks. One rat could control the situation because it was able to depress a lever to decrease the rate of shocks. Another rat received a shock whenever the first did, but could not control the frequency of the shocks. Subsequently, the second rat had elevated glucocorticoid secretion compared with the rat that could moderate the shocks (Weiss 1968).

Loss of Predictability

Predictable stress is easier to cope with than random presentation of stressors. In studies comparable to the one just described, predictability reduces the stress response of individuals. For example, if a warning stimulus is given prior to a shock, rats release less glucocorticoids as compared with rats that receive no warning stimuli (Sapolsky 1998).

Outlets for Frustration

The ability to engage in various behaviors can also ameliorate the effects of stress. When rats are shocked, there is relatively low glucocorticoid secretion if they can chew on a piece of wood or attack another rat (Sapolsky 1998). It is common knowledge that the effects of stress in humans can be ameliorated by engaging in a hobby or moderate exercise.

Presumably, psychological variables interact with fear, anxiety, or other emotions, which may, depending on the individual, increase cardiac tone and the need for glucocorticoid-induced release of stored energy. Only the hypothalamus, pituitary, and perhaps hippocampus directly mediate glucocorticoid secretion. However, the brain regions that *influence* this response include many, if not most, brain neural circuits: the limbic system, brain stem, amygdala, cortex, and thalamus, to name but a few. These brain regions may mediate fear, anxiety, or rage, and have important effects on the welfare of animals, including humans, but they only modulate the HPA system.

NEUROANATOMICAL PERSPECTIVE

Does the Brain Stem "Arousal System" Serve a Gating Function for Sensory Input, Which Facilitates Coping with Aversive Stimuli?

The brain stem is where the first central integration of much sensory information from both exteroceptive and visceroceptive sources occurs. First-order sensory information from the nonolfactory and nonauditory cranial nerves is processed in the brain stem. Projections from the parabrachial nucleus appear to mediate pain-related input from the dorsal horn of the spinal cord, as well as the cranial nerves. This pain information is ultimately relayed to the hypothalamus. The brain stem, or more precisely, the nucleus of the solitary tract (NST), receives input from the vagus nerve which monitors the viscera. Information is conveyed from the NST to the hypothalamus, including the paraventricular nucleus of the hypothalamus (PVN) and telencephalon (i.e., cerebral cortex, amygdala, hippocampal complex including the septum) that regulates the HPA axis directly. Vagal stimuli associated with energy imbalance are particularly potent signals for HPA activation. The dorsal medial nucleus of the vagus and other sets of preganglionic neurons in the brain stem mediate stress-evoked modulations to cardiac and other visceral responses.

The brain stem has three different neurotransmitter systems traditionally considered to be involved in arousal (Moruzzi and Magoun 1949): noradrenergic, acetylcholinergic, and serotonergic. These systems have wide-ranging connections that were traditionally called the "reticular activating system." The noradrenergic neurons of the locus coeruleus (LC), for example, connect to the neocortex, thalamus, hippocampus, pons, and medulla. These LC neurons have been shown to respond to stress by changing their electrophysiological properties and gene expression. However, a direct link between these brain stem nuclei and stress responses remains unspecified. The brain stem may sometimes help in coping with aversive stimuli.

Do the Hippocampus and Amygdala Participate in the Stress Response?

Hippocampus

It has been known for some time that lesions disrupting hippocampal output to the septal complex compromise adrenocorticotrophin (ACTH) and glucocorticoid secretion. Recent work has documented that the hippocampus is important in the negative feedback control systems of the HPA axis (McEwen 1999). The ventral subiculum (SUBv), but no other hippocampal region, appears to inhibit the gene expression of corticotrophin-releasing hormone (CRH) and ACTH blood concentrations (Watts 1996). The major function of the hippocampus is learning and memory, and several reports of chronic stress compromising learning and memory have been reported (Sapolsky 1992). Presumably, memory is used in arousal, behavioral planning (e.g., withdrawal or fighting), and other cognitive components of coping, but direct evidence remains elusive (Van de Kar and Blair 1999).

Amygdala

The role of the amygdala in fear-evoked stress responses is well documented (Fendt and Fanselow 1999). The amygdala receives both intrinsic and extrinsic sensory information, in both simple and highly processed forms, and may serve as an integrator of these sensory messages. An intact amygdala is necessary to show fear-induced stress responses in rats, rabbits, or humans (Fendt and Fanselow 1999). The amygdala can be activated by stimuli associated with stressors or by repeated exposure to the stressors themselves (i.e., learning), which permits anticipation and potentially facilitates coping with stressors.

Importantly, the amygdala projects to the PVN where CRH regulates ACTH from the anterior pituitary. This information arrives via the bed nucleus of the stria terminalis (BNST) in which information from the cortex, hippocampus, and amygdala is processed before arriving at the PVN and other regions of the hypothalamus.

The gating system of the brain stem is important for allocating attentional resources to stress-inducing sensory information from both the environment and the viscera. The emotional valences given to these stressful stimuli by the limbic system serve to categorize these stimuli into those that require action, those that can wait, and those that can be ignored. This information is integrated with previous experience, which interacts with the emotional impact of the stressful information and may feedforward to affect anxiety and fear, and ultimately affect coping. These non-HPA systems represent a stress modulatory system.

Is Sherrington's Notion of the Hypothalamus as the "Final Common Pathway" and the Integrator of the Brain Effector Response to the Endocrine and Autonomic Nervous System Still Valid?

In my view, the answer to this question is "yes." Some of the neural input to the PVN of the hypothalamus is described above. The PVN is important because that is where the majority of the neurohormone, CRH, is produced and stored. CRH is a peptide hormone consisting of 41 amino acids and serves as the major secretagogue of ACTH from the anterior pituitary. CRH is produced in the parvocellular division of the PVN, which projects to the median eminence where CRH is released into the hypothalamic-hypophyseal blood portal system. CRH may act as a neurotransmitter in the mediation of certain behavioral responses to stressors (Song et al. 1995). Intracerebroventricular (ICV) infusions of CRH mimic many of the common responses to stressors. A number of neurotransmitters can affect secretion of CRH and these are reviewed below.

CRH also receives input from the primary biological pacemakers in mammals, the suprachiasmatic nuclei (SCN). The SCN provide temporal coordination to the secretion of this peptide (i.e., a large bolus of CRH is secreted in anticipation to the onset of the activity phase of individuals' daily cycle). CRH provokes secretion of ACTH which, in turn, stimulates glucocorticoid secretion from the adrenal cortices. This preawakening secretion of glucocorticoids presumably provides energy prior to the first consumption of metabolic fuels of the day.

Importantly, other secretagogues of ACTH have been discovered. One of the most potent is vasopressin acting on V3 receptors in pituitary corticotrophic cells. It is possible that nonenergy stressors, such as low blood volume and dehydration, are mediated by this nanopeptide. Vasopressin activates a different signal transduction pathway than CRH when interacting with its receptor and can act synergistically with CRH to increase ACTH secretion two- to threefold (Turnbull and Rivier 1997). Indeed, some CRH-secreting cells actually "convert" during activity to secrete vasopressin. Other agents that can stimulate ACTH release from the pituitary include oxytocin, angiotensin II, cholecystokinin, noradrenaline, and adrenaline (= epinephrine) but are far less potent than CRH. The contribution of these neuropeptides and neurotransmitters to physiological adaptation to stressors is not known, but it remains plausible that the different hormones might mediate different types of stimulus effects.

Do Other Brain Regions Contribute to the Stress Response?

Many other brain regions are involved in the stress response. For example, dopaminergic neurons projecting from the midbrain are involved with movement and behavioral reward systems and are thus of critical importance in coping with various challenges. CRH and the noradrenergic neurons of the LC innervate and stimulate each other in a positive feedback system, which results in activation of one system in response to activation of the other (Calogero et al. 1988). LC is a major central mediator of the sympathetic system. It appears that CRH activates the stress response and that CRH is modulated by hormones, cytokines, and neurotransmitters (O'Connor et al. 2000). Additional regulation is maintained by CRH and norepinephrine collateral fibers themselves which inhibit presynaptic secretion of CRH and α_1-noradrenergic receptors, respectively (O'Connor et al. 2000). There are two classes of CRH receptors in the brain: dense CRH_1 receptor distributions are seen in the cortex, cerebellum, and sensory relay structures, whereas dense CRH_2 receptor distributions are found in the lateral septal nuclei, olfactory bulb, choroid plexus, and specific hypothalamic and amygdaloid nuclei (Chalmers et al. 1996). The separate CRH receptor subtype distributions suggest distinct functional roles in the brain (Chalmers et al. 1996).

CRH has additional functions in the brain beyond its regulatory effects on ACTH secretion (Chalmers et al. 1996). As noted, CRH can modulate the activity of the autonomic nervous system (Lehnert et al. 1998). CRH also can affect the amygdala (O'Connor et al. 2000), which in concert with the hypothalamus directly communicates with and regulates the autonomic nervous system (Berntson et al. 1998). ICV infusion of CRH in rats activates the HPA system to elevate blood glucose concentrations, heart rate, and blood pressure, as well as increased arousal, vigilance, and tolerance of pain and decreased food intake and sex drive (Sutton et al. 1982). CRH infusion directly into the LC has similar fear-enhancing and behavioral-activating effects (Butler et al. 1990).

Many of the adjustments to energy balance evoked by stressors also require accommodation by the thyroid gland. Although not as popular in studies of stress as the HPA axis, the hypothalamus-pituitary-thyroid (HPT) axis is involved in stress responses (Kelly 2000). Future studies of physiological and behavioral responses, in the context of welfare, will have to consider the HPT axis.

NEUROCHEMICAL PERSPECTIVE

What Neurotransmitters Modulate CRH Secretion?

Catecholamines

As mentioned, the PVN receives noradrenergic fibers from the brain stem. Adrenergic afferents from the caudal medulla innervate the parvocellular division of the PVN. Although traditionally believed that noradrenaline and adrenaline inhibit the HPA axis (Ganong 1980), recent studies indicate that both the brain stem noradrenergic system and the HPA axis are activated during stress responses. Both noradrenaline and adrenaline stimulate ACTH release from the pituitary *in vivo*, and ICV infusion of noradrenaline or adrenaline elevates CRH secretion in the HPA of rats in a dose-dependent manner. The enhancing effects of norepinephrine and epinephrine on CRH secretion could be blocked by prior treatment with a pharmacological α_1-adrenergic receptor antagonist. The enhancing effects of epinephrine on CRH secretion was also blocked by pretreatment with a pharmacological β-blocker. The PVN mainly has α-adrenergic receptor subtypes, and relatively few β-adrenergic subtypes. Both norepinephrine and epinephrine stimulate CRH secretion *in vitro*. The effects of these neurotransmitters are direct, because the stimulatory effects of the catecholamines on CRH secretion were unaffected by pretreatment with either serotonin or acetylcholine receptor antagonists. Thus, the catecholamines appear to have a facilitatory effect on CRH release *in vitro*, although the role of the catecholamines on CRH secretion *in vivo* remains controversial.

Acetylcholine

Although PVN neurons stain intensely for acetylcholinesterase (AChE), the enzyme that catabolizes acetylcholine, there is little indication that choline acetyltransferase (ChAT) activity occurs in the PVN (Calogero 1995). The source of cholinergic pathways into the PVN remains somewhat controversial. Direct assays of CRH in rodents have proved to be difficult for technical reasons. Thus, most assays of CRH are indirect, and the conclusions about CRH involvement must be made with caution. For example, treatment of freely moving, catherized rats with the muscarinic cholinergic agonist, arecoline, increased blood concentrations of ACTH and corticosterone in a dose-dependent manner. Atropine, pituitary stalk ablation, or pretreatment with anti-CRH rabbit serum reversed these responses. These results suggested, but did not establish, that acetylcholine engages the HPA axis via CRH. *In vitro* work showed that acetylcholine stimulated CRH secretion from hypothalamus explants in a dose-dependent manner. Acetylcholine agonists and antagonists affected CRH in the predicted directions. Additional evidence that acetylcholine stimulates the HPA axis via CRH is that pituitary explants did not show an increase in ACTH production when treated with acetylcholine (Calogero 1995).

Serotonin

It is well documented that serotonin (5-HT) elevates HPA activity. Treatment with 5-HT precursors, releasers, uptake inhibitors, or receptor agonists stimulated the secretion of ACTH and glucocorticoids *in vivo* (Fuller 1992). Serotonergic axons connect to CRH-containing neurons in the PVN. These serotonergic neurons probably originate in the midbrain raphe.

5-HT increased CRH activity in hypothalamic explants. A 5-HT receptor antagonist blocked this effect. Most *in vitro* pharmacological studies also support the stimulatory role of 5-HT in the direct regulation of CRH secretion. Generally, 5-HT and norepinephrine levels are increased (temporarily) above baseline during HPA activity and reduced below baseline during depression.

Gamma Aminobutyric Acid

A plexus of gamma aminobutyric acid (GABA) neurons terminate in the median eminence. GABA is the major inhibitory neurotransmitter in the brain. There are three major GABA receptor families, i.e., $GABA_A$, $GABA_B$, and $GABA_C$, with several subtypes within each family group. Although these GABA receptors share an affinity for the same neurotransmitter, they differ enormously in their molecular structure, pharmacology, and mechanisms of action (McKernan and Whiting 1996). For example, the $GABA_A$ receptor is a macromolecule with separate binding sites for GABA, benzodiazepines, and barbiturates, and activation of this receptor subtype causes an influx of chloride ions through an opened Cl^- channel, and subsequent hyperpolarization of the postsynaptic neuron. Muscimol and bicuculline are $GABA_A$ agonist and antagonist, respectively. In contrast, the $GABA_B$ receptor has no affinity for bicuculline, does not have binding sites for benzodiazepines, and is not linked to a chloride channel. Activation of $GABA_B$ receptors elicits an influx of calcium ions into the cell.

Both GABA and benzodiazepines suppress HPA axis function *in vivo* and *in vitro.* There is a negative correlation between GABA activity in the hypothalamus and blood corticosterone concentrations in rats. Indeed, ICV injection of GABA reduced CRH concentrations in the hypothalamic-hypophyseal-blood portal system directly.

In addition to barbiturates, $GABA_A$ receptors are also modulated by specific metabolites of steroid hormones from the gonads and adrenals, including progesterone and deoxycorticosterone, respectively (Costa 1998). These steroids are called neurosteroids when produced in the brain and, since they and other steroid metabolites act, in common with barbiturates, to potentiate GABA responses by prolonging the duration of the open channel time, they can have profound effects on the HPA axis. At high concentrations (e.g., during a stress response), barbiturates and steroid metabolites can "gate" the ionophore directly via a mechanism that does not involve the $GABA_A$ receptor. The steroid-gated Cl^- channel essentially blunts the inhibitory effects of GABA on CRH secretion and the system may positively feedforward resulting in yet higher glucocorticoid concentrations, which lead to still higher concentrations of the steroid metabolites.

One could consider other neurotransmitters in the mediation of the stress response. For example, dopamine neurons projecting from the midbrain are involved with movement and behavioral reward systems, and thus are involved in coping systems. The interaction of neurotransmitters with cytokines is also an important consideration when thinking about the brain stress system. The hippocampus and amygdala are important integrative sites, and the hippocampus is dense with receptors for hormones and cytokines (Dieterich et al. 1997). Importantly, the hippocampus unifies prior memories about the events and contributes to the early production of emotions during the neuroendocrine and behavioral response to stressors (Berntson et al. 1998; Butler et al. 1990; Wiersma et al. 1998).

FUNCTIONAL PERSPECTIVE

To summarize, the stress system of the brain consists of many subsystems; however, most of these subsystems ultimately feed into the HPA axis. Monitoring the HPA axis should, therefore, provide a useful means of assessing the welfare of individuals. By monitoring, I do not mean obtaining a single blood sample; animals can adapt to stress quickly, so a blood sample may not reveal any increased HPA activity. Rather, attempts should be made to monitor steroid byproducts in urine or feces in a noninvasive manner. It is not technically feasible to monitor CRH or other peptides in a noninvasive manner at this time, so further assessments await technological advances.

Although many neural systems interact with the HPA axis, it integrates environmental and visceral information as well as stimulates glucocorticoid secretion when warranted. Any adjustment in energy needs, either programmed by biological clocks or evoked by the environment, is mediated by the HPA axis. Well-documented negative effects of HPA activity appear to result from the balance of energy among competing costs, for example, it impairs reproductive and immune function. Generally in nature, individuals maintain the highest level of immune function that is energetically possible within the constraints of other survival needs, growth, and reproduction in habitats in which energy requirements and availability often fluctuate. The observation that immune function is generally compromised during specific energetically demanding times, such as winter, breeding (including pregnancy and lactation), migration, or molting, is consistent with the hypothesis that immune function is energetically optimized (John 1994; Zuk 1990). It appears that the hormones associated with stress are critical in the energy allocation decision-making process. Appreciation of the relationship between energetics and other physiological functions, such as immune function, is important in constructing an integrative view of stress.

Traditionally, the medical hypothesis for immunosuppression during pregnancy has been to protect the fetus from being attacked as foreign tissue by the maternal immune system (McCruden and Stimson 1991). However, reduced immune function during pregnancy may represent an energy-saving adaptation; this perspective may not be obvious in recent studies because most laboratory and clinical studies have been conducted on individuals with ad lib food availability (see below). Few women currently living in North America or Western Europe are energetically challenged during pregnancy. However, examination of the literature on pregnant women's health in the U.S.A. and Western Europe prior to 1920 or current studies of pregnant women in nonindustrialized nations reveal an elevated incidence of many diseases and immune dysfunctions during pregnancy, particularly during the third trimester (Sakamoto-Momiyama 1977). CRH is increased three- to fivefold during pregnancy and lactation in many species including humans. Glucocorticoids are similarly elevated and account for the immunosuppression. Although there may be offsetting endocrine adaptations, ultimately pregnancy evokes stress responses.

The HPA axis appears to be the primary stress system in the brain. Other important systems include the interaction of the CRH and LC-norepinephrine-sympathetic systems. Future studies should focus on specific environmental demands, specific physiological adjustments, and specific brain and behavioral responses in each instance. These numerous systems certainly influence the welfare of animals, including humans, and are generally linked to the HPA stress system. However, it is very difficult to monitor brain neurotransmitters or cytokines when assessing animal welfare. However, the HPA axis and its hormones can be monitored noninvasively as an indication of stress. An understanding of the adaptive significance of metabolic coping mechanisms is crucial to map out the neural systems that have evolved to mediate the perception, mobilization, and responses involved in effective coping with challenge.

ACKNOWLEDGMENTS

I thank Christopher Coe and an anonymous reviewer for providing organizational and critical suggestions. I am grateful for the support of the Dahlem Konferenzen der Freien Universität Berlin, the U.S. National Institutes of Health grant MH 22201, and U.S. National Science Foundation grant IBN 00-80745 during the preparation of this paper.

REFERENCES

Berntson, G.G., M. Sarter, and J.T. Cacioppo. 1998. Anxiety and cardiovascular reactivity: The basal forebrain cholinergic link. *Behav. Brain Res*. **94**:225–248.

Butler, P.D., J.M. Weiss, J.C. Stout, and C.B. Nemeroff. 1990. Corticotropin-releasing factor produces fear-enhancing and behavioral activating effects following infusion into the locus coeruleus. *J. Neurosci.* **10**:176–183.

Calogero, A.E. 1995. Neurotransmitter regulation of the hypothalamic corticotropin-releasing hormone neuron. *Ann. NY Acad. Sci.* **771**:31–40.

Calogero, A.E., W.T. Gallucci, G.P. Chrousos, and P.W. Gold. 1988. Catecholamine effects upon rat hypothalamic corticotrophin-releasing hormone secretion *in vitro*. *J. Clin. Invest.* **82**:839–846.

Chalmers, D.T., T.W. Lovenberg, D.E. Grigoriadis, D.P. Behan, and E.B. DeSouza. 1996. Corticotrophin-releasing factor receptors: From molecular biology to drug design. *TIPS* **17**:166–172.

Chrousos, G. 1998. Stressors, stress, and neuroendocrine integration of the adaptive response. The 1997 Hans Selye Memorial Lecture. *Ann. NY Acad. Sci.* **851**:311–335.

Costa, E. 1998. From $GABA_A$ receptor diversity emerges a unified vision of GABAergic inhibition. *Ann. Rev. Pharmacol. Toxicol.* **38**:321–350.

Dallman, M.F., S.F. Akana, A.M. Strack, E.S. Hanson, and R.J. Sebastian. 1995. The neural network that regulates energy balance is responsive to glucocorticoids and insulin and also regulates HPA axis responsivity at a site proximal to CRF neurons. *Ann. NY Acad. Sci.* **771**:730–742.

Dieterich, K.D., H. Lehnert, and E.B. DeSouza. 1997. Corticotropin-releasing factor receptors: An overview. *Ann. Rev. Pharmacol. Toxicol*. **105**:65–82.

Fendt, M., and M.S. Fanselow. 1999. The neuroanatomical and neurochemical basis of conditioned fear. *Neurosci. Biobehav. Rev*. **23**:743–760.

Fuller, R.W. 1992. The involvement of serotonin in the regulation of pituitary-adrenocortical function. *Frontiers Neuroendocrinol.* **13**:250–270.

Gatchel, R.J., A. Baum, and D.S. Krantz. 1989. An Introduction to Health Psychology. 2nd edition. New York: Newberry.

Ganong, W.F. 1980. Neurotransmitters and pituitary function: Regulation of ACTH secretion. *Fed. Proc.* **39**:2923–2930.

John, J.L. 1994. The avian spleen: A neglected organ. *Q. Rev. Biol*. **69**:327–351.

Kelly, G. 2000. Peripheral metabolism of thyroid hormones: A review. *Altern. Med. Rev.* **5**:306–333.

Kopin, I.J. 1995. Definitions of stress and sympathetic neuronal responses. *Ann. NY Acad. Sci*. **771**:19–30.

Lehnert, H., C. Schulz, and K. Dieterich. 1998. Physiological and neurochemical aspects of corticotropin-releasing factor actions in the brain: The role of the locus coeruleus. *Neurochem. Res.* **23**:1039–1052.

Mason, J.W. 1975. A historical view of the stress field. *J. Human Stress* **1**:6–36.

McCruden, A.B., and W.H. Stimson. 1991. Sex hormones and immune function. In: Psychoneuroimmunology, ed. R. Ader, pp. 475–493. New York: Academic.

McEwen, B.S. 1999. Stress and hippocampal plasticity. *Ann. Rev. Neurosci*. **22**:105–122.

McKernan, R.M., and P.J. Whiting. 1996. Which GABA-receptor subtypes really occur in the brain? *Trends Neurosci*. **19**:139–143.

Moruzzi, G., and H.W.Magoun. 1949. Brain stem reticular formation and activation of the EEG. *Electroencephalogr. Clin. Neurophysiol*. **1**:455–473.

O'Connor, T.M., D.J. O'Halloran, and F. Shanahan. 2000. The stress response and the hypothalamic-pituitary-adrenal axis: From molecule to melancholia. *Q. J. Med*. **93**:323–333.

Sakamoto-Momiyama, M. 1977. Seasonality in Human Mortality. Tokyo: Univ. of Tokyo Press.

Sapolsky, R.M. 1992. Stress, the Aging Brain, and the Mechanisms of Neuron Death. Cambridge, MA: MIT Press.

Sapolsky, R.M. 1998. Why Zebras Don't Get Ulcers: An Updated Guide to Stress, Stress-related Diseases, and Coping. San Francisco: W.H. Freeman.

Selye, H. 1936. A syndrome produced by diverse nocuous agents. *Nature* **138**:247–248.

Song, C., B. Early, and B.E. Leonard. 1995. Behavioral, neurochemical, and immunological responses to CRF administration: Is CRF a mediator of stress? *Ann. NY Acad. Sci.* **771**:55–72.

Sutton, R.E., G.F. Koob, M. LeMoal, J. Rivier, and W. Vale. 1982. Corticotrophin-releasing factor produces behavioural activation in rats. *Nature* **297**:331–333.

Tinbergen, N. 1951. The Study of Instinct. Oxford: Oxford Univ. Press.

Toates, F. 1995. Stress: Conceptual Biological Aspects. New York: Wiley.

Turnbull, A.V., and C. Rivier. 1997. Corticotropin-releasing factor (CRF) and endocrine responses to stress: CRF receptors, binding protein, and related peptides. *Proc. Soc. Exp. Biol. Med.* **215**:1–10.

Van de Kar, L.D., and M.L. Blair. 1999. Forebrain pathways mediating stress-induced hormone secretion. *Frontiers Neuroendocrinol.* **20**:1–48.

Watts, A.G. 1996. The impact of physiological stimulation on the expression of corticotropin-releasing hormone and other neuropeptide genes. *Frontiers Neuroendocrinol.* **17**:281–326.

Weiss, J.M. 1968. Effects of coping response on stress. *J. Comp. Physiol. Psychol.* **65**:251–260.

Wiersma, A., J.P. Konsman, S. Knollema, B. Bohus, and J.M. Koolhaas. 1998. Differential effects of central CRH infusion into the central nucleus of the amygdala in the Roman high-avoidance and low-avoidance in rats. *Psychoneuroendocrinology* **23**:261–274.

Zuk, M. 1990. Reproductive strategies and sex differences in disease susceptibility: An evolutionary viewpoint. *Parasitol. Today* **6**:231–233.

9

Motivation and Coping

P. JENSEN

Department of Animal Environment and Health, Swedish University of Agricultural Sciences, Section of Ethology, 532 23 Skara, Sweden

Keywords: behavior, control system, feedback, behavioral strategy, nest-building, pigs, fowl

ABSTRACT

When animals are kept in captivity, they will attempt to react to any internal or external stimuli according to motivation systems which were mostly shaped by evolution for a life in the wild. Domestication has modified only certain aspects of the natural behavior, mainly through slight alterations of behavioral strategies. In this chapter, different aspects of motivation and coping are examined. It is concluded that an important aspect of coping is the successful down-regulating of motivational systems. When this fails, for example, because of inadequate environment, the animal can be said to be in a state of stress. Motivation and its behavioral outputs are therefore important aspects of stress and coping. Accordingly, stress can arise as a consequence of a failure to respond appropriately to either internal stimuli, such as in the case of nest-building in sows, or to external stimuli, such as painful electric shocks. Whereas there will be some degree of poor welfare whenever animals are stressed, successful coping is no guarantee of good welfare. Animals coping at a high cost are also likely to have poor welfare.

INTRODUCTION

Animals on farms, in zoos and laboratories, and elsewhere in captivity often face environmental challenges which they may not have encountered during most of their evolution, and only lately during domestication. They may, therefore, be poorly suited to adapt to certain aspects of the environments where they are kept, and such problems with adaptation may sometimes be associated with clear signs of poor welfare (Fraser and Broom 1990). In some cases this may lead to reactions that many would describe as stress. When an animal manages to avoid stress in such a situation, it is sometimes referred to as coping. In this chapter, I argue

Coping with Challenge: Welfare in Animals including Humans
Edited by D.M. Broom

that an understanding of how animals cope requires a fundamental understanding of the evolutionary adaptations of the species.

Evolution works partly by shaping the motivational systems of animals, that is, by regulating when and how strongly an animal is prepared to use energy in achieving a particular goal or obtaining a certain commodity or simply performing a particular pattern of movement. Different motivational systems work in different ways, but they generally have many shared properties.

In this chapter, I present some definitions and models for coping and motivation in animals. I will attempt to demonstrate that coping, and thereby the state of stress itself, is to a large extent a behavioral and motivational phenomenon.

EVOLUTION AND DOMESTICATION OF BEHAVIOR

During evolution, animals have been selected for increased fitness in their natural environments over thousands of generations. Fitness in this context can roughly be defined as reproductive success (often measured at allelic level) and is not necessarily related to welfare; an animal with high fitness may have poorer welfare than one with low fitness (Barnard and Hurst 1996). One of the major achievements of behavioral research over the last decades is the realization that animals are equipped with flexible behavioral programs, usually referred to as strategies (Krebs and Davies 1991). A strategy involves a number of different possible behavioral outputs and a decision mechanism that helps the animal in selecting the best behavior under different conditions. If there is only one possible behavioral reaction to each specific stimulus, there is no need for the concept of strategy; for example, "response" would be adequate. The best strategy in this context is the one that gives the most fitness for the animal. For example, one specific social situation may require that the animal avoids conflict and assumes a subdominant role, while in a different situation, the same animal may be better off seeking conflict and attempting to become dominant. The best strategy will be the one that allows the animal to make the choice that will maximize its fitness in any group. Within a population of animals, different individuals may, therefore, show different behavior in a similar situation, even though they use the same strategy. Of course, different animals in a population may also use different strategies. The trait that is selected during evolution is net benefit, so animals are expected to assess the costs and benefits of each possible behavior and then select the one that provides the largest net benefit under prevailing conditions. Benefit is measured in fitness units but is often approximated by, for example, energy.

During domestication, animals have been genetically changed in relation to their ancestors, so as to affect various aspects of their adaptive traits, including behavior (Price 1997). However, as is obvious from many comparisons between domestic animals and their ancestors, behavior is usually modified in only a very limited manner. Generally, no new behaviors have developed and none has completely disappeared from the gene pool. The changes are usually referred to as altered release thresholds (Price 1997); however, a better description may be to say that the strategies have been modified. Since the domestic environment usually provides animals with food and protection, and involves selection for specific traits, the optimal behavior is likely to be different as compared with when food is scarce and predation pressure high. Strategies different from those that are most successful in the wild may therefore yield the highest fitness. We have found that domestic pigs and poultry use different strategies from those of their wild counterparts, for example, wild boar and jungle fowl, during food searching, and the differences appear to reflect an adaptation of strategies to the conditions existing during domestication (Schütz and Jensen 1999; Gustafsson et al. 1999b; Andersson et al. 2000). However, when comparing the maternal behavior of domestic pigs and wild boars, almost no differences were detected. This indicates that some fundamental behavior patterns are virtually unaffected by domestication (Gustafsson et al. 1999a).

It is therefore clear that domestic animals still possess behavior and motivational systems that are inherited from the ancestors, and have been little changed by domestication. Any understanding of how these systems function needs to be based on the assumption that they were developed to work in the environment of evolutionary adaptation, which is different in many ways from the domestic environment. Animals will therefore perceive stimuli and attempt to react to them in accordance with the best strategy in the environment of evolutionary adaptation, or sometimes with a slightly modified strategy.

STRESS AND COPING

Stress as a Physiological and Psychological Phenomenon

The traditional stress model, which is still common in many textbooks and scientific papers, describes stress mainly as a physiological state, induced by external, and usually aversive, stimuli. According to this traditional model, stress is defined by a cascade of physiological reactions, of which the activation of the hypothalamic-pituitary-adrenal (HPA) axis and the sympathetic nervous system are most well known. This traditional model is founded on the early work of Selye and Cannon (for a detailed account of the details, see Toates 1995). However, the concept of stress has been used in many different ways by different authors, and there is today no generally accepted definition. The ambiguities of the concept are amplified by the frequent use of qualifiers and modifiers, such as "social stress," "physiological stress," "eustress," or "distress."

At the end of the 1960s, Weiss conducted a series of well-designed experiments to investigate the role of the psychological perception of a stressful situation in the stress responses of rats. In the most central experiment (Weiss 1971), rats were given electric shocks in the tail while kept in single cages. One rat was yoked and subjected to the same shocks as the test animal. Test rats were given a predictive cue (sound) of each signal and were offered the possibility to terminate or postpone shocks by turning a wheel. Weiss found that when electric shocks were predictable to the animals, stomach ulcerations were smaller, and when the rats had some control and feedback, there were virtually no differences in ulceration between shocked animals and unshocked controls. This was attributed to the fact that when shocks could be predicted, and when the rats could exert some control and receive feedback, they were able to cope with the stressor. The conclusion of this experiment was that the physical nature of a stressor may have little relevance to how animals are affected; what matters is whether the animal can exert some behavioral control over the stressor.

This study has caused a strong interest in the psychological aspects of the stress response. It appears that exactly the same aversive stimuli, or stressors, can have very different effects on the physiology and pathology of the animal. Even when physiological reactions, according to the traditional stress model, are similar, the pathological consequences may differ depending on the psychological context. The results of Weiss have not always been replicated in similar and derived experiments, as reviewed by Toates (1995); however, there appears to be enough common ground for saying that the crucial measurement of stress may not be the physiological reaction per se, but rather how the animal perceives and is able to cope with the stimulus. Of course, physiological measures are essential parts of an overall stress assessment, providing indications of, for example, the intensity of a response.

Coping: The Way to Tackle Stressors

Coping is yet another concept that has been used in many different ways by different authors. At this workshop, it was used mainly as an equivalent to physiological and behavioral adaptation, as expressed in the background papers. Others have defined it as "having control of mental and bodily stability" (Fraser and Broom 1990), which is of course compatible with the present workshop use. As noted by Toates (1995), the concept is sometimes used in a way that approaches circularity while at other times it is used almost as a synonym for other behavioral concepts, such as reinforcement or, indeed, adaptation. There is a clear need to sharpen the language, and I argue that a motivation-based approach to understanding stress and coping offers advantages in this respect.

Recently, the concept of coping strategy has been used more commonly. The word *strategy* would imply that animals use different methods to cope and are equipped with decision mechanisms that allow them to select the best method under specific conditions. For example, in the classical studies of mice, some individuals attempted to cope by means of extrovert and active behavior, while others used more introvert and passive behavior (Sluyter et al. 1996). The concept may also imply that the coping pattern reflects a general behavioral strategy, which may be reflected in the ways animals react to different types of stressors, and perhaps also the way the animal behaves under unchallenged conditions. Whereas there appear to be good correlations between behavior in different situations in some species and in some studies (Ruis et al. 2000; Sluyter et al. 1996), others have failed to find evidence of such personality traits (Jensen 1995).

In the context of welfare studies, it is of course crucial to understand the details of how animals cope. If different individuals have differences in their preferred coping behavior, this is of central relevance. For example, farming systems that allow a good welfare for some animals, which have a successful coping strategy, may simultaneously be associated with a poor welfare for others, lacking such a strategy.

PROBLEMS WITH TRADITIONAL DEFINITIONS OF STRESS AND COPING

There are several problems with the concepts of stress and coping when they are applied to animal welfare studies. For example, the frequent use of qualifiers and alternative concepts such as "distress" and "eustress" distracts from the fact that under some conditions, animals develop pathologies because they are unable to cope. If we insist that stress is sometimes good for the animal ("eustress"), the wrong conclusion may arise, for example, that we should find farming systems which offer "optimal stress levels." However, one goal of welfare research is to identify systems and treatments where the risk of disease and poor welfare is high, so it seems most fruitful to reserve the use of the stress concept for those cases. It appears to me that there is no use for the notion that stress is sometimes good; stress is always negative, in the sense that it has adverse welfare effects. If we would accept that good stress is when an animal is successful in coping with a challenge and bad stress, or distress, when it has problems or fails, I would suggest that there is a better, and more established, word for it: life.

Traditional stress theory also appears quite mechanistic and inflexible, and this does not correspond to what we know about behavioral responses in animals. In focusing on well-specified physiological reactions, such as HPA activity, to well-defined and experimentally replicable stressors, such as electric shocks, the theory does not take adequate account of the fact that to animals, different stimuli may be aversive at different times. Furthermore, one stimulus may be irrelevant on one day, and extremely aversive, or even attractive, on another. Hence, traditional stress theory does not always take the motivation of the animals sufficiently into account. One way in which researchers can get away with this ignorance is by using stimuli that are almost always aversive, such as electric shocks.

Another major conceptual problem is that most stress research on animals has been conducted using aversive external stimuli. This may not, however, be relevant at all to the current animal welfare debate. Very rarely is the welfare debate concerned with aversive external events, such as pain, heat, or starvation, where stress research has a lot to offer. The reason is simply that these are not controversial welfare issues, since no one would question that animals experiencing pain or starving have a poor welfare. What causes controversy are questions like: are animals stressed when they lack opportunities and possibilities to perform species-specific behavior for which they are motivated? For example, is a tethered sow stressed before farrowing when she is unable to perform nest-building? Are animals stressed when they are unable to obtain certain commodities? For example, is a battery hen stressed when she is deprived of opportunities for dust-bathing or nesting?

Again, questions like these call for an examination of the motivations which guide the behaviors. There is no a priori reason to assume that stress responses to external aversive events would be similar to those evoked by internal drives.

WHAT IS MOTIVATION?

It may seem that motivation is just as poorly defined as coping. Whereas motivation is a common concept in ethology and psychology, there is no really universally accepted definition. In ethology there has been some tradition of using the concept to describe the internal events that affect an animal's readiness to engage in a particular behavior, and a recent textbook definition emphasizes precisely that "the internal state of the animal, which is the net result of stimuli arising from both inside and outside its body, constitutes its 'motivation' " (Manning and Dawkins 1998). One source of confusion is the formulation "internal state of the animal," which may be taken to mean any concentration of hormones in plasma, blood pressure, degree of gut fill, or neural states in the brain. As pointed out by Jensen and Toates (1997), what we seek to understand in motivational research are the processes by which an animal integrates information from different stimulus sources, both from inside and outside the body, and translates that into behavior. Thus, motivation is really concerned with the state of the nervous system, and other internal states are the stimuli arising from inside the body, which provide some of the necessary information to the brain. This was also in essence the meaning of the concept offered by Broom and Johnson (1993; see also Broom 1981).

A Conceptual Motivational Model

Much of the controversy in motivational research has been about the generality in nature of motivation systems. Some have liked to think that all or most motivational systems can be captured in relatively simple models, such as the old psychohydraulic model of Lorenz (see Manning and Dawkins 1998), whereas others reject the idea that any general model can ever apply to all motivational systems. Some attempts at reconciliation have been made (Toates and Jensen 1991), and it appears clear that the various models have different merits. Fred Toates and I came to the conclusion that both elements of homeostatic motivational models and "drive"-based models must be taken into account in any analysis of a specific behavior. On this basis, we (Jensen and Toates 1997) offered a simple model outlining the possible pathways whereby motivational states that result in tendencies to engage in particular behavior could be effected in the brain, and which therefore have to be considered in any analysis of a specific behavior pattern (Figure 9.1). Unlike many other models, this one does not claim that every motivational system works in this way; rather, it offers an account of the different pathways according to which any motivational system could work. It is therefore a conceptual model, to be used as a tool for specific analyses, and not a predictive model.

With respect to welfare, stress, and coping, the following two aspects of the model may be of particular interest:

1. There is no qualitative difference to an animal whether a particular motivational state is evoked by external or internal stimuli. In fact, both will usually be involved (Toates and Jensen 1991). For example, it is irrelevant to a sow which is motivated to nest-build, whether this motivation can be explained by a rise in certain hormone levels or by the perception of specific external stimuli, such as nest material. The state of the nervous system is likely to be the same in both cases.
2. Sometimes, the animal may be programmed to perform a specific motor program before it can detect any feedback effect on the stimuli that have aroused the motivation. An animal feeling hunger may be programmed to forage, ingest, chew, and swallow before it can detect that the gut is filling up; attempts to shortcut such pathways by tubular feeding have sometimes proven unsuccessful in reducing the feeding motivation (Toates and Jensen 1991).

The model is also relevant for the interpretation of a somewhat related subject, that of needs to show certain behaviors. Such needs are often assumed to be present for behavior that is normally released predominantly by the action of internal stimuli (see Jensen and Toates 1993). However, again it is difficult to see why it should matter to the animal whether a motivation has been caused by internal or external factors. It is therefore difficult to split the behavioral repertoire of an animal into some that are needs, and others that are not, based on the source of

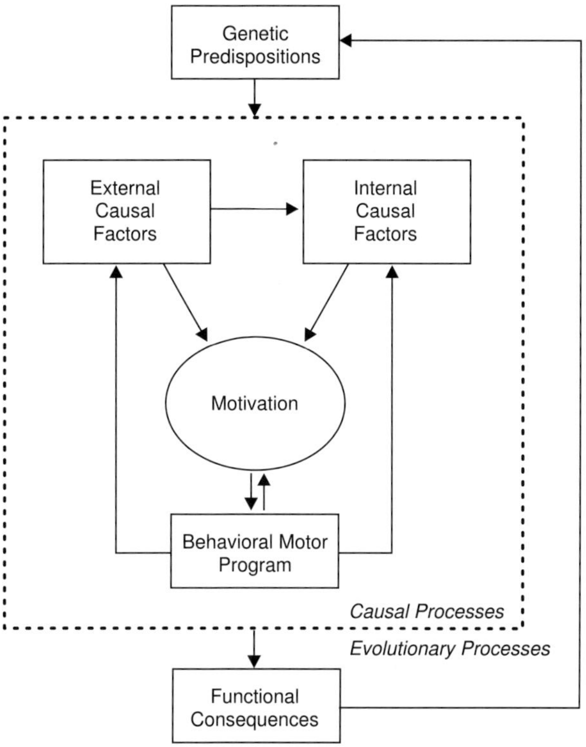

Figure 9.1 A conceptual motivational model, outlining the different pathways which need to be examined in any motivational analysis of a specific behavior. The model emphasizes that functional, evolutionary consequences of the behavior is not necessarily accessible to the causal control. Motivation, which is a state of the brain and the nervous system, is affected by both external and internal causal factors. A motivation is associated with specific motor programs, which may provide feedback along three pathways: by its consequences on the external stimuli, by the modifications the behavior cause in internal stimuli level, and directly on the motivation by proprioceptive feedback. The model is a modified version of the one presented by Jensen and Toates (1997).

stimuli arousing the motivation. However, a close understanding of the motivational system governing a particular behavior will be helpful in specifying the needs of the animal with respect to that behavior. These may be, for example, a need for specific substrates, or for secluded areas, or a specific social structure. Sometimes, there may simply be a need for sufficient space to carry out the motor patterns.

STRESS AND COPING IN RELATION TO MOTIVATION

As pointed out earlier, responses to stressful situations are modified by how the animal is able to act, or cope, with the stressor. How does this relate to the motivational model in Figure 9.1? To see whether it can offer any new insights into stress and coping, I shall examine a few cases in the context of the model.

Weiss' Experiment Interpreted in a Motivational Context

In the experiment by Weiss (1971), described earlier, the stressor used was an electric shock in the tail. This can be described as a painful external stimulus, and we could assume that the most relevant motivational system to invoke as a response to sudden pain is flight. Pain thresholds will probably differ depending on internal factors, and thus the motivation to flee will vary accordingly. However, a sudden and sufficiently intense pain sensation in the tail will most likely always cause the rat to be highly motivated to flee. In rats, the main behavior used for flight is running,that is, rapid movements of the legs. The type of response Weiss offered the rats to control the shock (to terminate or postpone it) was rotating a wheel with the help of the forepaws. It is striking that this response involves much the same behavior patterns as running. Interestingly enough, the yoked rats had the same setup of a turning wheel, but of course it was unconnected to the electric circuit. Nevertheless, Weiss described that the yoked rats were turning the wheel intensely after a shock, something we would expect if the shock had made the rats flight motivated.

We can therefore interpret the Weiss experiment from a motivational point of view in the following way: the rats were exposed to a sudden, strong external stimulus which aroused their flight motivation. In response to that, they recruited the species-specific flight motor program, running, and directed it towards whatever was available, which happened to be a turning wheel. The difference between the yoked and the "master" rats was that "masters" managed to achieve sufficient feedback from their behavior to eliminate the stimulus, whereas yoked rats received very little feedback from their activities. In control systems terminology, with reference to Figure 9.1, we may conclude that "master" rats were able to close the feedback loop and down-regulate the system, so flight motivation was turned off, while yoked rats were unable to do so. The yoked rats were left with the motivational system in an open loop state. This aspect of motivation apparently was the crucial factor in determining whether rats should develop pathologies or not when exposed to the aversive stimuli.

Nest-building in Sows

Consider another example, where the motivation is normally mainly caused by a sudden change in the internal environment. Under free-range conditions, sows build farrowing nests during the last day before giving birth. This is partly triggered by a preparturient rise in prolactin (Lawrence et al. 1994), and sows in all housing conditions attempt to carry out nest-building activities during this period (Jensen 1993). It has been argued that sows will not be motivated to nest-build if they are not stimulated by external stimuli, and that they may be unmotivated if they have a sufficiently comfortable lying place or a ready-made nest (i.e., are exposed to stimuli associated with the goal). This has been strongly refuted by experimental evidence (see Jensen 1993). It has also been suggested that it may be sufficient for the sows to carry out the motor patterns of nest-building, even if this does not result in a nest. This would be possible for sows that are tethered or confined in stalls, and they may therefore not experience any stress in this situation. However, sows in tethered conditions have considerably higher cortisol levels prior to farrowing than loose sows with straw (Lawrence et al. 1994), which strongly suggests that carrying out the movements is not sufficient.

A motivational interpretation of these data may look as follows: the preparturient increase in prolactin is a strong internal stimulus that evokes the nest-building motivation of the sow. In response to this, the sow recruits the nest-building behavior program. When the sow is loose and has access to nest material, there can be sufficient feedback from the behavior to allow the loop of the system to be closed. However, tethered or confined sows without straw are not able to close the loop, in spite of them performing similar motion patterns. A preparturient rise in cortisol seen in all sows is not in itself a valid sign of stress, but interestingly, the levels were significantly higher when the motivational system did not receive sufficient feedback.

A Motivational, Control-system Based Definition of Coping and Stress

Toates (1995) as well as Jensen and Toates (1997) have suggested that stress can be defined as the state an animal enters when a motivational system is locked in an open loop. According to the same definition, coping is when an animal is successful in closing the loop, i.e., it has the opportunity to act successfully according to its motivation by using relevant behavior. Coping therefore excludes stress; stress occurs only when there is a failure to cope. A rather similar definition is given by Broom and Johnson (1993), where stress is said to occur when one or several control systems are overtaxed and there is an adverse effect on the individual. While it is not clear what "overtaxed" actually implies, it would appear that a motivational system locked in an open loop is overtaxed. It should also be mentioned that stress, according to this definition, can occur under natural conditions, for example, during territorial conflicts.

As reviewed by Toates (1995), the positive effects of control only appear to work in experiments where the animals can use a coping behavior that is motivationally relevant, which strongly supports the present account. For example, using motivationally relevant running movements to control a shock seems to work, while controlling a fearful situation by approaching the fear-eliciting stimulus is not likely to be helpful to the animal.

If we accept these definitions, then we resolve the problems outlined earlier.

1. There is no need for good and bad stress. As long as the animal can cope with the challenge, there is no stress.
2. The coping responses and the situations that cause them may vary from individual to individual, depending on the strategy that the animal uses. Thus stress is not a mechanistic, inflexible reaction. It is as variable as animals.
3. There is no conceptual problem with the difference between external stressors and internal drives. What have been called stressors are simply stimuli that evoke specific motivations. Since stimuli may equally well be internal or external, animals can be stressed in either case.

Several studies, including those of Weiss, have shown that the pathological consequences of a stressor depend not only on controllability but also on the predictability of the stressor. It remains to be resolved how the concept of predictability can be presented in a motivational framework like the one offered here.

INDIVIDUAL VARIATION IN MOTIVATIONAL SYSTEMS AND STRESS RESPONSES

If stress occurs when a motivational system cannot function as it evolved in the environment of evolutionary adaptation, we may trace a number of different sources for variations between populations and individuals in stress responses. Again, using the model in Figure 9.1 as a conceptual support, we may first consider the input.

Individuals may differ in the way stimuli are perceived or in the way they are interpreted although this may prove to be difficult to distinguish. One could imagine that it would be possible to measure receptor activity in animals that differed in their responses to particular stimuli. In our own ongoing research, F_2 crosses between jungle fowl and a White Leghorn laying hen are exposed to series of different stimuli and tests. Crosses such as these show very high variation in genotype, with a corresponding phenotypic variation in most traits. When presented with a novel object, individual responses vary widely. Whereas some individuals are immediately investigative and apparently free from fear, others are agitated and make frequent escape attempts. There is, however, a significant sex effect, where males are considerably less investigative and more flight-prone (Schütz et al. 2001). It cannot be excluded that male and female fowl either perceive or interpret novel external stimuli differently.

The same line of reasoning would apply to internal stimuli. For example, in a comparison of maternal behavior between domestic pigs and wild boars, we induced farrowing by means of a single injection of $PGF_{2\alpha}$, a commonly used method for this purpose. Unexpectedly, wild boars farrowed several hours later after the injection than domestic pigs (Gustafsson et al. 1999a), which perhaps may be due to a domestication effect on sensitivity to either the injected substance or to prolactin. It is quite possible that individuals and populations may vary in their sensitivity to many hormones, which may be an important factor to consider when evaluating individual differences in stress responses.

Another source of individual variation would be strategies for coping. In this context, this would mean that individuals may differ either in how they weigh external and internal stimuli, or which behavior they decide to recruit based on the information they receive, or both. This may lead to more or less stable differences in coping behavior. In mice, it has been well documented that such differences are genetically determined and cause some individuals to adopt a more active pattern of coping while others adopt a more passive behavior (Sluyter et al. 1996). Such intrapopulation variation may be evolutionarily adaptive in some species, in particular those that have regular fluctuations in population size and territory stability, but not necessarily in others (Jensen 1995).

Domestication and selection may affect strategies, as mentioned earlier. In particular, behaviors that are energy demanding or risky may be selected against, both passively as a result of provisioning of food and protection and actively, as a correlated response when growth and reproduction are rewarded traits. The theory is that evolution favors an optimal allocation of finite resources to different life processes, and during domestication, production traits have to be gained at the expense of traits that compete for the energy (Beilharz et al. 1993). We have found that laying hens use less energy in searching for food than jungle fowl, which seems to be an adaptive reallocation of resources in a situation where food is provided by humans (Schütz and Jensen 1999). Furthermore, in the earlier mentioned crossing between jungle fowl and laying hens, there are indications that birds with high growth rates and high egg production are less active and less investigative in a novel object test (Schütz et al. 2001). It is therefore quite possible that the intense selection for increased production that takes place in all farm animal species may change coping strategies as a correlated, unintended side effect.

IMPLICATIONS FOR WELFARE

An animal that fails to cope and becomes stressed clearly has poor welfare. However, it does not necessarily follow that animals which can cope have good welfare. Even when coping is successful, it may be demanding for the animal, and at times coping may be achieved only at the price of very high energy expenditure. Hence, welfare has been defined as the state of an individual as regards its attempts to cope with the environment (Broom 1986, Broom and Johnson 1993). This is, of course, a much wider definition than the one advocated for stress and coping, and it clearly emphasizes that welfare assessments normally require a broader set of variables than measurements of stress.

Individual variation, which may be caused by different individuals using the same strategy but different behaviors, or by individuals using different strategies, call for caution in welfare assessments. It is essential that welfare is always assessed at the level of the individual. Whenever statistical measures are produced, it is necessary that care is taken not to lose central information. For example, in examining the occurrence of stereotypic behavior in sows, there is a profound difference between the following hypothetical statistical reports: (a) the sows spend on average 12% of their time performing stereotypies; (b) 50% of the sows were observed to perform stereotypies, and 10% of the sows performed it for more than 30% of the time. Only (b) allows for a scientific welfare assessment on the individual level.

CONCLUSIONS

Animals in captivity face all sorts of challenges. Some are caused by external stimuli and linked to the housing system or the husbandry routines. Others are caused by internal stimuli and linked to physiological systems evolved over thousands of generations in the wild. In any case, the animal shows behavioral or other responses, according to an evolutionarily optimized strategy. When an animal is able to respond to a potentially harmful stimulus, internal or external, by using its natural strategy and receives the biologically relevant feedback from its actions, it can be said to cope, and hence is not stressed. However, its welfare will be poor if it is forced to use a substantial amount of energy continuously in responding to potentially harmful stimuli.

REFERENCES

Andersson, M., P. Jensen, and E. Nordin. 2001. Domestication effects on foraging strategies in fowl (*Gallus gallus*). *Appl. Animal Behav. Sci.* **72**:51–62.

Barnard, C.J., and J.L. Hurst. 1996. Welfare by design: The natural selection of welfare criteria. *Animal Welfare* **5**:405–433.

Beilharz, R.G., B.G. Luxford, and J.L. Wilkinson. 1993. Quantitative genetics and evolution: Is our understanding of genetics suficient to explain evolution? *J. Animal Breed. & Genet.* **110**:161–170.

Broom, D.M. 1981. Biology of Behavior. Cambridge: Cambridge Univ. Press.
Broom, D.M. 1986. Indicators of poor welfare. *Br. Vet. J.* **142**:524–526.
Broom, D.M., and K.G. Johnson. 1993. Stress and Animal Welfare. London: Chapman and Hall.
Fraser, A.F., and D.M. Broom. 1990. Farm Animal Behaviour and Welfare. London: Baillière Tindall.
Gustafsson, M., P. Jensen, F.H. de Jonge, G. Illmann, and M. Spinka. 1999a. Maternal behaviour of domestic pigs and crosses between domestic pig and wild boar. *Appl. Animal Behav. Sci.* **65**:29–42.
Gustafsson, M., P. Jensen, F. de Jonge, and T. Schuurman. 1999b. Domestication effects on foraging strategies in pigs (*Sus scrofa*). *Appl. Animal Behav. Sci.* **62(4)**:305–317.
Jensen, P. 1993. Nest building in domestic sows: The role of external stimuli. *Animal Behav.* **45**:351–358.
Jensen, P. 1995. Individual variation in the behaviour of pigs: Noise or functional coping strategies? *Appl. Animal Behav. Sci.* **44**:2–4.
Jensen, P., and F.M. Toates. 1993. Who needs "behavioural needs"? Motivational aspects of the needs of animals. *Appl. Animal Behav. Sci.* **37**:161–181.
Jensen, P., and F.M. Toates. 1997. Stress as a state of motivational systems. *Appl. Animal Behav. Sci.* **54**:235–243.
Krebs, J.R., and N.B. Davies. 1991. Behavioural Ecology: An Evolutionary Approach. Oxford: Blackwell Scientific Publ.
Lawrence, A.B., J.C. Petherick, K.A. McLean, L.A. Deans, J. Chirnside, A. Vaughan, E. Clutton, and E.M.C. Terlouw. 1994. The effect of environment on behaviour, plasma cortisol and prolactin in parturient sows. *Appl. Animal Behav. Sci.* **39**:313–330.
Manning, A., and M.S. Dawkins. 1998. An Introduction to Animal Behaviour. Cambridge: Cambridge Univ. Press.
Price, E.O. 1997. Behavioural genetics and the process of animal domestication. In: Genetics and the Behaviour of Domestic Animals, ed. T. Grandin, pp. 31–65. San Diego: Academic.
Ruis, M.A.W., J.H.A. te Brake, J.A. van de Burgwal, I. de Jong, H.J. Blokhuis, and J.M. Koolhaas. 2000. Personalities in female domesticated pigs: Behavioural and physiological indications. *Appl. Animal Behav. Sci.* **66**:31–47.
Schütz, K., B. Forkman, and P. Jensen. 2001. Domestication effects on foraging strategy, social behaviour, and different fear responses: A comparison between the red junglefowl (*Gallus gallus*) and a modern layer strain. *Appl. Animal Behav. Sci.* **74**:1–14.
Schütz, K., and P. Jensen. 1999. Foraging behaviour and activity in red jungle fowl (*Gallus gallus*) and in domesticated breeds. In: Proc. 33rd Intl. Congress of the Intl. Soc. for Applied Ethology, ed. K.E. Böe, M. Bakken, and O.B. Bradstad, p. 92. Lillehammer: NLH, Agricultural Univ. of Norway.
Sluyter, F., G.A. van Oortsmerssen, A.J.H. de Ruiter, and J.M. Koolhaas. 1996. Aggression in wild house mice: Current state of affairs. *Behav. Genet.* **26**:489–496.
Toates, F. 1995. Stress — Conceptual and Biological Aspects. Chichester: Wiley.
Toates, F., and P. Jensen. 1991. Ethological and psychological models of motivation — Towards a synthesis. In: Intl. Conf. on Simulation of Adaptive Behavior, ed. J.A. Meyer and S.W. Wilson, pp. 194–205. Cambridge, MA: MIT Press.
Weiss, J.M. 1971. Effects of coping behaviour in different warning signal conditions on stress pathology in rats. *J. Comp. Physiol. Psychol.* **77**:1–13.

10

Is There a Major Stress System at the Periphery Other than the Adrenals?

S.W. PORGES

Department of Psychiatry, University of Illinois at Chicago
Chicago, IL 60612–7327, U.S.A.

Keywords: autonomic nervous system, evolution, polyvagal theory, vagus

ABSTRACT

This manuscript proposes that stress responses, which encompass complex behavioral and physiological adjustments, may be interpreted from a phylogenetic perspective. It is emphasized that the adrenals are not the sole stress response organs in the periphery. As the autonomic nervous system changed through the process of evolution, so did the interplay between the autonomic nervous system and the other physiological systems that respond to stress, including the hypothalamic-pituitary-adrenal (HPA) axis, the neuropeptides oxytocin and vasopressin, and the immune system. From this phylogenetic orientation, a model will be presented that emphasizes the integration among several systems that respond in the periphery during stress.

INTRODUCTION

Embedded in the mammalian nervous system are neuroanatomical structures related to the expression and experience of stress. Several of these structures are shared with other vertebrates and represent the product of phylogenetic development. Comparative research across vertebrate classes provides evidence that mammalian stress and coping response strategies are hierarchically ordered according to a phylogenetic stage (Porges 1997, 1998). Although research has demonstrated the relevance of neuroanatomical structures to stress and coping strategies, contemporary theories have not incorporated an "evolution" perspective into models of stress.

Coping with Challenge: Welfare in Animals including Humans
Edited by D.M. Broom

EVOLUTION AS AN ORGANIZING PRINCIPLE

Evolutionary forces have molded both contemporary physiology and behavior. The mammalian nervous system is a product of evolution. Via evolutionary processes the mammalian nervous system has emerged with specific features that react to challenge to maintain visceral homeostasis. These reactions change physiological state and, in mammals, limit sensory awareness, motor behaviors, and cognitive potentials. We can intuitively grasp the motor limitations of behavior in terms such as speed of movement and strength. However, we tend to neglect an important component of behavior that has been both product and contributor to our evolutionary adaptive success. This component is a functional unit that encompasses the regulation of neurobehavioral state to deal with the challenges that we collectively define as stress. Stress is the product of demands on the nervous system that result in a deviation from homeostasis. In general, the domains of homeostasis, which have been monitored, have focused on the visceral systems involved in cardiovascular, digestive, reproductive, and immune functions. Adaptive and successful coping to stress results in minimizing the magnitude and duration of this deviation. To survive, mammals must determine friend from foe, when an environment is safe, and communicate to their social unit. These survival-related behaviors limit the extent to which a mammal can be physically approached, whether vocalizations will be understood, and whether coalitions can be established. Moreover, these behavioral strategies, which are used to navigate through the "stress of life," form the bedrock upon which social behaviors and higher cognitive processes can be developed and expressed. Thus, learning and other expansive mental processes must be structured, manipulated, and studied within the context of how the environment fosters or ameliorates stress-related physiological states. Unless we appreciate these design features (i.e., vulnerabilities) of the mammalian nervous system, we will be creating stressful living environments that are not only inefficient, but may be harmful to both mental and physical health.

NEUROPHYSIOLOGY OF STRESS: LIMITATIONS OF AROUSAL THEORY

For over a century, researchers have measured autonomic variables (e.g., heart rate, palmar sweat gland activity) as indicators of emotional state related to perceived stress (e.g., fear, mental effort, workload, anxiety). Interest in measuring heart rate and sweat gland activity was theoretically supported by the acceptance of the "arousal" theory. Arousal theory made the assumption that peripheral physiological measures regulated by the sympathetic branch of the autonomic nervous system provided "sensitive" indicators of brain "arousal" or "activation." This view was based on a rudimentary understanding of the autonomic nervous system in which changes in easily measured peripheral organs (e.g., sweat glands, heart) were assumed to be accurate indicators of how the brain was processing emotional stimuli. Usually, the emotional states were associated with fight-flight behaviors and the sympathetic-adrenal system (e.g., increases in heart rate, sweat gland activity, and circulating catecholamines) as described by Cannon (1928). The current emphasis on cortisol as a dependent variable in stress research is consistent with historic views of stress-related arousal involving the adrenals.

Measures due, in part, to their availability and their neuroanatomical association with the target organs of the sympathetic nervous system (i.e., sweat, heart rate) and with secretions from the adrenal medulla (i.e., epinephrine, norepinephrine) and the adrenal cortex (i.e., cortisol), have become the primary physiological variables used to assess stress. Not by plan, but by default, an arousal theory emphasis created a research environment that neglected several important factors including: an understanding of the brain structures that regulate autonomic function; how these structures evolved from the most primitive vertebrates to mammals; how the autonomic nervous system interacts with the immune system, the HPA axis, and the neuropeptides, oxytocin and vasopressin; and the coevolution of stress and coping strategies with the increasing complexity of the autonomic nervous system.

POLYVAGAL THEORY: THREE PHYLOGENETIC SYSTEMS RELATED TO STRESS REACTIVITY

The "polyvagal theory" (Porges 1995, 1997, 1998) emphasizes the phylogenetic origins of brain structures that regulate social and defensive behaviors. The polyvagal theory proposes that the evolution of the mammalian autonomic nervous system provides the neurophysiological substrates for the emotional experiences and affective processes that are major components of social behavior. The theory proposes that physiological state limits the range of behavior and psychological experience. In this context, the evolution of the nervous system determines the range of emotional expression, quality of communication, and the ability to regulate bodily and behavioral state. The polyvagal theory links the evolution of the autonomic nervous system to affective experience, emotional expression, facial gestures, vocal communication, and contingent social behavior. Thus, the theory provides a plausible explanation of social, emotional, and communication behaviors and disorders. The theory also provides an explanation of stress-related responses. Relevant to adaptive stress responses, the theory makes the following assumptions:

- Evolution has modified the structures of the autonomic nervous system.
- The mammalian autonomic nervous system retains vestiges of phylogenetically older autonomic nervous systems.
- Emotional regulation and social behavior are functional derivatives of structural changes in the autonomic nervous system due to evolutionary processes.
- In mammals, the autonomic nervous system response strategy to challenge follows a phylogenetic hierarchy, starting with the newest structures and, when all else fails, reverting to the most primitive structural system.
- The phylogenetic stage of the autonomic nervous system determines the behavioral, physiological, and affective features of stress reactivity.

The polyvagal construct was introduced to emphasize and document the neurophysiological and neuroanatomical distinction between two branches of the tenth cranial nerve (i.e., vagus) and to propose that each vagal branch was associated with a different adaptive behavioral strategy. The vagus nerve, a primary component of the autonomic nervous system, exits the brain stem and has branches that regulate the striated muscles of the head and face (e.g., facial muscles, eyelids, middle ear muscles, larynx, pharynx, muscles of mastication) and in several

Table 10.1 The three phylogenetic stages of the neural control of the heart proposed by the polyvagal theory.

Phylogenetic stage	Autonomic nervous system component	Behavioral function	Lower motor neurons
III	Myelinated vagus	Social communication, self-soothing and calming, inhibit sympathetic-adrenal influences	Nucleus ambiguus
II	Sympathetic-adrenal	Mobilization (active avoidance)	Spinal cord
I	Unmyelinated vagus	Immobilization (death feigning, passive avoidance)	Dorsal motor nucleus of the vagus

visceral organs (e.g., heart, gut). The theory proposes that the different branches are related to unique adaptive behavioral strategies and articulates three phylogenetic stages of the development of the mammalian autonomic nervous system (see Table 10.1). These stages reflect the emergence of three distinct subsystems, which are phylogenetically ordered and behaviorally linked to communication (e.g., facial expression, vocalization, listening), mobilization (e.g., fight-flight behaviors), and immobilization (e.g., death feigning, behavioral "shutdown," and syncope). The mobilization system is dependent upon the functioning of the sympathetic nervous system (SNS). The most phylogenetically primitive component, the immobilization system, is dependent upon the unmyelinated or "vegetative" vagus, which is shared with most vertebrates. With increased neural complexity due to phylogenetic development, the organism's behavioral and affective repertoire is enriched. In this chapter, I expand the polyvagal theory to include response systems that interact with the autonomic nervous system (i.e., neuropeptides, HPA axis, immune) to mediate physiological state in response to stress.

Table 10.2 illustrates the phylogenetic differences in the structures that regulate the heart in vertebrates (Morris and Nilsson 1994; Santer 1994; Taylor 1992). The heart is selected because the regulation of the heart determines the availability of the metabolic resources required for mobilization as well as for growth and restoration. For example, cardiac output must be regulated to remain calm in safe environments, to mobilize for fight or flight behaviors, or to immobilize for death feigning or avoidance behaviors. To regulate cardiac output, several efferent structures have evolved. These structures represent two global and often opposing systems: (a) a sympathetic-catecholamine system including catecholamine secreting chromaffin tissue and spinal sympathetic nerves, and (b) a vagal system (a component of the parasympathetic nervous system) with branches originating in medullary source nuclei (i.e., dorsal motor nucleus of the vagus and nucleus ambiguus).

In the jawless fish, the neural control of the heart is very primitive. Some jawless fish, such as hagfish, rely on circulating catecholamines from diffuse chromaffin tissue to provide excitatory influences on the heart. Other jawless fish, such as lampreys, have a cardiac vagus. However, in contrast to all other vertebrates, which have a cardio-inhibitory vagus that acts via muscarinic cholinoceptors, the lamprey vagus is cardioexcitatory and acts via nicotinic receptors.

Table 10.2 Neural regulation of the heart as a function of vertebrate phylogeny.

Group	CHM	DVC	SNS	AD/m	VVC
Jawless fish	X+	(X+)			
Cartilaginous fish	X+	X–			
Bony fish	X+	X–	X+		
Amphibians	X+	X–	X+		
Reptiles	X+	X–	X+	X+	
Mammals	X+	X–	X+	X+	X–

CHM: chromaffin tissue;
DVC: dorsal vagal complex with vagal efferent pathways originating in the dorsal motor nucleus of the vagus and vagal afferents terminating in the nucleus tractus solitarius;
SNS: spinal sympathetic nervous system;
AD/m: adrenal medulla;
VVC: ventral vagal complex with efferent pathways originating in the nucleus ambiguus that regulate visceral structures (heart, bronchi, thymus) and striated muscles via special visceral efferents and afferents via the tractus solitarius, trigeminal, and facial nerve.

Cartilaginous fish, such as the sharks, rays, and skates, have an unmyelinated cardio-inhibitory vagus. Similar to the more recent vertebrates, the vagus has cells of origin in the dorsal motor nucleus of the vagus located in the medulla. The vagus in these fish is inhibitory and the cholinoceptors on the heart are muscarinic as they are in other vertebrates. The cardio-inhibitory vagus is functional in the cartilaginous fish. During conditions of hypoxia, the metabolic output is adjusted by lowering heart rate. This modification of neural regulation may provide a mechanism to enable the cartilaginous fish to increase their territorial range, by providing a neural mechanism that adjusts metabolic output to deal with changes in water temperature and oxygen availability. However, unlike the phylogenetically more recent bony fish and tetrapods, the cartilaginous fish do not have direct sympathetic input to the heart. Instead, cardiac acceleration and increases in contractility are mediated via β-adrenergic receptors stimulated by circulating catecholamines released from chromaffin tissue. Thus, since activation of metabolic output is driven by circulating catecholamines and not by direct neural innervation, once the excitatory system is triggered, the ability to self-sooth or calm is limited.

The bony fish are phylogenetically the first group of vertebrates in which the heart is regulated by both sympathetic and parasympathetic neural pathways. With opposing neural mechanisms from sympathetic and vagal pathways, rapid transitory changes in metabolic output are possible to support immediate changes in behavior from mobilization to immobilization. In bony fish this is observed as darting and freezing with direct neural components from the spinal cord via the sympathetic chain producing increases in heart rate and contractility, and direct neural pathways from the brain stem via the vagus producing cardio-inhibitory actions. Amphibians, similar to bony fish, have both sympathetic and parasympathetic neural innervation of the heart.

A distinct adrenal medulla, formed of chromaffin tissue, is present only in birds, reptiles and mammals (Santer 1994). The medulla is of ectodermal origin from which the epidermis, nervous tissue, and, in vertebrates, sense organs develop. Neural regulation by sympathetic nerves of the adrenal medulla provides a mechanism for rapid and controlled release of epinephrine and norepinephrine to increase cardiac output to match the metabolic demands of mobilization behaviors. In bony fish, chromaffin tissue is primarily related to parts of the cardiovascular system, but there also is chromaffin tissue associated with the kidney. However, in amphibians chromaffin tissue is primarily associated with the kidney and there are substantial aggregations of chromaffin cells located along the sympathetic chain ganglia. Thus, there is a phylogenetic shift in the location of chromaffin tissue and the concurrent evolution of a distinct adrenal medulla near the kidney.

Birds are conspicuously missing from Table 10.2; it represents mammals and their phylogenetic antecedents. Although mammals and birds are phylogenetic descendants of reptiles, mammals are not "direct" phylogenetic descendants of birds.

Unlike other vertebrates with a cardio-inhibitory vagus, the mammalian vagus contains two branches. One branch originates in the dorsal motor nucleus of the vagus and provides the primary neural regulation of subdiaphragmatic organs such as the digestive tract. However, at the level of the heart and unlike more primitive vertebrates, the efferent vagal pathways that originate in the dorsal motor nucleus of the vagus do not play a major role in the normal dynamic regulation of cardiac output. Rather, during embryological development in mammals, cells from the dorsal motor nucleus of the vagus migrate ventrally and laterally to the nucleus ambiguus (Schwaber 1986). There they form the cell bodies for the visceromotor myelinated axons that provide potent inhibition of the sinoatrial node, the pacemaker for the heart.

By transitory down-regulation of the cardio-inhibitory vagal tone to the heart (i.e., removing the vagal brake), the mammal is capable of rapid increases in cardiac output without activating the sympathetic-adrenal system. By engaging this system, rather than the sympathetic-adrenal system, mammals have an opportunity to increase metabolic output rapidly for immediate mobilization. Under prolonged challenge, the sympathetic system also may be activated. However, by rapidly reengaging the vagal system, mammals have the capacity to inhibit sympathetic input on the heart (Vanhoutte and Levy 1979) and rapidly decrease metabolic output to self-soothe and calm.

Three phylogenetic principles can be extracted from Table 10.2. First, there is a phylogenetic shift in the regulation of the heart from endocrine communication, to unmyelinated nerves, and finally to myelinated nerves. Second, there is a development of opposing neural mechanisms of excitation and inhibition to provide rapid regulation of graded metabolic output. Third, with increased cortical development, the cortex exhibits greater control over the brain stem via direct (e.g., corticobulbar) and indirect (e.g., corticoreticular) neural pathways originating in motor cortex and terminating in the source nuclei of the myelinated motor nerves emerging from the brain stem (e.g., special visceral efferent neural pathways embedded within cranial nerves V, VII, IX, X, XI).

These phylogenetic principles provide a basis for speculations regarding the behavioral and physiological responses associated with mammalian stress and coping strategies. In general, phylogenetic development results in increased neural control of the heart via the myelinated mammalian vagal system, which can promote transitory mobilization and the expression of sympathetic tone without requiring sympathetic or adrenal activation. With this new vagal system, transitory incursions into the environment or withdrawals from a potential predator can be initiated without the severe biological cost of the metabolic excitation associated with sympathetic-adrenal activation. Paralleling this change in neural control of the heart is an enhanced neural control of the face, larynx, and pharynx that enables complex facial gestures and vocalizations associated with social communication. This phylogenetic course results in greater central nervous system regulation of behavior, especially behaviors needed to engage and disengage with environmental challenges.

The Vagal Brake

The myelinated mammalian vagus can inhibit the influence of the SNS at the level of the heart. The mammalian vagus may function as an active vagal brake (see Porges et al. 1996) in which rapid inhibition and disinhibition of the vagal tone to the heart can rapidly mobilize or calm an individual. In addition, since the mammalian vagus has distinct pathways involved in the voluntary regulation of the striated muscles (e.g., corticobulbar pathways, afferents from face and mouth, efferents to larynx and pharynx), the regulation of heart rate is neuroanatomically and neurophysiologically linked to the function of the special visceral efferents.

Due to the tonic vagal influences to the sinoatrial node (i.e., the heart's pacemaker), resting heart rate is substantially lower than the intrinsic rate of the pacemaker. When the vagal tone to the pacemaker is high, the vagus acts as a brake on the rate the heart is beating. When vagal tone to the pacemaker is low, there is little or no inhibition of pacemaker. Thus, the vagal brake may be used as a construct to describe the functional modulation of heart rate by the myelinated vagal efferent pathways. The vagal brake provides a neural mechanism to change visceral state rapidly by slowing or speeding heart rate. Consistent with the assumptions of the polyvagal theory, the vagal brake contributes to the modulation of cardiac output by decreasing the inhibitory vagal control of the heart to speed heart rate and by increasing the inhibitory vagal control of the heart to slow heart rate. Thus, neurophysiologically the vagal brake provides a mechanism to support the metabolic requirements for mobilization and communication behaviors. Functionally, the vagal brake, by modulating visceral state, enables the individual to engage and disengage objects and other individuals rapidly and to promote self-soothing behaviors and calm behavioral states.

Difficulties in regulating the vagal brake result in the recruitment of other neural mechanisms (e.g., sympathetic regulation of the heart) and neural chemical mechanisms (e.g., stimulation of the HPA axis) to regulate physiological state. Thus, consistent with the polyvagal theory, if the vagal brake is not functioning or will not serve the survival needs of the organism, then the phylogenetically "older" systems (e.g., the sympathetic-adrenal system) will be recruited to regulate metabolic output to deal with environmental challenges. For example, if the vagal brake is not functioning, there is the potential for greater dependence on the sympathetic excitation of the cardiovascular system.

Evolution and Dissolution: A Hierarchical Stress Response Strategy

The evolution of the autonomic nervous system provides substrates for the emergence of three stress and coping subsystems, each linked to structures that evolved during identifiable phylogenetic stages. The polyvagal theory proposes that during danger or threat, the "older" less social are recruited. The older systems, although functional in short term, may result in damage to the mammalian nervous system when expressed for prolonged periods. Thus, the stress and coping neurophysiological strategies that are adaptive for reptiles (e.g., apnea, bradycardia, immobilization), may be lethal for mammals.

Although reminiscent of the triune brain proposed by MacLean (1990), the polyvagal theory emphasizes that even the phylogenetically more primitive structures have changed in structure and function. This phylogenetic adjustment of the autonomic nervous system represents an exaptation (i.e., a shift in function) of structures to express emotions and to adapt to the complex environment.

The polyvagal theory proposes a hierarchical response strategy to environmental challenges, with the most recent modifications employed first and the most primitive last. However, the response strategy is not all-or-none and may include transitional blends between the boundaries of the three hierarchical stages. These transitional blends may be determined by both visceral feedback and higher brain structures (including the HPA axis and vasopressinergic and oxytocinergic pathways that communicate between the hypothalamus and the dorsal vagal complex (DVC) consisting of the nucleus tractus solitarius and the dorsal motor nucleus of the vagus). Thus, the neurophysiological substrate of specific behavioral states and coping strategies may incorporate activation of a sequence of response systems representing more than one phylogenetic stage.

This phylogenetically based hierarchical response strategy is consistent with the concept of dissolution proposed by John Hughlings Jackson (1958) to explain disease of the nervous system. Jackson proposed that since the phylogenetically newer or "higher" components of the nervous system regulated or controlled the "lower" components (i.e., phylogenetically older), damage to the newer components would "disinhibit" the activity of the lower. The polyvagal theory (Porges 1995, 1997, 1998) proposes dissolution, not in response to disease or brain trauma, but as a response strategy to differential challenges (i.e., threats to survival) associated with stress. The ventral vagal complex (VVC), with its mechanisms for signaling and communication provides the initial response to the environment. The VVC inhibits, at the level of the heart, the strong mobilization responses of the spinal SNS. Withdrawal of the VVC, consistent with the Jacksonian principle of dissolution, results in a disinhibition of the sympathetic control of the heart. Similarly, withdrawal of sympathetic tone results in a disinhibition of the DVC control of the gastrointestinal tract and a vulnerability of the bronchi (i.e., constriction) and the heart (stopping or significantly slowly heart rate). There are several clinical consequences to unopposed DVC control, including defecation, due to a relaxation of the sphincter muscles and increased motility of the digestive tract; apnea, due to constriction of the bronchi; and bradycardia, due to stimulation of the sinoatrial node. Thus, when all else fails, the nervous system may elect a metabolically conservative course that is adaptive for primitive vertebrates, but lethal for mammals.

The Social Nervous System

Mammals and especially primates have evolved to have a well-defined social nervous system. Embryologically, components of several cranial nerves develop together to form the neural substrate of a social engagement system (see Porges 1998). This system provides the neural gatekeepers for social–emotional interactions. The social engagement system has a control component in the cortex (i.e., upper motor neurons) that regulates brain stem nuclei (i.e., lower motor neurons) to control eyelid opening (e.g., looking), facial muscles (e.g., emotional expression), middle ear muscles (e.g., extracting human voice from background noise), muscle of mastication (e.g., ingestion), laryngeal and pharyngeal muscles (e.g., vocalization and language), and head turning muscles (e.g., social gesture and orientation). Collectively, these muscles function as filters and control social engagement with the environment. The neural control of these muscles determines social experiences. In addition, the source nuclei (i.e., lower motor neurons) of these nerves, which are located in the brain stem, communicate directly with an inhibitory neural system that slows heart rate, lowers blood pressure, and actively reduces arousal to promote calm states consistent with the metabolic demands of growth and restoration of our neurophysiological systems. Direct corticobulbar pathways reflect the influence of frontal areas of the cortex (i.e., upper motor neurons) on the regulation of this system. Thus, the social nervous system is intimately related to stress reactivity. In addition, the anatomical structures involved in the social engagement system have neurophysiological interactions with the HPA axis, the neuropeptides of oxytocin and vasopressin, and the immune system. Thus, the social nervous system provides a theoretical model to explain the interactive and stress-related functions of several physiological systems that have central regulatory components, but are expressed in the periphery.

HOW DOES A PHYLOGENETIC ORIENTATION PROVIDE A THEORETICAL PLATFORM TO UNDERSTAND THE ADRENAL "STRESS" SYSTEMS AND THEIR FUNCTIONAL INTERACTIONS WITH THE AUTONOMIC NERVOUS SYSTEM?

The adrenal exhibits phylogenetic changes in structure and function. Above, because of its complementary function to the SNS, the adrenal medulla was discussed. However, since the adrenal cortex is treated as an endocrine organ and part of the HPA axis, it is traditionally studied and discussed independently of the adrenal medulla and the autonomic nervous system. Unlike the medulla, which is of ectodermal origin, the cortex is of mesodermal origin. The adrenal cortex secretes cortisol or corticosterone, which are frequently used to index stress. All vertebrates produce glucocorticoids and catecholamines. However, the structures that produce these secretions follow a phylogenetic trend that results in two anatomical changes. First, the chromaffin tissue and the adrenocortical cells, which are scattered through the viscera in the primitive vertebrates, cluster together to form functional structures. Second, these structures establish a close anatomical relation (the adrenal) with communication between the secretions of the two components of the adrenal and only in mammals does the adrenal have its own vascular supply and venous drainage.

Only mammals have adrenocortical cells clustered as a cortex of the adrenal. Interestingly, several reptiles (e.g., lizards and some snakes), which have a functioning adrenal medulla, have the adrenocortical cells partially encapsulated by the chromaffin cells. In the more primitive fish (jawless and cartilaginous), the chromaffin and adrenocortical cells are entirely separate. In the bony fish, both the adrenocortical cells and chromaffin tissue are scattered as independent groups within the "head" kidney. In amphibians, the chromaffin and adrenocortical tissues are scattered over the ventral surfaces of the kidney. In reptiles, the chromaffin and adrenocortical tissues become consolidated into distinct adrenal glands. Reptiles provide a phylogenetic marker in which the interrenals have an independent vascular supply and venous drainage and no longer (as in amphibians and bony fish) rely on the kidney and a renal portal system for the distribution of secretory products.

Several investigators have explored the adaptive function of the close proximity of the adrenocortical-cortisol-secreting adrenocortical cells (that subsequently forms the adrenal cortex) and the catecholamine-secreting chromaffin tissue (that subsequently forms the adrenal medulla) (Norris 1997). The medulla receives large quantities of glucocorticoids through the adrenal vascular system, and these hormones activate the enzyme system for converting norepinephrine into epinephrine. This effect would support sympathetic-adrenal medulla functions associated with mobilization and increased metabolic activity. Epinephrine has a potent effect on the elevation of blood glucose convergent with the influences of some of the adrenal steroids. Additionally, cortisol may directly enhance mobilization by converting lactate to glucose via gluconeogenesis in the liver. This process would contribute to the metabolic demands of mobilization by increasing the availability of glucose and also in reducing oxygen debt due to the accumulation of lactate.

Vagal activity has been implicated in the function of the adrenal cortex. Reports suggest that afferents originating in the subdiaphragmatic vagus (i.e., DVC) exhibit an inhibitory influence on HPA axis and reduce cortisol secretion (e.g., Miao et al. 1997; Bueno et al. 1989). Other research has demonstrated a covariation between increases in cortisol and decreases in cardiac vagal tone (Gunnar et al. 1995), which, consistent with the removal of the vagal brake and the stimulation of the SNS, would promote mobilization. Similarly, psychological stressors reduce cardiac vagal tone and increase cortisol plasma level (e.g., Cacioppo et al. 1995). Thus, in several situations there appears to be a coordinated response that functions to promote metabolic activity and mobilization behaviors by withdrawal of VVC tone and increasing both SNS activity and activation of the HPA axis.

In general, the functioning of the adrenal cortex and the secretion of cortisol appears to be integrated into the mobilization function of the autonomic nervous system by increasing sympathetic activation and circulating catecholamines. These effects suggest that, consistent with the phylogenetic approach described in the polyvagal theory, cortisol secretion may be related to maintenance of mobilization (i.e., the conversion of norepinephrine into epinephrine) for fight-flight behaviors and in the recovery from the lactate build up that may contribute to a functional oxygen debt (i.e., gluconeogenesis).

In addition, the reports of dysregulation of the HPA axis, low cortisol, or low cortisol reactivity in disorders such as schizophrenia (Jansen et al. 2000), posttraumatic stress disorder (Yehuda et al. 1996), and the consequences of neglect and abuse in children (De Bellis et al. 1994) may be explained within the context of the polyvagal theory. According to the theory, when mobilization strategies (i.e., fight-flight behaviors) are ineffective in removing the individual from the stressor and in modulating the effect of the stress, then the nervous system may degrade to a phylogenetically earlier level of organization. Thus, low cortisol or a hyporesponsive HPA axis may reflect a neural strategy associated with immobilization (i.e., passive avoidance) that would require a reduction in energy resources.

Immune Function

The vagus has been implicated in the immune system (e.g., Watkins et al. 1995). Recent studies have demonstrated the important role of subdiaphragmatic vagal afferents in conveying information regarding visceral state to the brain. The studies conceptualize the afferent vagus as providing a signal alerting the central structures regulating immune function. A few studies have described motor pathways via the vagus to the thymus (e.g., Bulloch and Pomerantz 1984). The link between the vagal function related to the stages of the polyvagal theory and the immune system is not clear. However, it might be plausible to speculate that the neural mediation of the myelinated vagus may, via direct influence on thymus and direct inhibition of the SNS, trigger a physiological state that would promote immune function. Likewise, withdrawal of vagal tone to the heart, increased sympathetic tone, and the release of cortisol have been associated with suppressed immune function.

Oxytocin and Vasopressin

Phylogenetically, although all vertebrates have peptides similar in structure to both oxytocin and vasopressin, only mammals have the specific receptors for both peptides. Oxytocin and vasopressin are synthesized primarily in the paraventricular and supraoptic nuclei of the hypothalamus and released centrally via parvocellular neurons and systemically via magnocellular neurons (Swanson and Sawchenko 1977). The central and systemic effects of these neuropeptides are different. Central release of oxytocin regulates the output of the dorsal motor nucleus of the vagus, usually maintaining output within levels optimal to support homeostasis. Thus, providing a proposed "antistress" function (see Carter 1998). Peripheral release of oxytocin is related to milk ejection, uterine contractions, and ejaculation. Central release of vasopressin appears to modulate afferent feedback from the viscera and to shift set points, independent of sensitivity, for vagal reflexes such as the baroreceptor reflex (Michelini 1994). The raising of the baroreceptor set point would, by increasing cardiac output, potentiate fight-flight behaviors and allow sympathetic excitation of the heart to be unopposed by homeostatic vagal reflexes. Thus, central levels of oxytocin have been assumed to be associated with vagal processes and central levels of vasopressin have been assumed to be associated with sympathetic processes (Uvnas-Moberg 1997).

Because the peripheral influences of oxytocin and vasopressin function through feedback, primarily via the sensory component of the DVC, the peripheral effects of these peptides are less clear and may be level dependent or differ as a function of acute versus chronic exposure. For example, it is possible that peripheral vasopressin may, by stimulating vagal afferents, trigger massive vagal responses via the dorsal motor nucleus of the vagus. In support of this speculation, it is known that in humans, peripheral vasopressin, and not oxytocin, is related to the nausea experienced during motion sickness (Koch et al. 1990). In addition, systemic vasopressin may induce a baroreceptor-mediated bradycardia and a fall in plasma concentration of norepinephrine (Buwalda et al. 1992; Michelini 1994).

Oxytocin may be part of a complex response profile related to the perception of the environment as safe. Consistent with this view, Uvnas-Moberg (1997) and Carter and Altemus (1997) propose that oxytocin promotes states resistant to stress (i.e., anti-stress). In contrast, vasopressin may be part of a complex response profile related to the perception that the environment is challenging or dangerous (i.e., stress). In fact, central vasopressin could potentiate mobilization responses via sympathetic excitation, while high levels of systemic vasopressin may potentiate a physiological shutdown associated with fear (e.g., bradycardia) via feedback to the dorsal motor nucleus and inhibition of sympathetic outflow (Ferguson and Lowes 1994). In addition, lesions of vagal afferents, which functionally block the visceral input to the sensory component of the DVC (areas sensitive to vasopressin), attenuate or abolish specific conditioned taste aversions (Andrews and Lawes 1992).

Based on the polyvagal theory, the mammalian vagus, with myelinated motor fibers originating in the nucleus ambiguus, provides a system for voluntary engagement with the environment with special features associated with the prosocial behaviors of communication. Paralleling this evolutionary shift in the vagus is a mammalian modification of the hypothalamic regulation of the DVC via both oxytocin and vasopressin. The advent of specific receptors for oxytocin and vasopressin increases the range of adaptive functions involving the DVC. In mammals, the dorsal motor nucleus of the vagus, the motor component of the DVC, is sensitive to oxytocin and insensitive to vasopressin. In contrast, the sensory components of the DVC, the nucleus of the solitary tract and area postrema, are most sensitive to vasopressin. Although the nucleus of the solitary tract has receptors for oxytocin (Landgraf et al. 1990), the area postrema may not be directly influenced by oxytocin (Carpenter 1990). The differential sensitivity of specific components of the DVC to these two neuropeptides, i.e., the differential effects of central and systemic release on visceral function and a potential level dependency, results in a wider range of response options including maximizing mobilization behaviors and the co-opting of the primitive vagal system associated with immobilization to support "anti-stress" functions such social engagement and growth and restoration (see Porges 1998).

A PHYLOGENETIC APPROACH TO THE STUDY OF STRESS

The phylogenetic orientation focuses our interest on the neural structures and neurobehavioral systems that we share with or have adapted from our phylogenetic ancestry. First, the three response systems proposed in the polyvagal theory (i.e., cranial nerves to regulate the face, sympathetic-adrenal system to increase metabolic output, and an inhibitory vagal system to decrease metabolic output and promote freezing and defecation) are the products of distinct neurophysiological systems. Second, these distinct neurophysiological systems represent a phylogenetically dependent hierarchy with the use of cranial nerves to regulate facial expression emerging in mammals, as system well developed in primates, the sympathetic-adrenal system shared with other vertebrates including reptiles, and the inhibitory vagal system shared with more primitive vertebrates including amphibians, bony fish, and cartilaginous fish (see Porges 1997, 1998). The three systems, which have been frequently studied in emotion research, represent different phylogenetic stages of neural development. This phylogenetic development starts with a primitive behavioral inhibition system, progresses to a fight-flight system, and, in humans (and other primates), culminates in a complex facial gesture and vocalization system. Thus, from a phylogenetic perspective, the nervous system of vertebrates evolved to support a greater range of behaviors and physiological states, including states that we often associate with stress.

The physiological responses associated with stress in mammals do not concisely fit into a single neurophysiological system. Although cortisol has often been labeled as the "stress" hormone, other systems clearly respond to stress. Moreover, there are situations of severe stress in which cortisol is neither responsive nor at a high level. By expanding the phylogenetic model proposed in the polyvagal theory to include the HPA axis, the hyporesponsive HPA can be interpreted as reflecting a primitive passive avoidance system. Thus, in response to the question proposed in the title of this paper, several systems with target organs at the periphery contribute to the organism's adaptation to challenge and stress. These neurobiological systems are not stress systems solely. Rather, there are intimately involved in the dynamic regulation of homeostasis, growth and restoration, metabolism, and mobilization. These systems form a complex set of mutually interacting neurophysiological pathways that communicate via nerves and neurally active chemicals (e.g., neurotransmitters, neuropeptides, hormones) to cope with survival challenges. This chapter has attempted to employ phylogeny as an organizing principle to illustrate the interactive and adaptive nature of several physiological systems that respond to stress.

ACKNOWLEDGMENTS

The preparation of this manuscript was supported in part by grants HD 22628 and MH 60625 from the National Institutes of Health. The ideas and concepts described in this paper were the product of discussions with C. Sue Carter during residency at the Rockefeller Foundation Conference Center in Bellagio, Italy.

REFERENCES

Andrews, P.L.R., and I.N.C. Lawes. 1992. A protective role for vagal afferents: A hypothesis. In: Neuroanatomy and Physiology of Abdominal Vagal Afferents, ed. S. Ritter, R.C. Ritter, and C.D. Barnes, pp. 280–302. Boca Raton: CRC.

Bueno, L., M. Gue, M.J. Fargeas, M. Alvinerie, J.L. Junien, and J. Fioramonti. 1989. Vagally mediated inhibition of acoustic stress-induced cortisol release by orally administered kappa-opioid substances in dogs. *Endocrinology* **124**:1788–1793.

Bulloch, K., and W. Pomerantz. 1984. Autonomic nervous system innervation of thymic-related lymphoid tissue in wildtype and nude mice. *J. Comp. Neurol.* **228**:58–68.

Buwalda, B., J.M. Koolhaas, and B. Bohus. 1992. Behavioral and cardiac responses to mild stress in young and aged rats: Effects of amphetamine and vasopressin. *Physiol. Behav.* **51**:211–216.

Cacioppo, J.T., W.B. Malaarkey, J.K. Kiecolt-Glaser, B.N. Uchino, S.A. Sgoutas-Emch, J.F. Sheridan, G.G. Berntson, and R. Glaser. 1995. Heterogeneity in neuroendocrine and immune responses to brief psychological stressors as a function of autonomic cardiac activation. *Psychosom. Med.* **57**:154–164.

Carpenter, D.O. 1990. Neural mechanisms of emesis. *Canad. J. Physiol. Pharmacol.* **68**:230–236.

Carter, C.S. 1998. Neuroendocrine perspectives on social attachment and love. *Psychoneuroendocrinology* **23**:779–818.

Carter, C.S., and M.Altemus. 1997. Integrative functions of lactational hormones in social behavior and stress management. *Ann. N.Y. Acad. Sci.* **807**:164–174.

Cannon, W.B. 1928. The mechanism of emotional disturbance of bodily functions. *New Engl. J. Med.* **198**:877–884.

De Bellis, M.D., G.P. Chrousos, L.D. Dorn, L. Burke, K. Helmers, M.A. Kling, P.K. Trickett, and F.W. Putnam. 1994. Hypothalamic-pituitary-adrenal axis dysregulation in sexually abused girls. *J. Clin. Endocrinol. Metab.* **78**:249–255.

Ferguson, A.V., and V.L. Lowes. 1994. Functional neural connections of the area postrema. In: Nucleus of the Solitary Tract, ed. I.R.A. Barraco, pp. 147–157. Boca Raton: CRC.

Gunnar, M.R., F.L. Porter, C.M. Wolf, J. Rigatuso and M.C. Larson. 1995. Neonatal stress reactivity: predictions to later emotional temperament. *Child Dev.* **66**:1–13.

Jackson, J.H. 1958. Evolution and dissolution of the nervous system. In: Selected writings of John Hughlings Jackson, ed. J. Taylor, pp. 45–118. London: Stapes Press.

Jansen, L.M.C., C.C. Gispen-de Wied, and R.S. Kahn. 2000. Selective impairments in the stress response in schizophrenic patients. *Psychopharmacology* **149**:319–325.

Koch, K.L., J. Summy-Long, S. Bingaman, N. Sperry, and R.M. Stern. 1990. Vasopressin and oxytocin responses to illusory self-motion and nausea in man. *J. Clin. Endocrinol. Metab.* **71**:1269–1275.

Landgraf, R., T. Mallkinson, T. Horn, W.L. Veale, K. Lederis, and Q.J. Pittman. 1990. Release of vasopressin and oxytocin by paraventricular stimulation in rats. *Am. J. Physiol.* **258**:155–159.

MacLean, P.D. 1990. The Triune Brain in Evolution. New York: Plenum Press.

Miao, F.J.-P., W. Janig, P.G. Green, and J.D. Levine. 1997. Inhibition of bradykinin-induced plasma extravasation produced by noxious cutaneous and visceral stimuli and its modulation by vagal activity. *J. Neurophysiol.* **78**:1285–1292.

Michelini, L.C. 1994. Vasopressin in the nucleus tractus solitarius: a modulator of baroreceptor reflex control of heart rate. *Braz. J. Med. Biol. Res.* **27**:1017–1032.

Morris, J.L., and S. Nilsson. 1994. The circulatory system. In: Comparative Physiology and Evolution of the Autonomic Nervous System, ed. S. Nilsson and S. Holmgren, pp. 193–246. Chur, Switzerland: Harwood Academic.

Norris, D.O. 1997. Vertebrate Endocrinology. New York: Academic.

Porges, S.W. 1995. Orienting in a defensive world: Mammalian modification of our evolutionary heritage. A polyvagal theory. *Psychophysiology* **32**:301–318.

Porges, S.W. 1997. Emotion: An evolutionary by-product of the neural regulation of the autonomic nervous system. *Ann. NY Acad. Sci.* **807**:62–77.

Porges, S.W. 1998. Love: An emergent property of the mammalian autonomic nervous system. *Psychoneuroendocrinology* **23**:837–861.

Porges, S.W., J.A. Doussard-Roosevelt, A.L. Portales, and S.I. Greenspan. 1996. Infant regulation of the vagal "brake" predicts child behavior problems: A psychobiology model of social behavior. *Dev. Psychobiol.* **29**:697–712.

Santer, R.M. 1994. Chromaffin systems. In: Comparative Physiology and Evolution of the Autonomic Nervous System, ed. S. Nilsson and S. Holmgren, pp. 97–117. Chur, Switzerland: Harwood Academic.

Schwaber, J.S. 1986. Neuroanatomical substrates of cardiovascular and emotional-autonomic regulation. In:Central and peripheral mechanisms of cardiovascular regulation, ed. A. Magro, W. Osswald, D. Reis, and P. Vanhoutte, pp. 353–384. New York: Plenum Press.

Swanson, L.W., and P.E. Sawchenko. 1977. Hypothalamic integration: Organization of the paraventricular and supraoptic nuclei. *Ann. Rev. Neurosci.* **6**:269–324.

Taylor, E.W. 1992. Nervous control of the heart and cardiorespiratory interactions. In: Fish Physiology: The Cardiovascular System, Vol. XII, ed. W.S. Hoar, D.J. Randall, and A.P. Farrell, pp. 343–387. New York: Academic.

Uvnas-Moberg, K. 1997. Physiological and endocrine effects of social contact. *Ann. NY Acad. Sci.* **807**:146–163.

Vanhoutte, P.M., and M.N.Levy. 1979. Cholinergic inhibition of adrenergic neurotransmission in the cardiovascular system. In: Integrative Functions of the Autonomic Nervous System, ed. C. McC. Brooks, K. Koizumi, and A. Sato, pp. 159–176. Tokyo: Univ. of Tokyo Press.

Watkins, L.R., L.E. Goehler, J.K. Relton, N. Tartaglia, L. Silbert, D. Martin, and S.F. Maier. 1995. Blockade of interleukin-1 induced hyperthermia by subdiaphragmatic vagotomy: Evidence for vagal mediation of immune–brain communication. *Neurosci. Lett.* **183**:27–31.

Yehuda, R., M.H. Teicher, R.K. Trestman, R.A. Levengood, and L.J. Siever. 1996. Cortisol regulation in posttraumatic stress disorder and major depression: A chronobiological analysis. *Biol. Psychiatry* **40**:79–88.

Standing left to right: Hans Reul, Robert Dantzer, Christopher Coe, Volker Stefanski, Steve Porges
Seated left to right: Per Jensen, Jeff Rushen, Adroaldo Zanella, Stafford Lightman

11

Group Report: Key Elements of Coping

J.M.H.M. REUL, Rapporteur

C.L. COE, R. DANTZER, P. JENSEN, S.L. LIGHTMAN, S.W. PORGES, J. RUSHEN, V. STEFANSKI, and A.J. ZANELLA

Keywords: defense systems, coping, motivation, autonomic nervous system, behavior

INTRODUCTION

We live in a world in which there is a continuous series of expected and unexpected challenges. How we respond to these challenges through behavioral, neuroendocrine, and autonomic systems reflects the mechanisms through which we cope with changes in our environment.

For many neurobiologists, concepts of "normality" are based on data from laboratory rodents, usually the rat. This animal is maintained in an artificial environment with little stimulation, next to no exposure to dangerous conditions, and no need to undergo difficult foraging expeditions for food. Laboratory rats have a very consistent rhythm of behavioral, autonomic, and neuroendocrine responses, including activity of the hypothalamic-pituitary-adrenal (HPA) axis and reproductive processes. We need to question, however, whether this is a fair portrayal of normality, whether it represents optimal conditions for good welfare and thus can serve as a paradigm for good animal husbandry. Although these animals are successful from the point of view of longevity, it would be presumptuous to assume that they have a better quality of life than their wild counterparts. However, much of the information presented in this report is necessarily based on work with rats. The brain structures mentioned refer to mammals, although some systems are identical in other vertebrates and many mechanisms are similar, even if the anatomical components differ.

Responses to challenge can be measured in several domains. The effects of these responses on the welfare, including the health, of the individual vary as a function of several factors, including the systems recruited to deal with the challenge, the species of the individual, the age of the individual, the context of the challenge, and the duration of the challenge.

No single response system provides an unambiguous window into the adaptive function of the challenge. Response systems such as the HPA axis, which has often been assumed to reflect a linear or unidimensional stress system, can be activated under normal adaptive situations not associated with stress, such as under conditions of increased metabolic demand or anticipatory to food intake. However, if the context, the adaptive behavior, and the interaction of a specific response system with other response systems (e.g., immune, autonomic, and endocrine system, brain) are observed, a different picture of functional adaptation emerges. If aspects of development and vulnerability, from both a behavioral and a neurobiological perspective, are introduced into the response profile, we start to see a complex collection of interactive response systems that coordinate to move the individual along a continuum from perceived threat to perceived safety. Insight into the phylogenetic history of species with regard to the development of their central nervous system (CNS) may help us to understand the challenge-specificity of, and the sequence in which, certain response systems are recruited. Possibly for some this can be even a *leitmotiv.* Moreover, it can provide insight into the interrelationships between the various physiological and behavioral response domains.

Our challenge is to unravel assumptions about quality of life and welfare from preconceptions of stress as a negative factor in the welfare of humans as well as wild and domestic animals. Acute challenges are part of the normal repertory of mammalian responses, and it seems that they are associated with negative or maladaptive responses in the organism when the capacity of coping responses of humans or other animals is exceeded.

ORGANIZATION OF DEFENSE SYSTEMS

To facilitate our discussion on the impact of challenges on the organism, we categorize challenges into different classes. The expression "challenge" was selected because, clearly, both negative and positive stimuli impact on the organism and recruit the same classes of response, that is, the autonomic, neuroendocrine, and motor systems.

The elicited adaptive process depends very much on the type of sensory and host defense system that is challenged. It is therefore important to understand how each defense system is structurally organized to account for the functional aspects of, for instance, the HPA axis response. In line with the distinction between teleceptors and interoceptors proposed a few decades ago by Garcia (Garcia and Ervin 1968), three different defense systems are described that are each characterized by specific afferent pathways to the CNS and a specific set of efferent pathways interconnected by challenge-specific integrative control mechanisms. Such systems serve not only to establish connections between specific stimuli and a given set of specific responses, they also allow the establishment of connections between several sensory attributes of the challenge and its consequences. Therefore, they structure the way the world is perceived and represented. The three defense systems are the immune defense system, the gut defense system, and the skin defense system (Dantzer 1991).

In the immune defense system, the immune system is challenged by a variety of antigens (e.g., viruses, bacteria, proteins, polysaccharides, dust). Activated cells of the immune system (e.g., macrophages) produce soluble mediators called cytokines that, above a certain threshold, activate specific circuits in the CNS. This activation occurs via afferent humoral and neural pathways, and results in a set of coordinated autonomic, neuroendocrine, and behavioral responses. These responses represent a reorganization of the organism's priorities to cope with pathogens. Furthermore the adaptive responses elaborated in the CNS are controlled by cytokines that are produced locally in the brain and act via various mediators, including prostaglandins and nitric oxide. In this context, the main purpose of the elevated levels of glucocorticoids is to inhibit the production of cytokines, both in the periphery and in the brain.

Another example of an interoceptive sensory system can be observed in the gut and is geared to enable the organism to handle edible and nonedible foods, based on their immediate gustatory characteristics and their long-term effects. Harmful or beneficial effects of food are monitored via neural (vagal) and humoral pathways, and they need to be related to the sensory qualities of the ingested food. In this context, the HPA axis activation in response to poison has a very different range of functions than the hormonal response to immune stimuli. It aims mainly at down-regulating the production of endogenous mediators of sickness. Apart from aversive effects exerted by ingested poisonous compounds, the afferent vagal and sensory nerves also signal pleasantly tasting compounds, precipitating positive sensations at the higher limbic and neocortical levels. In addition to this example observed in the gut, there are a variety of other interoceptive systems, that is, those monitoring the PO2, PCO2, glucose, osmolarity, Na^+, K^+, and other ions in the blood, plasma pH, filling of the stomach and gut, blood pressure, core body temperature, and others.

The skin defense system, an example of an exteroceptive sensory system, is activated by stimuli that threaten the integrity of the skin barrier. Its output consists of a set of species-specific defense reactions that are hierarchically organized, depending on the nature and intensity of challenges and the organism's ability to predict and control them. Processing of information within this system is extremely rapid, and associative abilities conform to the laws of classical conditioning. In the context of the activation of the skin defense system, activation of the HPA axis is necessary for mobilizing energy and modulating the strength of associative relationships.

In the real world, these systems are not independent. For instance, breaking down the skin barrier is likely to be associated with entry of pathogen microorganisms and activation of the immune system. Vagal afferents can modulate pain thresholds on the skin through solitary-spinal pathways. Chemical, electrical, and physiological activation of cardiopulmonary vagal, diaphragmatic vagal, and subdiaphragmatic vagal afferents have been reported either to facilitate or inhibit nociception in mammals (Randich and Gebhart 1992). Stimuli that are detected by any of the sensory organs are often highly specific, which was captured in the early ethological notion of key stimuli. Without previous training or experience, animals will react to different stimuli by recruiting specific behavioral responses (see, e.g., Goodenough et al. 1993). In addition, activation of these three different systems is certainly not the only way in which organisms can respond to demands in their internal and external milieu. In this context, some doubt whether it makes sense to make any classification system. However, classification is helpful in facilitating insight into how one defense system (e.g., the skin defense system in response to the threat of a predator) impacts on the functioning of another system (e.g., the immune defense system). Such insight cannot evolve without detailed knowledge of the organization of each defense system. In other words, it is not possible to describe influences of stress on the immune system without knowing how and why the immune system has established reciprocal relationships with the CNS. Similarly, it is not possible to describe the commonalities between these systems if there is no knowledge of the resources that need to be allocated to each of these systems and the way in which they can be modified by experience, including sensitization, habituation, and conditioning. This allocation strategy is optimized by evolution, so that gene propagation is maximized.

NEURAL, NEUROENDOCRINE, BEHAVIORAL, AND PSYCHOLOGICAL SUBSTRATES OF COPING

To enable an organism to cope with acute and chronic — aversive and positive — challenges as well as daily ongoing patterns of activities (e.g., feeding, foraging, grooming), it is equipped with a number of neural, physiological, and behavioral psychological systems. We felt that key elements of coping include the viscero-motor reflexes at the level of the vagal complex, the corticotrophin-releasing factor (CRF) driven systems (including the HPA axis), behavioral systems, and psychological mechanisms. This sequence of presentation has been chosen deliberately, as it reflects phylogenetically the rise in complexity, both in a multidimensional and integrative sense.

The Autonomic Nervous System, Focusing on the Ventral Vagal Complex

The phylogeny of the mammalian nervous system provides us with an important organizing principle to investigate and understand the mammalian response to challenges. The order in which neural control systems evolved to deal with challenges from within and outside of the organism, should help in understanding the sequence in which these adaptive systems are recruited.

Most models of reactivity have been based on the characteristics of the stimulus or the response systems and have not addressed the priorities and hierarchical integration of neural regulation. In addition, since most indices of challenge have relied on changes in visceral state, such as activation of either HPA axis or the sympathetic nervous system (SNS) and the adrenal medulla (SAM), little interest has been directed at the integrative nature of the neural regulation of the autonomic nervous system (ANS) and the integration with endocrine and immune functions. Thus, an understanding of the physiology of the nervous system and, especially, the phylogeny of the ANS becomes an initial starting point in our quest to understand stress, coping, and welfare.

An investigation into the specific phylogenetic history of the mammalian ANS provides insights into priorities, hardwiring, and flexibility of neurally mediated visceral reactivity to challenge. In global terms, although the regulation of the vertebrate ANS has progressed through several stages, vestigial functional response systems, with their requisite neuroanatomical structures, remain embedded in our nervous system. These ancient systems are capable of responding when the newer structures of the highly developed mammalian nervous system do not "remove" the organism from a potentially life-threatening event. Thus, implicit in the phylogenetic description is a survival-driven hierarchy of responses (Porges, this volume) identifiable by the reverse order in which the specific neural systems evolved.

The phylogenetic approach is applied to the neural regulation of the heart in the polyvagal theory (Porges, this volume). This approach emphasizes the phylogenetic origins of brain structures and neural systems that regulate the visceral state to promote both social and defensive behaviors. The model proposes that the physiological state limits the range of behavior and psychological experience.

The phylogenetic model provides an alternative entree to trigger the neural dissolution proposed by Jackson (1958). The polyvagal theory proposes that the sequence of dissolution can be triggered by a lack of success in removing the individual organism from a life-threatening event or situation. Thus, the nervous system is being driven by a neural system that detects whether the organism is safe or threatened. The physical characteristics of threat need not be real, but the nervous system must perceive the event as life-threatening on a metaphorical level. Thus, believing that one will be shot, thinking that a loved one has died, or even having one's laboratory taken away might produce a similar degrading of neural regulation. To simplify the model and put it into a human perspective, the first response to a challenge will be to use the neural regulation of the special visceral efferents and the myelinated vagus. This integrated system promotes (a) limited mobilization by modulating the vagal input to the heart (i.e., removing the vagal brake to increase metabolic output without sympathetic excitation), (b) calm states by re-engaging the vagal influence on the heart, and (c) the neural regulation of the face and head via special visceral efferent pathways related to vocalizations, intonations, ingestion, facial cueing, looking (eyelid opening), and head gestures. Thus the infants of humans (and several other mammalian species) look at their mothers, vocalize, and suckle as mechanisms to self-soothe and maintain a calm state.

The ventral vagal system, via myelinated pathways to the viscera and the striated muscles of the head, promotes social communication and self-soothing behaviors. If the ventral vagal system fails to cope with challenge, then a phylogenetically older system, the SNS, is triggered. The SNS and the adrenal medulla and cortex are recruited to promote mobilization behaviors such as fight–flight. If mobilization fails to move the individual into a safe environment, then the lowest level of this phylogenetically ordered hierarchy is triggered. The oldest system involves the dorsal vagal system, which via unmyelinated pathways creates a physiological state characterized by massive bradycardia and apnea to support immobilization. This immobilization system is characterized by a relatively hardwired visceromotor system that functions to reduce metabolic demands to conserve resources. This system is very adaptive and successful for reptiles. However, since the mammalian brain has a great dependency on oxygenated blood, the reduced cardiac output during this state will, within a short period, place the mammalian brain at great risk. Thus, although this system functions well for reptiles, who have relatively small brains, it is potentially lethal for mammals.

Although the lower motor neurons of both vagal systems are located in the brain stem, the new vagal system (i.e., at the level of the nucleus ambiguus) may be influenced by monosynaptic inputs via corticobulbar pathways from neocortex. Additionally, the dorsal vagal complex, consisting of the dorsal motor nucleus of the vagus and the nucleus of the tractus solitarius, receives direct and indirect influences from limbic and neocortical forebrain areas, such as those mediated via the neuropeptides, oxytocin and vasopressin. It is important to note that the oldest vagal system is capable of reflexive and adaptive conservation responses when not inhibited by higher level systems. Thus, visceroception does not always result in the same viscero-motor response.

The Forebrain Corticotropin-releasing Factor System and Glucocorticoid Hormone-binding Receptors: Significance for Autonomic, Neuroendocrine, and Behavioral Regulation

A major neuropeptidergic system residing in the forebrain of higher vertebrates including mammals is the CRF system. Neurons expressing the highest levels of CRF are located in parvocellular portions of the paraventricular nucleus (PVN) of the hypothalamus. Other nuclei containing CRF-expressing neurons include the bed nucleus of the stria terminalis (BNST) and the central nucleus of the amygdala (CeA). Apart from CRF, a related peptide, urocortin, is being expressed in the brain with a distinct localization pattern. CRF and urocortin bind to two types of CRF receptors, type 1 and type 2, the type 1 receptor being most relevant with regard to CRF actions. Data collected over the last twenty years from a broad array of experiments have provided conclusive evidence that CRF is the principal mediator of the HPA axis, autonomic, and behavioral response to stressful and other challenges, irrespective of immunoceptive, interoceptive, or exteroceptive nature. CRF neurons driving the HPA axis response reside in the PVN, whereas those stimulating sympathetic outflow and behavioral response are primarily located in the BNST, CeA, and dispersed throughout the neocortex. Paraventricular CRF neurons also project caudally to pontine and brain stem nuclei, including the sympathetic motor and raphe nuclei. Furthermore, CRF plays an important role in maintaining the diurnal rhythm in the activity of the HPA axis, which is driven by a vasopressinergic projection originating in the suprachiasmatic nucleus (SCN).

Following a stressful challenge and possibly also as a result of pleasant events, such as sexual activity, that are less studied, CRF release in distinct areas of the CNS leads to a coordinated sympathetic (elevated circulating levels of noradrenaline and adrenaline), HPA axis (elevated plasma glucocorticoid levels), and behavioral (freezing if in an unfamiliar environment or escape if stressed in the home cage) response. The responses run in a "coordinated" fashion as the fast sympathetic response supports the flight–freeze response, and the sympatho-medullary response facilitates the glucocorticoid response by rendering the adrenocortical cortex more sensitive to the action of ACTH. Thus, the glucocorticoid response is much slower (peak at 20–30 min post-stress onset) than the fast (seconds) autonomic and behavioral responses. Nevertheless, the glucocorticoid response is important to replenish energy stores or, if the challenge lasts for a longer time period (hours), to support metabolic capacity. In higher limbic and neocortical regions, glucocorticoids are critically involved in the acquisition of adaptive behavior (see below). Glucocorticoids shut down the HPA response through negative feedback actions, mainly at the level of the anterior pituitary and PVN.

Glucocorticoids also have a complex role in behavioral regulation. Glucocorticoid hormones bind to two types of receptors in the CNS: the mineralocorticoid receptor (MR) and the glucocorticoid receptor (GR) (Reul et al. 2000). MRs are mainly localized in the hippocampus whereas GRs are distributed ubiquitously in the brain with especially high concentrations in the PVN, hippocampus, and neocortex. The two glucocorticoid-binding receptors differ by at least a factor of 10 in binding affinity for the natural glucocorticoids, that is, cortisol and corticosterone, which is of major significance for the functional properties of these receptors. Because of their high affinity for binding natural glucocorticoids, MRs are already fully occupied by the minute levels of glucocorticoid (<10 ng/ml) circulating under baseline conditions during the circadian trough of HPA activity (in rats and mice, early morning; humans, late afternoon). GRs, which exhibit lower affinity for binding natural glucocorticoids than MRs, only become occupied extensively after stress and during the circadian peak of glucocorticoid secretion. These pharmacological differences form the basis for the different functional properties of MR and GR in the brain (Reul et al. 2000). The differential functional properties of MR and GR can be observed in HPA axis regulation, electrophysiological processes, autonomic output, and behavioral regulation (de Kloet et al. 1998).

Recent findings add another level of complexity to the role of CRH and MR in the control of the HPA axis. It was observed that psychological stressors, such as forced swimming and novelty, induce an increase in the density of MRs in the hippocampus of rats within 8 to 24 h (Gesing et al. 2001). Importantly, the effect of these stressors was mediated by CRH in the brain. Moreover, the stress-induced rise in MRs was associated with an enhanced MR-mediated inhibitory tonus on HPA axis activity (Gesing et al. 2001). These findings demonstrate not only that the regulatory set points of the HPA axis are dynamically organized, they also show a new role of CRH in stress regulation. Apparently, apart from its role in mediating the acute response to stress in terms of sympathetic and HPA activation and inducing behavioral changes, CRH activates an adaptive mechanism by increasing hippocampal MR levels, which leads to enhanced inhibitory control of the HPA axis (Reul et al. 2000; Gesing et al. 2001). Given that hippocampal MRs, apart from HPA control, are involved in a variety of other, stress-responsive, neurobiological mechanisms, including electrophysiological properties, long-term potentiation, serotonergic neurotransmission, and several types of behavior, the altered MR status induced by CRH may bring about adaptive adjustments in set points in many of these systems.

With regard to behavior, MRs play an important role in determining strategy in challenging situations, whereas GRs in the hippocampus are of critical importance in memory consolidation processes executed post-challenge (de Kloet et al. 1998). The latter property of GR is nicely illustrated by the phenomenon of acquired immobility in the forced swimming test. In this test, rats or mice are forced to swim for 15 minutes in 25°C warm water on the first day and reexposed to forced swimming for 5 minutes 24 hours later. During the 5 minute retest, the time is scored during which the animal remains "immobile" (a state of hardly any leg movements) in the water. The acquired immobility response is currently regarded as a behavioral adaptive response of the animal, presumably to save energy, because it "remembers" from the previous day that escape is impossible. This immobility response is highly dependent on the presence of glucocorticoids at concentrations sufficient to bind to GRs, because it is highly impaired in adrenalectomized animals as well as in animals treated with GR antagonists before or immediately after the first test. Thus, GRs are instrumental for the consolidation of this behavioral adaptive response. This is an interesting example of the important role of glucocorticoid hormones in attempting to cope with stressful challenge.

BEHAVIORAL PERSPECTIVES OF THE COPING RESPONSE

Behavioral systems are one of the primary means that animals have of responding to external challenges, especially challenges that are acute, for example, risk of predation. Behavioral responses are often highly sensitive and can be seen in response to those less severe challenges that may not provoke physiological responses. However, it is now widely accepted that behavioral responses are not simple expressions of an emotional state, as was once thought, but represent highly organized systems that have evolved to nullify or remove the challenge. The animals' responses to the likely presence of a predator, rather than being simple expressions of a "fear" state, involve behaviors aimed at detecting and avoiding the predator. For example, grazing ungulates intersperse their grazing with periods of scanning the environment, known in the literature as vigilance. For a number of species, it has been shown that the relative proportion of time spent grazing and time spent vigilant is fairly sensitive to the degree of predation risk. Increases in predation risk assessment resulting from actual sighting of a predator, previous encounters with predators in that place, or distance from cover, result in a reduction in grazing and an increase in vigilance. These low energy cost, antipredator responses are replaced by behaviors such as flight or attack following the close approach of the predator. Thus, the behaviors reflect the activation of an organized motivational system that has evolved to reduce the incidence of predation and cannot, in its entirety be interpreted as reflecting different levels of "fear."

One consequence of the complexity of response to challenge is that behavioral responses are usually highly specific to the particular type of challenge, and it seems unlikely that we shall find general, behavioral stress responses that are given in response to all types of challenge. For example, social isolation of domestic ungulates such as sheep, cattle, or pigs results in the performance of specific types of vocalization, presumably with the aim of reestablishing social contact. Such vocalizations are not necessarily given in response to other stressors that do not involve social isolation. In some species, vocalizations are part of antipredator behavior where they function as alarm calls. With regard to some primate species, such as vervet monkeys, the calls given are specific to a particular type of predator; different calls are given to ground predators as opposed to aerial predators. Moreover, playback experiments show that these alarm calls elicit different types of behavioral responses in the monkeys hearing these calls, suggesting that they provide information to other monkeys about the type of the predator. The same has been found in chickens, who emit specific calls for ground predators and aerial predators (Evans et al. 1993). In addition, high specificity in signaling has been found in other chick vocalizations, such as the food call (Evans and Marler 1994).

Behavioral responses to chronic stress are much less well understood; however, there is evidence for specificity here as well. For example, mice confined for a long period in a small cage by a door that they have observed to open, address much of their stereotypy to that door. Disruptions of daily rhythms, especially reductions in sleep, have been proposed as more general behavioral responses to chronic stress, but more data are needed in this research area.

COGNITIVE ELEMENTS OF IMPORTANCE FOR THE COPING PROCESS

From a psychological perspective, coping in humans has traditionally been described primarily as a cognitive process with significant ramifications for behavior and emotion regulation, that is, the emphasis has been placed on the appraisal or perceptual dimension of the individual's reaction. Most researchers consider coping to involve actions and brain function in order to manage or minimize problems. Here the goals of coping are quite clear: (a) behavioral responses that are oriented to resolving the problem, and (b) emotion regulation to enable the individual to remain in an emotional state conducive to solving the problem or continuing to gather new information for eventual resolution.

In human research, we must place equal emphasis on the resources that the individual brings to the situation and, in particular, the personality traits or more enduring dispositions that one has from prior experiences with challenging situations. Thus, for example, some have suggested that a prior history of successful experiences with coping leads to more vigorous and sustained attempts at coping in the future. This complex topic is often summarized under the term of having a high sense of self-efficacy, or a feeling that one is capable of being effective. Others have suggested that a dispositional bias toward believing that things will most likely turn out well is also a critical element of coping. This theoretical framework is often captured by the terms of optimism and pessimism, a belief system leading to positive or negative views about the future. These ideas are elaborated further in formulations that account for resilience in the face of adversity and sustained challenge, including the concept of hardiness. Hardiness is attributed by some, such as Kobasa (1979), to be dependent upon three primary traits: (a) control, an attribution of internal locus of control, (b) commitment, a persevering style of continually seeking the goal, often because of a higher order belief in the value of the goal, and (c) challenge, a cognitive bias toward interpreting demanding events as challenges rather than threats.

Such complex cognitive models and constructs cannot and probably should not be readily applied to how animals react and adapt to challenging situations, but they underscore the importance of including a cognitive framework in animal studies. We know that prior experiences of animals, for example, a history of winning or losing fights, continues to affect their reactions to new situations. Similarly, experience with a severe, uncontrollable event may have very persistent influence over responses to future challenges, a perspective often summarized by the term "learned helplessness." In some cases, it has been possible to establish neurobiological alterations that underlie these long-term biases in how animals react, such as a decrease in brain monoamine levels, which may then lead us to reevaluate whether the shift is just at the cognitive level or also includes lower-level brain stem-based changes. However, the importance of cognitive perspectives in animal studies can be found in other types of observations as well. For example, members of gregarious species, such as monkeys, react very differently to challenges when living in a social group than when confronting challenges alone. A specific example is the behavioral and hormonal reaction evinced by monkeys when seeing a predator such as a snake. When squirrel monkeys were presented with a snake as individuals or in pairs, there were strong behavioral reactions as well as a marked elevation of cortisol levels. However, when monkeys in social groups of six animals were presented with the same snake stimulus, there was no increase in adrenal activity, despite similar manifestations of fear reactions to the snake. Further exploration of this attenuation of adrenal responses to challenging situations revealed that the beneficial effects of social companions emerged when the group size exceeded three. Briefly, three seemed to be the defining element for a group in this species. One can imagine that the sense of being a part of group might be different across species, especially in those species that aggregate in herds.

Several decades of research have also emphasized the importance of three other factors in evaluating how animals react to challenging situations: (a) control, the ability to react in a way that changes the situation, (b) predictability, the capacity to anticipate the future and immediate consequences of one's actions, and (c) information feedback, the capacity of the animal to get information about the success or failure of it's actions. From this perspective, challenging situations are ones that are less controllable and predictable, ones where there is little feedback about the consequences of behavioral responses. Conversely, the challenges and demands of the situations are reduced progressively as the sense of control and predictability is enhanced. These ideas can be readily translated into a discussion about good welfare in animals and help to explain the reactions of animals. Novelty, at least temporarily, presents an animal with a situation that is less predictable. As familiarity with the situation grows, it ceases to be a challenging situation. This phenomenon is immediately recognized in a commonsense way when one talks about habituating or adapting an animal to a new cage or housing situation. Since animals in natural settings often encounter some degree of novelty on a regular basis, there is often debate about how much stimulation we should provide in captive environments controlled and regulated by humans. This concern about the possible deleterious effects of too little stimulation, that is, boredom, is often one of the more controversial items considered in discussions about animal welfare.

MOTIVATIONAL CONCEPTS OF COPING

Motivation constitutes a central aspect of coping with challenges in animals. When an animal is challenged by an external or internal stimulus, it will often need to perform behavior in order to cope with the challenge. The way in which animals perceive and interpret a particular stimulus, the way in which this is transformed into a behavioral output, and the way this behavior in its turn affects the tendency to perform different behaviors is studied within the framework of motivation. A more thorough description of motivation is given by Jensen (this volume).

It is imperative that motivational aspects of behavior are on a different explanatory level than neurally based aspects. Where possible, it is necessary to include available neurophysiological information to reach a more complete explanation of motivation; however, sometimes we have limited knowledge of the neural correlates and must use a black-box approach to the actual brain processes. Proper understanding of motivation, such as why the animal recruits a particular behavior in response to some stimuli rather than a different behavior, or why some species recruit one behavior in response to a specific stimulus and another species recruits a different behavior to the same stimulus, also requires knowledge of the evolutionary and ecological aspects of behavior.

The term motivational state is usually taken to mean the state at any moment of all the motivational systems of an animal (McFarland and Sibly 1975). A motivational system integrates stimuli from both external and internal sources and organizes behavior output towards a specific goal. In this context, the term goal can be given a very wide meaning, covering everything from a physiological set-point to a cognitively represented goal, sometimes referred to as an "expectation copy." During evolution, animals have been equipped with limited set-ups of behavioral programs (often consisting partly of motor patterns which are more or less inflexible) and mechanisms whereby such programs (the way the motor patterns are assembled to adaptive behavior) can be adaptively modified in response to experience. These behavioral programs can be recruited in the attainment of the goal. The thwarting of such programs may in itself concern pose challenges to animals.

There are at least two ways in which behavioral programs may be thwarted and thus coping be made difficult: (a) by not allowing the animal to perform the motivationally relevant motion patterns, and (b) by not allowing the animal to obtain the relevant goal. The extent to which each of these coping problems is relevant will vary in different motivational systems. Often, down-regulation of motivation may require the performance of the relevant motion patterns, and sometimes the mere performance may actually be sufficient (through proprioceptive feedback). For example, hens appear to be able to down-regulate dust-bathing motivation by performing dust-bathing movements on a wire floor. In other cases, attaining the goal seems to be the important part of down-regulation, whereas the actual motor patterns performed are less motivationally relevant. An example of this is aggressive behavior, for example, in territory defense. This involves two extreme types of motivational control systems: (a) those that are controlled exclusively by internal factors, where external stimulation and feedback plays a small role; (b) those that are controlled principally by external factors and where there are no important internal causal factors and no major physiological variables that are controlled by performing the behavior.

The concept of needs, in particular needs to show certain behaviors, was developed to emphasize the importance of internal factors. This means that the behavior may be elicited by, for example, hormones, and that the feedback regulation may be primarily proprioceptive. However, there is a risk in categorizing and splitting the behavioral repertoire into needs and non-needs, since not expressing a particular motivated behavior in specific situations may well be a source of poorer welfare, regardless of what has released the behavior. For example, antipredatory flight behavior is, in most species, not dependent on internal causal factors but is released exclusively as a result of external stimuli from a predator. There is no reason to expect that animals would have a need to perform antipredatory flight behavior in the absence of a predator and, therefore, this would not qualify as a need according to this classification. However, an animal may encounter stimuli which are interpreted as a predator, for example, a human visitor on a farm or a cleaning brush in a battery cage. Once the flight behavior is motivated, the animal may have a strong need to perform it. It would therefore be wrong to refer to needs only when they are triggered by internal stimuli. A proper determination of the needs of animals therefore requires understanding of the motivational systems. Some needs will only be possible to fulfill by performing motivationally relevant behavior.

In neural terms, motivational systems can be seen as neurobiological systems that integrate sensory information, visceral input, and behavioral responses, including the proprioceptive components of behavioral activities. They are therefore similar to the emotional brain circuits that are described by Panksepp (this volume) and include the basic control systems that have been described in the first part of this chapter. Although the information that is processed by motivational systems is specific to the motivational system under consideration, there are certain cognitive elements of this information that are common to different motivational systems. They are based on the basic dimensions through which the brain, as an active information processing system, represents the external world. These basic dimensions include novelty, predictability, controllability, the hedonic valence of the situation and, at least in human beings, the conformity to social and individual mores of the type of response that the subject emits.

STEREOTYPY: AN ATTEMPT TO COPE WITH CHALLENGE RESULTING IN POOR WELFARE

The discrepancies concerning the use of stereotypies as indicators of poor welfare can be partially explained by the diversity of behavioral sequences associated with these behavioral abnormalities. Repetitive behavioral patterns, which have no specific goal or function, can vary widely. Some stereotypies may involve oral manipulation (e.g., sham chewing in sows, cribbing behavior in horses); other stereotypies are primarily locomotor (e.g., pacing in polar bears or autistic infants). Scientific work has been carried out to understand the factors that play a role in the development and maintenance of stereotypies. Stereotypies are manifested as a disruption of behavioral sequences and also as a simplification of complex sequences that are repeated in an apparently functionless way. In some animal species, maternal deprivation has been identified as a factor promoting the development of stereotypies. Our understanding of the neurobiological basis of environmentally induced stereotypies is emerging from several lines of research. Barren social environment in horses, as well as early weaning in pigs and mink, have been implicated in inducing the development of stereotypies. Animals that develop high levels of stereotypies showed a down-regulation of mu and kappa (not delta) opioid receptor sites in the frontal cortex (Zanella et al. 1996), hypothalamus, and hippocampus (Loyens). Animals selected to develop high levels of stereotypies also showed a depletion of dopamine in the frontal cortex when compared with control animals (Zanella and Mason 1997). Higher levels of the kappa-opioid agonist dynorphin 1-17 have been found in prefrontal cortex of sows kept in individual crates than in loose housed sows (Zanella et al. 1998). Thus, dopamine and opioids seem to be neurobiological correlates of stereotypic behavior.

MEASURES OF COPING

Measurement of Autonomic Nervous System Function

The ANS is phylogenetically organized and that organization provides a model for identifying which response systems would change during challenge. The ANS as a response system reacting to challenge is neither a linear arousal system characterized by degrees of sympathetic activation nor a balance system characterized by the cooperative interaction of sympathetic and parasympathetic inputs. However, there are aspects of linear arousal and autonomic balance in the relation of homeostatic function and in the maximization of specific visceral states. By contrast, autonomic responses to challenge recruit specific neural pathways, which regulate and prioritize autonomic function in the service of acute and chronic survival needs. The polyvagal theory (Porges 1995, 1998) emphasizes the phylogenetically ordered hierarchy as metaphor for the hierarchy of responses to challenges. Thus, specific measures can be identified that are mapped into the neural sequence of the phylogenetic order, which, starting with the most recent system in mammals, is characterized by a myelinated vagus, preceded by a spinal sympathetic system, and initiated by an unmyelinated vagus.

According to this model, specific physiological processes can be measured that map into each of the three global phylogenetic "stages" of neural regulation of the ANS. The model works best when discussing the functioning of the heart and the gut. At the level of the heart, the myelinated vagus inhibits the sympathetic input. The output of the myelinated vagus can be accurately and dynamically evaluated by quantifying the amplitude of a respiratory rhythm observed in the heart rate pattern. This mechanistic output is determined by the impact of the myelinated vagal efferent fibers to the sinoatrial node of the heart, and the rhythm in this output is determined by the interneuronal communication between the nucleus ambiguus and the nucleus of the solitary tract. The amplitude and rhythm of this efferent input to the sinoatrial node is determined by brain stem structures monitoring the visceral state via the nucleus tractus solitarius. This rhythm is known as respiratory sinus arrhythmia. Interestingly, upper motor neurons on cortex, as well as subcortical areas, can inhibit and regulate the lower motor neurons of this system. The system provides a rapid and controlled modulation of the predominant vagal input to the heart and an inhibition of the sympathetic input to the heart. Functionally, this system works as a brake that can modulate the cardiac output in a rapid and controlled manner.

The vagal brake described above is functional for mild acute challenges. However, when increased metabolic activity is required for prolonged mobilization strategies, the vagal brake is removed and the SNS is stimulated. This system increases cardiac contractility and peripheral blood flow changes. In addition, at the level of the gut, the SNS inhibits the tonic vagal actions required for appropriate digestion. When mobilization does not fulfil the need of the organism and move the organism from a threatened environment to a state of safety, a primitive unmyelinated vagal system is initiated. This immobilization system is reflected in conservation actions, such as bradycardia, apnea, and defecation. Thus, a sequence can be observed in the ANS, from a protective and regulated vagal brake that inhibits sympathetics at the level of the heart, to a system of mobilization that recruits increased cardiac output and inhibits gut functions (i.e., diverts metabolic resources to mobilization strategies), to a conservative response that is adaptive for more primitive organisms such as reptiles (which do not require much oxygen) but may be potentially lethal for mammals (i.e., death associated with death feigning).

Measurement of HPA Axis Activity

The HPA axis displays two modes of activity profiles. First, in humans and most other animals, the HPA axis shows a circadian variation in activity during the day with higher levels of ACTH and, depending on the species, corticosterone or cortisol circulating during the behaviorally active period. An exception to this is the dog, in which no diurnal but rather episodic elevations in HPA activity are observed. Normally, in humans and most other animals, there is a marked difference between the amplitudes of ACTH and glucocorticoid hormones attained at the diurnal peak: the amplitude of glucocorticoids is much higher than that of ACTH. The reason for this is mainly that the increased sympathetic input to the adrenal medulla occurring during the behaviorally active period makes the adrenal cortex more sensitive to adrenocorticotrophic signals. Second, superimposed on the diurnal variation in HPA activity are elevations in hormonal output caused by stress, metabolic demands, inflammatory and immune challenges, and positive events, including sexual activity.

In humans and other animals living under chronically aversive conditions ("chronic stress"), continuously elevated levels of HPA hormones have often been observed, primarily during the trough of the circadian cycle. Hypersecretion of HPA hormones has also been observed in patients suffering from major depression and other psychiatric diseases. To address at which level(s) the regulation of the HPA axis is disturbed and to discern the mechanism(s) underlying the chronically elevated HPA output, a number of approaches have been developed:

1. Circadian levels of ACTH and glucocorticoid hormones in plasma can be monitored. In addition, the biologically relevant "free" fraction of glucocorticoids can be determined in saliva, in blood by *in vivo* microdialysis, or in tissue such as brain (Linthorst et al. 1994, 1995; Reul et al. 2000).
2. Reactivity of the HPA axis to adverse events can be addressed with a potentially stressful challenge test. For humans, for instance, public speaking may be a stressful event. Experimental animals, such as rats and mice, can be subjected to a variety of challenges. Here, a distinction is often made between so-called physical (e.g., exposure to ether or a cold environment), psychological (e.g., exposure to a novel environment, exposure of a mouse to a rat or cat), social (e.g., exposure of an adult male rat to an unfamiliar resident male), or mixed physical/psychological (e.g., forced swimming) challenges. Larger animals, for instance sheep, can be much disturbed by a barking dog.
3. The status of the glucocorticoid negative feedback at the level of the anterior pituitary can be addressed with the dexamethasone suppression test. Dexamethasone is a synthetic glucocorticoid hormone that accumulates poorly in the brain.
4. The status of CRF receptor function at the anterior pituitary level can be tested with a CRF challenge test in which CRF is administered intravenously and the response in plasma ACTH and glucocorticoids is assessed (Holsboer et al. 1986; Rothuizen et al. 1993). Typically, depressed patients show a blunted hormonal response to a CRH challenge, most likely due to desensitization of CRF receptors at the anterior pituitary level.
5. The combined dexamethasone/CRF test is often used to investigate the HPA axis in patients suffering from major depression or other stress-related psychiatric diseases. In this test, patients or probands receive a single oral dose (1–1.5 mg) of dexamethasone at 23.00 h followed by an intravenous injection of 100 μg human CRF at 15.00 h the following day (Holsboer et al. 1987). Basically, this allows the hypothalamic drive of the HPA axis in the face of glucocorticoid negative feedback at the pituitary level to be tested.

The advantage of the combined dexamethasone/CRF test compared with the dexamethasone suppression test is that it is much more sensitive in disclosing HPA axis disturbances in depressed patients: more than 80% of depressed patients show an abnormal dexamethasone/CRF test outcome, whereas only 45–50% of the patients display an early escape in the dexamethasone suppression test. During the course of antidepressant treatment, the dexamethasone/CRF test outcome normalizes just prior to the amelioration of the clinical symptoms, suggesting that normalization of the HPA axis disturbance is a prerequisite for the clinical success. Indeed, studies have shown that a positive test outcome is strongly associated with the long-term remittance of the depressive psychopathology (Zobel et al. 2001). Thus, the combined dexamethasone/CRF test could be used as a tool to monitor the efficacy of antidepressant treatment. Studies in rats have shown that a pivotal step in the effects of antidepressants on the HPA axis is an upregulation of MRs in the hippocampus, which leads to an enhanced tonic inhibitory influence on CRF-containing parvocellular neurons in the PVN. The secondary consequences of this effect on MR include a dampened reactivity of the HPA axis to stress, an upregulation of brain and anterior pituitary glucocorticoid receptor, and a reduction in adrenal size (Reul et al. 1993, 1994; for reviews, see Reul et al. 2000; Holsboer 2000).

Measurement of the Immune System

Significant physiological alterations associated with coping take place in the immune system. The immune system is a complex defense system consisting of many different subsystems. The immune system can roughly be divided into specific (also called "adaptive") and unspecific ("innate") as well as in humoral and cellular defense subsystems. Since pathogens exist in many different forms, a wide variety of defense mechanisms have evolved to combat infection. An appropriate assessment of immune competence therefore requires quantitative (e.g., cell numbers) as well as qualitative (e.g., activity levels) information on each of the subsystems. It is important to recognize that the blood (the most common source of specimen in many mammalian species including humans) is usually not the site of immunological reactivity. Most immunological reactions take place in lymphatic organs, such as the spleen and the lymph nodes. Nevertheless, because immune cells are transported through the blood to lymphatic organs, skin, and other body compartments, the blood can provide a relevant diagnostic window into the functional state and migration pattern of many important immune cells.

A large number of studies in animals and humans indicates that challenging social and nonsocial conditions (often associated with low predictability or loss of control) are linked to major immunological changes. For example, studies by Laudenslager and colleagues (1983) showed that inescapable electroshock reduces lymphocyte proliferation in laboratory rats but escapable shock does not. In social confrontation studies in rats and other mammals, profound immunological changes were observed in the defeated (subdominant) animals (Raab et al. 1986; Stefanski 2000; Stefanski and Engler 1999; von Holst 1998). In monkeys, profound immunological changes were recorded after disruption of social relationship (e.g., separation from companions or mother; Coe 1993). Reduced proliferative response of lymphocyte to mitogenic challenge, decreased activity levels of NK cells, altered cytokine production pattern, or lowered numbers of blood T cells (indicating an altered migration pattern) have been reliably observed. In addition to these rather suppressive immunological effects, an activation of immune subsystems can also occur on the cellular level, for instance, by the release of neutrophiles into circulation (Stefanski and Engler 1998) and enhancement of phagocyte activity of monocytes or macrophages. It can, however, also occur on the molecular level via the induction of cytokine receptors and CD4 molecules on the cell surface (Wiegers et al. 1995, 2000, 2001; Wiegers and Reul 1998). As with many other physiological responses to environmental challenges, the temporal dynamics (e.g., duration of confrontation or separation) may crucially determine the immunological outcome (Stefanski and Engler 1998). In many mammalian species, challenging social and nonsocial conditions are associated with increased activity levels of the HPA and/or SAM systems. Both systems produce molecules, such as CRF, ACTH, glucocorticoids (corticosterone, cortisol), norepinephrine, and epinephrine that are capable of interacting with immune cells. Lymphocytes also bear receptors for, and respond to, many other molecules that are released during stressful conditions, such as endogenous opioids or prolactin. In addition, sympathetic innervation of lymphoid organs may directly influence lymphocyte function. The complexity of neuroendocrine mediators that interact with the immune system may provide the basis for understanding the variability in immune function in challenging social and nonsocial conditions.

Measurement of Cognitive Abilities

Assessment of coping processes requires measurement not only of behavioral, physiological, and immune responses to challenges but also of the way the challenge is appraised. The challenging aspect of the situation can be assessed, based on the degree of predictability and controllability that is accessible to the individual exposed to it. Such assessment is important in view of the well-demonstrated consequences of prediction and control in psychological and physiological aspects of welfare.

FUTURE RESEARCH AND RECOMMENDATIONS

Future studies should rely on multiple measures when trying to gauge how an individual reacts in a stressful or challenging situation. This conclusion reflects the inherent difficulty in specifying a single representative or index response for the overall psychological and physiological reaction. In addition, the term stress continues to be used so broadly, and often inappropriately, that it will ultimately be more useful to describe carefully the specific circumstances that elicit specific changes in behavior or physiology.

Ultimately, the goal must be to understand better the organizing principles that integrate these separate elements of the organism's response to challenging stimuli and situations. This understanding will involve an appreciation of how individuals prioritize behavioral responses to the demands of the situation, as well as how the nervous system prioritizes the body's physiological responses. During times of challenge, the normal hierarchy of needs and responses may also become disrupted, initially to facilitate coping, but subsequently through the disorganization that occurs in the context of pathology and maladaptive reactions.

To appreciate the adaptive or maladaptive nature of behavioral and physiological adjustments during the response to challenge, we still need to learn more about the relationship to energy balance, resource allocation, and various motivational states. That is, to what extent do certain behavioral and physiological responses reflect a trade-off between short- and long-term gain?

One important area of study is the role of experience in shaping future responses, especially in the developing organism. It has been known for a long time that coping strategies are strongly influenced by prior success or failure. In addition, there are critical periods in development when the individual is particularly sensitive to such experiences. Further, the very young and the old individual may have fewer coping options available, both at the behavioral and physiological level. For example, the young infant in a social species may have to rely on recruiting aid from others. The regulatory tolerable range determination for the critical physiological systems may also be less stable than in the healthy adult.

The previous points underscore the need to consider acute responses within a larger context. Sometimes the value may be apparent only when appreciated with respect to long-term outcomes. Depending upon the response, a life span perspective or comparison of individual versus group needs may be needed. Here one should anticipate marked species differences, such as between solitary and gregarious animals, or between short- and long-lived ones.

There is an obvious need for those interested in how individuals react to challenge to continue these attempts at communicating across disciplines. In particular, this communication is needed to develop common terms and concepts. Progress in the past has sometimes been hampered by borrowing terms and applying them in either simplistic or inappropriate ways.

REFERENCES

Coe, C.L. 1993. Psychosocial factors and immunity in nonhuman primates: A review. *Psychosom. Med.* **3**:298–308.

Dantzer, R. 1991. Stress and disease: A psychobiological perspective. *Ann. Behav. Med.* **13**:205–210.

De Kloet, E.R., E. Vreugdenhil, M.S. Oitzl, and M. Joels. 1998. Brain corticosteroid receptor balance in health and disease. *Endocr. Rev.* **19**:269–301.

Evans, C.S., J.M. Macedonia, and P. Marler. 1993. Effects of apparent size and speed on the response of chickens, *Gallus gallus*, to computer-generated simulations of aerial predators. *Animal Behav.* **46**:1–11.

Evans, C.S., and P. Marler. 1994. Food calling and audience effects in male chickens, *Gallus gallus*: Their relationships to food availability, courtship, and social facilitation. *Animal Behav.* **47**:1159–1170.

Garcia, J., and F.R. Ervin. 1968. Gustatory-visceral and telereceptor-cutaneous conditioning. Adaptation in the internal and external milieux. *Comm. Behav. Biol.* **1 (Part A)**:389–415.

Gesing, A., A. Bilang-Bleuel, S.K. Droste, A.C.E. Linthorst, F. Holsboer, and J.M.H.M. Reul. 2001. Psychological stress increases hippocampal mineralocorticoid receptor levels: Involvement of corticotropin-releasing hormone. *J. Neurosci.* **21**:4822–4829

Goodenough, J., B. McGuire, and R. Wallace. 1993. Perspectives on Animal Behaviour. New York: Wiley.

Holsboer, F. 2000. The corticosteroid receptor hypothesis of depression. *Neuropsychopharmacology* **23**:477–501.

Holsboer, F., A. Gerken, U. Von Bardeleben, W. Grimm, H. Beyer, O.A. Müller, and G.K. Stalla. 1986. Human corticotropin-releasing hormone (CRH) in patients with depression, alcoholism, and panic disorder. *Biol. Psychiatry* **21**:601–611.

Holsboer, F., U. von Bardeleben, K. Wiedemann, O.A. Müller, and G.K. Stalla. 1987. Serial assessment of corticotropin-releasing hormone response after dexamethasone in depression: Implications for pathophysiology of DST nonsuppression. *Biol. Psychiatry* **22**:228–234.

Jackson, J.H. 1958. Evolution and dissolution of the nervous system. In: Selected Writings of John Hughlings Jackson, ed. J. Taylor, pp. 45–118. London: Stapes Press.

Kobasa, S.C. 1979. Stressful life events, personality, and health: On inquiry into hardiness. *J. Personality Soc. Psychol.* **37**:1–11.

Laudenslager, M.L., S.M. Ryan, R.C. Drugan, R.L. Hyson, and S.F. Maier. 1983. Coping and immunosuppression: Inescapable but not escapable shock suppresses lymphocyte proliferation. *Science* **221**:568–570.

Linthorst, A.C.E., C. Flachskamm, F. Holsboer, and J.M.H.M. Reul. 1994. Local administration of recombinant human interleukin-1β in the rat hippocampus increases serotonergic neurotransmission, hypothalamic-pituitary-adrenocortical axis activity and body temperature. *Endocrinology* **135**:520-532.

Linthorst, A.C.E., C. Flachskamm, P. Müller-Preuss, F. Holsboer, and J.M.H.M. Reul. 1995. Effect of bacterial endotoxin and interleukin-1β on hippocampal serotonergic neurotransmission, behavioral activity, and free corticosterone levels: An *in vivo* microdialysis study. *J. Neurosci.* **15**:2920–2934.

McFarland, D.J., and R.M. Sibly. 1975. The behavioural final common path. *Phil. Trans. R. Soc. Lond. B* **270**:265–293.

Porges, S.W. 1995. Orienting in a defensive world: Mammalian modifications of our evolutionary heritage. A polyvagal theory. *Psychophysiology* **32**:301–318.

Porges, S.W. 1998. Love: An emergent property of the mammalian autonomic nervous system. *Psychoneuroendocrinology* **23**:837–861.

Raab, A., R. Dantzer, B. Michaud, P. Mormède, K. Taghzouti, H. Simon, and M. Le Moal. 1986. Behavioural, physiological, and immunological consequences of social status and aggression in chronically coexisting resident-intruder dyads of male rats. *Physiol. Behav.* **2**:223–228.

Randich, A., and G.F. Gebhart. 1992 Vagal afferent modulation of nociception. *Brain Res. Rev.* **17**:77–99.

Reul, J.M.H.M., A. Gesing, S. Droste, I.S.M. Stec, A. Weber, C. Bachmann, A. Bilang-Bleuel, F. Holsboer, and A.C.E. Linthorst. 2000. The brain mineralocorticoid receptor: Greedy for ligand, mysterious in function. *Eur. J. Pharmacol.* **405**:235–249.

Reul, J.M.H.M., M.S. Labeur, D.E. Grigoriadis, E.B. De Souza, and F. Holsboer. 1994. Hypothalamic-pituitary-adrenocortical axis changes in the rat after long-term treatment with the reversible monoamine oxidase-A inhibitor moclobemide. *Neuroendocrinology* **60**:509–519.

Reul, J.M.H.M., I. Stec, M. Söder, and F. Holsboer. 1993. Chronic treatment of rats with the antidepressant amitriptyline attenuates the activity of the hypothalamic-pituitary-adrenocortical system. *Endocrinology* **133**:312–320.

Rothuizen, J., J.M.H.M. Reul, F.J. Van Sluijs, J.A. Mol, A. Rijnberk, and E.R. De Kloet. 1993. Increased neuroendocrine reactivity and decreased brain mineralocorticoid receptor-binding capacity in aged dogs. *Endocrinology* **132**:161–168.

Stefanski, V. 2000. Social stress in laboratory rats: Hormonal responses and immune cell distribution. *Psychoneuroendocrinology* **4**:389–406.

Stefanski, V., and H. Engler. 1998. Effects of acute and chronic social stress on blood cellular immunity in rats. *Physiol. Behav.* **5**:733–741.

Stefanski, V., and H. Engler. 1999. Social stress, dominance, and blood cellular immunity. *J. Neuroimmunol.* **94**:144–152.

Von Holst, D. 1998. The concept of stress and its relevance for animal behavior. *Adv. Study Behav.* **27**:1–131.

Wiegers, G.J., M.S. Labeur, I.E.M. Stec, W.E.F. Klinkert, F. Holsboer, and J.M.H.M. Reul. 1995. Glucocorticoids accelerate anti-T cell receptor-induced T cell growth. *J. Immunol.* **155**:1893–1902.

Wiegers, G.J., and J.M.H.M. Reul. 1998. Induction of cytokine receptors by glucocorticoids: Functional and pathological significance. *Trends Pharmacol. Sci.* **19**:317–321.

Wiegers G.J., I.E.M. Stec, W.E.F. Klinkert, A.C.E. Linthorst, J.M.H.M. Reul. 2001. Glucocorticoid hormones enhance or inhibit anti-T-cell receptor-induced T-cell proliferation as a function of cell density. *Neuroendocrinology* **173**:139–148.

Wiegers, G.J., I.E.M. Stec, W.E.F. Klinkert, J.M.H.M. Reul. 2000. Glucocorticoids regulate TCR-induced upregulation of CD4: Functional implications. *J. Immunol.* **164**:6213–6220.

Zanella, A.J., D.M. Broom, J. Hunter, and M. Mendl. 1996. Brain opioid receptors in relation to stereotypies, inactivity and housing in sows. *Physiol. Behav.* **59**:769-775.

Zanella, A.J., P. Brunner, J. Unshelm, D.M. Broom, and M.T. Mendl. 1998. The relationship between housing and social rank on cortisol, β-endorphin and dynorphin (1-13) secretion in sows. *Appl. Animal Behav. Sci.* **59**:1–10.

Zanella, A.J., and G. Mason. 1997. Neurophysiological correlates associated with behavioral traits in mink. 3rd North Am. Regional Meeting of the Intl. Soc. for Applied Ethology, Lennoxville, Canada. www.usask.ca/wcvm/herdmed/applied-ethology/ISAE/ISAEabstrt/zanella.html 25 October 2001.

Zobel, A.W., T. Nickel, A. Sonntag, M. Uhr, F. Holsboer, and M. Ising. 2001. Cortisol response in the combined dexamethasone/CRH test as predictor of relapse in patients with remitted depression: A prospective study. *J. Psychiatr. Res.* **35**:83–94.

12

Intrinsic and Extrinsic Factors Modulating Physiological Coping Systems during Development

P.M. PLOTSKY[1], M.M. SANCHEZ[1], and S. LEVINE[2]

[1]Stress Neurobiology Laboratory, Department of Psychiatry and Behavioral Sciences, Emory University School of Medicine, Atlanta, GA 30322, U.S.A.
[2]Department of Psychiatry, Center for Neuroscience, University of California at Davis, Davis, CA 95616, U.S.A.

Keywords: neuroplasticity, stress, hippocampus, glucocorticoids, rats, monkeys

ABSTRACT

The nature–nurture debate has matured into an appreciation that the genomic blueprint for constructing a central nervous system does not unfold in a vacuum but is instead constantly subjected to feedback by gene products as well as by external influences. Intrinsic factors inherent in the normal developmental process are likely to define the envelope or phase space of coping strategies available to an individual. Extrinsic factors acting in an experience-dependent and experience-expectant fashion elicit patterns of synaptic activity as well in humoral factors that potentially limit the range of actual coping strategies available to an individual. In many species, the primary caregiver serves as the interface between dependent offspring and the external environment for a variable, species-typical period of time. Thus, perinatal factors may serve to "bias" or "program" developing neurocircuits involved in the perception and evaluation as well as the response to external events. Such environmentally based biasing, acting via mechanisms involved in learning and memory processes, potentially alters the coping strategies of the offspring in a way which confers some survival advantage.

INTRODUCTION

Successful adaptation of an individual organism to its environment (i.e., coping) occurs through multiple processes beginning in the prenatal period, continuing through the neonatal

Coping with Challenge: Welfare in Animals including Humans
Edited by D.M. Broom © 2001 Dahlem University Press

period, and extending throughout life. Underlying the development and implementation of coping strategies is the ability of the individual to perceive, evaluate, and respond to its environment. It is long past time to replace the age-old nature–nurture controversy with the realization that the genetic or biological substrate with which an organism is born is modifiable by nongenetic factors or, put another way, the full expression of a particular phenotype is shaped by a complex interaction of genetic and environmental factors. This thesis flows from the following underlying assumptions on which the remainder of our treatise is based. First, although a given behavior, neuroendocrine response, or coping strategy (e.g., temperament) may have a heritable component, it does not necessarily follow that a complex output in any of these domains is entirely attributable to an inherited gene or set of genes. Second, it is imperative to recognize that even though genes probably do not code for behavior, neuroendocrine response, or coping strategy in a direct way, this does not exclude the possibility that an individual gene cannot play a critical role in the expression of an output in any of these domains. Individual genes code for specific proteins (e.g., enzymes, structural proteins, trophic factors, neurotransmitters, receptors) while a given behavioral or other output is generated by neural circuits composed of multiple, interconnected neuronal elements made of these proteins. Third, it should be appreciated that most complex outputs and behavioral dysfunctions are likely to be polygenetic and not regulated by a single gene. The development, differentiation, function, and connectivity of each neuron is dependent on the coordinated activity of numerous genes. The products of these genes interact with one another, their local environment, and extrinsic stimuli giving rise to coordinated neuronal activation subserving some particular output or suite of outputs underlying coping with environmental challenges.

In this brief treatise, we will discuss intrinsic and extrinsic factors underlying this adaptive process, with an emphasis on extrinsic factors, as it relates to anatomical and functional changes in the developing central nervous system and the long-term consequences of these adaptations on relevant outputs — behavioral, autonomic, and endocrine — underlying coping. Finally, we will explore the potential costs and benefits to the individual, sometimes via the social group, of these adaptations. This is not meant to be a comprehensive document, but merely a working roadmap of our current state of knowledge, shortcomings of past and current formulations, and potential avenues for future explorations. To that end, we will focus on selective illustrative examples utilizing references to broader reviews for further edification. We must begin, however, with the basic housekeeping task of providing a defensible operational definition of what we mean when we use the terms stress and coping.

OPERATIONAL DEFINITIONS OF STRESS AND COPING

Stress and coping are rather fluidly defined in the spheres of biology and psychology. The general concept of stress in biology as a "nonspecific response of the body to any demand" was introduced by Selye (1936, 1974). This conceptualization was subsequently modified by J.W. Mason (1971, 1975), among others, who successfully argued that there was not a general, nonspecific stress response. A stress or stressor involves an actual or perceived threat to an organism, a failure of expectations, prediction, or the absence of a sense of control (Levine and Ursin 1991; Plotsky 1991; von Holst 1998). The dependent variables, usually measured as indices of the stress response, include pituitary–adrenal hormonal secretion, behavioral changes, and autonomic nervous system measures such as heart rate, blood pressure, and systemic catecholamine secretion.

Coping has been defined as the utilization of available strategies and resources to reduce the impact of a stressor successfully or, more globally, to meet daily environmental challenges (Koolhaas et al. 1999; Wechsler 1995). Failure or "bad" coping is not an option, as the alternative to coping is death. Interestingly, many of the same dependent measures are employed as indices of coping, and it is clear that coping depends heavily upon building an accurate internal representation of the world to enable the organism to develop a sense of predictability and controllability (Bohus et al. 1987; Weiss 1968, 1971a,b). The physiological and psychological costs of coping have been termed allostatic load, which implies that repeated activation of allostatic systems over a lifetime involves constant trade-offs in the utilization of an individual's physiological and psychological resources (McEwen 1998; Sterling and Eyer 1988). Numerous factors influence the ability of organisms to interpret environmental events and to marshal appropriate coping responses to them. It is important to appreciate that what might appear to be maladaptive in one situation may prove to be quite adaptive in another situation, suggesting that those individuals with the widest envelope of potential coping strategies from which to choose when confronting an environmental event may exhibit increased fitness (narrowly defined as increased reproductive success).

NEUROPLASTICITY IN THE IMMATURE AND ADULT NERVOUS SYSTEM

During the perinatal period, external signals from the environment interact with the concurrently unfolding genetic blueprint for construction of the central nervous system via processes including activity-dependent plasticity, thus giving rise to a stable, individual phenotype governing perception of and responsiveness to salient features of the environment (McGue and Bouchard 1998; Sroufe 1997; Rutter et al. 1997). This basic role of experience-expectant and experience-dependent stimuli in the "programming" of the developing nervous system has been best illustrated in developing and adult cortical sensory processing regions (Ben-Ari 1995; Constantine-Paton et al. 1990; Diamond et al. 1993; Katz and Shatz 1996; Kolb and Whishaw 1998; Shatz 1990; Wheal et al. 1998).

It is important to develop a conceptualization of how a changing palette of environmental stimuli are interpreted by an individual and converted to stable alterations in the structure and function of the nervous system. Of particular interest are findings of the repeated involvement of the cyclic adenosine monophosphate (cAMP) and neurotrophic factor cascades in transduction of extrinsic stimuli into alterations in patterns of gene expression, remodeling neurocircuit structure and function (i.e., neuroplasticity), and in neuronal survival (Duman et al. 1999; Gould et al. 2000; McEwen 1999, 2000; Thoenen 1995). Additional candidate inter- and intracellular signaling cascades and details of their molecular mechanisms are being identified at an increasing rate, thus it would be premature to emphasize any particular cascade at this time. Among the most well-studied structures in the central nervous system is the hippocampus, a region involved in spatial mapping and construction of an internal representation of the physical environment as well as in basic learning and memory processes. As might be expected on the basis of its role, the hippocampus appears to exhibit a high degree of developmental plasticity and may retain a considerable margin of structural plasticity into adulthood via continued neurogenesis as well as more classical mechanisms of neuroplasticity (Gould et al. 2000; McEwen 2000). This plasticity is reversible and appears to be modulated by conditional actions of (a) glucocorticoids, (b) excitatory amino acids via the N-methyl-D- aspartate (NMDA), (c) the serotonin, and (d) the GABA-benzodiazepine systems. The concept of conditional actions of extracellular, and perhaps intracellular, signaling molecules is central to the ability to orchestrate spatial and temporal information transfer and processing. Within this concept, the final action of any localized signaling cascade is, in part, determined by the concurrent activity of other signaling cascades in that cell and may involve, but is not limited, to: (a) the formation of homo- or hetero-dimers of transcriptional regulatory factors thus altering their ability to translocate to the nucleus, their site of binding, their activation or repressive actions on gene expression, and the half-life of recycling these factors for reuse or degradation, (b) competition among components of signaling cascades potentially expressed in stoichiometric interactions between limited intermediates of these signaling cascades.

The general process of adaptation, illustrated in Figure 12.1, may be viewed as "biasing" or "environmental programming" of neural circuitry on the basis of intrinsic and extrinsic

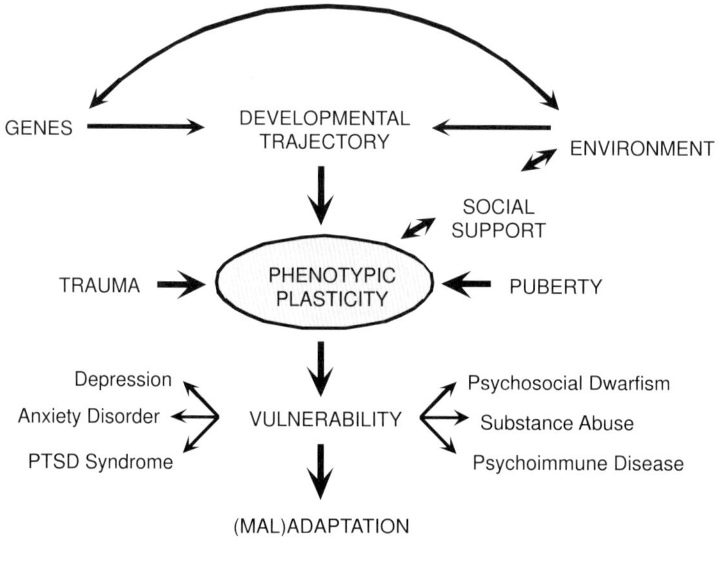

Figure 12.1 The interaction between an individual's genome and early environmental stimuli leads to individual differences representing either adaptation or maladaptation in interpreting subsequent environmental stimuli. During the perinatal period, such interactions may temporarily or permanently alter levels of gene expression, thus initiating a cascade resulting in stable adaptations of numerous neurocircuits affecting behavior, neuroendocrine, and autonomic nervous system responsivity as well as coping responses throughout life. During the perinatal period, social buffering may mute the effects of adverse early experience and thus lead to resiliency. Other factors, including puberty and trauma occurring throughout the life span, may exacerbate the underlying vulnerabilities resulting in expression of physio- or psychopathology. (Modified and used with permission from Ladd et al. 2000.)

factors. The overall effect of these processes, illustrated in Figure 12.2, is the alteration of an individual organism's coping responses in a manner adaptive to a particular environmental niche leading to resiliency and reproductive success (i.e., fitness). However, these adaptations may prove to be maladaptive in rapidly changing environmental circumstances leading to reduced reproductive success and increased disease susceptibility.

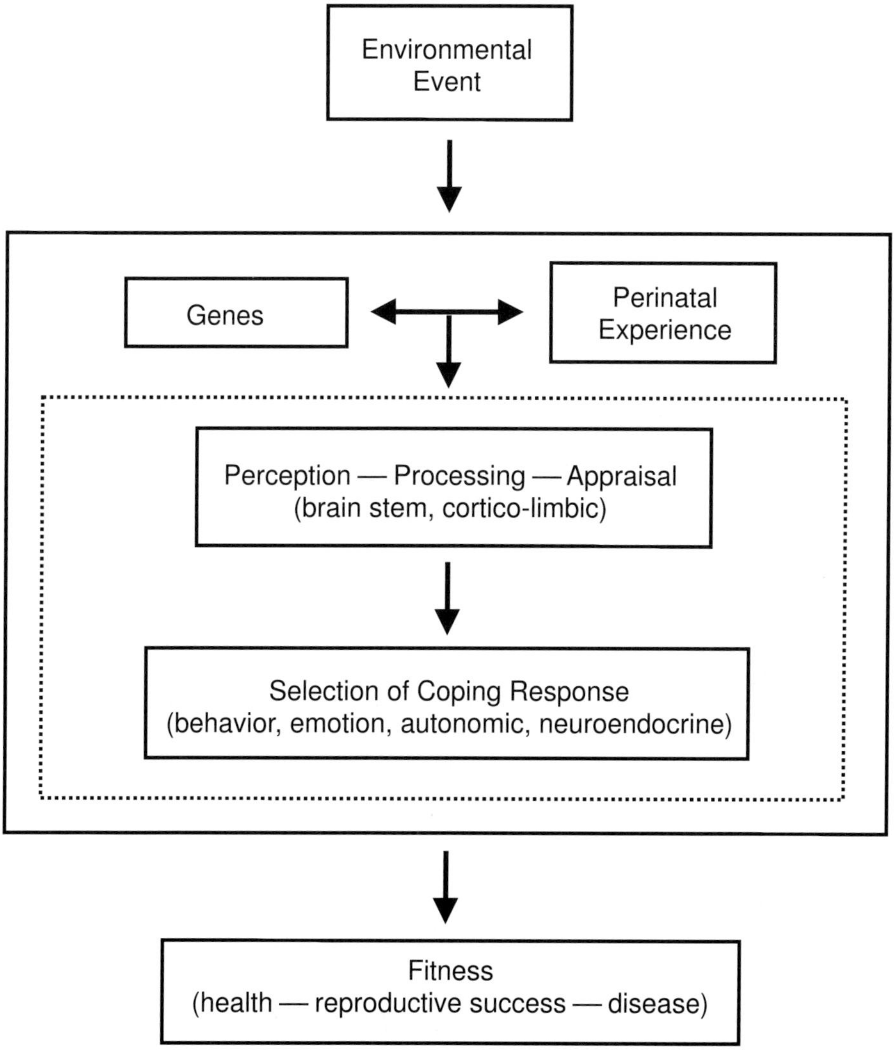

Figure 12.2 Interactions between genome and environmental events to determine fitness. Exposure to adverse environmental stimuli during the perinatal period sculpts neurocircuits such that the individual's perception and processing of subsequent environmental stimuli becomes biased eliciting individual adaptive or maladaptive coping responses. The degree by which these responses are inappropriate exacts an allostatic load, which may be expressed as lowered reproductive success, increased disease susceptibility or, in the case of adaptive responses, resiliency.

THE ROLE OF EXTRINSIC FACTORS: PARENTAL INVESTMENT AND THE MOTHER–INFANT RELATIONSHIP

The parental investment theory, currently the primary evolutionary model of parenting, postulates that the likelihood of maltreatment of the offspring increases when the survival costs of providing parental care outweigh the survival benefits (Kriegman 1999; Mann 1992). Caring for offspring increases the chances that the progeny will survive and, thus, protects the parental contribution to the species gene pool and, presumably, fitness. However, when the current offspring has a limited chance of survival and subsequent reproduction as a result of, for example, birth defects, illness, injury, food shortages, environmental degradation, or other factors, continued parental investment in the offspring may be counterproductive from an evolutionary standpoint. Continuing to expend energy to nurture offspring who will likely never reproduce may serve to undermine the parent's opportunity to make a lasting contribution to the survival of the species. This may best be realized by abandoning or even actively destroying the current offspring, thereby preserving parental energy to be invested in subsequent reproductive opportunities (Dawson 1996).

The developmental influence of the mother–infant dyad has long been a subject of interest. Bowlby (1977a,b) formalized this relationship using the concept of "security of attachment." Both the mother and the infant cooperatively set and then learn the rules necessary for a continued and productive relationship (Alberts 1994; Lewis and Goldberg 1969). Indeed, Hofer (1994) has characterized the primary caregiver as a "hidden regulator" of the development of regulatory systems in the offspring. It may be possible to translate the concept of "security of attachment" into a broader concept applicable to lower species by postulating that the basis for "security of attachment" is the development of a sense of predictability and controllability on both sides of the dyad and, furthermore, that this basis for the relationship is necessary for appropriate development of systems underlying perception and evaluation of environmental cues as well as the regulation of the primary output systems (i.e., autonomic nervous system, neuroendocrine, behavior). Numerous neurochemical substrates appear to participate in mediation of infant–mother attachment (Nelson and Panksepp 1998). The implication that follows from this series of postulates, then, is that an abnormal, neglectful, or abusive relationship between the primary caregiver and the progeny will result in abnormal development in the progeny of these critical systems for interpreting and responding to the surrounding environment. Numerous support for this thesis is available in the epidemiological, psychiatric, and primary research literature (see, for instance, Andrews and Rosenblum 1994; De Bellis et al. 1999a, b; Heim et al. 1997; Kaufman et al. 2000; Kraemer et al. 1991; Ladd et al. 2000; McCauley et al. 1997; Pollak et al. 2000; Suomi 1997).

THE HYPOTHALAMIC-PITUITARY-ADRENAL AXIS

Relative maintenance of the internal milieu and development of effective coping strategies are essential to survival. The synthesis and secretion of glucocorticoids represents the final step in a neuroendocrine cascade beginning in the central nervous system designed to ensure adequate energy substrates and coordination of other response systems when faced with the initial exposure to most categories of stressors.

Activation of the hypothalamic-pituitary-adrenal (HPA) axis by stressors, circadian drive, and humoral influences initiates this cascade by releasing multiple hypothalamic adrenocorticotrophic hormone (ACTH) secretagogues into the hypophysial-portal circulation, which provides the functional link between the central nervous and endocrine systems (Antoni 1986; Plotsky 1991). Corticotrophin-releasing factor (CRF), elaborated by perikarya in the hypothalamic parvocellular paraventricular nuclei (PVN) (Swanson et al,. 1983), plays an integral and obligatory role in regulation of adenohypophysial ACTH secretion and gene expression of the precursor protein proopiomelanocortin (Autelitano et al. 1990). Intrinsically weaker secretagogues, such as arginine vasopressin (AVP), modulate ACTH secretion; however, their actions are dependent upon the presence of CRF (Antoni 1983; Plotsky 1991). ACTH, in turn, stimulates the *de novo* production and release of glucocorticoids from the adrenal cortex. The glucocorticoids, primarily cortisol in primates and corticosterone in rodents, act to mobilize energy substrates by inhibiting glucose uptake in nonexercising muscle, facilitating gluconeogenesis as well as stimulating catabolic processes such as lipogenolysis and proteolysis. Additionally, glucocorticoids suppress anabolic reactions and inflammation, and synergize with catecholamines released from the adrenal medulla and sympathetic nerve terminals to elevate circulating glucose levels as well as to increase the heart rate and blood pressure.

Prolonged elevation of circulating glucocorticoids can have negative physiological and psychological consequences (McEwen 2000; Munck et al. 1984; Sapolsky 2000). Therefore, the levels of circulating glucocorticoids are tightly controlled by the process of glucocorticoid-mediated negative feedback (Dallman et al. 1987; De Kloet 1991; Jacobson and Sapolsky 1991; Munck et al. 1984). Glucocorticoids are necessary for survival and have positive as well as negative consequences (De Kloet et al. 1999). Glucocorticoid feedback processes are mediated by Type I or mineralocorticoid receptors (MR) primarily localized within the septal and hippocampal regions and by Type II or glucocorticoid receptors (GR) localized in the adenohypophysis and widely distributed throughout the central nervous system including the hypothalamic area, corticolimbic system, brain stem, and spinal cord (De Kloet 1991; Jacobson and Sapolsky 1991). Importantly, MR and GR serve as transcriptional regulators which, upon activation, translocate to the nucleus where they bind in homo- or hetero-dimeric complexes to DNA recognition sites to alter levels of gene expression. It is believed that the ratio between MR and GR in the hippocampus may determine the set-point of the HPA axis during the trough period of the diurnal rhythm and may also participate in determining the maximal amplitude of stressor-induced activation (De Kloet et al. 1998). Together, these mechanisms permit rapid adjustment of the HPA axis in response to the demands of the environment.

Interoceptive and exteroceptive stimuli are funneled to the hypothalamic PVN via visceral and somatosensory pathways from: (a) noradrenergic nuclei in the brain stem including the ventral tegmental area (VTA), nucleus of the solitary tract (NTS), and the locus coeruleus, (b) limbic input from the bed nucleus of the stria terminalis (BNST) including projections from the ventral hippocampus, amygdala, and prefrontal cortex, (c) internal regulatory input via the intrahypothalamic nuclei, and (d) neurohumoral information from the circumventricular organs such as the subfornical organ (Cullinan et al. 1995; Gray and Bingaman 1996; Plotsky et al. 1989; Sawchenko et al. 1993). Each of these pathways mediates signals from a specific class of stressors, although there may be cases of overlap in pathway activation patterns. This is reflected in different patterns of ACTH secretagogue release (Plotsky 1991) and in different temporal patterns of the stress response (Koolhaas et al. 1997).

Development of the HPA axis has been recently reviewed in Walker et al. (2000). There is great variation among species in the development of the various components of the HPA axis and the onset of their functional activity. Depending on the degree of maturity of the fetus at birth, developmental stages in central function are differentially expressed during neonatal life. While studying the ontogeny of the pituitary–adrenal axis, Jailer (1950) observed that neonatal rats exhibited resistance to mount an adult-like corticosterone response to an acute stressor to environmental stimuli from postnatal days (PND) 4–14, which he called the "stress nonresponsive period." Subsequent studies of the neonatal stress response in rats demonstrated that the axis was capable of stressor-specific activation during this period that was generally of lower amplitude than observed in adults, thus it was renamed the stress hypo-responsive period (SHRP) (Walker et al. 1986, 1991; Witek-Janusek 1988). The SHRP has also been described in other species, including the rabbit, guinea pig, and dog, with a duration of only a few days at the most. The presence of such a pituitary–adrenal "refractory period" in nonhuman primates and humans remains controversial.

Researchers originally believed that the mechanisms underlying the SHRP resided solely at the level of the adrenal gland, perhaps serving to protect the developing brain against elevated concentrations of circulating glucocorticoids. However, as compared with adults, neonatal rats exhibit low concentrations of hypothalamic CRF and AVP, decreased efficiency of POMC cleavage, reduced adrenal sensitivity to ACTH, U-shaped ontogenetic curves of corticosterone and corticosterone-binding globulin concentrations, and adult-like GR density in the anterior pituitary (Rosenfeld et al. 1992) suggesting involvement of multiple loci in the regulation of the SHRP. However, it must also be noted that recent evidence indicates that the hypothalamic component of the HPA axis does appear to respond to stressors during at least a portion of the SHRP (Dent et al. 2000).

INTERACTIONS BETWEEN CORTICOTROPHIN-RELEASING FACTOR AND NORADRENERGIC CIRCUITS

Administration of exogenous CRF and related peptides in the central nervous system, such as urocortin, appear to enhance behavioral arousal as measured by locomotor activation and responses to stressors (Dunn and Berridge 1990; Koob 1999; Owens and Nemeroff 1991). These CRF effects appear to be independent of HPA axis activation and can be reversed by administration of specific and selective CRF antagonists. More importantly, administration of CRF antagonists can reverse behavioral responses to many stressors, suggesting that the endogenous CRF systems may play a role in mediating behavioral responses to stressors.

Forebrain norepinephrine primarily originates from catecholaminergic neurons located in the brain stem. Norepinephrine pathways emanating from the locus coeruleus have long been hypothesized to be involved in mediating alertness, vigilance, arousal, and stress (for a review, see Koob 1999; Valentino et al. 1998). Neuroanatomical, pharmacological, and electrophysiological evidence supports an important interaction between central norepinephrine and CRF systems. For instance, the locus coeruleus is innervated by extra-hypothalamic CRF circuits originating in the BNST and CEA (Koob 1999; Valentino et al.

1993; Van Bockstaele et al. 1999). In addition, stereotaxic injection of CRF adjacent to the locus coeruleus elicits changes in neuronal burst activity and increased release of norepinephrine in terminal fields of the locus coeruleus (for a review, see Koob 1999; Valentino et al. 1993). This interaction may be particularly important in situations where an organism must mobilize not only the HPA axis, but also the central nervous system in response to an environmental challenge. Consequently, dysregulation of either the extra-hypothalamic CRF system or the locus coeruleus noradrenergic system could lead to a feedforward cascade resulting in serious dysregulation of vigilance, autonomic nervous system responses, neuroendocrine responses, and behavioral responses to environmental stimuli and may play a key role in a variety of pathophysiologic conditions involving abnormal responses to stressors such as anorexia nervosa, anxiety, and affective disorders (see Francis et al. 1999a).

LONG-TERM CONSEQUENCES OF ADVERSE EARLY EXPERIENCE ON THE HPA AXIS, BEHAVIORAL RESPONSES, AND COPING

Rat Studies

There is considerable individual variability in the regulation of the HPA axis and anxiety- or fear-like behavior in rats and monkeys, particularly with respect to rearing conditions (Ladd et al. 2000; Levine 1987; Liu et al. 1997; Meaney et al. 1996; Schanberg et al. 1984; Suomi 1997). Ader, Denenberg, Levine, and others (Ader and Grota 1969; Denenberg 1964; Levine 1957) first reported that early handling (i.e., infantile stimulation) reduced novelty-induced fearfulness and adrenal responsiveness to stressors. In contrast, maternal separation exerted the opposite effect on adrenal responsivity and, when assessed, anxiety-like behavior (for a review, see Francis et al. 1999a ; Ladd et al. 2000). Prenatal stress of the mother also leads to alterations in the behavioral and neuroendocrine responsiveness of the offspring (reviewed in Weinstock (1997). Furthermore, this response is not limited to rats but appears to represent a general response to withdrawal of maternal care (Hennessy 1997). The lack of particular aspects of maternal care (e.g., contact, feeding, licking) appears to be the primary mediator of numerous developmental effects, including neuroendocrine responsiveness, behavioral responsiveness, and growth (Evoniuk et al. 1979; Liu et al. 1997; Schanberg et al. 1984; Suchecki et al. 1993; Van Oers et al. 1998). These classic studies demonstrated that the development of rudimentary adaptive responses to stressors could be influenced by postnatal environmental events.

In the rat, maternal behavior consists of several rituals associated with feeding that are vital for proper infant development. During each nursing bout, the dam may assume an arched-back posture to encouraging suckling or a blanket posture, which is less conducive to suckling. Each nursing bout concludes with anogenital licking or grooming to stimulate urination and defecation. Studies have shown that, in the rat, feeding (suckling) stimulates the infant heart rate (see Hofer 1994), while stroking (licking) stimulates the secretion of growth hormone and expression of ornithine decarboxylase (Evoniuk et al. 1979; Schanberg et al. 1984). Suchecki and colleagues (1993) reported that these maternal behaviors suppress the HPA axis at different levels, thus preventing a pup from mounting an adult-like HPA response to a moderate stressor. In addition, Van Oers and colleagues (1998) demonstrated that, at least in the case of a single, 24-hour separation, the outcome was not mediated by an increase of corticosterone. Thus, in the neonatal rat, feeding desensitizes the adrenal gland to circulating ACTH, while stroking inhibits the release of ACTH from the anterior pituitary. These observations may explain why maternal separation is one of the few stressors that can induce a substantial HPA response in rat pups during the SHRP.

To gain a better understanding of these processes and their consequences in the development of coping strategies and resistance or vulnerability to subsequent medical and psychiatric illness, Plotsky and Meaney have developed a series of animal rodent and primate animal models involving exposure to pre- or neonatal adverse events. Our most extensive knowledge of the long-term consequences of adverse early experience has been derived from studies of Long Evans hooded rats exposed to brief handling and neonatal maternal separation for 180 minutes per day (HMS180) on PND 2 to 14 as compared to briefly handled rats separated for 15 minutes on each day from PND 2–14, or to animal facility-reared controls (AFR). It is important to note that due to the immature state of rat pup development during this period, these animals are entirely dependent upon their mothers for thermoregulation, nutrition, stimulation of micturition, and protection for at least the first 10 days of life. Brain development in these pups at birth is roughly equivalent to a human gestational age of 23–24 weeks.

When tested as adults (> PND 60), offspring exposed to neonatal maternal separation displayed hyper-responsiveness psychological/emotional stressors when compared to AFR or HMS15 progeny. Thus, when exposed to the mild stressor of a series of airpuff startles, the HMS180 animals showed a two- to threefold increase in ACTH and corticosterone secretion as well as a delayed shut-off of secretion of these hormones as compared to AFR or HMS15 rats. The HMS180 animals also exhibited resistance to glucocorticoid-mediated negative feedback when challenged with dexamethasone. This HPA axis hyper-responsiveness persists throughout the remainder of these animals' lives and is associated with increased anxiety-like behavior, development of depression-like syndrome, anhedonia, proclivity for increased ethanol intake, an increased susceptibility to the development of insulin resistance and diabetes, as well as a decrease in immune surveillance with increased rates of tumor growth and metastases later in life. A greater portion of HMS180 than AFR or HMS15 rats also exhibit cognitive difficulties as they age. These studies are reviewed in Ladd et al. (2000) and Meaney et al. (1994) and specific alterations are discussed below.

Interestingly, neonatal maternal separation for periods of 180 minutes profoundly disrupted maternal behavior upon reunion of the pups and the dam in their home cage. Compared with the dams of the HMS15 protocol, dams in the HMS180 protocol exhibited an apparent disorganization of maternal behavior with behavioral anxiety, delay in pup retrieval as well as in the proportion of time spend licking and grooming pups. These observations support the maternal-mediation hypothesis (Denenberg et al. 1964). However, it must be stressed that in other species and, perhaps, even in rodents, social buffering plays an important role. Maternal behavior can be normalized by providing the dams with an age- and gender-matched litter during the period of separation. Furthermore, cross-fostering studies demonstrate the importance of maternal behavior as a regulator of the development of regulatory systems in their pups. Finally, what appears as disrupted maternal behavior may, indeed, be the dam's adaptive attempt to find a safe location to which to move her nest and pups after they were disturbed.

Much of this research has focused on factors underlying the increased responsiveness of the HPA axis and in anxiety in the HMS180 animals. By PND 14, the HPA axis response to psychological, but not physical, stressors was elevated by 2.5-fold in HMS180 offspring and this hyper-responsiveness was evident even in 24-month-old rats. Underlying these behavioral and neuroendocrine outcomes are changes in the characteristics and function of multiple neurocircuits, including the CRF, norepinephrine, and GABA systems. Briefly, these behavioral and neuroendocrine changes were accompanied by:

1. up-regulation of CRF gene expression in the bed nucleus of the BNST, the central nucleus of the amygdala, and the hypothalamic PVN with a consequent increase in CRF peptide content in terminal fields of these neurons,
2. down-regulation of hippocampal GR binding and mRNA levels as well as up-regulation of MR binding and mRNA levels in the hippocampus,
3. decreased GR mRNA expression in the medial prefrontal cortex,
4. increased expression of tyrosine hydroxylase mRNA in the locus coeruleus,
5. increased basal and stress-induced norepinephrine release in locus coeruleus terminal fields,
6. down-regulation of α_2-adrenergic receptors in the locus coeruleus,
7. alterations in electrophysiological characteristics of the locus coeruleus and the ventral tegmental area,
8. changes in GABA-A receptor binding as well as in receptor subunit composition in selected brain regions,
9. reduced mossy fiber density,
10. reductions in oxytocin receptor binding in the basomedial nucleus of the amygdala, and
11. reduced neurogenesis in the hippocampal dentate gyrus (Caldji et al. 2000a, b; Cirulli et al. 2000; Francis et al. 1999a; Gould et al. 2000; Huot et al., 2001; Ladd et al. 2000; Liu et al. 2000; Liu et al. 1997; Meaney et al. 1996).

Thus, animals experiencing moderate periods of maternal separation during the neonatal period show fundamental reorganization of neurocircuits involved in neuroendocrine regulation (hypothalamic PVN, hippocampus), regulation of alerting and vigilance behavior (locus coeruleus), regulation of autonomic nervous system tone (locus coeruleus, nucleus of the solitary tract), and regulation of fear- or contextual-conditioning (CEA, BNST). Alterations in cortical and dopaminergic circuits are also apparent but are less well defined at the current time. Interestingly, Stout et al. (2000) has reported elevations of CRF mRNA in the CEA of anhedonic rats. Neither the precise factors mediating these changes nor the order of the cascade of changes has been elucidated. However, testable hypotheses have been generated that are within reach of experimental techniques. Unraveling this cascade may have profound implications for our understanding of factors involved in organizing, coping, and, subsequently, in factors determining health and disease.

Studies in our laboratory have demonstrated multiple methods for ameliorating the effects of these changes associated with adverse early experience. For instance, cross-fostering pups subjected to HMS180 conditions to dams whose pups were only separated for 15 min prevented development of the "separation phenotype" described above, while fostering HMS15 pups to dams whose pups were separated for 180 minutes yielded offspring that appear to have experienced maternal separation. In addition, merely providing an age- and gender-matched litter to the dam during the period of separation from her own litter prevented the expression of disturbed maternal behavior and yielded normal offspring. Treatment of adult offspring expressing the "maternal separation phenotype" with antidepressants, such as the serotonin reuptake inhibitor paroxetine or the norepinephrine reuptake inhibitor reboxetine, for 21–28 days reversed most aspects of neonatal maternal separation. However, these effects were transient with "symptoms" returning within 10–14 days after cessation of treatment. Comparison of these observations with observations from nonhuman primate studies as well as clinical studies of patients with posttraumatic stress disorder or major depressive disorder suggests the importance of perinatal and early childhood factors in organizing critical neurocircuits in the brain that may influence coping skills and lead to disease susceptibility. These results also suggest opportunities for early intervention as well as later treatment.

Nonhuman Primate Studies

Although the role of prenatal maternal stress has been studied in relation to adverse birth outcomes and problems in human development, a reliable chain of cause-and-effect has not been definitively identified thus far (Lobel 1994). In nonhuman primates, exposure of pregnant females to stressful situations resulted in an increased miscarriage rate (Sackett 1981) as well as increased dermatoglyphic asymmetry, which is indicative of altered fetal development and is associated with increased perinatal mortality (Newell-Morris et al. 1989). There seem to be periods of vulnerability for the effects associated with prenatal stress exposure in nonhuman primates. Schneider and colleagues (1999) demonstrated that although exposure to mild psychological stress during any point of gestation resulted in deficits of attention and neuromotor maturation in macaque infants, early gestational stress (E45–90) resulted in infants with more pronounced and pervasive motor effects than did mid-late gestational (E90–E145) stress. However, even daily exposure to stressors during mid-late gestation resulted in lower birth weights, impaired neuromotor development, and attention deficits in the offspring including delays on learning tasks, reduced motor/exploratory behavior in a novel environment, enhanced behavioral responses to stressors, higher levels of abnormal behavior, and deficits in social behavior (Clarke and Schneider 1993; Schneider 1992a,b,c; Schneider and Coe 1993). HPA axis function was also affected (Clarke et al. 1994) as were CSF concentrations of biogenic amines (Schneider et al. 1998).

Although interesting, these results must be interpreted with caution as factors other than prenatal stressors may have affected the outcome. For example, in some cases, both the control and the prenatally stressed infants were separated from the mother on PND 2, nursery-reared until 30 days old, and then peer-reared until testing at 6 or 12 months of age (Schneider 1992a,b; Clarke et al. 1994). This may be significant because monkeys exhibit long-term consequences of maternal separation and maternal–infant dyad disruption (see below).

Perhaps the most common procedure used to investigate attachment relationships in nonhuman primates has been to disrupt this relationship by differing periods of separation. In contrast to the early work of Harlow and coworkers, which consisted of total separation of the mother and infant, the maternal deprivation paradigm, as used in the studies to be described, imposes intermittent periods of separation. During the intervening periods the infant is normally with its mother or with its mother and other mother–infant dyads. When the studies were initiated, the primary purpose was to examine the behavioral and endocrine responses of squirrel monkey (*Saimiri sciureus*) mothers following the removal of their infants. The rationale for examining the mother was based upon the theories of attachment as postulated by Bowlby (1969). One criterion for attachment in humans is that both members of the mother–infant dyad respond to the loss of the attachment object. Although there is a considerable literature that describes the behavior of the infant following the loss of the mother, there is a paucity of studies that have examined the maternal response to loss of the infant. It is difficult to determine whether the lack of information concerning the maternal response to separation is based on the fact that: (1) the infant had been the primary focus of most separation studies, or (2) whether the behavior of the mother was not changed in any significant way and, therefore, was not reported. There are numerous reports that the infant responded to separation from the mother by a marked increase in " distress vocalizations." These vocalizations are idiosyncratic to the infant and have a distinct sonographic pattern that is unique to infant distress and are often associated with specific neurochemical and endocrine changes (Levine and Wiener 1988).

If one assumes that increased distress calls are an indication of increased arousal then a pattern of physiological changes should accompany the behavioral manifestations of maternal separation in the infants. In studies that examined the physiological manifestations of disruptions of the mother–infant relationship, the activity of the HPA axis was measured in both mothers and infants. Hennessy et al. (1979) hypothesized that increased activity of the HPA axis was a sensitive measure of arousal. Their initial study demonstrated that there was a marked increase in the plasma levels of cortisol in both members of the dyad following separation. For reasons not germane to this discussion, squirrel monkeys secrete exceptionally high levels of cortisol. For the infant, the magnitude of the elevations of cortisol was directly proportional to the time separated. Thus, at the end of a 24-hour period of separation, the levels of cortisol in the infant were among the highest observed in any primate. Interestingly, the cortisol response to separation did not appear to habituate. Repeated separations continued to elicit elevations of cortisol that were equal to that observed during the initial separation. Further, age of the offspring did not appear to modify the response to separation (Levine 1993).

In the early studies of the hormonal responses to separation, the monkeys were maintained as mother–infant dyads in an individual primate cage. However, squirrel monkeys, like most primates, normally rear their infants in social groups. It must be noted that in the wild, females and their offspring remain in their natal social groups while the males do not interact with either the females or the infants and are excluded from the group except during the breeding season. The effects of social factors on the behavioral and hormonal responses to the absence of the mother were examined in response to three different separation paradigms. The first procedure replicated the initial experiment. Thus, mother–infant pairs were reared in social isolation. Under these conditions the infant was either removed or placed in a different but identical cage, or the mother was removed and the infant remained in the familiar home cage. The data indicated that the critical variable associated with the marked elevation of cortisol was the loss of the mother. The magnitude of the cortisol response was similar whether the infant was placed in a novel environment during separation or remained in familiar surroundings. The second separation procedure was to remove the mother from the social group and leave the infant with familiar conspecifics. Although loss of the mother still elicited a significant elevation of cortisol, the magnitude of the response was markedly reduced. Further, distress vocalizations were also reduced compared to monkeys separated in the absence of social partners. In the final condition, infants and mothers were both removed from their home cage and placed adjacent to each other with visual, auditory, and olfactory cues available but were unable to make physical contact. Under these conditions, infant vocalizations were higher and persisted for much longer. While the elevations of cortisol were higher than in social group-reared monkeys, they were lower than in isolation-reared animals. The behavior of the infant did change following separation. Immediately following reunion, the infant spent the majority of its time either in direct contact with the mother or in very close proximity. Reunion also appears to terminate the cortisol response.

The initial studies reported an inverse relationship between vocalizations and the magnitude of the cortisol response. This relationship was not apparent in this study or in studies with mother–infant separation in marmosets (*Callithrix jacchus*; Norcross and Newman 1999). Subsequent research (Wiener et. al. 1988), in which vocalizations were manipulated pharmacologically, indicated that there was no systematic relationship between distress vocalizations and the hormonal response to separation. Although vocalizations can be increased by naltrexone and decreased by morphine pretreatment, both groups showed similar elevations of cortisol following separation that did not differ from saline-injected controls. Thus, it appears that the behavioral and endocrine responses to separation are controlled by different neural mechanisms.

Although the effects of disrupting the ongoing mother–infant relationship on the HPA axis have been emphasized, other physiological systems also reflect the impact of maternal separation. Evaluation of the major metabolites of serotonin, dopamine, and norepinephrine in the cerebrospinal fluid of infant monkeys following separation revealed a change in only 3-methoxy-4-hydroxyphenylglycol (MHPG), a norepinephrine metabolite, which was significantly elevated by maternal separation. Furthermore, the extent of the increases in MHPG was determined by the social milieu during the period of separation. Disruption of the mother–infant relationship also affects immune function (Coe and Erickson 1997).

Little information on the long-term effects of repeated intermittent separations is available in this species. One study did examine two-year-old squirrel monkeys that had experienced postnatal separations (Levine et al. 1997). These previously separated and nonseparated juveniles were isolated from their social groups and placed in individual cages. On all behavioral, hormonal, and noradrenergic assessments, the repeatedly separated monkeys showed a reduced response compared with the nonseparated controls. These data are reminiscent of the data on the HPA responses of early handled rodents when exposed to novelty.

The limited information on the long-term effects of intermittent postnatal separations is intrinsic to the problems of conducting early experience research with primates. Given the high cost of maintaining nonhuman primates, these animals are frequently subjected to numerous experiments, thus adding a confounding element to any long-term experiments. Not unlike human research, it is often difficult to determine in what way these prior experiences effect the outcome measures in any single experiment. That separations during the primate's infancy are very salient events has been abundantly documented. In addition to experiential factors, individual differences in vulnerability to emotional distress may also, in part, be heritable. Combining paternal half-sibling analysis with three postnatal rearing protocols that altered aspects of maternal availability, Lyons and colleagues (1999) found significant postnatal rearing effects with the lowest heritabilities for peak elevations of cortisol 24 hours following removal of the mother in otherwise undisturbed groups. In contrast, individual differences in cortisol 3–7 days later revealed the highest heritabilities and negligible postnatal rearing effects. Overall, paternal half-siblings that responded with long lasting increases in cortisol spent more time near their mother in undisturbed groups and exhibited long-lasting increases in separation-induced peep-calls. These findings are in general agreement with human twin studies in suggesting an interaction between genetic and environmental influences.

In primates, the magnitude and duration of these responses are influenced by social factors that can modify and shape the responses. For the infant primate, the primary social relationship is between the mother and infant or an appropriate interactive mother surrogate. One of the functions of this relationship, in addition to its obvious nurturing aspects, appears to be to enable the infant to modulate its levels of stress and arousal. Thus the infant learns that it can, by means of maintaining contact and proximity to the mother, control and regulate the physiological and behavioral responses to stress.

Although we have thus far limited our discussion to HPA responses in infant monkeys following brief periods of separation, there are additional data (Levine and Coe 1988, Levine 2000) that support the role of familiar social partners in modulating the endocrine response to stress in adult monkeys. The cortisol responses to a natural predator and to stimuli associated with fear conditioning were significantly reduced when presented to monkeys in a stable social group. Hamburg and Anderson (1967) emphasized the need for continuity of personal relationships when individuals are in crisis situations. Other investigators have pointed to social support as a major factor in specific health-related issues (Elliot 1989). Thus the phenomena of social buffering appears to be of importance as a stress modulator across many species. Although patterns of social organization vary considerably among species, primates spend most of their lives interacting socially and the importance of these social interactions at multiple levels cannot be overemphasized. When dealing with issues related to welfare in animals and humans, the value and importance of social factors must be taken into account and ample opportunity for social relationships should be one of the dimensions of welfare that cannot be overlooked.

The effects of social environment on development have also been studied in Old World monkeys such as the rhesus monkey (*Macaca mulatta*). Rhesus monkeys are highly social animals, living in troops with moderately stable social organization and clear dominance hierarchies. Apart from the relative physical maturity of the rhesus infant at birth, a difference between the early development of infant humans and rhesus monkeys is the amount of physical contact between mother and neonate (Seay and Gottfried 1975). The macaque infant remains closely attached to the mother's ventral surface almost constantly for several weeks and it sleeps in close ventral contact with the mother for nearly a year, when weaning occurs. This intimate contact provides warmth, comfort, and essential social guidance and training.

The spontaneous occurrence of infant abuse and neglect among nonhuman primates is the closest similarity to human child maltreatment and, therefore, a likely model for the effects of this kind of early adverse experience on development. Within large social groups of rhesus monkeys and pigtail macaques, 2–10% of the infants born every year are physically abused by their mothers (Maestripieri 1999). Examples of infant physical abuse in monkeys include violent behaviors such as biting, hitting, dragging the infant on the ground by the tail or legs, throwing or stepping on it (Troisi and D'Amato 1983; Troisi et al. 1982). The fact that abuse is more likely to occur within certain families of monkeys suggests that infant abuse "traits" may be transmitted along the maternal line for generations (Maestripieri et al. 1997a,b; Maestripieri and Carroll 1998). However, it is unclear if this intergenerational pattern of abuse is heritable via aspects of temperament associated with anxiety, impulsivity, and low threshold for aggression that have been related to low serotonergic function (Brown et al. 1979; Higley, King et al. 1996; Higley, Mehlman et al 1996), or to repetition of learned behaviors.

Altering environmental variables, such as predictability of food presentation, has also been used to disturb normal social as well as maternal behavior in macaques (Rosenblum and Andrews 1994). Variable foraging demand, which is the unpredictable alternation of periods of strenuous foraging with others with free access to food, altered female bonnet macaque social behavior and reduced the amount of time available for responding to infant solicitations for contact and attention, resulting in insecure patterns of attachment behavior (Rosenblum and Andrews 1994). Although the variable foraging demand allowed for apparently normal physical and social development of the offspring, the progeny become more fearful and responsive to stressful stimuli, both behaviorally and physiologically, and less socially competent. Physiological markers of this experience in the offspring include persistent alterations in cerebrospinal fluid levels of CRF, somatostatin and biogenic amines (Coplan et al. 1996, 1998), and a reduced growth hormone response to the α-adrenergic agonist clonidine (Coplan et al. 2000). Elevations in cerebrospinal fluid CRF levels and a blunted growth hormone response to clonidine challenge are frequently observed in patients with mood and anxiety disorders (Heim et al. 1997).

Separation of the rhesus infant from its mother results in protest by the infant characterized by increased activity, vocalization, defecation (Harlow et al. 1971), and increased HPA axis activity (Bayart et al. 1990). Rearing rhesus infants in partial social isolation, in which animals have auditory, visual, and olfactory, but not tactile contact with conspecifics, has profound long-term consequences for these animals including the occurrence of autistic-like behaviors, the inability to develop positive social relationships within the social group, hyper-aggressiveness, disoriented sexual behavior, and lack of recognition of social cues. More severe isolation rearing results in correspondingly more severe autistic-like self-clasping, self-orality, rocking and huddling, as well as lack of exploration and social interaction (Harlow et al. 1965, 1971; Mason 1960; Seay and Gottfried 1975). Interestingly, if animals are exposed to a social environment prior to or by 6 months of age, the species typical repertoire of social and nonsocial behaviors develop, even in the infants reared in total isolation (Harlow and Suomi 1971).

In addition to the long-term effects that prolonged periods of maternal and/or social deprivation early in life have in nonhuman primates, these early adverse experiences can also have long-lasting effects on ingestive behaviors, for example, polydipsia and polyphagia, without affecting the body weight (Miller et al. 1969), immune system function (Lubach et al. 1995), the sympathetic response to stressful stimuli (Martin et al. 1988), central catecholamine levels or innervation patterns (Ginsberg et al. 1993a,b; Higley et al. 1992; Kraemer et al. 1989; Lewis et al. 1990; Martin et al. 1991), and HPA axis function. Although higher basal levels of plasma cortisol have been reported in singly-reared rhesus monkeys at 19 months of age (Sackett et al. 1973; Champoux et al. 1989), no differences in the HPA axis response to stressful stimuli were detected in these animals or in 4-year-old animals (Meyer and Bowman

1972). In peer-reared rhesus macaques, either reduced (Clarke 1993; Meyer et al. 1975) or increased (Higley, Hasert et al. 1991; Higley, Suomi et al. 1991; Higley et al. 1992) basal HPA function has been reported. Interestingly, when increased HPA function is detected in peer-reared young rhesus monkeys, either in basal conditions (Higley, Hasert et al. 1991) or in response to separation stress (Fahlke et al. 2000), the levels of plasma cortisol were positively correlated with alcohol consumption. Loss of the normal circadian rhythm or elevation of salivary cortisol levels have been reported in maltreated children (Hart et al. 1995, 1996)

Early maternal and social deprivation interferes with the normal cognitive development of primates (Gluck et al. 1973; Beauchamp et al. 1991; Beauchamp and Gluck 1988; Gluck and Sackett 1976; Sanchez et al. 1998). The cognitive impairments have been partially attributed to the strong perseverative response tendencies exhibited by the deprived animals, which reflect a failure of systems involved in the cessation and inhibition of ongoing behavior. They have also been attributed to attentional or motivational deficits. However, the neurochemical and neuroanatomical changes underlying such impairments are still unclear. Some of the plastic events observed in the CNS have been actually attributed to somatosensory deprivation and restriction of movement, such as the decreased dendritic branching of Purkinje cells in the cerebellum (Floeter and Greenough 1979) or in cells of the somatosensory and motor cortex (Struble and Riesen 1978). Siegel and colleagues (1993) demonstrated the increased expression of neurofilament protein in the hippocampal dentate gyrus. Abnormal electrophysiological recordings from hippocampus, frontal cortex, and sensory relay nuclei for vestibular and proprioceptive function in the cerebellum and somatosensory thalamus have also be reported in rhesus monkeys reared in social isolation (Heath 1972), which may underlie the putative aberrations in emotion and perception in these animals.

Structural magnetic resonance imaging (MRI) has been used as a tool to detect gross neuroanatomical changes in rhesus monkeys separated from the mother at 2 months of age and singly-housed until 12 months old (Sanchez et al. 1998). The scans performed at 18 months of age revealed a reduction in the midsagittal size of the corpus callosum, in parallel with a decrease in white but not gray matter content in the parietal and prefrontal cortices in the separated animals. However, the different rearing conditions were not associated with differences in the total brain, cerebellar or hippocampal volumes, or the size of the anterior commissure, another structure containing interhemispheric fibers. This lack of effect suggested specific reductions in cross-hemispheric projections, possibly resulting from a reduction in the number, myelination, or growth of callosal axons from the same number of neurons. These alterations were not reversed by introduction of the isolates into a peer-group for six months and occurred in parallel with cognitive impairments.

POTENTIAL COSTS AND BENEFITS TO THE INDIVIDUAL AND THE GROUP

The case has been made that the developing central nervous system is malleable in the face of perinatal experience. This neuroplasticity undoubtedly serves valuable functions such as enabling learning and memory, limited repair of damaged circuits, and adaptation to the environment into which the individual is born. The interaction between the genome and the environment may lead to a wide range of coping strategies within a population, ensuring survival of the species over a wide range of environmental circumstances. Importantly, many of the neurocircuits that are "programmed" by these extrinsic factors play a role in shaping an individual's perceptions of and responses to its environment. Thus, any early effects will be predicted to have cascading effects.

It is particularly salient that cortico-limbic circuits are so heavily affected by these processes. For instance, the central nucleus of the amygdala and the bed nucleus of the stria terminalis play central roles in mediation of fear and context-dependent potentiation of startle (Davis 1986; Kaufman et al. 2000; LeDoux 2000). Both of these nuclei receive abundant noradrenergic input from the locus coeruleus and both send CRF-containing fibers to the locus coeruleus. Exposure to stressors activating the HPA axis results in increased exposure to glucocorticoids. At the level of the central nucleus of the amygdala, glucocorticoids enhance CRF gene expression, thus elevating CRF mRNA (Makino et al. 1994), and CRF peptide content in this region and its terminal fields including the locus coeruleus (Plotsky, in preparation).

Perhaps the adaptive capability of these neurocircuits is a double-edged sword. Affective disorders may have their roots in the adaptive ability of neurocircuits that have evolved to organize counter-regulatory responses to stressors (Nesse 1999). Indeed, these diseases may be characterized as conditions of an inability to utilize effective coping strategies. Dysregulation of these cortico-limbic systems has been implicated in the etiology or pathophysiology of psychiatric diseases with an emphasis on CRF, norepinephrine, and glucocorticoids (Heim et al. 1997; Kaufman et al. 2000; Nemeroff 1996) While enormous progress has been made in unraveling the proximate physiological mechanisms that account for anxiety, stress, and affective disorders, these conditions continue to cause conceptual confusion. It is in this vein that studies of animal models will be most valuable.

Mechanisms involved in development of the brain and in mediating neuroplasticity are obvious targets for further study and pharmacological targeting. As discussed by Duman (2000), Gould et al. (2000) as well as de la Rosa and de Pablo (2000), factors affecting the balance between neurogenesis and apoptosis may have long-term consequences, especially if these events occur during critical periods of brain development. Handling, exercise, and environmental enrichment appear to stimulate neurogenesis and the up-regulation of neurotrophic factors, while environmental impoverishment, abuse, and neglect may have the opposite effect (Cirulli et al. 2000; Gould et al. 1999; Kempermann and Gage 1999; van Praag et al. 1999). These mechanisms are sensitive to differences within the normal range of behavior. For example, Liu et al. (2000) recently reported that rats pups exposed to natural extremes in maternal care (e.g., licking and grooming, arched back nursing) one standard deviation above the mean maternal behavior scores of a large cohort of dams showed improved cognitive performance, as well as increased hippocampal synaptogenesis, increased expression of brain-derived neurotrophic factor critical and NMDA subunits in the hippocampus, and increased cholinergic fiber density as compared to those rats reared by dams showing maternal behavior one standard deviation below the cohort mean. This result is similar to studies by Huot and colleagues (2001) showing that maternal separation and subsequent reunion with the dam resulted in elevations of plasma corticosterone levels during the SHRP and, as adults, maternally separated rats exhibited slight, but significant impairments on acquisition in the Morris water maze task, elevated indices of anxiety, and stress-induced pituitary–adrenal hyper-responsiveness. In addition, maternally separated rats exhibited significant mossy fiber hypoplasia in the stratum oriens region of the hippocampus as measured by Timm's staining (Huot et al. 2001). These changes may be the result of neonatal exposure to elevated glucocorticoids and/or changes in other signaling systems in response to maternal separation. Overall, the results suggest that neonatal maternal separation and associated alterations in circulating glucocorticoids during critical periods of hippocampal development can disrupt hippocampal structure and function in a stable manner which may then contribute to the overall stress hyper-responsive phenotype.

Related to these studies, it is interesting to note that Meaney's laboratory has shown that natural variations in maternal behavior have profound and long-lasting effects on the offspring (Liu et al. 2000). Furthermore, this laboratory has recently demonstrated nongenomic transmission of maternal behavior and stress responsiveness in laboratory rats (Francis et al. 1999b). In these experiments, female pups reared by high- versus low-licking and grooming dams were permitted to reach adulthood and then mated. The maternal behavior of three generations of these offspring resembled that of their own dams; offspring of low-licking and grooming dams could be rescued from exhibiting the same type of maternal behavior by cross-fostering them to high-licking and grooming dams. Mechanistically, our laboratory (unpublished) has evidence of down-regulation of the developing oxytocin receptor system in the amygdala of maternally separated rats. The oxytocin receptor system develops during the neonatal period and has been associated with affiliative and maternal behavior in females. Thus the inappropriate development of this system in the offspring of low lickers and groomers may predispose them to exhibit similar maternal behavior. Experimentally, one could inject an oxytocin antisense construct bilaterally into the amygdala during the first week of life in normal pups, then observe their dam–infant interactions and subsequent development relative to pups receiving a control construct; alternatively, rescue of the low licker and groomer offspring might be attempted by injection of an oxytocin receptor overexpression vector.

We are in exciting times. Mechanistically, the studies cited above suggest that the quality of the perinatal environment is a critical factor in the development of coping flexibility. This has profound implications for programs designed to alleviate suffering, poverty, and ignorance. It suggests a clear pathway to reduce disease susceptibility and suffering in animals and humankind. The major implication is that these changes in central neurocircuits are neither unavoidable nor permanent as long as early intervention occurs in the form of increased availability of prenatal and postnatal counseling and care, reduction of substance abuse, and increased environmental stability and predictability. However, timing of these interventions is important: the later one begins, the less flexibility there is to alter the developmental trajectory of these critical neurocircuits. Both our depth of basic neuroscience knowledge as well as our armamentarium of sophisticated experimental tools has increased exponentially. A multitude of previously unthinkable experiments is now within our grasp. What we are lacking is the ability to think outside the confines of the dogma in which we have been steeped in order to advance the field. We must pay attention to our data and our insights. By following these simple prescriptions, we may yet make progress in unraveling the balance between intrinsic and extrinsic factors affecting the development of mechanisms underlying coping strategies.

ACKNOWLEDGMENTS

I wish to thank K.V. Thrivikraman, senior members of my laboratory group, Charles B. Nemeroff, and members of the NIMH Early Experiences and Glucocorticoid Network: Linking Animal and Human Developmental Psychopathology (MH60766) for helpful discussions and comments. I dedicate this brief treatise to the memory and work of Bela Bohus with whom I had many stimulating discussions during the last decades of his life. This research was supported by U.S. National Institutes of Health grants MH50113 and MH58922.

REFERENCES

Ader, R., and L.J. Grota. 1969. Effects of early experience on adrenocortical reactivity. *Physiol. Behav.* **4**:303–305.

Alberts, J.R. 1994. Learning as adaptation of the infant. *Acta Paediatr.* **Suppl. 397**:77–85.

Andrews, M., and L. Rosenblum. 1994. Developmental consequences of altered dyadic coping patterns in bonnet macaques. In: Current Primatology: Social Development, Learning and Behavior, ed. J. Roeder, B. Thierry, J. Anderson, and N. Herrenschmidt, pp. 265–271. Strasbourg: Univ. Louis Pasteur.

Antoni, F.A. 1986. Hypothalamic control of adrenocorticotropin secretion: Advances since the discovery of 41-residue corticotropin-releasing factor. *Endocr. Rev.* **7**:351–357.

Antoni, F.A., M.C. Holmes, and M.T. Jones. 1983. Oxytocin as well as vasopressin potentiate ovine CRF *in vitro*. *Peptides* **4**:411–415.

Autelitano, D.J., M. Blum, M. Lopingco, R.G. Allen, and J.L. Roberts. 1990. Corticotropin-releasing factor differentially regulates anterior and intermediate pituitary lobe proopiomelanocortin gene transcription, nuclear precursor RNA and mature mRNA *in vivo*. *Neuroendocrinology* 51:123–130.

Bayart, F., K.T. Hayashi, K.F. Faull, J.D. Barchas, and S. Levine. 1990 Influence of maternal proximity on behavioral and physiological responses to separation in infant Rhesus Monkeys (*Macaca mulatta*). *Behav. Neurosci.* **104**:98–107.

Beauchamp, A.J., and J.P. Gluck. 1988. Associative processes in differentially reared monkeys (*Macaca mulatta*): Sensory preconditioning. *Dev. Psychobiol.* **21**:355–364.

Beauchamp, A.J., J.P. Gluck, and M.H. Lewis. 1991. Associative processes in differentially reared rhesus monkeys (*Macaca mulatta*): Blocking. *Dev. Psychobiol.* **24**:175–189.

Ben-Ari, Y. 1995. Activity-dependent forms of plasticity. *J. Neurobiol.* **26**:295–298.

Bohus, B., R.F. Benus, D.S. Fokkema, .M. Koolhaas, C. Nyakas, G.A. Van Oortmerssen, A.J.A. Prins, A.J.H. De Ruiter, A.J.W. Scheurink, and A.B. Steffens. 1987. Neuroendocrine states and behavioral and physiological stress responses. In: Progress in Brain Research, ed. E.R. De Kloet, V.M. Wiegant, and D. De Wied, pp. 57–70. Amsterdam: Elsevier.

Bowlby, J. 1977a. The making and breaking of affectional bonds. I. Aetiology and psychopathology in the light of attachment theory. An expanded version of the Fiftieth Maudsley Lecture, delivered before the Royal College of Psychiatrists, November 19, 1976. *Br. J. Psychiatry* **130**:201–210.

Bowlby, J. 1977b. The making and breaking of affectional bonds. II. Some principles of psychotherapy. The Fiftieth Maudsley Lecture. *Br. J. Psychiatry* **130**:421–431.

Bowlby, J. 1969. Attachment: Attachment and Loss, Vol. 1. New York: Basic Books.

Brown, G.L., F.K. Goodwin, J.C. Ballenger, P.F. Goyer, and L.F. Major. 1979. Aggression in humans correlates with cerebrospinal fluid amine metabolites. *Psychiatr. Res.* **1**:131–139.

Caldji, C., J. Diorio, and M.J. Meaney. 2000a. Variations in maternal care in infancy regulate the development of stress reactivity. *Biol. Psychiatry* **48**:1164–1174.

Caldji, C., D. Francis, S. Sharma, P.M. Plotsky, and M.J. Meaney. 2000b. The effects of early rearing environment on the development of GABAA and central benzodiazepine receptor levels and novelty-induced fearfulness in the rat. *Neuropsychopharmacology* **22**:219–229.

Champoux, M., C.L. Coe, S.M. Schanberg, C.M. Kuhn, and S.J. Suomi. 1989. Hormonal effects of early rearing conditions in the infant rhesus monkey. *Am. J. Primatol.* **19**:111–117.

Cirulli, F., E. Alleva, A. Antonelli, and L. Aloe. 2000. NGF expression in the developing rat brain: Effects of maternal separation. *Dev. Brain Res.* **123**:129–134.

Clarke, A.S. 1993 Social rearing effects on HPA axis activity over early development and in response to stress in rhesus monkeys. *Dev. Psychobiol.* **26**:433–446.

Clarke, A.S., and M.L. Schneider. 1993. Prenatal stress has long-term effects on behavioral responses to stress in juvenile rhesus monkeys. *Dev. Psychobiol.* **26**:293–304.

Clarke, A.S., D.J. Wittwer, D.H. Abbott, and M.L. Schneider. 1994. Long-term effects of prenatal stress on HPA axis activity in juvenile rhesus monkeys. *Dev. Psychobiol.* **27**:257–269.

Coe, C.L., and C.M. Erickson. 1997. Stress decreases lymphocyte cytolytic activity in the young monkey even after blockade of steroid and opiate hormone receptors. *Dev. Psychobiol.* **30**:1–10.

Constantine-Paton, M., H.T. Cline, and E. Debski. 1990. Patterned activity, synaptic convergence, and the NMDA receptor in developing visual pathways. *Ann. Rev. Neurosci.* **13**:129–154.

Coplan, J.D., M.W. Andrews, L.A. Rosenblum, M.J. Owens, S. Friedman, J.M. Gorman, and C.B. Nemeroff. 1996. Persistent elevations of cerebrospinal fluid concentrations of corticotropin-releasing factor in adult nonhuman primates exposed to early-life stressors: Implications for the pathophysiology of mood and anxiety disorders. *Proc. Natl. Acad. Sci. USA* **93**:1619–1623.

Coplan, J.D., E.L.P. Smith, R.C. Trost, B.A. Scharf, M. Altemus, L. Bjornson, M.J. Owens, J.M. Gorman, C.B. Nemeroff, and L.A. Rosenblum. 2000. Growth hormone response to clonidine in adversely reared young adult primates: Relationship to serial cerebrospinal fluid corticotropin-releasing factor concentrations. *Psychiatry Res.* **95**:93–102.

Coplan, J.D., R.C. Trost, M.J. Owens, T.B. Cooper, J.M. Gorman, C.B. Nemeroff, and L.A. Rosenblum. 1998. Cerebrospinal fluid concentrations of somatostatin and biogenic amines in grown primates reared by mothers exposed to manipulated foraging conditions. *Arch. Gen. Psychiatry* **55**:473–477.

Cullinan, W.E., J.P. Herman, D.L. Helmreich, and S.J. Watson. 1995. A neuroanatomy of stress. In: Neurobiological and Clinical Consequences of Stress: From Normal Adaptation to PTSD, ed. M.J. Friedman, D.S. Charney, and A.Y. Deutch, pp. 3–25. Philadelphia: Lippincott-Raven.

Dallman, M.F., S. Akana, C.S. Cascio, D.N. Darlington, L. Jacobson, and N. Levin. 1987. Regulation of ACTH secretion: Variations on a theme of B. *Rec. Prog. Horm. Res.* **43**:113–173.

Davis, M. 1986. Pharmacological and anatomical analysis of fear conditioning using the fear-potentiated startle paradigm. *Behav. Neurosci.* **100**:814–824.

Dawson, K. 1996. Evolutionary consequences of a trade-off between parental effort and mating effort. *J. Theor. Biol.* **183**:139–158.

De Bellis, M.D., A.S. Baum, B. Birmaher, M.S. Keshavan, C.H. Eccard, A.M. Boring, F.J. Jenkins, and N.D. Ryan. 1999a. Developmental traumatology Part I: Biological stress systems. *Biol. Psychiatry* **45**:1259–1270.

De Bellis, M.D., M.S. Keshavan, B. Clark, B.J. Casey, J.N. Giedd, A.M. Boring, K. Frustaci, and N.D. Ryan. 1999b. Developmental traumatology. Part II: Brain development. *Biol. Psychiatry* **45**:1271–1284.

De Kloet, E.R. 1991. Brain corticosteroid receptor balance and homeostatic control. *Frontiers Neuroendocrin.* **12**:95–164.

De Kloet, E.R., M.S. Oitzl, and M. Joels. 1999. Stress and cognition: Are corticosteroids good or bad guys? *Trends Neurosci.* **22**:422–426.

De Kloet, E.R., E. Vreugdenhil, M.S. Oitzl, and M. Joels. 1998. Brain corticosteroid receptor balance in health and disease. *Endocr. Rev.* 19:269–301.

De La Rosa, E.J., and F. De Pablo. 2000. Cell death in early neural development: Beyond the neurotrophic theory. *Trends Neurosci.* **23**:454–458.

Denenberg, V.H. 1964. Critical periods, stimuli input, and emotional reactivity: A theory of infantile stimulation. *Psychol. Rev.* **71**:335–351.

Dent, G.W., M.A. Smith, and S. Levine. 2000. Rapid induction of corticotropin-releasing hormone gene transcription in the paraventricular nucleus of the developing rat. *Endocrinology* **141**:1593–1598.

Diamond, M.E., M. Armstrong-James, and F.F. Ebner. 1993. Experience-dependent plasticity in adult rat barrel cortex. *Proc. Natl. Acad. Sci. USA* **90**:2082–2086.

Duman, R.S., J. Malberg, S. Nakagawa, and C. D'Sa. 2000. Neuronal plasticity and survival in mood disorders. *Biol. Psychiatry* **48**:732–739.

Duman, R.S., J. Malberg, and J. Thome. 1999. Neural plasticity to stress and antidepressant treatment. *Biol. Psychiatry* **46**:1181–1191.

Dunn, A.J., and C.W. Berridge. 1990. Physiological and behavioral responses to corticotropin-releasing factor administration: Is CRF a mediator of anxiety and of stress responses? *Brain Res. Rev.* **15**:71–100.

Elliot, G. 1989. Stress and illness. In: Psychosomatic Medicine, ed. S. Cherin, pp. 45–90. Madison: Intl. Univ. Press.

Evoniuk, G.E., C.M. Kuhn, and S.M. Schanberg. 1979. The effect of tactile stimulation on serum growth hormone and tissue ornithine decarboxylase activity during maternal deprivation in rat pups. *Commun. Psychopharmacol.* **3**:363–370.

Fahlke, C., J.G. Lorenz, J. Long, M. Champoux, S.J. Suomi, and J.D. Higley. 2000. Rearing experiences and stress-induced plasma cortisol as early risk factors for excessive alcohol consumption in nonhuman primates. *Alcoholism Clin. Exp. Res.* **24**:644–650.

Floeter, M.K., and W.T. Greenough. 1979. Cerebellar plasticity: Modification of Purkinje cell structure by differential rearing in monkeys. *Science* **206**:227–229.

Francis, D.D., C. Caldji, F. Champagne, P.M. Plotsky, and M.J. Meaney. 1999a. The role of corticotropin-releasing factor-norepinephrine systems in mediating the effects of early experience on the development of behavioral and endocrine responses to stress. *Biol. Psychiatry* **46**:1153–1166.

Francis, D.D., J. Diorio, D. Liu, and M.J. Meaney. 1999b. Nongenomic transmission across generations in maternal behavior and stress responses in the rat. *Science* **277**:1659–1662.

Ginsberg, S.D., P.R. Hof, W.T. McKinney, and J.H. Morrison. 1993a. Quantitative analysis of tuberoinfundibular tyrosine hydroxylase- and corticotropin-releasing factor-immunoreactive neurons in monkeys raised with differential rearing conditions. *Exp. Neurol.* **120**:95–105.

Ginsberg, S.D., P.R. Hof, W.T. McKinney, and J.H. Morrison. 1993b. The noradrenergic innervation density of the monkey paraventricular nucleus is not altered by early social deprivation. *Neurosci. Lett.* **158**:130–134.

Gluck, J.P., H.F. Harlow, and K.A. Schiltz. 1973. Differential effect of early enrichment and deprivation on learning in the rhesus monkey (*Macaca mulatta*). *J. Comp. Physiol. Psychol.* **81**:598–604.

Gluck, J.P., and G.P. Sackett. 1976. Extinction deficits in socially isolated rhesus monkeys (*Macaca mulatta*). *Dev. Psychol.* **12**:173–174.

Gould, E., A. Beylin, P. Tanapat, A. Reeves, and T.J. Shors. 1999. Learning enhances adult neurogenesis in the hippocampal formation. *Nature Neurosci.* **2**:260–265.

Gould, E., P. Tanapat, T. Rydel, and N. Hastings. 2000. Molecular and cellular hypothesis of antidepressant action: Regulation of hippocampal neurogenesis in adulthood. *Biol. Psychiatry* **48**: 715–720.

Gray, T.S., and E.W. Bingaman. 1996. The amygdala: Corticotropin-releasing factor, steroids, and stress. *Crit. Rev. Neurobiol.* **10**:155–168.

Hamburg, D., and J.E. Anderson. 1967. Perspective on coping behavior. *Arch. Gen. Psychiatry* **17**:277–284.

Harlow, H.F., R.O. Dodsworth, and M.K. Harlow. 1965. Total social isolation in monkeys. *Proc. Natl. Acad. Sci. USA* **54**:90–97.

Harlow, H.F., M.K. Harlow, and S.J. Suomi. 1971. From thought to therapy: Lessons from a primate laboratory. *Am. Sci.* **59**:538–549.

Harlow, H.F., and S.J. Suomi. 1971. Social recovery by isolation-reared monkeys. *Proc. Natl. Acad. Sci. USA* **68**:1534–1538.

Hart, J., M. Gunnar, and D. Cicchetti. 1995. Salivary cortisol in maltreated children, evidence of relations between neuroendocrine activity and social competence. *Dev. & Psychopath.* **7**:11–26.

Hart, J., M. Gunnar, and D. Cicchetti. 1996. Altered neuroendocrine activity in maltreated children related to symptoms of depression. *Dev. & Psychopath.* **8**:201–214.

Heath, R.G. 1972. Electroencephalographic studies in isolation-raised monkeys with behavioral impairment. *Diseases Nerv. Sys.* **33**:157–163.

Heim, C., M.J. Owens, P.M. Plotsky, and C.B. Nemeroff. 1997. The role of early adverse life events in the etiology of depression and posttraumatic stress disorder: Focus on corticotropin-releasing factor. *Ann. NY Acad. Sci.* **821**:194–207.

Hennessy, M.B. 1997. Hypothalamic-pituitary-adrenal responses to brief social separation. *Neurosci. Biobehav. Rev.* **21**:11–29.

Hennessy, M.B., J.P. Heybach, J. Vernikos, and S. Levine. 1979. Plasma corticosterone concentrations sensitively reflect levels of stimulus intensity in the rat. *Physiol. Behav.* **22**:821–825.

Higley, J.D., M.F. Hasert, S.J. Suomi, and M. Linnoila. 1991. Nonhuman primate model of alcohol abuse: Effects of early experience, personality, and stress on alcohol consumption. *Proc. Natl. Acad. Sci. USA* **88**:7261–7265.

Higley, J.D., S.T. King, Jr., M.F. Hasert, M. Champoux, S.J. Suomi, and M. Linnoila. 1996a. Stability of interindividual differences in serotonin function and its relationship to severe aggression and competent social behavior in rhesus macaque females. *Neuropsychopharmacology* **14**:67–76.

Higley, J.D., P.T. Mehlman, S.B. Higley, B. Fernald, J. Vickers, S.G. Lindell, D.M. Taub, S.J. Suomi, and M. Linnoila. 1996. Excessive mortality in young free-ranging male nonhuman primates with low cerebrospinal fluid 5-hydroxyindoleacetic acid concentrations. *Arch. Gen. Psychiatry* **53**:537–543.

Higley, J.D., S.J. Suomi, and M. Linnoila. 1991. CSF monoamine metabolite concentrations vary according to age, rearing, and sex, and are influenced by the stressor of social separation in rhesus monkeys. *Psychopharmacology* **103**:551–556.

Higley, J.D., S.J. Suomi, and M. Linnoila. 1992. A longitudinal assessment of CSF monoamine metabolite and plasma cortisol concentrations in young rhesus monkeys. *Biol. Psychiatry* **32**:127–145.

Hofer, M.A. 1994. Early social relationships as regulators of infant physiology and behavior. *Acta Paediatr.* **Suppl. 397**:9–18.

Huot, R.L., P.M. Plotsky, R.H. Lenox, and R.K. McNamara. 2001. Neonatal maternal separation impairs hippocampal mossy fiber development and spatial learning in adult Long Evans rats. *Brain Res.*, in press.

Jacobson, L., and R.M. Sapolsky. 1991. The role of the hippocampus in feedback regulation of the hypothalamic-pituitary-adrenal axis. *Endocr. Rev.* **12**:118–134.

Jailer, J.W. 1950. The maturation of the pituitary-adrenal axis in the newborn rat. *Endocrinology* **46**:420–425.

Katz, L.C., and C.J. Shatz. 1996. Synaptic activity and the construction of cortical circuits. *Science* **274**:1133–1138.

Kaufman, J., P.M. Plotsky, C.B. Nemeroff, and D.S. Charney. 2000. Effects of early adverse experience on brain structure and function: Clinical implications. *Biol. Psychiatry* **48**:778–790.

Kempermann., G., and F.H. Gage. 1999. Experience-dependent regulation of adult hippocampal neurogenesis: Effects of long-term stimulation and stimulus withdrawal. *Hippocampus* **9**:321–332.

Kolb, B., and I.Q. Whishaw. 1998. Brain plasticity and behavior. *Ann. Rev. Psychol.* **49**:43–64.

Koob, G.F. 1999. Corticotropin-releasing factor, norepinephrine and stress. *Biol. Psychiatry* **46**:1167–1180.

Koolhaas, J.M., S.M. Korte, S.F. De Boer, B.J. Van Der Vegt, C.G. Van Reenen, H. Hopster, I.C. De Jong, M.A.W. Ruis, and H.J. Blokhuis. 1999. Coping styles in animals: Current status in behavior and stress-physiology. *Neurosci. Biobehav. Rev.* **23**:925–935.

Koolhaas, J.M., P. Meerlo, S.F. De Boer, J.H. Strubbe, and B. Bohus. 1997. The temporal dynamics of the stress response. *Neurosci. Biobehav. Rev.* **21**:775–782.

Kraemer, G., M. Ebert, D. Schmidt, and W. McKinney. 1991. Strangers in a strange land: A psychobiological study of infant monkeys before and after separation from real or inanimate mothers. *Child Dev.* **62**:548–566.

Kraemer, G.W., M.H. Ebert, D.E. Schmidt, and W.T. McKinney. 1989. A longitudinal study of the effects of different social rearing conditions on cerebrospinal fluid norepinephrine and biogenic amine metabolites in rhesus monkeys. *Neuropsychopharmacology* **2**:175–189.

Kriegman, D. 1999. Parental investment, sexual selection, and evolved mating strategies: Implications for psychoanalysis. *Psychoanalytic Psychol.* **16**:528–553.

Ladd, C.O., R.L. Huot, K.V. Thrivikraman, C.B. Nemeroff, M.J. Meaney, and P.M. Plotsky. 2000. Long-term behavioral and neuroendocrine adaptations to adverse early experience. In: Progress in Brain Research: The Biological Basis for Mind Body Interactions, vol. 122, ed. E.A. Mayer and C.B. Saper, pp. 81–103. Amsterdam: Elsevier.

LeDoux, J.E. 2000. Emotion circuits in the brain. *Ann. Rev. Neurosci.* **23**:155–184.

Levine, S. 1957. Infantile experience and resistance to physiological stress. *Science* **126**:405–406.

Levine, S. 1987. Psychobiologic consequences of disruption in mother–infant relationships. In: Perinatal Development: A Psychological Perspective, ed. N.A. Krasnegor, E.M. Blas, M.A. Hofer, and P.W. Smotherman, pp. 359–376. New York: Academic.

Levine, S. 1993. Influence of social factors on the response to stress. *Psychother. & Psychosom.* **60**: 33–38.

Levine, S. 2000. Influence of psychological variables on the activity of the hypthalamic-pituitary-adrenal axis. *Eur. J. Pharmacol.* **405**:149–160.

Levine, S., and C.L. Coe. 1988. Psychosocial modulation of neuroendocrine activity. In: Biorhythms and Stress in the Physiopathology of Reproduction, ed. P. Pancheri and L. Zichella, pp. 41–52. New York: Hemisphere Publ. Corp.

Levine, S., D.M. Lyons, and A.F. Schatzberg. 1997. Psychobiological consequences of social relationships. *Ann. NY Acad. Sci.* **807**:210–218.

Levine, S., and H. Ursin. 1991. What is stress? In: Stress: Neurobiology and Neuroendocrinology, ed. M.R. Brown, G. Koob, and C. Rivier, pp. 3–21. New York: Marcel Dekker.

Levine, S., and S.G. Wiener. 1988. Psychoendocrine aspects of mother–infant relationships in nonhuman primates. *Psychoendocrinology* **13**:143–154.

Lewis, M.H., J.P. Gluck, A.J. Beauchamp, M.F. Keresztury, and R.B. Mailman. 1990. Long-term effects of early social isolation in *Macaca mulatta*: Changes in dopamine receptor function following apomorphine challenge. *Brain Res.* **513**:67–73.

Lewis, M., and S. Goldberg. 1969. Perceptual-cognitive development in infancy: A generalized expectancy model as a function of the mother–infant interaction. *Merrill-Palmer Qtly.* **15**:81–100.

Liu, D., J. Diorio, J.C. Day, D. Francis, and M.J. Meaney. 2000. Maternal care, hippocampal synaptogenesis and cognitive development in rats. *Nature Neurosci.* **3**:799–806.

Liu, D., B. Tannenbaum, C. Caldji, D. Francis, A. Freedman, S. Sharma, D. Pearson, P.M. Plotsky, and M.J. Meaney. 1997. Maternal care, hippocampal glucocorticoid receptor gene expression and hypothalamic-pituitary-adrenal responses to stress. *Science* **277**:1659–1662.

Lobel, M. 1994. Conceptualizations, measurement, and effects of prenatal maternal stress on birth outcomes. *J. Behav. Med.* **17**:225–272.

Lubach, G.R., C.L. Coe, and W.B. Ershler. 1995. Effects of early rearing environment on immune responses of infant rhesus monkeys. *Brain Behav. Immun.* **9**:31–46.

Lyons, D.M., F.L. Martel, S. Levine, N.J. Risch, and A.F. Schatzberg. 1999. Postnatal experiences and genetic effects on squirrel monkey social affinities and emotional distress. *Horm. Behav.* **36**:266–275.

Maestripieri, D. 1999. The biology of human parenting: Insights from nonhuman primates. *Neurosci. Biobehav. Rev.* **23**:411–422.

Maestripieri, D., and K.A. Carroll. 1998. Risk factors for infant abuse and neglect in group-living rhesus monkeys. *Psych. Sci.* **9**:143–145.

Maestripieri, D., K. Wallen, and K.A. Carroll. 1997a. Genealogical and demographic influences on infant abuse and neglect in group-living sooty mangabeys (*Cercocebus atys*). *Dev. Psychobiol.* **31**:175–180.

Maestripieri, D., K. Wallen, and K.A. Carroll. 1997b. Infant abuse runs in families of group-living pigtail macaques. *Child Ab. & Negl.* **21**:465–471.

Makino, S., P.W. Gold, and J. Shulkin. 1994. Effects of corticosterone on CRH mRNA and content in the bed nucleus of the stria terminalis: Comparison with the effects in the central nucleus of the amygdala and the paraventricular nucleus of the hypothalamus. *Brain Res.* **657**:141–149.

Mann, J. 1992. Nurturance or negligence: Maternal psychology and behavioral preference among preterm twins. In: Evolutionary Psychology and the Generation of Culture, ed. J. Barkow and L. Cosmides, pp. 367–390. New York: Oxford Univ. Press.

Martin, L.J., D.M. Spicer, M.H. Lewis, J.P. Gluck, and L.C. Cork. 1991. Social deprivation of infant rhesus monkeys alters the chemoarchitecture of the brain: I. Subcortical regions. *J. Neurosci.* **11**: 3344–3358.

Martin, R.E., G.P. Sackett, V.M. Gunderson, and B.M. Goodlin-Jones. 1988. Auditory-evoked heart rate responses in pigtailed macaques (*Macaca nemestrina*) raised in isolation. *Dev. Psychobiol.* **22**:251–260.

Mason, J.W. 1971. Motivational factors in psychosocial development. In: Nebraska Symposium on Motivation, vol 18, ed. J.K. Cole, pp. 35–67. Lincoln, NE: Univ. of Nebraska Press

Mason, J.W. 1975. A historical view of the stress field. *J. Human Stress* **1**:6–36.

Mason, W.A. 1960. The effects of social restriction on the behavior of rhesus monkeys. I: Free social behavior. *J. Comp. Physiol. Psychol.* **53**:582–589.

McCauley, J., D. Kern, and K. Kolodner. 1997. Clinical characteristics of women with a history of childhood abuse. *J. Am. Med. Assoc.* **277**:1362–1368.

McEwen, B.S. 1998. Stress, adaptation, and disease. Allostasis and allostatic load. *Ann. NY Acad. Sci.* **840**:33–44.

McEwen, B.S. 1999. Stress and hippocampal plasticity. Ann. Rev. Neurosci. 22:105–122.

McEwen, B.S. 2000. Effects of adverse experiences for brain structure and function. *Biol. Psychiatry* **48**:721–731.

McGue, M., and T.J. Bouchard, Jr. 1998. Genetic and environmental influences on human behavioral differences. *Ann. Rev. Neurosci.* **21**:1–24.

Meaney, M.J., J. Diorio, D. Francis, J. Widdowson, P. LaPlante, C. Caldji, S. Sharma, J. Seckl, and P.M. Plotsky. 1996. Early environmental regulation of forebrain glucocorticoid receptor gene expression: Implications for adrenocortical responses to stress. *Dev. Neurosci.* **18**:49–72.

Meaney, M.J., B. Tannenbaum, D. Francis, S. Bhatnagar, N. Shanks, V. Viau, D. O'Donnell, and P.M. Plotsky. 1994. Early environmental programming hypothalamic-pituitary-adrenal responses to stress. *Sem. Neurosci.* **6**:247–259.

Meyer, J.S., and R.E. Bowman. 1972. Rearing experience, stress and adrenocorticosteroids in the rhesus monkey. *Physiol. Behav.* **8**:339–343.

Meyer, J.S., M.A. Novak, R.E. Bowman, and H.F. Harlow. 1975. Behavioral and hormonal effects of attachment object separation in surrogate-peer-reared and mother-reared infant rhesus monkeys. *Dev. Psychobiol.* **8**:425–435.

Miller, R.E., I.A. Mirsky, W.F. Caul, and T. Sakata. 1969. Hyperphagia and polydipsia in socially isolated rhesus monkeys. *Science* **165**:1027–1028.

Munck, A., P.M. Guyre, N.J. Holbrook. 1984. Physiological functions of glucocorticoids in stress and their relations to pharmacological actions. *Endocr. Rev.* **5**:25–44.

Nelson, E.E., and J. Panksepp. 1998. Brain substrates of infant–mother attachment: Contributions of opioids, oxytocin, and norepinephrine. *Neurosci. Biobehav. Rev.* **22**:437–452.

Nemeroff, C.B. 1996. The corticotropin-releasing factor (CRF) hypothesis of depression: New findings and new directions. *Mol. Psychiatry* **1**:336–342.
Nesse, R.M. 1999. Proximate and evolutionary studies of anxiety, stress and depression: Synergy at the interface. *Neurosci. Biobehav. Rev.* **23**:895–903.
Newell-Morris, L.L., C.E. Fahrenbruch, and G.P. Sackett. 1989. Prenatal psychological stress, dermatoglyphic asymmetry, and pregnancy outcome in the pigtailed macaque (*Macaca nemestrina*). *Biol. Neonate* **56**:61–75.
Norcross, J.L., and J.D. Newman. 1999. Effects of separation and novelty on distress vocalizations and cortisol in the common marmoset (*Callithrix jacchus*). *Am. J. Primatol.* **47**:209–222.
Owens, M.J., and C.B. Nemeroff. 1991. The physiology and pharmacology of corticotropin-releasing factor. *Pharmacol. Rev.* **43**:425–473.
Plotsky, P.M. 1991. Pathways to the secretion of adrenocorticotropin: A view from the portal. *J. Neuroendocrinol.* **3**:1–9.
Plotsky, P.M., E.T. Cunningham, and E.P. Widmaier. 1989. Catecholaminergic modulation of corticotropin-releasing factor and dexamethasone secretion. *Endocr. Rev.* **10**:437–458.
Pollak, S.D., D. Cicchetti, K. Hornung, and A. Reed. 2000. Recognizing emotion in faces: Developmental effects of child abuse and neglect. *Dev. Psychol.* **36**:679–688.
Rosenblum, L.A., and M.W. Andrews. 1994. Influences of environmental demand on maternal behavior and infant development. *Acta Paediatr.* **Suppl. 397**:57–63.
Rosenfeld, P., D. Suchecki, and S. Levine. 1992. Multifactorial regulation of the hypothalamic-pituitary-adrenal axis during development. *Neurosci. Biobehav. Rev.* **16**:553–568.
Rutter, M., J. Dunn, R. Plomin, E. Simonoff, A. Pickles, B. Maughan, J. Ormel, J. Meyer, and L. Eaves. 1997. Integrating nature and nurture: Implications of person-environment correlations and interactions for developmental psychopathology. *Dev. Psychopathol.* **9**:335–364.
Sackett, G.P. 1981. A nonhuman primate model for studying causes and effects of poor pregnancy outcomes. In: Preterm Birth and Psychological Development, ed. S. Friedman and M. Sigman, pp. 41–63. New York: Academic.
Sackett, G.P., R.E. Bowman, J.S. Meyer, R.L. Tripp, and S.A. Grady. 1973. Adrenocortical and behavioral reactions by differentially raised rhesus monkeys. *Physiol. Psychol.* **1**:209–212.
Sanchez, M.M., E.F. Hearn, D. Do, J.K. Rilling, and J.G. Herndon. 1998. Differential rearing affects corpus callosum size, and cognitive function of rhesus monkeys. *Brain Res.* **812**:38–49.
Sapolsky, R.M. 2000. The possibility of neurotoxicity in the hippocampus in major depression: A primer on neuron death. *Biol. Psychiatry* **48**:755–765.
Sawchenko, P.E., T. Imaki, E. Potter, K. Kovics, J. Imaki, and W. Vale. 1993. The functional neuroanatomy of corticotropin-releasing factor. In: Corticotropin-releasing Factor. CIBA Found. Symp. 172, ed. D.J. Chadwick, S. Marsh, and K. Ackrill, pp. 5–29. Chichester: Wiley.
Schanberg, S.M., G. Evoniuk, and C.M. Kuhn. 1984. Tactile and nutritional aspects of maternal care: specific regulators of neuroendocrine function and cellular development. *Proc. Soc. Exp. Biol. Med.* **175**:135–146.
Schneider, M.L. 1992a. Delayed object permanence development in prenatally stressed rhesus monkeys infants (*Macaca mulatta*). *Occ. Therapy J. Res.* **12**:96–110.
Schneider, M.L. 1992b. Prenatal stress exposure alters postnatal behavioral expression under conditions of novelty challenge in rhesus monkey infants. *Dev. Psychobiol.* **25**:529–540.
Schneider, M.L. 1992c. The effect of mild stress during pregnancy on birth weight and neuromotor maturation in rhesus monkey infants (*Macaca mulatta*). *Infant Behav. Dev.* **15**: 389–403.
Schneider, M.L., A.S. Clarke, G.W. Kraemer, E.C. Roughton, G.R. Lubach, S. Rimm-Kaufman, D. Schmidt, and M. Ebert. 1998. Prenatal stress alters brain biogenic amine levels in primates. *Dev. & Psychopathol.* **10**:427–440.
Schneider, M.L., E.C. Roughton, A.J. Koehler, and G.R. Lubach. 1999. Growth and development following prenatal stress exposure in primates: An examination of ontogenetic vulnerability. *Child Dev.* **70**:263–274.

Schneider, M.L., and C.L. Coe. 1993. Repeated social stress during pregnancy impairs neuromotor development of the primate infant. *J. Dev. Behav. Pediatr.* **14**:81–87.

Seay, B., and N.W. Gottfried. 1975. A phylogenetic perspective for social behavior in primates. *J. Gen. Psychol.* **92**:5–17.

Selye, H. 1936. A syndrome produced by diverse noxious agents. *Nature* **138**:32.

Selye, H. 1974. Stress without Distress. New York: J.B. Lippincott.

Shatz, C.J. 1990. Impulse activity and the patterning of connections during CNS development. *Neuron* **5**:745–756.

Siegel, S.J., S.D. Ginsberg, P.R. Hof, S.L. Foote, W.G. Young, G.W. Kraemer, W.T. McKinney, and J.H. Morrison. 1993. Effects of social deprivation in prepubescent rhesus monkeys: Immunohistochemical analysis of the neurofilament protein triplet in the hippocampal formation. *Brain Res.* **619**:299–305.

Sroufe, L.A. 1997. Psychopathology as an outcome of development. *Dev. Psychopathol.* **9**:251–268.

Sterling, P., and J. Eyer. 1988. Allostasis: A new paradigm to explain arousal pathology. In: Handbook of Life Stress, Cognition, and Health, ed. S. Fisher and J. Reason, pp. 629–649. New York: Wiley.

Stout, S.C., P. Mortas, M.J. Owens, C.B. Nemeroff, and J.-L. Moreau. 2000. Increased corticotropin-releasing factor in the bed nucleus of the stria terminalis of anhedonic rats. *Eur. J. Pharmacol.* **401**:39–46.

Struble, R.G., and A.H. Riesen. 1978. Changes in cortical dendritic branching subsequent to partial social isolation in stumptailed monkeys. *Dev. Psychobiol.* **11**:479–486.

Suchecki, D., D. Rosenfeld, and S. Levine. 1993. Maternal regulation of the hypothalamic-pituitary-adrenal axis in the infant rat: The roles of feeding and stroking. *Dev. Brain Res.* **75**:185–192.

Suomi, S.J. 1997. Early determinants of behavior: Evidence from primate studies. *Br. Med. Bull.* **53**:170–184.

Swanson, L.W., P.W. Sawchenko, J. Rivier, and W. Vale. 1983. Organization of ovine corticotropin-releasing factor (CRF)-immunoreactive cells and fibers in the rat brain: An immunocytochemical study. *Neuroendocrinology* **36**:165–186.

Thoenen, H. 1995. Neurotrophins and neuronal plasticity. *Science* **270**:593–598.

Troisi, A., and F.R. D'Amato. 1983. Is monkey maternal abuse of offspring aggressive behavior? *Aggr. Behav.* **9**:167–173.

Troisi, A., F.R. D'Amato, R. Fuccillo, and S. Scucchi. 1982. Infant abuse by a wild-born group-living Japanese macaque mother. *J. Abn. Psychol.* **91**:451–456.

Valentino, R.J., A.L. Curtis, M.E. Page, L.A. Pavcovich, and S.M. Florin-Lechner. 1998. Activation of the locus ceruleus brain noradrenergic system during stress: Circuitry, consequences, and regulation. *Adv. Pharmacol.* **42**:781–784.

Valentino, R.J., S.L. Foote, and M.E. Page. 1993. The locus coeruleus as a site for integrating corticotropin-releasing factor and noradrenergic mediation of stress responses. *Ann. NY Acad. Sci.* **697**:173–188.

Van Bockstaele, E.J., J. Peoples, and R.J. Valentino. 1999. Anatomic basis for differential regulation of the rostrolateral peri-locus coeruleus region by limbic afferents. *Biol. Psychiatry* **46**:1352–1263.

Van Oers, H.J., E.R. De Kloet, and S. Levine. 1998. Maternal deprivation effect on the infant's neural stress markers is reversed by tactile stimulation and feeding but not by suppressing corticosterone. *J. Neurosci.* **18**:10,171–10,179.

Van Praag, H., G. Kempermann, and F.H. Gage. 1999. Running increases cell proliferation and neurogenesis in the adult mouse dentate gyrus. *Nature Neurosci.* **2**:266–270.

von Holst, D. 1998. The concept of stress and its relevance for animal behavior. *Adv. Study Behav.* **27**:1–31.

Walker, C.-D., K.J.S. Anand, and P.M. Plotsky. 2000. Development of the hypothalamic-pituitary-adrenal axis and the stress response. In: Handbook of Physiology: Coping with the Environment, ed. B.S. McEwen, pp. 237–270. New York: Oxford Univ. Press.

Walker, C.D., M. Perrin, W. Vale, and C. Rivier. 1986. Ontogeny of the stress response in the rat: Role of the pituitary and the hypothalamus. *Endocrinology* **80**:1177–1179.

Walker, C.D., K.A. Scribner, C.S. Cascio, and M.F. Dallman. 1991. The pituitary-adrenocortical system of the neonatal rat is responsive to stress throughout development in a time-dependent and stressor-specific fashion. *Endocrinology* **128**:1385–1395.

Wechler, B. 1995. Coping and coping strategies: A behavioural view. *Appl. Animal Behav. Sci.* **43**:123–134.

Weinstock, M. 1997. Does prenatal stress impair coping and regulation of the hypothalamic-pituitary-adrenal axis? *Neurosci. Biobehav. Rev.* **21**:1–10.

Weiss, J.M. 1968. Effects of coping response on stress. *J. Comp. Physiol. Psychol.* **65**:251–260.

Weiss, J.M. 1971a. Effects of coping behavior in different warning signal conditions on stress physiology in rats. *J. Comp. Physiol. Psychol.* **77**:1–13.

Weiss, J.M. 1971b. Effects of coping behavior with and without a feedback signal on stress pathology in rats. *J. Comp. Physiol. Psychol.* **77**:22–30.

Wheal, H.V., Y. Chen, J. Mitchell, M. Schachner, W. Maerz, H. Wieland, D.V. Rossum, and J. Kirsch. 1998. Molecular mechanisms that underlie structural and functional changes at the postsynaptic membrane during synaptic plasticity. *Prog. Neurobiol.* **55**:611–640.

Wiener, S.G., C.L. Coe, and S. Levine. 1988. Endocrine and neurochemical sequelae of primate vocalizations. In: The Physiological Control of Mammalian Vocalization, ed. J. Newman, pp. 367–394. New York: Plenum.

Witek-Janusek, L. 1988. Pituitary-adrenal response to bacterial endotoxin in developing rats. *Am. J. Physiol.* **255**:E525–E530.

13

How and Why Coping Systems Vary among Individuals

J.M. KOOLHAAS, S.F. DE BOER, B. BUWALDA, B.J. VAN DER VEGT, C. CARERE, and A.G.G. GROOTHUIS

Department of Animal Physiology, University of Groningen
9750 AA Haren, The Netherlands

Keywords: Individual variation, coping strategies, fitness, population ecology, neurobiology, physiology

ABSTRACT

This chapter summarizes the behavioral, neuroendocrine, and neurobiological differences between proactive and reactive coping strategies in rats and mice. Evidence indicates that one of the main differences between the two coping strategies is the degree of behavioral plasticity. With repeated experience, proactive coping males develop rather rigid, routine-like behavior whereas reactive coping males seem to remain more flexible. This seems to be due to the degree in which behavior is guided by feedback signals. Proactive coping males tend to reduce the contribution of feedback, in particular after repeated positive outcome of behavior. Behavioral differentiation in coping strategies is correlated with a differentiation in a variety of neuroendocrine systems, including the hypothalamic-pituitary-adrenal (HPA) axis and the autonomic nervous system. At the level of the brain, coping strategies are associated with a differentiation in serotonergic neurotransmission and in a number of peptidergic brain mechanisms, such as the sexually dimorphic vasopressinergic system. Studies of feral populations of mice and birds indicate that the two coping strategies may have a function in the population dynamics of the species. Depending on the environmental conditions, such as population density, food availability, etc., the two coping strategies may have a differential fitness. Indeed a survey of the scarce literature on the evolutionary ecology of individual differentiation within a species suggests that, under certain conditions, natural selection might result in bimodality of individuals in a population. Coping strategies may be less distinct or even absent in laboratory animals and farm animals due to the absence of natural selection pressure, artificial selection, domestication, or inbreeding.

Coping with Challenge: Welfare in Animals including Humans
Edited by D.M. Broom

INTRODUCTION

Welfare, health, and stress-related disease are generally considered to be influenced by a complex function of the actual environmental demands and the individual's capacity to cope with these demands. A wide variety of medical, psychological, and biological studies, both in humans and other animals demonstrate that individuals may differ in their capacities to cope with environmental demands. Factors that have been shown to affect individual coping capacity include, for example, genotype, ontogeny, adult experience, age, and social support. Researchers have long tried to determine the individual vulnerability to stress-related diseases using estimates of the individual coping capacity. These attempts go back to the times of Hippocrates, who distinguished between four temperaments, that is, choleric, sanguine, phlegmatic, and melancholic. Each of these temperaments was supposed to reflect a general attitude to deal with everyday problems. More recent approaches use the concept of coping and try to classify coping responses into distinct coping strategies. Psychologists define coping in humans as the cognitive, behavioral, and emotional ways of managing stressful situations. According to Lazarus and Folkman (1984), coping responses are determined by the appraisal of the degree of control over important resources available to the individual. In their conceptualization, psychologists divide coping responses into emotion-focused and problem-focused coping strategies. The concept of coping is also used by biologists to understand the behavioral and physiological ways animals manage stressful conditions. We prefer to use the concept with this meaning. A coping strategy can be defined therefore as a coherent set of behavioral and physiological stress responses which is consistent over time and context, and which is characteristic of a certain group of individuals. Most studies describe individual differences in behavior and physiology as trait characteristics that may determine the individual vulnerability to stress related diseases.

While the studies mentioned above are in the realm of the biomedical sciences, there is a growing interest in individual differentiation in behavior and physiology in the science of ecology and evolutionary biology. Individual variation in coping with challenges in the natural habitat is not only considered as the origin of speciation, but may be an important factor in the regulation of populations as well. It seems that coping strategies have been shaped by evolution and form general adaptive response patterns in reaction to everyday challenges in the natural habitat. The aim of this chapter is to describe coping strategies in animals as they might be observed in nature and to summarize the available evidence of their possible function. We will focus on aggressive behavior as an important indicator and component of coping strategy. We argue that the individual variation in aggressive behavior is related to a differentiation in behavioral flexibility. This differential degree of flexibility seems to be one of the major factors involved in the differential fitness of individuals under various environmental conditions in nature.

COPING STRATEGIES: HOW?

Much of our current thinking on coping strategies is derived from the work of Jim Henry (Henry and Stephens 1977). He suggested, on the basis of social stress research in humans and other animals, that two different stress response patterns may be distinguished. The first type, the active response, was originally described by Cannon (1915) as the fight–flight response. Behaviorally, territorial control and aggression characterize the active response. Engel and Schmale (1972) originally described the second type of stress response as the conservation–withdrawal response. This response pattern is characterized behaviorally by immobility and low levels of aggression. These authors consider the degree to which animals react with an aggressive response to a stressor as an important discriminating factor of the two coping strategies. These ideas led to the hypothesis that the individual level of aggressive behavior, that is, the tendency to defend the home territory, is related to the way individual males react to environmental challenges in general. The hypothesis was tested by Benus (Benus et al. 1991b) using male house mice that were genetically selected for either short attack latency (SAL) or a long attack latency (LAL). These selection lines of wild house mice were derived from a completely different line of research aimed at the genetic basis of aggression. Also, when other indices of aggressive behavior are taken into account, the SAL males are considered extremely aggressive whereas the LAL males have very low levels of intermale aggressive behavior. The results of a series of experiments in mice and rats, which are summarized in Table 13.1, demonstrate that the individual tendency to initiate aggressive behavior is indeed predictive for the individual reaction to other, nonsocial environmental challenges. This pattern of behavioral responses is consistent with the concept of coping strategies.

Table 13.1 shows that the individual level of aggressive behavior is related to the way in which the animals react to a wide variety of environmental challenges. It seems that aggressive males have a strong tendency to take the initiative, that is, attack, active avoidance, and nest building. Nonaggressive males seem to accept the situation more easily as it is, responding only when absolutely necessary. This difference in response initiation forms the basis of the terminology we currently use for the different coping strategies. In our view, high levels of aggression are a reflection of a more general proactive coping strategy, whereas a low level of aggression reflects a reactive coping strategy.

An important fundamental question is whether the two types of behavior patterns can be considered to represent strategies of coping in the sense that they are both aimed at successful

Table 13.1 Summary of the behavioral differences between high-aggressive and low-aggressive male rats and mice (Koolhaas, Everts et al. 1999).

Behavioral Characteristics		
	High aggressive	*Low aggressive*
Tests for proactivity:		
Attack latency	Low	High
Active avoidance	High	Low
Defensive burying	High	Low
Nest-building	High	Low
Behavioral flexibility tests:		
Routine formation	High	Low
Cue dependency	Low	High

environmental control. Several experiments indicate that the different behavior patterns can indeed be considered as coping strategies. A nice example is the shock-prod defensive burying test. In this test, the animal is confronted with a small, electrified prod in its home cage. Because this prod is a novel object, the experimental animal will explore it by sniffing at the object. Consequently, the animal receives a mild but aversive shock. As soon as it has experienced the shock, the animal has two options to avoid further shocks. It may either hide in a corner of the cage to avoid further contact with the shock prod, or it may actively bury the shock prod with the bedding material of the cage. Under these free choice conditions, aggressive males spend most of the 10-min test-time burying, while nonaggressive males show immobility behavior. However, both types of response are equally successful in avoiding further shocks. In this particular test, successful coping can be defined operationally as avoidance of further shocks. In that sense, both response types in the defensive burying test can be considered as successful coping. However, Treit et al. (1981) described the defensive burying test as an anxiety test originally. High levels of burying behavior would indicate a high level of anxiety. Therefore, one may argue that the differences in burying behavior reflect a difference in baseline emotional state, such as anxiety. We think that this interpretation is incorrect. When high- and low-aggressive males (rats and mice) are tested in a well-validated anxiety test, the elevated plus maze, no differences were observed (see Figure 13.1). The difference between the anxiety tests is that the elevated plus maze is more a test for baseline anxiety while the burying test is a test for fear induced by the electric shock. In line with the concept of proactive and reactive coping, this fear can be expressed either as burying behavior or as immobility behavior.

It is important to emphasize that the differentiation in coping strategies may not be expressed equally clearly in all challenging situations. In particular, tests that measure aspects of initiative or proactivity seem to be most discriminative. This holds, for example, for latency measures such as the attack latency test in males or the defensive burying test, which allow the animal a choice between proactive and reactive coping. Although female mice usually do not show territorial aggression, females of the SAL selection line show much more

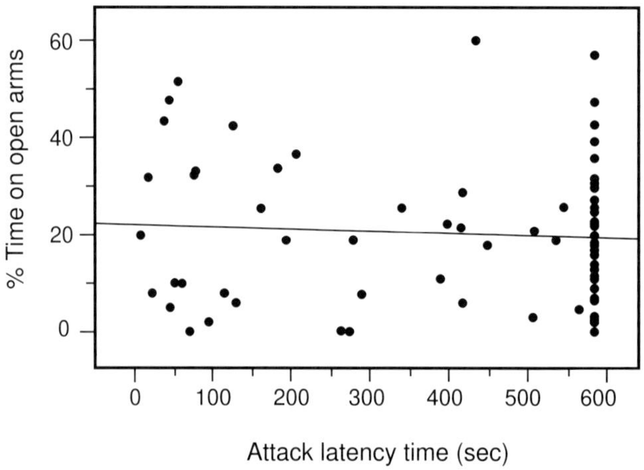

Figure 13.1 Absence of a correlation between attack latency of wild type male rats as measured in the resident intruder test and time spent in the open arm of the elevated plus maze.

defensive burying than female mice of the LAL selection line. This supports our view that aggression is only one of a larger set of behavioral characteristics that make up the proactive coping strategy.

Behavioral Flexibility

The concept of coping strategies is basically descriptive, that is, it describes the correlations and consistencies of the behavior of individual animals under different environmental conditions. The question is to what extent these differences might be causally related to differential cognitive abilities. Several experiments indicate that proactive and reactive coping strategies differ in the degree in which behavior is guided by environmental cues (Koolhaas, Korte et al. 1999). Aggressive males easily develop routines, that is, a rather intrinsically driven rigid type of behavior. Nonaggressive males are more flexible and react to environmental stimuli all the time. This can be demonstrated, for example, by the way in which aggressive and nonaggressive males react to a small change in an otherwise stable environment. Animals can be trained to run a maze for a food reward. Rats and mice can learn such a task easily. When they reached a stable task performance, the reaction to a small change in the maze was studied. In one experiment, a small piece of tape was put on the floor in one of the alleys of the maze, while in another experiment, the maze was turned ninety degrees with respect to the extramaze cues. In both experiments, aggressive males paid little or no attention to the change, that is, there was no increase in time to complete the task and no increase in the number of errors made in the maze. Nonaggressive males on the other hand started exploring the maze again and hence took much more time to get to the goal box and make more errors in the task. This suggests that the behavior of the nonaggressive male may be much more guided by environmental stimuli, while aggressive males seem to develop routines. This has also its consequences for behavior in a social setting. Benus tested to what extent male mice are able to distinguish a male mouse from a female mouse when they were introduced into their territory. Both highly aggressive and low-aggressive male mice that have no experience with conspecifics introduced into their home cage distinguish a male from a female intruder. A male intruder elicits aggressive behavior while a female elicits sociosexual behavior. However, when these males were trained with a male intruder on nine consecutive days, the first reaction of the high-aggressive mice was to attack the female when presented on the tenth day. The low-aggressive males still discriminate between a male and a female as before (Benus et al. 1990). This shows that the aggressive animal tends to rely on its previous experience rather than on a detailed account of its environment. A similar difference in behavioral plasticity can be demonstrated in the response to a twelve-hour shift in light–dark cycle. Aggressive male mice stay in their original day–night rhythm for a few days after which their rhythm gradually shifts to the new cycle. Nonaggressive males, on the other hand, shift their rhythm immediately; they are twice as fast in adapting to the new light–dark cycle as the aggressive males (Benus et al. 1988).

These experiments show that the two coping strategies differ more generally in the use of feedback and feedforward behavioral control. The proactive animal acts primarily on the basis of earlier experience, that is, feedforward control. The reactive coping animal seems to rely on feedback information. This fundamental difference in behavioral control may causally explain why the animals differ in such a wide variety of behavioral tests. One may conclude that the concept of coping strategies can be reduced to an underlying difference in a limited number of behavioral control mechanisms. Moreover, this differential use of behavioral control mechanisms also relates to the adaptive character of the two coping strategies. A proactive coping animal may be adapted to stable environmental conditions. After all, a feedforward behavioral control works best under highly predictable conditions. The reactive coping strategy may do better under variable and unpredictable environmental conditions. Indeed, field studies on feral mouse populations indicate that aggressive males are more successful under stable colony conditions, whereas nonaggressive males do better during migration (see below).

Neuroendocrinology of Coping Strategies

Coping strategies are not only characterized by differences in behavior but also by differences in physiology and neuroendocrinology. As mentioned earlier, tests that measure aspects of initiative or proactivity seem to be most discriminative. The defensive burying test in rodents is such a test; it allows the animal a choice between proactive and reactive coping. In general, high plasma noradrenaline and relatively low plasma adrenaline and corticosterone accompany defensive burying, while freezing behavior is associated with relatively low plasma noradrenaline and high plasma corticosterone levels. In a strain of wild-type rats, the more aggressive males showed the highest levels of burying behavior and showed a larger catecholaminergic (both plasma noradrenaline and adrenaline) reactivity after electrified prod exposure than the nonaggressive rats did (Sgoifo et al. 1996). Also during social defeat, the more competitive proactive male rats react with higher responses of blood pressure and catecholamines than the more reactive rats. In addition, these competitive males had higher baseline levels of noradrenaline (Fokkema et al. 1988). The same can be observed in a comparison between strains. The aggressive wild-type rats responded to social defeat with larger sympathetic (plasma noradrenaline levels) reactivity and concomitantly lower parasympathetic reactivity (as measured by increased heart rate response and decreased heart rate variability) than the less aggressive Wistar rats (Sgoifo et al. 1997). Thus, proactive coping rodents show in response to stressful stimulation a low HPA axis reactivity (low plasma corticosterone response), but high sympathetic reactivity (high levels of catecholamines). In contrast, reactive coping rodents show higher HPA axis reactivity and higher parasympathetic reactivity (Table 13.2). Differences in endocrine activity have also been observed for HPA axis and gonadal axis activity under baseline conditions. In aggressive mice, reduced circadian peak plasma corticosterone levels have been observed as compared with nonaggressive mice (Korte et al. 1996). In mice of the SAL selection line and in wild-type male rats, high baseline levels of testosterone have been observed, while the proactive coping male is also more sensitive to the behavioral effects of testosterone (De Ruiter et al. 1992).

There is a growing body of evidence that farm animals show the same differences in coping strategy and the associated neuroendocrinology as well. Hessing and colleagues showed that male castrated pigs could be characterized as high resistant or low resistant at an early age (1–2 weeks) by means of a back-test (manual restraint) (Hessing 1994). In this back-test, a piglet is put on its back and the number of bouts of resistance is used to characterize the animal. The high-resistant pigs made more escape attempts and mean heart rate frequency was higher than in low-resistant pigs. At three and eight weeks of age, the high-resistant ones were

Table 13.2 Summary of the physiological and neuroendocrine differences between proactive and reactive animals.

Physiological and Neuroendocrine Characteristics		
	Proactive	*Reactive*
HPA axis activity	Low	Normal
HPA axis reactivity	Low	High
Neurosympathetic reactivity	High	Low
Adrenomedullary reactivity	High	Medium
Parasympathetic reactivity	Low	High
Testosterone production	High	Low
Testosterone sensitivity	High	Low

less inhibited in approaches to novel objects in an open field. However, high-resistant pigs spent less time in exploring the novel object than low-resistant pigs. Heart rate frequency of high-resistant pigs was also substantially increased in reaction to a falling novel object, while heart rate frequency of low-resistant animals was only slightly increased or even decreased (bradycardia), suggesting that parasympathetic reactivity was higher in low-resistant pigs (Hessing et al. 1994). Hessing and colleagues did not find clear differences in HPA axis reactivity between the high- and low-resistant animals although basal plasma cortisol levels were higher in low-resistant than in high-resistant pigs, and this was accompanied by adrenal hypertrophy. Recently, however, Ruis et al. (2001) showed clear differences in HPA axis reactivity in high- and low-resistant female pigs. The low-resistant animals had higher HPA axis reactivity than the high-resistant ones. This was shown by higher salivary cortisol responses to a novel environment test, to routine weighing at 25 weeks of age, and to administration of a high dose of adrenocorticotrophic hormone (ACTH). Interestingly, the low-resistant animals with high HPA axis reactivity at 24 weeks of age showed less aggression in group-feeding competition tests, hesitated longer to leave their home pens and to contact a human than did high-resistant animals. Altogether, pigs that showed high resistance in the back-test and low HPA axis reactivity and high sympathetic reactivity in response to stressful stimulation are thought to be representatives of the proactive coping strategy. In contrast, pigs that showed low resistance in the back-test and high HPA axis reactivity and high parasympathetic reactivity are thought to be representatives of the reactive coping strategy. It should be pointed out that several other studies did not indicate two coping strategies in pigs.

Recently, it was shown that laying hens, from two lines with high or low propensity to feather peck, also show individual differences in physiological and behavioral responses to stress that are similar to the described coping strategies. During manual restraint (keeping the bird on its side by hand for 8 min), the high feather pecking line showed more resistance and higher mean heart rate frequency and lower parasympathetic reactivity than the low feather pecking line (Korte et al. 1999). HPA axis reactivity (plasma corticosterone levels) was highest in the low feather pecking line, while the sympathetic reactivity (plasma noradrenaline levels) was the highest in the high feather pecking line. These data suggest that chickens of the high feather pecking line are representatives of the proactive coping strategy, whereas birds of the low feather pecking line are representatives of the reactive coping strategy.

Neurobiology of Coping Strategies

Several studies in our rats and mice show a widespread central nervous differentiation between the two coping strategies, for example, at the level of the peptidergic modulation of the central nucleus of the amygdala (Roozendaal et al. 1992), the vasopressinergic neurons in the bed nucleus of the stria terminalis and its innervation of the lateral septum (Compaan et al. 1992), the suprachiasmatic nucleus (Bult et al. 1993), postsynaptic 5-HT1a receptor sensitivity, the hippocampal mossy fiber system (Sluyter et al. 1994), and striatal dopaminergic mechanisms (Benus et al. 1991a). These differences reflect differences in the state of brain mechanisms in terms of number of neurons, degree of arborization of neurons, hormonal and neurotransmitter receptor binding capacity, etc., which in concert may determine the tendency to cope either proactively or reactively with environmental challenges.

It is interesting to notice that the vasopressinergic system of the lateral septal area is well known to be sexually dimorphic and testosterone dependent. Males are characterized by a considerably higher density of LS-AVP fibers than females. However, the differences in density of LS-AVP fibers, as observed in the proactive and reactive coping strategies within the male gender turn out to be as large as the differences between the sexes. This indicates that the differentiation in coping strategies is somehow related to the differentiation between the sexes. Indeed, there is some evidence that the perinatal processes involved in the sexual differentiation are also involved in the development of the two coping strategies within the male gender (Koolhaas et al. 1999a) (Table 13.3).

COPING STRATEGIES: WHY?

Most of the data mentioned thus far have been obtained using males under rather limited experimental conditions in the laboratory. To allow conclusions about the adaptive significance of coping strategies, one needs data from more complex natural populations as well. There are several examples in mammals, birds, and fish showing that phenotypic variation is somehow maintained within a single natural population. These field studies support the general view as outlined above on the adaptive significance of individual variation in behavior and physiology of animals. On the basis of extensive ecological studies in a sunfish species (*Lepomis gibbosus*), Wilson (1998) found evidence that two morphologically different phenotypes occupy different habitats while both seem to have an advantage over morphologically intermediate individuals. Subsequently, he considers the possibility that the same might be true for the individual differentiation in behavior, in particular for the differentiation in shy and bold animals. The way in which the distinction between shy and bold is described in a number of species, including monkeys, octopus, and fish, suggests that the shy individual might be similar to the reactive coping male, while the bold individual might be similar to the proactive coper. This leads on to the question whether there is any evidence from field studies in rats and mice that the proactive and reactive coping strategy have a differential adaptive significance. In feral mouse populations, there is some evidence that different behavioral

Table 13.3 Summary of the central nervous differences between aggressive and nonaggressive male rats and mice.

Neurochemical Characteristics		
	Proactive	*Reactive*
Septum AVP-ir fibers	Low density	High density
SCN AVP-ir fibers	Low density	High density
AVP infusion in ACE	Insensitive	Immobility
AVP infusion in ACE	Insensitive	Bradycardia
CRH infusion in ACE	Behavioral activity	Insensitive
CRH infusion in ACE	Tachycardia	Insensitive
Hippocampal mossy fibers	Small	Large
Striatal dopamine	Sensitive	Insensitive
5-HT turnover	High	Low
5-HT 1a binding	High	Low
5-HT 1a receptor mRNA	High	Low
5-HT 1a receptor sensitivity	High	Low

phenotypes may have a differential fitness depending on the environmental conditions. In nature, mouse populations are known to go through phases of growth and phases of decline. Such population cycles may have a period from four to seven years and can be so extreme that the population suddenly collapses at the end of a cycle and becomes extinct. Chitty hypothesized that the cyclic nature of rodent populations might be due to a disruptive selection for aggressive behavior in the course of the population cycle. Evidence that this might be the case indeed was obtained in a study by Van Oortmerssen and Busser (1989) in seminatural populations of house mice consisting of both males and females. Phenotypic characterization of the laboratory bred male offspring (F1) of fathers caught from these colonies, revealed a bimodal distribution of attack latencies as measured in a standardized resident intruder paradigm. Subsequent selective breeding for high- and low-attack latencies resulted within five generations in a stable SAL selection line. After a number of failures, due to infertility of the offspring, we managed to obtain a LAL selection line as well. Cross-fostering and back-cross experiments show that the phenotypic differentiation in aggressive behavior as observed in the colonies has a strong genetic basis.

An analysis of the mortality reveals a strong increase in females, juveniles, and pre-weanling juveniles just before the crash of the population (Figure 13.2). The idea that this increase in mortality is due to enhanced levels of aggression in the males is supported by an index of territoriality. Figure 13.3 shows the number of males present in the nests. In the growth phase of the population, on average four males were found together in each nest. Towards the crash however, males were generally found alone, indicating a reduced tolerance for additional males in the nest. During the whole population cycle, animals are migrating from the population. It seems that the nonaggressive phenotype is more successful in

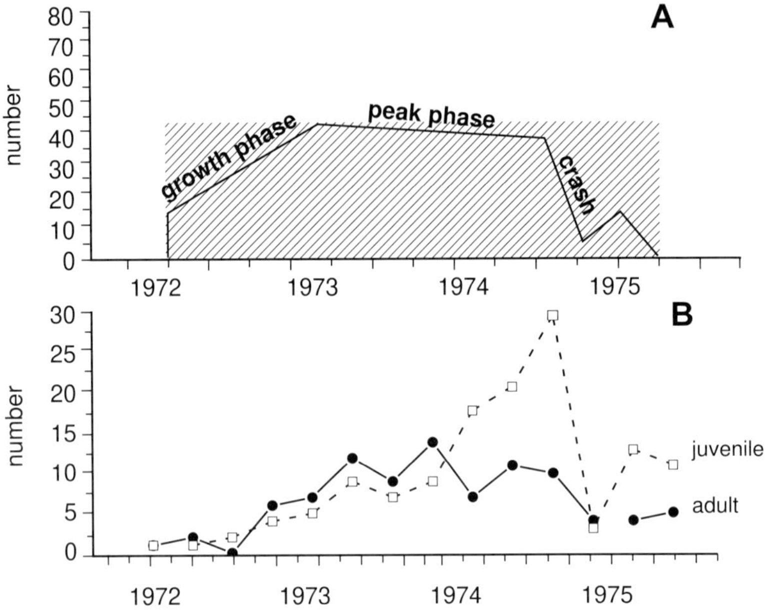

Figure 13.2 A: Change in population density of a feral population of house mice. B: Mortality of adult and juvenile house mice during the various phases of the population cycle.

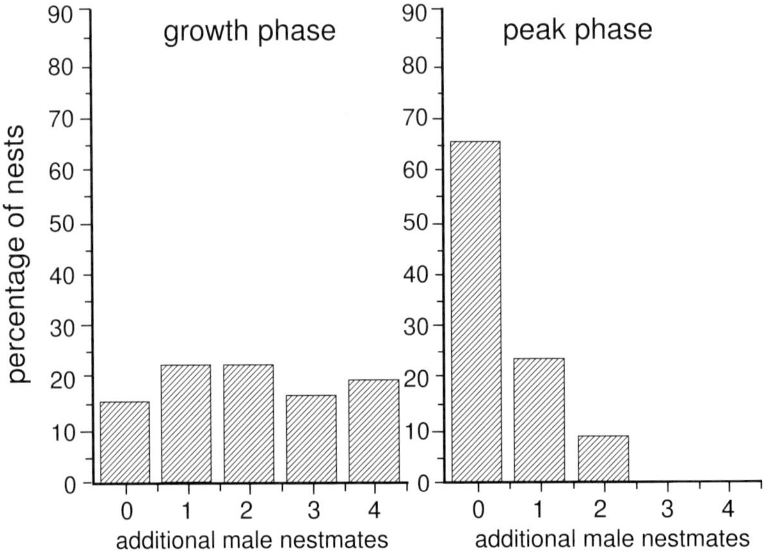

Figure 13.3 Number of males found together in one nest during the growth phase and at the end of the peak phase of a colony of wild house mice.

establishing a new colony than the highly aggressive phenotype. This leads to the more general view that the two phenotypes observed in seminatural populations of house mice might have a differential fitness depending on the phase of the population cycle.

Recent studies in the great tit (*Parus major*), a small song bird, show a similar differentiation of phenotypes like in our rodent studies. Bold animals are relatively aggressive, superficial explorers and more prone to develop behavioral routines than shy birds. In a field study, Verbeek observed a bimodal distribution of aggressive behavior of individual birds. Subsequent experiments showed that this differentiation has a genetic basis and can be considered behaviorally as a differentiation in proactive and reactive coping strategy (Verbeek et al. 1994). The maintenance of two different coping strategies within the same population may be explained by either habitat or frequency selection. In the first, different habitats favor different strategies. High-density areas with ample food supply may favor the bold type, while low-density areas with low food availability favor the shy strategy, which is relatively good in exploration. Frequency-dependent selection occurs when the relative frequency of the strategies determine fitness. In this sense, the concept of coping strategies addresses a fundamental issue in evolutionary biology.

CONCLUDING REMARKS

Individual differentiation in behavior and physiology is a well-known phenomenon in many animal species. Unfortunately, there are relatively few studies using a more systematic approach to characterize individual response patterns across various environmental conditions. However, the scarce literature suggests that the dimension of proactive and reactive coping strategies can be distinguished in a wide variety of animal species (Koolhaas, Korte et al. 1999). Authors may use different terms to characterize phenotypes, such as shyness and boldness or proactive and reactive, but they all seem to share the same basic characteristics. Detailed analysis of coping strategies in rats and mice indicate that the most fundamental difference between proactive and reactive coping is the degree in which behavior is guided by feedback and feedforward control. This results in a differential degree of behavioral flexibility. This degree of flexibility may have its origin in a differential survival value in nature. The challenge for the future is to integrate ethological, stress physiological, and ecological approaches in the study of coping strategies.

From a biomedical point of view, the concept of coping strategies implies that different animals have a differential capacity to adapt to the same environmental conditions. Negative health consequences might arise if an animal cannot cope with the stressor or needs very demanding coping efforts. In view of the differential neuroendocrine reactivity and neurobiological makeup, one may expect different types of stress pathology to develop under conditions in which a particular coping strategy fails. Although there are only a limited number of studies performed concerning pathology in relation to the type of coping strategy adopted, there are some indications that the two coping strategies differ in susceptibility to develop cardiovascular pathology, ulcer formation, stereotypies and infectious disease (Koolhaas, Korte et al. 1999).

If we accept the idea that nature somehow favors the existence of different phenotypes within one species, one may wonder how this relates to animals bred by human beings in laboratory or animal husbandry conditions. Both in wild house mice and in a small bird, the great tit (*P. major*), latency measures seem to have a bimodal distribution (Verbeek et al. 1994). However, one has to realize that the distribution is not truly bimodal because latency measures are generally finite, i.e., above a certain time the latency is set to the maximum value. This leads to an accumulation of individual scores in the distribution curve at this maximum value. Nevertheless, the individual behavioral scores in wild populations are certainly not normally distributed. In a laboratory-bred population of wild rats, we find three peaks in the distribution of attack latency times with a clear intermediate group. This difference with the wild situation may be explained by the fact that there is little or no selection pressure in the laboratory. It is tempting to consider the possibility that intermediate animals are less successful in nature. They may either not survive or reproduce in nature or be forced to choose either of the extreme coping strategies. Many studies use heavily domesticated animals that are usually selected for specific traits as well. In general, individual behavioral scores are normally distributed in these animals. Moreover, it is hard to tell how a certain inbred or domesticated strain relates to the original and presumably functional distribution of its wild ancestors. Nevertheless, it is intriguing that the extremes of this normal distribution often still fulfill the criteria for proactive and reactive coping strategies, both behaviorally and physiologically (Ruis et al. 2001). Although the discussion on the shape of the distribution curve is important from an evolutionary point of view, it does not seem to matter much when the individual vulnerability to stress related diseases is concerned. After all, it has been repeatedly shown that the extremes in a population, irrespective of the detailed distribution curve, may differ not only quantitatively, but also qualitatively in their behavioral and physiological response pattern to stress.

REFERENCES

Benus, R.F., B. Bohus, J.M. Koolhaas, and G.A. van Oortmerssen. 1991a. Behavioural differences between artificially selected aggressive and nonaggressive mice: Response to apomorphine. *Behav. Brain Res.* **43**:203–208.

Benus, R.F., B. Bohus, J.M. Koolhaas, and G.A. van Oortmerssen. 1991b. Heritable variation in aggression as a reflection of individual coping strategies. *Experientia* **47**:1008–1019.

Benus, R. F., S. Den Daas, J.M. Koolhaas, and G.A. van Oortmerssen. 1990. Routine formation and flexibility in social and nonsocial behaviour of aggressive and nonaggressive male mice. *Behaviour* **112**:176–193.

Benus, R.F., J.M. Koolhaas, and G.A. van Oortmerssen. 1988. Aggression and adaptation to the light-dark cycle: Role of intrinsic and extrinsic control. *Physiol. Behav.* **43**:131–137.

Bult, A., L. Hiestand, E.A. van der Zee, and C.B. Lynch. 1993. Circadian rhythms differ between selected mouse lines: A model to study the role of vasopressin neurons in the suprachiasmatic nuclei. *Brain Res.Bull.* **32**:623–627.

Cannon, W.B. 1915. Bodily Changes in Pain, Hunger, Fear, and Rage. New York: Appleton.

Compaan, J.C., R.M. Buijs, C.W. Pool, A.J.H. De Ruiter, and J.M. Koolhaas. 1992. Differential lateral septal vasopressin innervation in aggressive and nonaggressive male mice. *Brain Res. Bull.* **30**:1–6.

De Ruiter, A.J.H., J.M. Koolhaas, J.N. Keijser, G.A. van Oortmerssen, and B. Bohus. 1992. Differential testosterone secretory capacity of the testes of aggressive and nonaggressive house mice during ontogeny. *Aggr. Behav.* **18**:149–157.

Engel, G.L., and A.H. Schmale. 1972. Conservation withdrawal: A primary regulatory process for organic homeostasis. In: Physiology, Emotions, and Psychosomatic Illness, ed. R. Porter and J. Knight, pp. 57–95. New York: Elsevier.

Fokkema, D., K. Smit, J. van der Gugten, and J.M. Koolhaas. 1988. A coherent pattern among social behavior, blood pressure, corticosterone, and catecholamines measures in individual male rats. *Physiol. Behav.* **42**:485–489.

Henry, J.P., and P.M. Stephens. 1977. Stress, Health and the Social Environment: A Sociobiological Approach to Medicine. Berlin: Springer.

Hessing, M. 1994. Individual behavioural characteristics in pigs and their consequences for pig husbandry. Ph.D. diss., Wageningen Agricultural Univ.

Hessing, M.J.C., A.M. Hagelso, W.G.P. Schouten, P.R. Wiepkema, and J.A.M. van Beek. 1994. Indiviual behavioral and physiological strategies in pigs. *Physiol. Behav.* **55**:39–46.

Koolhaas, J.M., H.G.J. Everts, A.J.H. De Ruiter, S.F. De Boer, and B. Bohus. 1999. Coping with stress in rats and mice: Differential peptidergic modulation of the amygdala-lateral septum complex. In: Progress in Brain Research, vol. 119, ed. I.J. Urban, J.P.H. Burbach, and D. De Wied, pp. 437–438. New York: Elsevier.

Koolhaas, J.M., S.M. Korte, S.F. De Boer, B.J. van Der Vegt, C.G. van Reenen, H. Hopster, I.C. de Jong, M.A. Ruis, and H.J. Blokhuis. 1999. Coping styles in animals: Current status in behavior and stress physiology. *Neurosci. Biobehav. Rev.* **7**:925–935.

Korte, S. M., O.C. Meijer, E.R. De Kloet, B. Buwalda, J. Keijser, F. Sluyter, G. van Oortmerssen, and B. Bohus. 1996. Enhanced 5-HT1: A receptor expression in forebrain regions of aggressive house mice. *Brain Res.* **1–2**:338–343.

Korte, S.M., W. Ruesink, and H.J. Blokhuis. 1999. Heart rate variability during manual restraint in chicks from high- and low-feather pecking lines in laying hens. *Physiol. Behav.* **65**:649–652.

Lazarus, R.S., and S. Folkman. 1984. Coping and adaptation. In: Handbook of Behavioral Medicine, ed. W.D. Gentry. New York: Guilford Press.

Roozendaal, B., A. Wiersma, P. Driscoll, J.M. Koolhaas, and B. Bohus. 1992. Vasopressinergic modulation of stress responses in the central amygdala of the Roman high-avoidance and low-avoidance rat. *Brain Res.* **596**:35–40.

Ruis, M.A.W., J.H.A. te Brake, J.A. van de Burgwal, I.C. de Jong, H.J. Blokhuis, and J.M. Koolhaas. 2001. Personalities in domesticated pigs: Behavioral and physiological indications for an active–passive continuum. *Appl. Animal Behav. Sci.,* in press.

Sgoifo, A., S.F. De Boer, J. Haller, and J.M. Koolhaas. 1996. Individual differences in plasma catecholamine and corticosterone stress responses of wild-type rats: Relationship with aggression. *Physiol. Behav.* **60**:1403–1407.

Sgoifo, A., S.F. De Boer, C. Westenbroek, F.W. Maes, H. Beldhuis, T. Suzuki, and J.M. Koolhaas. 1997. Incidence of arrhythmias and heart rate variability in wild-type rats exposed to social stress. *Am. J. Physiol.* **42**:H1754–H1760.

Sluyter, F., L. Jamot, G.A. van Oortmerssen, and W.E. Crusio. 1994. Hippocampal mossy fiber distribution in mice selected for aggression. *Brain Res.* **646**:145–148.

Treit, D., J.P.J. Pinel, and H.C. Fibiger. 1981. Conditioned defensive burying: A new paradigm for the study of anxiolytic agents. *Pharmacol. Biochem. Behav.* **15**:619–626.

Van Oortmerssen, G.A., and J. Busser. 1989. Studies in wild house mice. III: Disruptive selection on aggression as a possible force in evolution. In: House Mouse Aggression: A Model for Understanding the Evolution of Social Behavior, ed. P.F. Brain, D. Mainardi, and S. Parmigiani, pp. 87–117. Chur: Harwood Academic.

Verbeek, M.E.M., P.J. Drent, and P.R. Wiepkema. 1994. Consistent individual differences in early exploratory behavior of male great tits. *Animal Behav.* **48**:1113–1121.

Wilson, D.S. 1998. Adaptive individual differences within single populations. *Phil. Trans. R. Soc. Lond.* **353**:199–205.

14

How Do Animals Cope with Social Problems?

M. MENDL

Centre for Behavioural Biology, Department of Clinical Veterinary Science, University of Bristol, Langford, BS40 5DU, U.K.

Keywords: social status, dominance, coping strategies, individual differences, stress, welfare, hypothalamic-pituitary-adrenal system, sympathetic-adrenal-medullary system

ABSTRACT

The problem of competing with others to establish and maintain social status, faced by most group-living and territorial animals, is considered here. In captivity, the benefits that usually accompany high rank are often small or absent because reproduction and access to resources such as food is organized and controlled by humans. Therefore, measures of physiological stress responses are important indicators of how captive animals cope with their social environment, and hence of their welfare. Under natural conditions, these measures inform us about the physiological costs of different positions in the social hierarchy. Classic models of social stress suggest that high-ranking individuals who actively maintain control of the social group are characterized by activation of the sympathetic-adrenal-medullary (SAM) system. In contrast, low-rankers who fail to control the environment show avoidance and withdrawal responses and elevated hypothalamic-pituitary-adrenal (HPA) activity. Both responses are viewed as adaptive attempts to cope with different sorts of social problems, but prolonged activation of both physiological systems can lead to pathology. A straightforward relationship between rank and coping response is, however, not borne out by recent research, which shows that individuals of similar rank may actually exhibit quite different responses to social situations. One possibility is that they perceive, evaluate, and respond to similar situations in quite different ways. Evidence for underlying predispositions to show particular coping responses is considered alongside evidence which suggests that perception and evaluation of the social situation may be an equally important source of individual differences. Another possibility is that the nature of the social situation faced by similar-ranking individuals can be quite different. Changes in social dynamics, as the result of group formation, social instability, presence or absence of social support, and opportunities for assessment and avoidance, can lead to marked alterations in the coping responses of both high- and low-ranking animals. The individual's actual experiences and perceptions of the social

---

Coping with Challenge: Welfare in Animals including Humans
Edited by D.M. Broom

environment are, therefore, more important than its rank per se in determining how it responds. Some of the factors that influence how successfully individuals cope with the social environment are summarized and implications for welfare and areas for further research discussed.

INTRODUCTION

In many mammalian and bird species, including those kept in captivity, the social environment can be a source of positive, affiliative relationships and social support; however, it can also be a source of "social problems." These include parent–offspring conflict, sibling conflict, the problem of mate selection, and social isolation. My main focus in this chapter is on the problem of competing with others to establish and maintain social status or dominance rank. I will concentrate on this for a number of reasons.

First, it is a very common problem. Most group-living and territorial species form social structures in which individuals differ in their ability to displace others and in their priority of access to resources. These abilities can be used as indicators of an individual's social status. Establishing a social position, maintaining it, and dealing with its consequences are problems that all individuals have to face. Second, there is a growing understanding of the physiological, as well as the behavioral, consequences and correlates of competitive interactions and relationships. This allows us to make more informed assessments of how well individuals are able to cope with their position in the dominance hierarchy. Third, research has shown that the physiological stress responses induced by competitive and dominance interactions can be among the most potent that an individual will experience (Koolhaas et al. 1997). This means that these interactions and the individual's ability to cope with them can have potentially profound effects on an animal's fitness and welfare.

My aim in this paper is to provide a brief overview of the behavioral and physiological responses shown by animals of differing social status. I will then consider how these responses vary between individuals and in different social conditions.

SOCIAL STATUS, FITNESS, AND COPING IN NATURAL AND CAPTIVE ENVIRONMENTS

Conventional measures of dominance rank usually reflect the number or proportion of animals in a group that each individual is able to displace in competition over resources, such as food, or when no resource is being contested directly. Other types of measure may focus on each individual's experience of winning or losing agonistic interactions, irrespective of who they are against. In both cases, individuals can be ranked in terms of their relative success. In a review of over 700 studies, Ellis (1995) suggests that animals who hold high rank usually also achieve a greater level of reproductive success, an important indicator of fitness, than those lower in the hierarchy. This is especially so for males in carnivore, ungulate, and rodent species, although a less straightforward relationship is observed sometimes in male primates and, in general, in females where variation in reproductive success is usually lower than in males. Nevertheless, females also show extreme examples of this relationship in which low-rankers are completely reproductively suppressed (e.g., Abbott et al. 1997).

If we view coping as a response to a challenging environment which minimizes reductions in fitness (cf. Broom and Johnson 1993), then achieving a high rank involves successful coping with a competitive social environment under natural conditions, as it not only minimizes reductions in fitness but often strongly enhances fitness. However, in captive animals, the major fitness benefits of obtaining high rank are often removed because reproduction is organized and controlled by humans rather than being linked directly to social order. Likewise, access to resources such as food, water, and preferred lying areas will be less rank dependent in captive conditions where husbandry procedures should prevent individuals monopolizing these resources. In these conditions, the effectiveness of high rank as a way of coping with the social environment will depend strongly on the costs incurred in achieving high rank. The same is true for captive lower-ranking individuals, who usually lack the reproductive disadvantages they experience under natural conditions. In terms of the welfare of captive animals, then, it is important to examine how animals of different social rank cope when the reproductive consequences of their behavior are removed from the equation. This information is also relevant in the natural situation in order to demonstrate whether individuals with similar reproductive performance also incur similar costs.

How should costs be assessed? Costs such as severe injury or death are easy to measure but comparatively uncommon in most within-species social interactions. A much more common phenomenon is the activation of physiological stress responses. These can be used as indicators of how hard the animal is working to cope with its environment, and hence of its welfare (Broom and Johnson 1993) and may, in the long term, lead to various pathological consequences (Henry and Stephens 1977; von Holst 1998). Over the last 30–40 years, there have been laboratory and field studies in which the physiological stress responses shown by individuals of different rank and those shown by individuals experiencing various forms of social conflict have been examined. Most information has been collected on the activity of the HPA and the SAM systems, with some information also available on the activity of gonadal and immune systems. I will focus in particular on the HPA and SAM systems.

RELATIONSHIPS BETWEEN SOCIAL STATUS, STRESS, AND COPING

If we use the HPA and SAM systems as indicators of socially induced stress, it is important to understand their correlates and functional significance in this context. In their classic model of stress responses to social challenges, Henry and colleagues (Henry and Stephens 1977; Ely and Henry 1978) suggested that individuals who showed high levels of locomotor activity, aggression, and active defense of resources or territory were characterized by activation of amygdala and SAM pathways and elevated gonadal activity (especially testosterone production in males). In contrast, individuals who showed low locomotor activity, decreased competitive behavior, and apparent withdrawal from social interactions exhibited activation of hippocampus–septum and HPA pathways and decreased gonadal activity.

Henry and colleagues postulated that the two response profiles reflect an individual's perception of its social situation and have different functions and consequences. Individuals perceiving a threat to their control of the social environment show the behavioral responses associated with SAM activation. These function to facilitate maintenance of social status and territorial control. However, if responses are sustained for a long period of time, repeated activation of the SAM system can lead to cardiovascular pathology (Henry and Stephens 1977).

Individuals who perceive that they are failing to control their social environment (e.g., receiving repeated attack, unable to escape) show behavior associated with HPA activation. This link between a perceived lack of control or predictability and activation of the HPA system is well known from many studies of nonsocial stress (Levine et al. 1989). The withdrawal responses facilitate avoidance of further social defeat thus enhancing survival chances (Ely and Henry 1978). If the responses have to be maintained for a long time, repeated or prolonged activation of the HPA system can lead to immune and reproductive suppression (von Holst 1998).

Both responses can thus be viewed as adaptive attempts to cope with different sorts of social predicament, or "coping strategies." If they do not alter the individual's perception of its predicament and, consequently, are sustained indefinitely, they can lead to long-term pathology and even death (see von Holst 1998). In these circumstances, they clearly fail to cope (*sensu* Broom and Johnson 1993).

In their model, Henry and colleagues emphasized that these two response profiles could be differentially activated according to an individual's perception of its control of the social environment. However, in their research on captive colonies of mice, they characterized the SAM activation response as being typical of high-ranking territorial males and the HPA activation response as typifying subordinate males. Individuals could switch their response type if their rank position was manipulated, suggesting that response was a function of rank rather than an attribute of the individual itself (Henry and Stephens 1977; Ely and Henry 1978).

The idea that animals of different social status show predictably different stress response profiles is attractive, and some evidence for the rank-stress response relationship reported by Henry and coworkers has been obtained in other species. For example, high-ranking wolf cubs have higher heart rate and metabolic expenditure than their lower-rank siblings (Folk et al. 1970). High-ranking free-living male baboons appear to have a strong cardiovascular stress response to SAM activation (Sapolsky and Share 1994). Studies of captive rodents, primates, ungulates, and free-living baboons (see Sapolsky 1994; von Holst 1998; Mendl et al. 1992) have found that low-ranking individuals have higher levels of HPA activation than high-rankers.

However, as emphasized in von Holst's (1998) comprehensive review of this area, a general straightforward relationship between social status and measures of the activation of HPA, SAM, immune, and gonadal systems does not exist. There are some methodological reasons for this including variation between studies in the type and range of measurements taken, and exactly how data are collected. These are important issues, but I will not consider them in any detail here. More interesting and relevant to this symposium is the emerging picture that an individual's actual experiences and perceptions of the social environment are more important than its rank per se in influencing its physiological responses.

In the rest of this chapter, I will first consider the possibility that individual differences in responses to similar social challenges are the result of predispositions to perceive, evaluate, or respond to the social environment in certain ways. Thereafter, I identify several features of the social environment that can have marked effects on how individuals cope with social challenge. An understanding of these has implications for appropriate housing and husbandry design to enhance the welfare of captive animals.

INDIVIDUALS DIFFER IN HOW THEY PERCEIVE AND RESPOND TO SOCIAL CHALLENGE

In a number of studies, individual conspecifics of similar rank and in apparently similar social circumstances show very different behavioral and physiological responses. For example, Sapolsky (1994) and coworkers found that not all dominant free-living male baboons have low basal cortisol levels. In fact, this was only the case for those males who were good at monitoring, predicting, and controlling the social situation. Sachser et al. (1998) showed that male guinea pigs lacking the ability to use nonaggressive behavior to develop relationships with unfamiliar conspecifics exhibited high levels of HPA activation during group formation in stark contrast to the remarkably low levels of activation normally observed.

A number of studies, principally of captive animals, have identified different response types among low-ranking animals. Low-ranking individuals may continue to compete actively and aggressively, or they may apparently accept their social position and avoid encounters with others. In some studies, it appears that the former response is associated with increased HPA or SAM activation relative to the latter response (e.g., female pigs, Mendl et al. 1992; male rats, Fokkema et al. 1995).

In studies of tree shrews, von Holst (1998) further subdivided "avoider" individuals into those who showed active avoidance and defensive behavior ("subdominants"), and those who showed withdrawal or hiding behavior and low levels of activity ("submissives"). He showed that subdominant males displayed elevated SAM activity, including elevations in baseline heart rate, and were able to survive in the presence of a dominant male. Submissive males, on the other hand, showed elevated HPA activation, reduced SAM activity, and reduced measures of immune function, and often died. Stefanski (1998) identified similar, but not identical, response types in male rats defeated in fights by a resident dominant.

An important question is what underlies these individual differences. In theory, one can conceptualize an individual's response to social challenge as a function of its perception and evaluation of the situation, and of the range of response types that it is able show (see Figure 14.1). This schema suggests that there are at least two different ways by which individual differences in responses to apparently similar social situations can be achieved. Individuals may be predisposed to show only one or a limited range of response types such that they tend to behave in similar ways irrespective of the precise details of the situation. Their responses will thus be relatively fixed. Alternatively, individuals may be capable of showing a wide range of responses; however, differences in how they perceive and evaluate apparently similar situations lead to differences in the response that they select. Such individuals might show a number of different responses over time and in different situations.

Individuals Exhibit Fixed Response Profiles: The Example of Active and Passive Copers

There is some evidence that there are different types of individuals in the population who are predisposed to show a particular fixed response to social challenge. Koolhaas and coworkers (1999) have suggested that the two coping response profiles described by Henry and Stephens (1977) may be individual specific. Active (or proactive) copers display a suite of behavioral and physiological characteristics, which includes high levels of locomotor

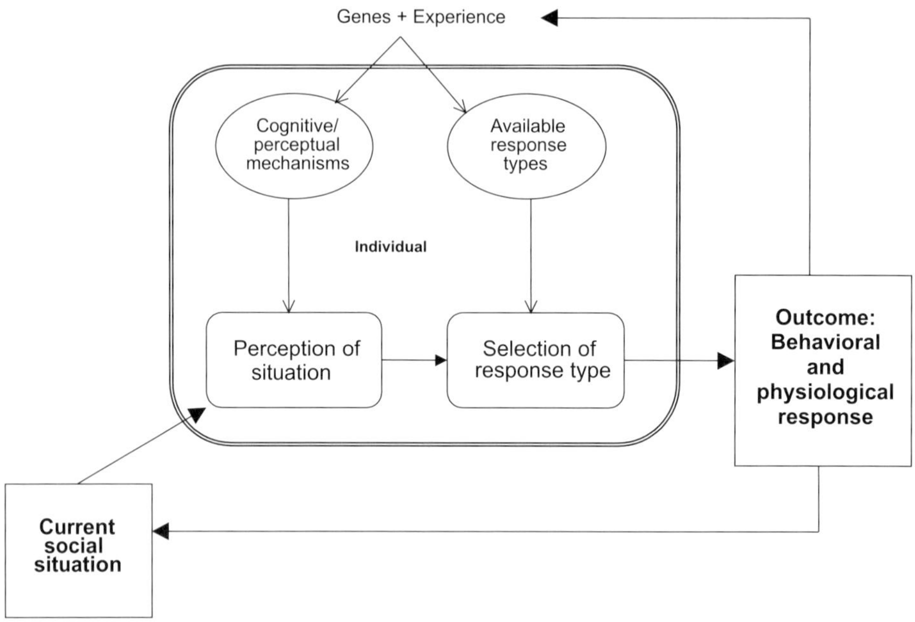

Figure 14.1 Theoretical framework representing the processes that may underlie an individual's response to social challenge. Perception and evaluation of the situation is influenced by cognitive or perceptual mechanisms which, in turn, influence the type of response selected. Cognitive and perceptual mechanisms and available response types are shaped by genetic and experiential factors. Individuals may thus vary in one or both of these parameters, and hence in the response selected and its outcome.

activity, rapid attack, a high propensity to develop routine-like behavior, low responsiveness to changes in environmental cues, low HPA and high SAM reactivity, as well as high testosterone activity. In contrast, passive (or reactive) copers show the opposite characteristics (Benus et al. 1991; Koolhaas et al. 1999). Fixed responses of this sort could underlie the differences between submissive and subdominant individuals observed by von Holst (1998) and Stefanski (1998).

Much of the initial work in this area was done on males from two lines of wild house mice from a single population (van Oortmerssen and Bakker 1981) under specific testing conditions (see Benus et al. 1991). In this species, van Oortmerssen and colleagues suggested that active coper males are more successful in fitness terms when in a familiar environment, while passive coper males do better during dispersal in unfamiliar environments. Thus, there are reasons to believe that disruptive selection linked to population cycles may have driven the evolution of the two phenotypes (van Oortmerssen and Busser 1989; see also Wilson 1998).

Is the active/passive dichotomy an important determinant of individual differences in response to social challenge? This depends on how widespread the dichotomy is across species; whether it is a cause rather than a consequence of individual differences and, if so, whether individuals remain consistent in their response strategies in complex and changing social environments.

We should only expect a similar dichotomy of responses in other species if appropriate disruptive selection pressures exist. In general, selection pressures may favor individuals showing one type of response in one situation, and a different type in another situation (context-specific individual differences, Wilson 1998), rather than one response type in all situations. However, constraints on neurophysiological organization may mean that individuals are inevitably biased to show one or other response profile, irrespective of context. The existence and influence of such constraints needs to be investigated.

Evidence is most convincing if the *clustering* of behavioral and physiological characteristics outlined above can be reproduced in other species and if it is clearly bimodally distributed with individuals consistently exhibiting one or other response cluster (Jensen 1995; Mendl and Deag 1995). Although distinct individual differences in relevant characteristics (e.g., activity) may be common, this does not mean that they necessarily form part of an active/passive cluster. This must be demonstrated empirically. There is some evidence for similar response clusters in male rats, tree shrews, and great tits (Fokkema et al. 1995; von Holst 1998, Verbeek 1998), while there has been little research on females of these species, and studies of pigs remain equivocal (e.g., Hessing et al. 1993; Forkman et al. 1995). In general, evidence that individuals are predisposed to show one or other response is at present limited to a few species.

In species which tend to show more active or more passive responses, to provide the most convincing evidence that active/passive personalities underlie variation in responses to social groupings, it is necessary to demonstrate that individuals characterized as active or passive show the expected behavior and physiological responses in a social context. This requires following *individuals* from group formation onwards. Merely observing the active/passive phenotypes in a social setting does not indicate whether they are *consequences* of the individual's social experience, as suggested by Henry and Stephens (1977), or manifestations of consistent individual characteristics.

There have been only a limited number of studies following active/passive individuals from the establishment of a social group onwards. For example, Fokkema et al. (1995) showed that rats predefined as active/aggressive individuals in individual tests behaved aggressively and displayed SAM activation and elevated blood pressure when placed in a social group. Interestingly, only one of these individuals became dominant in each group and he did not show elevated blood pressure. Verbeek (1998) demonstrated that male great tits classified as active explorers (a trait with similar behavioral features to the active phenotype described in mammals) were more aggressive when initially introduced into a social group. However, many of these individuals subsequently failed to achieve high-ranking status, perhaps due to their rather indiscriminate fighting behavior, and showed increased inhibition of aggression following lost fights, whereas passive explorer males gradually rose up the hierarchy as they settled into the group. Other studies have also shown that individual characteristics such as high aggressiveness or elevated SAM function can predict aggressive behavior (e.g., pigs, Hessing et al. 1993) and the acquisition of high rank once animals enter a group (e.g., guinea pigs and tree shrews, see von Holst 1998; marmosets, Saltzman et al. 1996). However, a causal relationship between aggressiveness and dominance is often not observed (e.g., Verbeek 1998).

These findings indicate that there is some predictive value of the initial categorization of individuals. However, it is clear that behavioral and probably physiological profiles of the individuals may change according to their experience of success or failure in the social group (see also Benus et al. 1990). An active individual who succeeds in establishing high rank may show low locomotor and SAM activity relative to an active individual who becomes subdominant (Fokkema et al. 1995). It is as if achievement or nonachievement of the "goal" of a particular phenotype, rather than the characteristics of that phenotype per se, is the key factor influencing the resulting physiological response (e.g., pigs, Mendl et al. 1992; rats, Hurst et al. 1996). This may mean that, once a group is formed, it is not possible to identify the initial coping strategy of an individual from its behavior or physiology in the group situation.

In general, it seems likely that, within the complexities of social environments, "core" coping strategies may become masked by behavior that is influenced by the immediate social context and has been shaped by prior experience with other individuals (Mendl and Deag 1995). An important question is whether the "core" coping strategies remain intact or are completely altered by social experience. Suomi (1987) notes that stable individual differences in behavioral and physiological reactivity of rhesus monkeys are only revealed in threatening or challenging situations, but cannot be detected in stable social situations. There seems to be some underlying consistency which is not altered by social experience but is only revealed in certain circumstances. Longitudinal studies are needed to examine whether this is also true for active and passive copers.

Individuals Perceive/Evaluate the Same Social Conditions Differently

Individual variation in response to the same social conditions may result from differences in how an individual perceives and evaluates that situation, and responds according to these perceptions, rather than from any predisposition to show a particular form of coping response. Here I shall use the term perceive to refer to any mechanisms involved in the perception and evaluation of a situation and course of action which result in the selection of a particular response (see Figure 14.1). Differences in perception and appraisal of a situation may arise for a number of reasons. For example, an individual's size, strength, and weaponry, how much it values the contested resource, and its current social status and previous experience of victory or defeat may all influence how highly motivated it is to compete and hence how vigorously it competes and how readily it accepts defeat (see Archer 1988). Thus, low-ranking or defeated individuals may persist in competing with high-rankers if they value the contested resource highly, or accept defeat if they do not attach much value to the resource. These differences in perception may vary across time and situation within the same individual and are thus "state dependent."

It is tempting to speculate that state-dependent differences in perception and evaluation underlie some of the cases in which individuals appear to accept low social status readily (e.g., low-ranking female pigs, Mendl et al. 1992; low-ranking and reproductively suppressed female marmosets, Abbott et al. 1997). It is possible that these individuals are operating a "biding-time" tactic in which they perceive that the most beneficial course of action is to accept low social status now and challenge for higher status later. This may be particularly evident in female mammals, which generally stay in their natal groups. In some cases, helping related individuals to reproduce in the natal group may provide similar fitness benefits to attempt to breed elsewhere (Emlen 1991), and accepting low rank may thus be an equally successful alternative strategy to competing for high rank (see Saltzman et al. 1996). The relative benefits of these strategies may depend on how many other individuals are adopting them (frequency dependence). Perceptual or evaluative mechanisms allowing animals to assess this will facilitate choice of the most appropriate response type. The nature of such mechanisms is unknown but is likely to involve an ability to monitor and accumulate information about the quality of social interactions experienced by oneself and others in the group. There is evidence for the existence of such apparently sophisticated abilities (Hogue et al. 1996).

In addition to state-dependent differences in how individuals perceive and monitor social conditions, there may also be more consistent individual differences in these abilities. For example, Sapolsky (1994) suggested that the ability of dominant baboons to monitor and respond effectively to the behavior of others differed consistently between individuals and hence appeared to be a type of personality characteristic which had a strong influence on how the individuals coped with their high social status.

Interestingly, one important feature of the active/passive dichotomy is the marked difference in the propensity to monitor and respond to environmental changes. For example, Benus et al. (1990) showed that resident male mice displaying the active phenotype rapidly developed routine-like responses in repeated intruder-attack tests and failed to alter their aggressive behavior when their own female partner was used as the intruder. Passive male mice noticed the change and did not attack their own female. In a more complex group situation, Verbeek (1998) suggested that male great tits showing the passive phenotype were more sensitive to the social interactions going on around them. They more often attacked a dominant individual after it had been defeated several times by another bird than did males showing the active phenotype. This behavior increased their chances of moving up the hierarchy, which they did more successfully than many of the active-type birds.

These findings suggest that both active and passive copers may be able to display high-aggressive and low-aggressive type responses (e.g., Benus et al. 1990), but that the passive phenotype more readily inhibits aggressive behavior in inappropriate conditions. Thus, differences in the ability to monitor the environment may be the critical factor underlying some or all of the differences in response shown by active and passive animals in the social context (Koolhaas et al. 1999). Precisely how such differences are manifest is unclear, but it is possible that they result in variation in the ability of individuals to form and retrieve memories of the identity and social relevance of other individuals, and of their past interactions. If active animals are poorer at attending to and memorizing information concerning the identity of individuals who defeated them, this could explain their propensity to persist in attacking these animals in a group situation (e.g., Fokkema et al. 1995; Verbeek 1998).

Summary

The schema illustrated in Figure 14.1 draws attention to two possible processes underlying individual differences in responses to similar social situations. It is likely that both are involved in many cases, but disentangling their roles will give us a better idea of exactly how individuals "decide" to respond to a social challenge and how flexible their responses are likely to be. Any predispositions to show a specific response type may influence responses to novel social situations but may be poor predictors of subsequent social rank and general behavior and physiology in a settled group. However, in novel or challenging social conditions, they may be revealed once more. From a functional viewpoint it makes sense for individuals to revert to underlying behavioral predispositions in these situations, provided that this allows them to show rapid responses or direct full attention to the surroundings, rather than to organizing and moderating their behavior according to various learnt rules (Mendl and Deag 1995). The implications are that if individuals do have specific coping responses, they will be good predictors of their responses in novel social situations, for example, when animals are mixed together for the first time (e.g., Erhard et al. 1997), but less so in established groups. The causes of variation in how individuals perceive and monitor the social environment could include both genetic factors, as may be the case in some active and passive phenotypes, and also experiential factors as appears to be the case for male guinea pigs with different levels of social skill (Sachser et al. 1998).

THE CURRENT SOCIAL SITUATION AFFECTS RESPONSES TO SOCIAL CHALLENGE

A review of the literature shows that several key aspects of the social environment appear to influence how animals respond to social challenge.

Formation of New Groups: Establishing Social Status

When social status is initially being acquired, animals face quite different challenges from those experienced in established groups. Under natural conditions, individuals may grow up in a group and experience age-related changes in social status, or inherit rank from their parents, or slowly become assimilated into a new group and acquire rank through associations with others or by regular and measured challenges to dominants. Acquisition of rank may thus be a gradual and relatively benign process.

However, in captivity, unfamiliar individuals are often abruptly mixed together with limited opportunity to assess or avoid each other. In these circumstances, all individuals usually experience elevated HPA and SAM activity during initial encounters (see von Holst 1998). These encounters are often highly aggressive and their outcomes influence subsequent stress responses. If individuals succeed in removing the social threat by winning encounters and establishing dominance, they may show a return to baseline activity (e.g., tree shrews, von Holst 1998). Continued elevation of SAM activity may occur in individuals that perceive ongoing social challenge (e.g., they win encounters but face further threat from other individuals (Ely and Henry 1978); they lose encounters but maintain their challenge to dominants (Fokkema et al. 1995)). Defeated individuals, or those who fail to integrate in a group, may show prolonged HPA activation and related pathological consequences (e.g., rats, Koolhaas et al. 1997; rabbits, von Holst 1998), especially if they are unable to predict or control their exposure to aggression or other forms of social threat (tree shrews, von Holst 1998).

The abrupt formation of new groups thus usually results in high levels of aggression and a variety of physiological stress responses which depend mainly on the individual's experience of success or failure.

Opportunities to Use "Nonfighting Assessment" Rather than Fighting to Settle Disputes

One way of minimizing the high levels of aggression and stress observed during group formation is to use some form of assessment, which does not involve overt aggression, to establish status or settle disputes. Under natural conditions, fighting is itself an important way of assessing and obtaining information about the relative competitive ability of an opponent (see Enquist and Leimar 1983). However, "nonfighting assessment" of relative social status and competitive strength through displays or other cues is predicted to occur when costs of escalated aggression, such as injury, are high (Parker 1974). Therefore, these nonfighting assessment abilities (subsequently referred to here as assessment) should have evolved in species and sexes where fighting can be damaging. In addition, resolution of disputes through assessment should be prevalent when there are asymmetries in the abilities of competitors, and environmental conditions must also facilitate assessment by allowing individuals time and space to monitor each other.

The ability to use assessment during pairwise encounters has been demonstrated in various species under both natural (e.g., red deer, Clutton-Brock and Albon 1979) and captive conditions (e.g., tree shrews, von Holst 1998). There are also a number of studies indicating that individuals are able to acquire and use information about the relative status of group members by observing their interactions with each other (e.g., baboons, Kummer 1981; chickens, Hogue et al. 1996). This "eavesdropping" ability suggests that animals are capable of quite complex modification of their behavior depending on the interactions going on around them.

Assessment abilities may thus be quite widespread across species, including those kept in captivity. However, under captive conditions, the opportunity to use these abilities is often limited. Unfamiliar animals are frequently mixed together abruptly and in very confined and densely populated pens. They lack the time and space required for some forms of assessment (e.g., Clutton-Brock and Albon 1979). Furthermore, they may be housed with animals of the same sex and similar size and weight such that asymmetries between competitors are minimized. Both these factors increase the likelihood of fights occurring rather than assessment, and the detrimental consequences that accompany damaging physical interactions. Captive conditions that facilitate the use of assessment are likely to enhance the ability of individuals to cope with social mixing, and improve their welfare.

Social Stability

Group formation is an extreme example of social instability; however, established groups may also go through unstable periods when members join or leave, rank reversals occur, or resources become scarce and competition increases. Under these circumstances, high-ranking individuals may experience stronger physiological stress responses than when in settled groups. For example, Kaplan and Manuck (1997) showed that high-ranking male cynomolgus monkeys experiencing regular group membership changes developed high levels of coronary atherosclerosis. This is likely to have resulted from recurrent SAM activation accompanying the males' responses to repeated challenge to their social position. Such effects were not observed in high-ranking males in stable social groups or in low-ranking males. Sapolsky (1994) noted that in unstable groups of free-living baboons, dominant males had higher basal levels of glucocorticoids than dominants in stable groups. In wild rabbits living in large enclosures, dominants in larger and more unstable social groupings had similar HPA activity to subordinates whereas in smaller stable groups their HPA activity was lower (von Holst 1998).

Taken together these and other data indicate that when the social environment is unpredictable and dominant individuals have less control over it and need to assert their status, they are more likely to experience elevations in the activity of the HPA and/or SAM systems. HPA activation may rise to such a level that it does not differ between high- and low-ranking animals. Under these conditions, dominant individuals need to work harder to cope both behaviorally and through activation of stress physiology systems, and this increases the chances of stress-related pathological consequences. In contrast, in stable conditions when dominants are easily able to control access to resources, often with little overt aggression, they may exhibit minimal activation of either the HPA or SAM systems (von Holst 1998). When social instability and increased competition result from limitations in resources, subordinate individuals may be particularly susceptible to receiving high levels of aggression from others, and this may result in further elevations in their HPA activity (cf. Sapolsky 1994).

Presence or Absence of Social Support

A variety of studies have indicated that the presence of a companion can influence the extent to which an individual exhibits socially related stress responses. This is best documented in studies of social separation. For example, Levine et al. (1997) summarize work showing that an infant squirrel monkey's stress response to separation from its mother can be minimized if other mothers and/or familiar peers are present. Sachser et al. (1998) observed similar effects in young guinea pigs even if the animal present was completely unfamiliar. This phenomenon of "social buffering" has received less attention in the context of social subordination. However, Ruis et al. (1999) showed that the effects of a single defeat, which include strong HPA and SAM responses and bear some resemblance to features of human depression (Koolhaas et al. 1997), were considerably greater if a rat was left in isolation following defeat as opposed to being allowed to return to its social group. The extent to which this effect was due to social isolation per se is not clear.

Work on some primate species indicates that, following an agonistic interaction, the contestants themselves may exhibit apparent reconciliation behavior such as embracing, mouth to mouth contact, and holding hands. It appears that this behavior may act to repair or prevent damage to social relationships and emphasizes the potential importance of social support in allowing animals to cope with conflict situations (de Waal and Aureli 1997). The presence of social support clearly has the potential to ameliorate the detrimental consequences of being a low-ranking or defeated individual but has received little investigation in this context.

Escape and Avoidance Opportunities

The success of coping responses involving avoidance of aggressive conspecifics depends entirely on the availability of escape opportunities. In captive, paired tree shrew males, the prolonged activation of the HPA system of individuals exhibiting a passive avoidance response and its sometimes lethal consequences are likely to be the result of an inability to escape from the dominant male. This occurs even if both males are physically but not visually separated by a transparent barrier, indicating the potency of the psychological as opposed to the physical component of this stress response (von Holst 1998).

Creel et al. (1996) pointed out that a link between social subordination and elevated HPA activity has primarily been demonstrated in studies of captive animals where subordinates are usually unable to avoid aggressive dominants. In the wild, such avoidance is possible and the consequence may be that subordinates suffer considerably less socially induced stress. In their studies of free-living wild dogs and dwarf mongooses, subordinates indeed appeared to have lower HPA activity than dominants.

Summary: Experience of Specific Social Events Is More Important than an Individual's Nominal Rank

The above points demonstrate that measures of an individual's social status or rank may provide relatively little information about its precise experience of social events. Individuals, ostensibly with the same rank but living in different social conditions, may vary greatly in exactly how they experience the social environment, and this appears to be a major determinant of how they cope. A high-ranking individual may dominate calmly without having to display aggressive behavior, or may rush around chaotically quelling repeated challenges to its status. A low-ranking individual may successfully avoid aggression, or be repeatedly attacked and have no affiliative relationships with other group members to help it cope. The calm high-ranker and the successful low-rank avoider may even show similar low levels of physiological stress (cf. Mendl et al. 1992).

Species differences in social organization may also mean that individuals in apparently similar rank positions experience quite different social conditions. For example, Sachser et al. (1998) demonstrated that captive guinea pigs could be housed at high densities without any apparent difference in SAM and HPA activation in high- and low-ranking individuals. They suggested that, unlike many species, guinea pigs are able to form social structures characterized by tolerance and low levels of aggression even at high densities in captivity. In both free-living wild dogs and dwarf mongooses (Creel et al. 1996), basal glucocorticoid levels are higher in dominant than in subordinate individuals, while the reverse is true in free-living baboons (Sapolsky 1994). This difference may occur because dominants in the former two species are regularly involved in overt aggressive encounters to maintain their status, while this is less common in baboons (Creel et al. 1996). Another possible explanation is that animals of different status take different roles concerning the outside world (e.g., defense of the group, decision making), and that these roles have an important influence on HPA activity and glucocorticoid levels. Thus, the demands of different rank positions may be quite species specific, and the precise nature of these demands may be more important than nominal social rank in determining how well an individual copes.

CONCLUSIONS, WELFARE IMPLICATIONS, AND AREAS FOR FURTHER RESEARCH

Figure 14.2 provides a summary of how animals can respond to the challenges of establishing and living in a social group and what the physiological correlates of different responses are. It is based on the information presented in this chapter and in Table III in von Holst (1998). The figure can apply to pairwise encounters in which one dominant and one subordinate animal

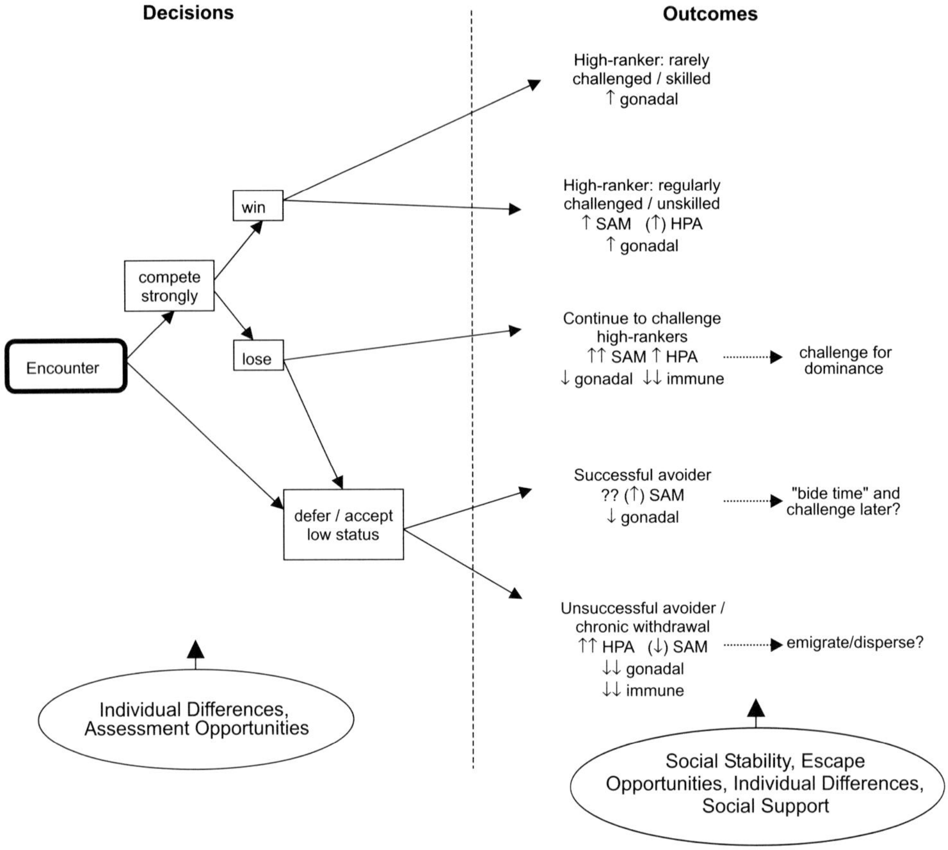

Figure 14.2 Summary of some of the types of response (outcome) to social challenge that individuals show and their physiological correlates, based on this chapter and Table III in von Holst (1998). Putative functions of the "non-high-ranker" outcomes are indicated (e.g., "emigrate/disperse"). The outcomes can be viewed as consequences of a series of decisions taken by individuals during encounters with others. The decisions made may be influenced by individual differences in perception or coping responses, and the availability of assessment opportunities. The outcome achieved may also be affected by individual differences, and by the stability of the social environment and the availability of escape opportunities and social support.

emerge, or it can be taken to represent the progress of an individual through the establishment of a larger social group.

Each outcome's physiological and behavioral correlates are important indicators of its relative effectiveness as a coping response especially when, as in captivity, rank, reproduction, and resource access are not strongly linked. It could be argued that a high-ranking animal with high control of the social environment and the lowest level of physiological stress copes most successfully, while a low-ranker with no social control and high physiological stress fails to cope. However, a high-ranker with poor control of the social environment may cope less well, and thus have poorer welfare, than a low-ranker who accepts its status. The use of HPA and SAM activity as indicators of physiological stress and hence of coping can, of course, be criticized. For example, in low-ranking and reproductively suppressed marmosets, levels of reproductive hormones may play as important a causal role as social stress in influencing HPA activity (Abbott et al. 1997). Information on other measures of stress responses, immune and gonadal function, access to and use of resources, alterations in behavior patterns (e.g., sleep, activity), weight changes and growth rates, and injuries or disease should strengthen future studies of coping in social situations.

Figure 14.2 also emphasizes that the outcome that is reached depends on various decisions that individuals make during social encounters, and on the success or failure of these decisions. As should be clear from this chapter, current social conditions and differences between individuals in how they perceive and respond to these conditions are critical determinants of the decisions made, their success or otherwise, and the outcome achieved.

From a welfare perspective, enhancing group stability, facilitating the use of assessment as opposed to fighting to resolve disputes, and providing escape or avoidance opportunities will help to decrease the amount of work that individuals need to do to cope and hence improve their welfare. The beneficial effects of social support and other affiliative behavior require further investigation. The observation that, in some species, subdominant individuals use behavior which partially or completely offsets any reproductive advantage of being behaviorally dominant, should encourage us to examine whether we can create group structures in which the usual asymmetries in relative costs and benefits experienced by individuals of different status are minimized. There is some evidence that groups containing both high- and low-aggressive individuals (characterized in tests) are more harmonious than groups containing only one or other type, at least in the short term (see Mendl and Deag 1995). However, longitudinal analysis of how *individuals* perform in these groups is needed.

More information is required on exactly how individuals perceive the social environment and on individual variation in these abilities. Studies of assessment, eavesdropping, and social recognition memory may all reveal mechanisms and individual differences that could enhance our understanding of how individuals cope with the social environment. They will be relevant to studies of very large groups, typical of some farm animal housing systems, in which social organization based on recognition of individuals may become unfeasible. The factors affecting social stress responses under these conditions are poorly understood. Any further search for active and passive coping strategies in other species should be guided by a consideration of which species and sexes are likely to have experienced appropriate disruptive selection. It should also be accompanied by more open-ended research for other types of individual difference, such as those linked to perceptual abilities. More generally, the cross-time and cross-situation consistency of individual differences, particularly in changing social environments, and the effects of genes and experience on their development, need further research. If genetic variation contributes to individual differences in, for example, the propensity to exhibit aggressive behavior, then selection for particular types of characteristic (e.g., low aggressiveness) might lead to lower levels of socially induced stress in captive groups. Similarly, if experiences during early life enhance or disrupt the development of appropriate social skills (see Sachser et al. 1998), then by respectively maximizing or minimizing exposure to these experiences, we will be able to ensure that few individuals develop to become socially inept and/or hyper-aggressive.

There is a notable lack of information on how females cope with social challenges and this should be addressed. Females may show quite different responses and associated consequences to males (see Hurst et al. 1996). Coping responses should also be viewed from a lifetime perspective. Responses shown at one age need not be at all predictive of what happens later. For example, captive low-ranking individuals who appear to accept their status may be showing a "biding-time" response or an "escape/emigration" response. The former should be followed by a challenge for higher rank at a later date while the latter may not be, and it may only be possible to distinguish these through longer-term study. Similarly, highly aggressive or domineering behavior during early life may be beneficial to an individual at that time but may lead to a more rapid decline in social competitive abilities, for example, as a result of injury or other physical damage, as the individual grows older.

Finally, the large amount of data available on primates and rodents should be complemented by studies of other species. These should look for some of the subtleties of social interaction and their effects on coping abilities that have been revealed in studies of primates in particular. Research is accumulating on pigs, but further work is needed on this and other captive species in which the social environment may be the single most important influence on animal welfare.

ACKNOWLEDGMENTS

I thank the U.K. Biotechnology and Biological Sciences Research Council for support, and the members of the Dahlem Workshop for comments on this chapter.

REFERENCES

Abbott, D.H., W. Saltzman, N.J. Schultz-Darken, and T.E. Smith. 1997. Specific neuroendocrine mechanisms not involving generalized stress mediate social regulation of female reproduction in cooperatively breeding marmoset monkeys. *Ann. NY Acad. Sci.* **807**:219–238.

Archer, J. 1988. The Behavioural Biology of Aggression. Cambridge: Cambridge Univ. Press.

Benus, R.F., B. Bohus, J.M. Koolhaas, and G.A. van Oortmerssen. 1991. Heritable variation for aggression as a reflection of individual coping strategies. *Experientia* **47**:1008–1019.

Benus, R.F., S. den Daas, J.M. Koolhaas, and G.A. van Oortmerssen. 1990. Routine formation and flexibility in social and non-social behaviour of aggressive and non-aggressive male mice. *Behaviour* **112**:176–193.

Broom, D.M., and K.G. Johnson. 1993. Stress and Animal Welfare. London: Chapman and Hall.

Clutton-Brock, T.H., and S.D. Albon. 1979. The roaring of red deer and the evolution of honest advertisement. *Behaviour* **69**:145–170.

Creel, S., N.M. Creel, and S.L. Monfort. 1996. Social stress and dominance. *Nature* **379**:212.

De Waal, F.B.M., and F. Aureli. 1997. Conflict resolution and distress alleviation in monkeys and apes. *Ann. N.Y. Acad. Sci.* **807**:317–328.

Ellis, L. 1995. Dominance and reproductive success among nonhuman animals: A cross-species comparison. *Ethol. Sociobiol.* **16**:257–333.

Ely, D.L., and J.P. Henry. 1978. Neuroendocrine response patterns in dominant and subordinate mice. *Horm. Behav.* **10**:156–169.

Emlen, S.T. 1991. Evolution of cooperative breeding in birds and mammals. In: Behavioural Ecology, ed. J.R. Krebs and N.B. Davies, pp. 301–337. Oxford: Blackwell Scientific.

Enquist, M., and O. Leimar. 1983. Evolution of fighting behaviour: Decision rules and assessment of relative strength. *J. Theor. Biol.* **102**:387–410.

Erhard, H.W., M. Mendl, and D.D. Ashley. 1997. Individual aggressiveness can be measured and used to reduce aggression after mixing. *Appl. Animal Behav. Sci.* **54**:137–151.

Fokkema, D.S., J.M. Koolhaas, and J. van der Gugten. 1995. Individual characteristics of behavior, blood pressure, and adrenal hormones in colony rats. *Physiol. Behav.* **57**:857–862.

Folk, Jr, G.E., M.W. Fox, and M.A. Folk. 1970. Physiological differences between alpha and subordinate wolves in a captive sibling pack. *Am. Zool.* **10**:487.

Forkman, B., I.L. Furuhaug, and P. Jensen. 1995. Personality, coping patterns and aggression in piglets. *Appl. Animal Behav. Sci.* **44**:31–42.

Henry, J.P., and P.M. Stephens. 1977. Stress, Health, and the Social Environment. New York: Springer-Verlag.

Hessing, M.J.C., A.M. Hagelsø, J.A.M. van Beek, P.R. Wipekema, W.G.P. Schouten, and R. Krukow. 1993. Individual behavioural characteristics in pigs. *Appl. Animal Behav. Sci.* **37**:285–295.

Hogue, M.E., J.P. Beaugrand, and P.C. Lague. 1996. Coherent use of information by hens observing their former dominant defeating or being defeated by a stranger. *Behav. Proc.* **38**:241–252.

Hurst, J.L., C.J. Barnard, R. Hare, E.B. Wheeldon, and C.D. West. 1996. Housing and welfare in laboratory rats: Time-budgeting and pathophysiology in single-sex groups. *Animal Behav.* **52**:335–360.

Jensen, P. 1995. Individual variation in the behaviour of pigs — Noise or functional coping strategies. *Appl. Animal Behav. Sci.* **44**:245–255.

Kaplan, J.R., and S.B. Manuck. 1997. Using ethological principles to study psychosocial influences on coronary atherosclerosis in monkeys. *Acta Physiol. Scand.* **Suppl. 640**:96–99.

Koolhaas, J.M., S.M. Korte, S.F. de Boer, B.J. van der Vegt, C.G. van Reenen, H. Hopster, I.C. de Jong, M.A.W. Ruis, and H.J. Blokhuis. 1999. Coping styles in animals: Current status in behavior and stress physiology. *Neurosci. Biobehav. Rev.* **23**:925–935.

Koolhaas, J.M., P. Meerlo, S.F. de Boer, J.H. Strubbe, and B. Bohus. 1997. The temporal dynamics of the stress response. *Neurosci. Biobehav. Rev.* **21**:775–782.

Kummer, H. 1982. Social knowledge in free-ranging primates. In: Animal Mind — Human Mind, ed. D. Griffin, pp. 113–130. Dahlem Workshop Report. Berlin: Springer-Verlag.

Levine, S., C. Coe, and S.G. Wiener. 1989. Psychoneuroendocrinology of stress: A psychobiological perspective. In: Psychoendocrinology, ed. F.R. Brush and S. Levine, pp. 341–377. San Diego: Academic.

Levine, S., D.M. Lyons, and A.F. Schatzberg. 1997. Psychobiological consequences of social relationships. *Ann. NY Acad. Sci.* **807**:210–218.

Mendl, M., and J.M. Deag. 1995. How useful are the concepts of alternative strategy and coping strategy in applied studies of social behaviour? *Appl. Animal Behav. Sci.* **44**:119–137.

Mendl, M., A.J. Zanella, and D.M. Broom. 1992. Physiological and reproductive correlates of behavioural strategies in female domestic pigs. *Animal Behav.* **44**:1107–1121.

Parker, G.A. 1974. Assessment strategy and the evolution of fighting behaviour. *J. Theor. Biol.* **47**:223–243.

Ruis, M.A.W., J.H.A. te Brake, B. Buwalda, S.F. de Boer, P. Meerlo, S.M. Korte, H.J. Blokhuis, and J.M. Koolhaas. 1999. Housing familiar male wildtype rats together reduces the long-term adverse behavioural and physiological effects of social defeat. *Psychoneuroendocrinology* **24**:285–300.

Sachser, N., M. Dürschlag, and D. Hirzel. 1998. Social relationships and the management of stress. *Psychoneuroendocrinology* **23**:891–904.

Saltzman, W., N.J. Schultz-Darken, and D.H. Abbott. 1996. Behavioural and endocrine predictors of dominance and tolerance in female common marmosets, *Callithrix jacchus*. *Animal Behav.* **51**:657–674.

Sapolsky, R.M. 1994. Individual differences and the stress response. *Sem. Neurosci.* **6**:261–269.

Sapolsky, R.M., and L.J. Share. 1994. Rank-related differences in cardiovascular function among wild baboons — Role of sensitivity to glucocorticoids. *Am. J. Primatol.* **32**:261–275.

Stefanski, V. 1998. Social stress in loser rats: Opposite immunological effects in submissive and subdominant males. *Physiol. Behav.* **63**:605–613.

Suomi, S.J. 1987. Genetic and maternal contributions to individual differences in rhesus monkey biobehavioral development. In: Perinatal Development: A Psychobiological Perspective, ed. N.A. Krasnegor, E.M. Blass, A.M. Hofer, and W.P. Smotherman, pp. 397–419. New York: Academic.

van Oortmerssen, G.A., and Th.C.M. Bakker. 1981. Artificial selection for high and low attack latencies in wild *Mus musculus domesticus*. *Behav. Genet.* **11**:115–126.

van Oortmerssen, G.A., and J. Busser. 1989. Studies in wild house mice. 3: Disruptive selection on aggression as a possible force in evolution. In: House Mouse Aggression: A Model for Understanding the Evolution of Social Behaviour, ed. P.F. Brain, D. Mainardi, and S. Parmigiani, pp. 87–117. Chur: Harwood Academic.

Verbeek, M. 1998. Bold or Cautious. Behavioural Characteristics and Dominance in Great Tits. Ph.D. diss., Univ. of Wageningen.

von Holst, D. 1998. The concept of stress and its relevance for animal behavior. *Adv. Stud. Behav.* **27**:1–131.

Wilson, D.S. 1998. Adaptive individual differences within single populations. *Phil. Trans. R. Soc. Lond. B* **353**:199–205.

15

Protective and Damaging Effects of Stress Mediators

Lessons Learned from the Immune System and Brain

B.S. McEwen

Harold and Margaret Milliken Hatch Laboratory of Neuroendocrinology,
The Rockefeller University, New York, N.Y. 10021, U.S.A.

Keywords: stress, allostasis, homeostasis, allostatic load, immune, delayed-type hypersensitivity, brain, hippocampus, adaptive plasticity

ABSTRACT

The hormones and other physiological agents that mediate the effects of stress on the body have protective and adaptive effects in the short run and yet can accelerate pathophysiology when they are overproduced or mismanaged. Here the protective and damaging effects of these mediators as they relate to the immune system and brain are considered. "Stress" is a principle focus, but this term is rather imprecise, as are uses of "homeostasis." Therefore, two new terms are used, allostasis and allostatic load, to supplement and clarify the meaning of "stress" and "homeostasis." For the immune system, acute stress enhances immune function whereas chronic stress suppresses it; these effects can be beneficial for some types of immune responses and deleterious for others. For the brain, acute stress enhances the memory of events that are potentially threatening to the organism. Chronic stress, on the other hand, causes adaptive plasticity in the brain in which local neurotransmitters as well as systemic hormones interact to produce structural as well as functional changes. Only under extreme conditions does permanent damage ensue. Indeed, adrenal steroids tell only part of the story as far as how the brain adapts or shows damage. Moreover, comparison of the effects of experimenter-applied stressors and psychosocial stressors show that what animals do to each other is often more potent than what experimenters do to them. Even then, the brain is resilient and capable of adaptive plasticity, and these aspects have turned out to have unexpected clinical ramifications. Finally, a scheme is presented to connect the protective and damaging effects of the mediators of allostasis and allostatic load.

Coping with Challenge: Welfare in Animals including Humans
Edited by D.M. Broom

INTRODUCTION

Stress is an aspect of our daily lives and a word often used in our conversations, and yet there is ambiguity in what we mean when we talk about "stress." The brain is the master controller of the interpretation of what is stressful and the behavioral and physiological responses that are produced. The brain is also significantly affected by stress, along with the immune system, metabolism and cardiovascular systems as well as other systems of the body. Hypothalamic-pituitary-adrenal cortex (HPA) hormones play a major role in mediating both adaptive and maladaptive responses and do so by interacting with specific aspects of each tissue. What is often overlooked is that the HPA hormones are protective in the short run and yet can participate in damage when they are overproduced or not shut off when no longer needed. In this chapter, I discuss the circumstances and mechanisms that differentiate between protection and damage.

Animals, including humans, continually learn, and some experiences are classified as "stressful" in part because HPA hormones are produced. Contrary to Selye, who mainly emphasized physical stressors, psychological and experiential factors are among the most powerful of stressors: novelty, withholding of reward, and anticipation of punishment rather than the punishment itself are among the most potent activators of HPA and autonomic nervous system (ANS) activity (Mason 1959, 1975).

Although environmental impacts producing HPA activity are usually thought about as bad and damaging, recent studies paint a different picture as far as the brain is concerned. The first is that the brain appears to handle repeated impacts of this kind over weeks by showing adaptive plasticity in which local neurotransmitters as well as systemic hormones interact to produce structural as well as functional changes. Thus, adrenal steroids do not tell the whole story as far as how the brain adapts. Moreover, comparison of the effects of experimenter-applied stressors and psychosocial stressors show that what animals do to each other is often more potent than what we, as experimenters, do to them. Even then, there is reason to believe that the brain is resilient and capable of adaptive plasticity. A similar story applies to the immune system, which will be described. First, however, it is important to consider a number of important concepts and terms before discussing stress, the immune system, and the brain.

PROTECTIVE AND DAMAGING EFFECTS OF STRESS MEDIATORS: HOMEOSTASIS AND ALLOSTASIS

"Stress" is often defined as a threat, real or implied, to homeostasis, while "homeostasis" refers to the maintenance of a narrow range of vital physiological parameters necessary for survival. In common usage, "stress" usually refers to an event or succession of events that cause a response, often in the form of "distress" but also, in some cases confusingly, referring to a challenge that leads to a feeling of exhilaration, as in "good" stress. The term "stress," however, is full of ambiguities. It is often used to mean the event (stressor) or the response (stress response). It is normally used in a negative sense, and sometimes it is used to describe a chronic state of imbalance in response. In this chapter, "stress" will be used to describe an event or events that are interpreted as threatening to an individual and which elicit physiological and behavioral responses. The brain is the key organ involved in interpreting and responding to potential stressors. Before considering its role, the following are some additional key terms and the way they are used in this article.

Stress Response

The most commonly studied physiological systems that respond to stress are the HPA axis and the ANS, particularly the sympathetic response of the adrenal medulla and sympathetic nerves. These systems respond in daily life according to stressful events as well as to the diurnal cycle of rest and activity. Thus, these systems do more than respond to "stressors" even though they are frequently identified as "stress response systems." Behaviorally, the response to stress may consist of fight or flight reactions or potentially health-related behaviors such as eating, alcohol consumption, smoking, or other forms of substance abuse. Another type of reaction to a potentially stressful situation is an increased state of vigilance, accompanied, at least in our own species, by enhanced anxiety and worrying, particularly when the threat is ill-defined or imaginary and when there is no clear alternative behavioral response that would end the threat. The behavioral responses to stress and these states of anxiety are both capable of exacerbating and potentiating the production of the physiological mediators of health outcomes.

Homeostasis

Homeostasis, in a strict sense, applies to a limited number of systems like pH, body temperature, and oxygen tension, i.e., components of the internal milieu that are truly essential for life and are therefore maintained over a narrow range, as a result of their critical role in survival. These systems are not activated or varied in order to help the individual adapt to its environment. In contrast, systems that show "variation to meet perceived/anticipated demands" (Sterling and Eyer 1988) characterizes the state of the organism in a changing world and reflects the operation of most body systems in meeting environmental challenges, e.g., through fluctuating hormones, heart rate and blood pressure, cytokines of the immune system, and other tissue mediators like neurotransmitters and hormones. Those mediators are most certainly not held constant, although their levels may usually operate within a range, and they participate in processes leading to adaptation as well as contributing to pathophysiology when they are produced insufficiently or in excess, i.e., outside of the normal range.

Allostasis

Introduced by Sterling and Eyer in 1988, allostatis is used to characterize how blood pressure and heart rate responses vary with experiences and time of day and also to describe changes in the set point of these parameters in hypertension. They referred to the change in set point as the primary example that distinguishes allostasis from homeostasis. Yet there is a much broader implication of what they wrote. In their paper, they state: "Allostasis emphasizes that the internal milieu varies to meet perceived and anticipated demand." This led us (McEwen and Stellar 1993) to define allostasis more broadly than the idea of a changing set point, namely, as the process for actively maintaining homeostasis. This is important because, in our view, the systems that vary according to demand, like the HPA axis and ANS, actually help maintain those systems that are truly homeostatic. Moreover, large variations in the HPA axis and ANS do not lead directly to death as would large deviations in oxygen tension and pH.

Therefore, we propose that allostasis is a much better term for physiological adaptation mechanisms than is homeostasis, which should be reserved for the parameters that are essentially maintained for survival. Therefore, allostasis is the process that keeps the organism alive and functioning, that is, maintaining homeostasis or "maintaining stability through change" and promoting adaptation and coping, at least in the short run (McEwen 1998, 2000).

We note, however, that another view of homeostasis is that it can also refer to the operation of coordinated physiological processes that maintain most of the steady states of the organism (Cannon 1929). In this interpretation, homeostasis and allostasis might seem to mean almost the same thing. The problem with the use of "homeostasis" in this context is that it does not distinguish between those systems essential for life and those that maintain them.

What are some examples of allostasis? Sterling and Eyer (1988) used variations in blood pressure as an example: in the morning, blood pressure rises when we get out of bed and blood flow is maintained to the brain when we stand up to keep us conscious. This type of allostasis helps maintain oxygen tension in the brain. Other examples include catecholamine and glucocorticoid elevations during physical activity to mobilize and replenish, respectively, energy stores needed for brain and body function under challenge. These adaptations maintain essential metabolism and body temperature.

Examples of allostasis go beyond the immediate control of body temperature and pH to broader aspects of individual survival, for example, from pathogens or physical danger. As we will see below, in the immune system, acute stress-induced release of catecholamines and glucocorticoids facilitates the movement of immune cells to parts of the body where they are needed to fight an infection or to produce other immune responses (Dhabhar and McEwen 1999). Finally, in the brain, glucocorticoids and catecholamines act in concert to promote the formation of memories of events of potentially dangerous situations so that the individual can avoid them in the future (Roozendaal 2000). Yet, each of these adaptive processes has a potential cost to the body when allostasis is either called upon too often or is inefficiently managed, and that cost is referred to as "allostatic load."

Allostatic Load

Whereas allostasis refers to the process of adaptation to challenges, "allostatic load" refers to the price that the body pays for being forced to adapt to adverse psychosocial or physical situations. It also represents either the presence of too much allostasis or the inefficient operation of the allostasis response systems, which must be turned on and then turned off again after the stressful situation is over.

What are some examples of the damaging as well as the adaptive effects in different systems? For example, glucocorticoids, so-named because of their ability to promote conversion of protein and lipids to usable carbohydrates, serve the body well in the short run by replenishing energy reserves after a period of activity, like running away from a predator. Glucocorticoids also act on the brain to increase appetite for food and to increase locomotor activity and food-seeking behavior (Leibowitz and Hoebel 1997), thus regulating behaviors which control energy input and expenditure. This is very useful when we do manual labor or play active sports, but it is not beneficial when we grab a pizza and a beer while watching television or writing a paper. Inactivity and lack of energy expenditure creates a situation where chronically elevated glucocorticoids resulting from poor sleep, ongoing stress, or as side effects of a rich diet can impede the action of insulin to promote glucose uptake. One of the results of this interaction is that insulin levels increase and, together, insulin and glucocorticoid elevations promote the deposition of body fat, and this combination of hormones promotes the formation of atherosclerotic plaques in the coronary arteries (Brindley and Rolland 1989).

For the heart, we see a similar paradox. Getting out of bed in the morning requires an increase in blood pressure and a reapportioning of blood flow to the head so that we can stand up without fainting (Sterling and Eyer 1988). Our blood pressure rises and falls during the day as physical and emotional demands change, providing adequate blood flow as needed. Yet repeatedly elevated blood pressure promotes generation of atherosclerotic plaques, particularly when combined with a supply of cholesterol and lipids and oxygen-free radicals that damage the coronary artery walls (Manuck et al. 1995). Beta adrenergic receptor blockers are known to inhibit this cascade of events and to slow down the atherosclerosis that is accelerated in dominant male cynomologus monkeys exposed to an unstable dominance hierarchy (Manuck et al. 1991). Thus catecholamines and the combination of glucocorticoids and insulin can have dangerous effects on the body, besides their important short-term adaptive roles (Brindley and Rolland 1989).

We now move to consider protective and damaging effects of mediators of allostasis in the immune system and brain, where the beneficial and injurious aspects of this process are not as well appreciated.

STRESS AND IMMUNE FUNCTION

The immune system is regulated by neural input from sensory, sympathetic, and parasympathetic nerves (Bulloch 2000) as well as by circulating hormones, of which the glucocorticoids are among the most prominent (McEwen et al. 1997; Spencer et al. 2000). Long regarded as inhibitors of immune function, adrenal steroids have now been recognized as having biphasic effects upon immune function, as shown recently in studies of delayed-type hypersensitivity (DTH) (Dhabhar and McEwen 1999, 1997). This makes more sense, because the organism's response to acute challenge is in other respects protective. Under acute stress, energy reserves are mobilized, vegetative processes and reproduction are suppressed, and the body is made ready for fight or flight, which may involve wounding. Thus the immune defense system should acutely gear up to protect the organism from infections and accelerate wound healing.

A primary underlying mechanism for these effects is the translocation or "trafficking" of immune cells between the blood and different primary, secondary, and tertiary immune tissues (see Bulloch 2000; Spencer et al. 2000). Elevations of stress hormones, both glucocorticoids and catecholamines, direct the movement of various cell types of the immune system. As shown in Figure 15.1, lymphocytes, monocytes, and natural killer cells are all reduced in number in blood and increased in number in tissues such as the skin as a result of acute stress; the same thing is seen after acute glucocorticoid administration (Dhabhar and McEwen 1996; Dhabhar et al. 1995).

Once immune cells have marginated and begun to enter the tissue, other factors become involved as local mediators of further activation of immune function. Interferon gamma

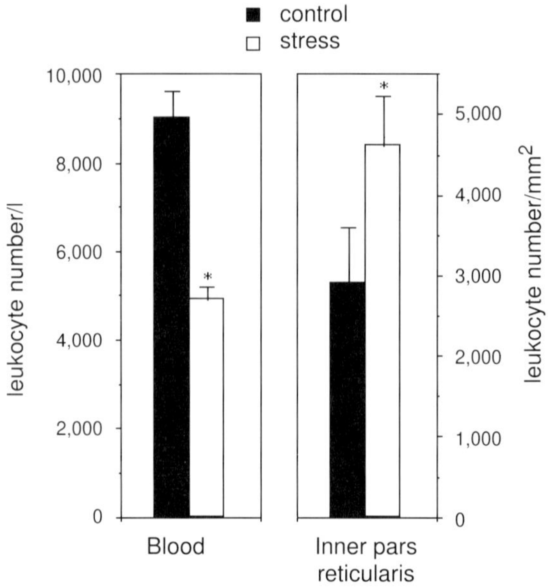

Figure 15.1 Immune cells are depleted reversibly from the blood during stress-induced elevation of glucocorticoids and marginate on blood vessel walls and enter tissues such as the skin (Dhabhar 1996).

(IFN) is an important factor in the stress-enhancement of the DTH response, and this has been shown by a lack of a stress effect in mice lacking the receptor for IFN gamma (Dhabhar et al. 2000). Furthermore, immunoneutralization of IFN gamma in normal mice also blocks the stress effect. Thus, although IFN gamma is not required for the baseline DTH response, it is evidently important for the manifestations of the effects of acute stress. IFN gamma is known to induce expression of antigen-presenting and cell adhesion molecules on endothelial cells and macrophages and cell adhesion molecules on leukocytes. It is also significant that glucocorticoids induce IFN gamma receptors on monocytes (for discussion and references, see Dhabhar et al. 2000).

What about the effects of chronic stress? Acute stress shows a dose dependency to activate the DTH response, and this is related to the magnitude of glucocorticoid secretion (Dhabhar and McEwen 1997). As shown in Figure 15.2, exogenous glucocorticoids mimic this dose dependency, but at higher glucocorticoid doses there is an inverted U dose response curve and suppression of the DTH response is seen (Dhabhar and McEwen 1999).

Dexamethasone, which does not bind to serum corticosteroid binding globulin, mimics the suppressive effects of high dose glucocorticoids (Dhabhar and McEwen 1999). As shown in Figure 15.3, chronic stress over 3–5 weeks produces a suppression of the DTH response and also suppressed the initial sensitization of the response (Dhabhar and McEwen 1997).

An important factor in this suppression is the lack of immune cell trafficking, and this may be due, at least in part, to habituation of the corticosterone response to stress (Dhabhar and McEwen 1997). However, there are likely to be other effects of repeated stress or high dose glucocorticoids to reduce production or actions of the cytokines and cell adhesion molecules that make the immune system work. Thus the immune system exemplifies the contrasting

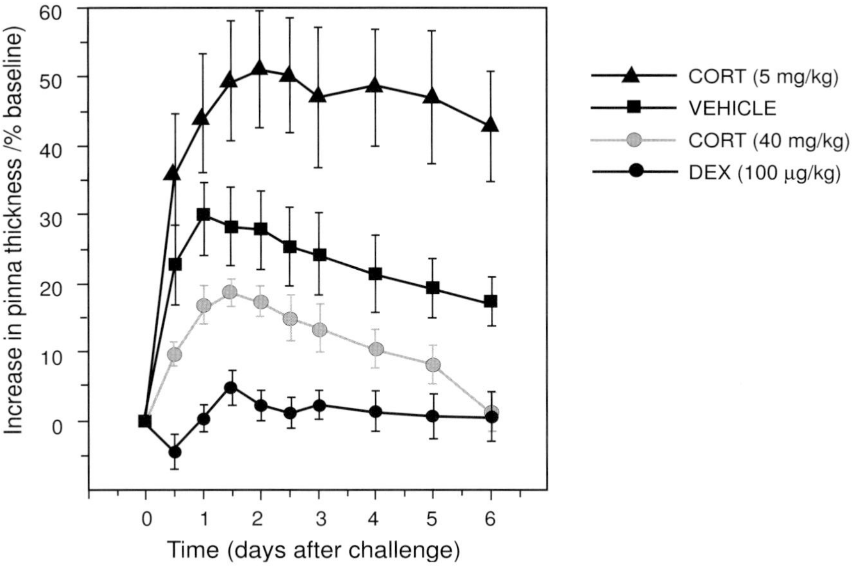

Figure 15.2 Physiological doses of corticosterone mimic effects of stress in enhancing the DTH response in the ear, whereas high doses of corticosterone and dexamethasone, which does not bind to CBG, suppresses the DTH response (from Dhabhar and McEwen 1999).

aspects of "protection" and "damage" even if we have a long way to go in understanding the underlying mechanisms for these actions. Now we turn to the brain and consider how it handles repeated stress.

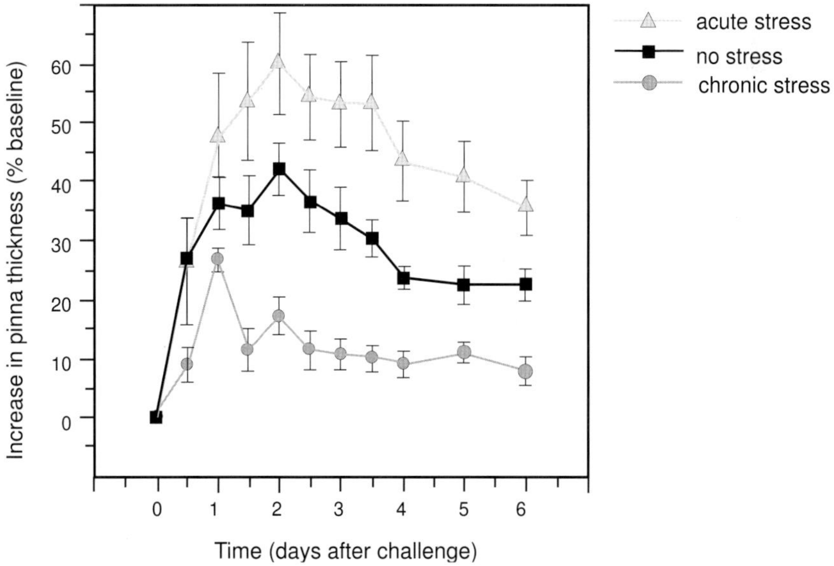

Figure 15.3 Acute stress enhances, whereas chronic stress suppresses, the DTH response (from Dhabhar and McEwen 1997, 1999).

STRESS, ADAPTIVE PLASTICITY, AND THE HIPPOCAMPUS

The brain is the key to interpreting and responding to potentially stressful events; it is also a target for the actions of stress hormones, particularly glucocorticoids. In the short run, acute elevation of both glucocorticoids and catecholamines facilitates the formation of memories of events associated with strong emotions (McGaugh 2000; Roozendaal 2000). Chronically, however, stress hormones and glucocorticoids, in particular, contribute to impairment of cognitive function and promote damage to brain structures such as the hippocampus (Lupien and McEwen 1997; McEwen and Sapolsky 1995; Sapolsky 1992).

The response of the brain to both acute and chronic stress must be regarded in terms of its capacity to show adaptive plasticity. The adult brain is more plastic than previously believed. Remodeling of synaptic contacts and dendrites in the hypothalamus with the onset of lactation (Michaloudi et al. 1997; Stern and Armstrong 1998) and growth and branching of dendrites of cerebrocortical neurons in an enriched environment and after training (Greenough and Bailey 1988; Withers and Greenough 1989) are two examples of such plasticity. Recent studies on the hippocampal formation of the brain provides further examples of adult brain plasticity, which is regulated by hormones in adult life and during brain development. The hippocampus is involved in episodic, declarative, contextual and spatial learning, and memory, as well as a component in the control of autonomic and vegetative functions such as ACTH secretion (Eichenbaum and Otto 1992; Jacobson and Sapolsky 1991; Phillips and LeDoux 1992). The hippocampus is also vulnerable to damage by stroke and head trauma and susceptible to damage during aging and repeated stress (Sapolsky 1992).

Hippocampal neurons express receptors for circulating adrenal steroids (McEwen et al. 1968), and work in many laboratories has shown that the hippocampus has two types of adrenal steroid receptors, Type I (mineralocorticoid) and Type II (glucocorticoid), which mediate a variety of effects on neuronal excitability, neurochemistry, and structural plasticity (DeKloet et al. 1998). These effects, which involve hormone-mediated effects on gene expression, include the regulation of branching and length of dendrites in the pyramidal cells of Ammon's horn and the replacement of nerve cells in the dentate gyrus. The hippocampus is also sensitive to gonadal hormones and expresses both intracellular androgen and estrogen receptors (Kerr et al. 1995; Weiland et al. 1997). Gonadal and adrenal hormones participate in functional and structural changes in adult life, as well as in developmental events, which include sexual differentiation and influences of early stressful life experiences.

Many of these hormone effects do not occur alone but rather in the context of ongoing neuronal activity. In particular, excitatory amino acids and NMDA receptors, as well as serotonin, play an important role in the functional and structural changes produced in the hippocampal formation by steroid hormones. With regard to stress effects, two types of adaptive plasticity will be considered in this article, namely, the remodeling of dendrites of hippocampal pyramidal neurons in response to repeated stress and chronic elevation of glucocorticoids and the inhibition of neurogenesis of granule cells in the dentate gyrus (see Figure 15.4).

As reviewed elsewhere (McEwen 1999), elevated levels of adrenal steroids participate, along with excitatory amino acid neurotransmitters, in promoting the remodeling of dendrites of CA3 pyramidal neurons in the hippocampus, as well as the suppression of neurogenesis of dentate gyrus granule neurons. Recent evidence indicates that the

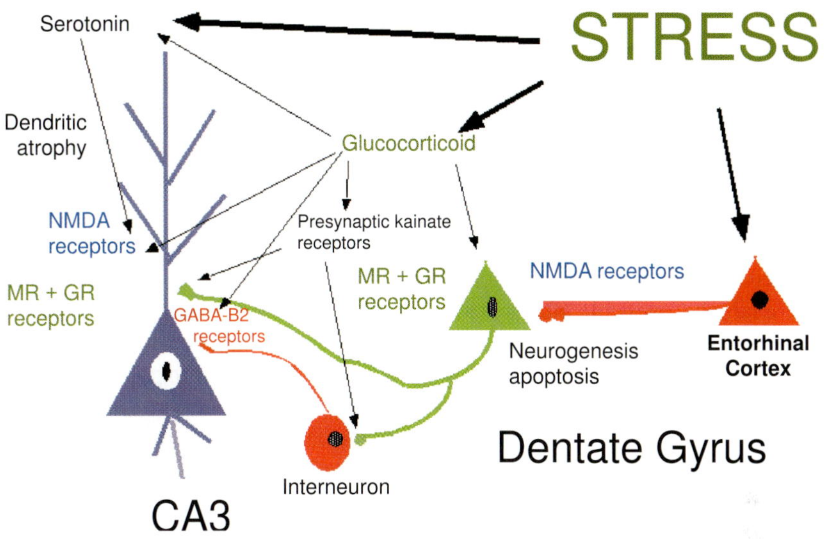

Figure 15.4 Schematic diagram of the role of neurotransmitters and glucocorticoids in regulating neurogenesis and dendritic remodeling in the dentate gyrus-CA3 system of the hippocampal formation. Granule neurons are replaced in adult life, and neurogenesis as well as apoptotic neuronal death are regulated by stress as well as by seizure-like activity. Granule neurons send mossy fibers to both the CA3 pyramidal neurons and to interneurons in the hilus, which, in turn, send inhibitory projections to the CA3 pyramidal neurons. Balance between excitatory input and inhibitory tone from the interneurons is presumed to be very important to the excitability of CA3 neurons. Evidence summarized in the text indicates that excitatory amino acid release during repeated stress, aided by circulating glucocorticoids, leads to a reversible remodeling of apical dendrites over 3–4 weeks in rats and tree shrews. Serotonin also participates, possibly by aiding the excitatory amino acid activity at the NMDA receptor, and reduced GABA-benzodiazepine-mediated inhibitory activity at synapse from the interneurons on CA3 pyramidal neurons may also exacerbate the remodeling. Excitatory input to the dentate granule neurons from the entorhinal cortex acts via NMDA receptors in concert with circulating adrenal steroids to regulate the rate of neurogenesis and apoptotic cell death, and both acute and chronic stress appear to be capable of inhibiting neurogenesis in the dentate gyrus.

remodeling of dendrites after repeated stress occurs in other parts of the hippocampus besides CA3 (Sousa et al. 2000). As will be seen below, the degree of dendritic remodeling and suppression of neurogenesis depend on the nature of the stressor. Furthermore, recent evidence points to a considerable degree of reversibility of the structural changes in the brain that are caused by repeated stress (Conrad et al. 1999; Sousa et al. 2000). Moreover, there are some surprises as far as the magnitude of these effects in relation to the apparent degree of stress, measured from the standpoint of the secretion of traditional stress hormones.

RESTRAINT AND MULTIPLE STRESS EFFECTS

A closer examination of the effects of stressors on structural plasticity in the hippocampus and neurochemical changes in other brain regions reveals some important properties of both experimenter-applied stressors and natural psychosocial stressors which emphasize the adaptive nature of these changes in brain structure and highlight the complex interactions between circulating stress hormones and neurotransmitters.

Contrary to expectations, there is not a consistent relationship between stress-induced dendritic remodeling in hippocampus and elevated glucocorticoid secretion or other signs of chronic stress such as thymus atrophy, adrenal hypertrophy, and body weight reduction. A recent study showed that a chronic multiple stress paradigm, which is designed to avoid habituation of glucocorticoid responses to the same stressor, produced dendritic remodeling not only in the CA3 region of the hippocampus but also in the dentate gyrus and CA1 (Sousa et al. 2000). Glucocorticoid treatment mimicked the effects of stress, and the effects of both treatments were largely reversible after termination of the treatment (Sousa et al. 2000).

We showed previously that a similar multiple stress paradigm, consisting of daily 1 h restraint, 1 h shaking, 30 min swim in body temperature water, did indeed produce continued elevation of glucocorticoids throughout the 21-d paradigm, in contrast to 21 d of 6 h/day restraint stress which produced habituation of the glucocorticoid stress response (Magarinos and McEwen 1995a). Both treatments, however, caused similar degrees of dendritic remodeling in CA3 and produced nonsignificant atrophy of the thymus, hypertrophy of the adrenal glands, and moderate decrease in body weight gain over 21 d. In spite of the minimal changes in the expected parameters of chronic stress, blockade of glucocorticoid synthesis with cyanoketone prevented dendritic remodeling (Magarinos and McEwen 1995b) and oral administration of glucocorticoids, which does not involve the stress of injection, causing dendritic remodeling (Magarinos et al. 1996b, 1999).

It is important to emphasize that dendritic remodeling is reversible following the termination of stress or corticosterone treatment (Conrad et al. 1999; Sousa et al. 2000). Moreover, both dilantin and tianeptine prevent dendritic remodeling caused by repeated restraint stress and glucocorticoid injections. Treatment with tianeptine is able to reverse the remodeling produced by 21 d oral corticosterone and to do so within 2 weeks, even while the corticosterone treatment is continuing (Magarinos et al. 1999).

A final note on the structural plasticity produced by repeated restraint stress is that dentate gyrus neurogenesis is not inhibited by a single acute restraint stress but is impaired after 21 d of daily restraint (Pham, Nacher, Hof, McEwen, unpublished). The reduction in neurogenesis is not as large as has been reported in rats after natural predator odor (Galea et al. 1996) or in tree shrews or marmosets after acute or chronic psychosocial stress (Gould et al. 1997, 1998).

We interpret these effects as further evidence of the importance of endogenous neurotransmitters along with glucocorticoids in the regulation of structural plasticity in the hippocampus. Serotonin and NMDA receptors are both implicated as mediators of dendritic remodeling, and GABA-benzodiazepine receptors play an opposing role. In the regulation of neurogenesis, serotonin acting via 5-HT1A receptors plays an opposing role to glucocorticoids and NMDA receptors.

The multiplicity of regulatory factors for structural plasticity in the hippocampus is consistent with the interpretation of these changes as reflecting adaptive plasticity, rather than as a pathway leading to damage. Behavioral changes accompanying repeated restraint stress fit with this view and provide additional surprises. In spite of the apparent habituation of the HPA response to repeated restraint stress, these animals develop increased fear in an open field apparatus and showed increased fear conditioning to both tone and contextual cues (Conrad et al. 1999). Moreover, repeatedly restrained rats show progressively increasing aggression towards their cage mates after they are released from their restrainers (Wood and McEwen, unpublished).

Finally, repeated stress that produces dendritic atrophy in the CA3 region impairs hippocampal-dependent learning. Rats that received 21 d of restraint stress were impaired in performance on an 8-arm radial maze when they were trained starting one day after the end of stress but not when trained 18 d later (Luine et al. 1994). We now know that dendritic atrophy is reversible within 7–10 d after the end of stress (Conrad et al. 1999). The impairment was in the same direction, but not as great as impairment found in aging rats. Moreover, stress effects were prevented by prior treatment of rats with phenytoin or with tianeptine under the same conditions in which both drugs prevented the stress-induced atrophy of CA3c pyramidal neurons (Luine et al. 1994; Watanabe et al. 1992a, b). A subsequent study showed that 21 d of repeated restraint stress impaired the short-term (4 h) retention of a spatial recognition memory in a hippocampus-dependent Y-maze task; again, stress impairment was prevented by tianeptine treatment during the stress regimen (Conrad et al. 1996).

STRESS FROM CONSPECIFICS

The theme of adaptive plasticity in the face of repeated stress is further documented by studies of more naturalistic types of stressors, and these studies provide additional clues as to the multiplicity of neurochemical pathways affected by chronic stressors. In the tree shrew, a resident-intruder paradigm was used to produce chronic stress in the intruder over 28d. The effect of the intruder is largely psychological so it is described as psychosocial. This procedure caused the same type of dendritic remodeling of CA3 neurons in the tree shrew as was found in the rat; moreover the effects of this type of stress were prevented by concurrent treatment of intruder tree shrews with Dilantin (Magarinos et al. 1996a). Dilantin treatment did not prevent the stress-induced elevation in catecholamines and glucocorticoids, and the HPA response was very robust and showed no signs of habituating over time, in contrast to the story with repeated restraint stress (see above) (Magarinos et al. 1996a).

Neurogenesis was inhibited by both acute and chronic psychosocial stress in the intruder tree shrew (Gould et al. 1997) and by acute psychosocial stress in a resident-intruder paradigm in the marmoset (Gould et al. 1998), and these effects were substantially larger than that seen in rats after chronic restraint stress (see above). Moreover, the dentate gyrus is 30% smaller in the chronically stressed tree shrew, although granule neuron number shows a smaller reduction (Gould, Fuchs and McEwen, unpublished). This finding suggests that there may be other changes such as atrophy of dendritic branching to account for the decrease in dentate gyrus volume. It should also be noted that neurogenesis is inhibited in the marmoset after acute psychosocial stress .

Another important animal model to study stress resulting from intraspecific competition is the visible burrow system (VBS) in the rat (Blanchard 1990). Five males and two females are placed in an apparatus that has an open chamber and tunnels and compartments, and there is a videocamera to record the animals' behavior. A dominant male usual emerges within several days and controls access to the food and water cups and to the females. Within the two-week span of this study, it is not uncommon for one or two of the subordinate males to die, usually not so much from wounds as from autonomic collapse related to defeat. A complex set of neuroendocrine and neurochemical changes has been found for the dominants and subordinates, which are both experiencing stress compared with control animals living in the normal cage environment (see Table 15.1.) The most striking difference is among the subordinates, in which two types have been recognized. The non-responders are so identified because they show an attenuated HPA response to acute restraint stress and also show a reduced level of CRH mRNA in the hypothalamus (Albeck et al. 1997).

For rats in the VBS and the intruder tree shrew there is a parallel involvement of serotonin. In the VBS, both dominant and subordinate rats show both dendritic remodeling, and they also show a downregulation of 5-HT transporter expression in the CA3 region, indicating either a reduced density of serotonin terminals or a reduced expression of the transporter. Moreover, repeated restraint stress and psychosocial stress in rats suppresses expression of the inhibitory 5-HT1A receptor in the hippocampus, and this is also true in the intruder tree shrews described above (Flugge 1995; McEwen et al. 1995; McKittrick et al. 1996). These stress-induced changes are relevant to the suspected role of 5-HT1A receptors in anxiety disorders and depression (Lopez et al. 1998). For example, deletion of the 5-HT1A receptor in mice results in increased anxiety, due in part to impairment of the GABA-benzodiazepine system (Ramboz et al. 1998; Sibille et al. 2000).

In the VBS, there is also an upregulation of 5-HT2A receptors in the cerebral cortex (McKittrick et al. 1995). This also has relevance to depressive illness, since the inactivation of the 5-HT2A receptor by antisense induces an antidepressant-like effect in mice (Sibille et al. 1998). Thus, animal models involving severe psychosocial stress of one animal by other

Table 15.1 Summary of changes in dominant and subordinate rats under psychosocial stress in the visible burrow.

	Endocrine Parameters	**Neural Parameters**
Dominants and subordinates show:	Elevated corticosterone Adrenal hypertrophy Thymic involution	CA3 dendritic remodeling Downregulation of 5HTT in hippocampus Decreased 5HT1A receptors Increased CRF mRNA in hypothalamus
Dominants show:		Somewhat more dendritic remodeling Somewhat greater downregulation of 5HTT in hippocampus
Subordinates show:	Progressive weight loss Greater thymic involution Decreased CBG Decreased testosterone Decreased insulin and glucose	Increased CRF mRNA levels in CeAmygdala Increased tyrosine hydroxylase mRNA and protein in locus coeruleus
Nonresponsive subordinates show:	Largest body weight loss Lowest testosterone Lowest CBG No response to novel stressor	Depressed CRF mRNA levels in the hypothalamus

conspecifics produce brain changes that are relevant to human psychiatric conditions such as anxiety and depression.

CONCLUSIONS

Study of the influence of acute and chronic stress on the immune system and brain has highlighted the dual role of the mediators of allostasis in adaptation and protection, on one hand, and pathophysiology, damage and disease, on the other. What is the connection between these opposing processes? The concepts of allostasis and allostatic load are intended to supplement and clarify traditional concepts of homeostasis and stress by emphasizing the interconnected nature of the mechanisms that promote adaptation or lead to pathophysiological change.

A key aspect is the temporal pattern of the allostatic response. Figure 15.5 shows a number of scenarios in which the typical hormonal response to challenge is shown to be either overused or disregulated, leading to allostatic load. In each case, the result is overexposure to the mediators of allostasis. The first type of allostatic load reflects a lot of repeated challenges, that is, many stressful events. The second type is a failure to habituate over time to the same stressor, something that most, but not all, individuals are able to do (Kirschbaum et al. 1995). The third type of allostatic load reflects a failure of each response to challenge to shut off in a normal manner. This may be true of the response to stressful events, and it also may be true of the diurnal rhythm. Failure to shut off the blood pressure response to stress is recognized in individuals with a familial history of hypertension (Gerin and Pickering 1995). Elevated, diurnal cortisol associated with depression and inadequate sleep is associated with bone mineral loss (Michelson et al. 1996) and elevated glucose levels (Spiegel et al. 1999). The final type of allostatic load is an inadequate response of the primary mediators of allostasis. In the case of glucocorticoids, the lack of adequate glucocorticoids leads to overreactivity of the cytokines that are normally held in check by glucocorticoids. In the Lewis compared to the Fischer rat, relative glucocorticoid insufficiency is associated with increased autoimmune and inflammatory disturbances.

Let us see how the scheme shown in Figure 15.5 may apply to the examples given above for the effects of repeated stress on the immune system and brain. In the case of the immune system, it is likely that glucocorticoids bring about the suppression of the DTH response described above. However, the exact sites and processes that show effects of repeated stress remain to be elucidated, that is, whether these involve downregulation, for example, of cytokines or cell surface adhesion molecules. In the case of dendritic remodeling in the hippocampus, the patterns of response of the HPA axis only tell part of the story, and the excitatory amino acids must be studied to show patterns of release and reuptake that may help explain the occurrence of dendritic remodeling even after HPA response has become habituated. Some evidence for a disregulation of extracellular levels of glutamate in aging rat hippocampus has been presented (Lowy et al. 1995), and this points to a prolongation of extracellular levels of glutamate as a possibly important factor in neuronal remodeling. As far as the suppression of neurogenesis, the variation in the ability of different stressors to suppress neurogenesis, as described above, will require studies on patterns of HPA response and excitatory amino acid responses to each of these stressors.

Thus stress can produce adaptive changes in the brain and other parts of the body, and we do not see the monolithic and invariant response to stressors that is implied by the General

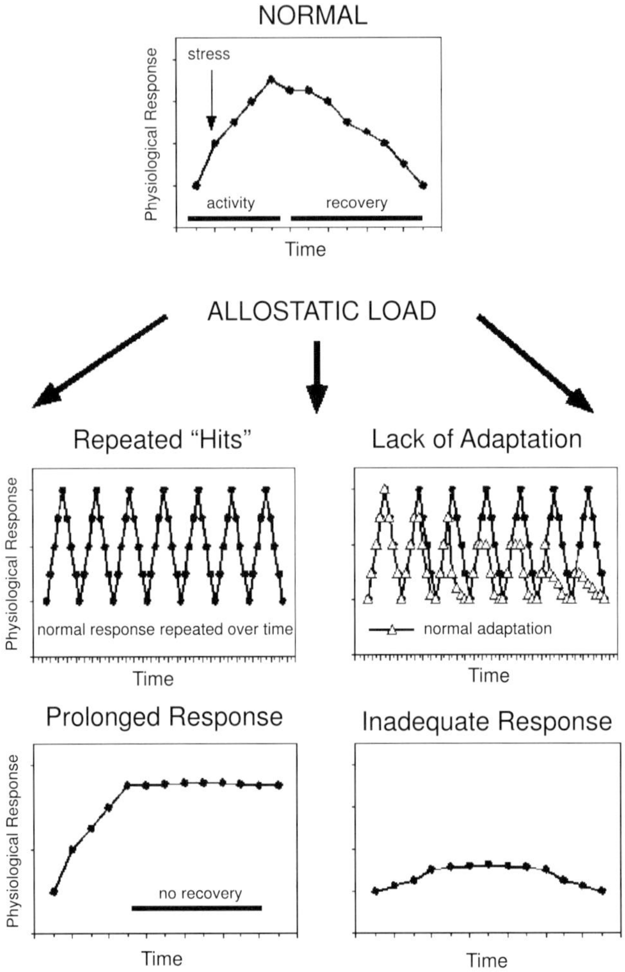

Figure 15.5 Three types of allostatic load. The top panel illustrates the normal allostatic response, in which a response is initiated by a stressor, sustained for an appropriate interval, and then turned off. The remaining panels illustrate three conditions that lead to allostatic load: *repeated "hits"* from multiple stressors; *lack of adaptation; prolonged response* due to delayed shut down; and *inadequate response* that leads to compensatory hyperactivity of other mediators (e.g., inadequate secretion of glucocorticoid, resulting in increased levels of cytokines that are normally counterregulated by glucocorticoids). Figure drawn by Dr. Firdaus Dhabhar, Rockefeller University. From McEwen (1998) by permission.

Adaptation Syndrome of the late Hans Selye but, rather, a range of adaptations to stressors by individuals who differ widely in the patterns of their responses. We are also aware of a range of mechanisms in tissues that subserve the processes of adaptation. In addition, individual differences occur, in part, because of differences in the interpretation of the threat and, in part, because of the unique nature of each stressful situation and the individuality of responses.

We have also seen that the brain and other organs of the body are rather more resilient and adaptable than we might think, and the key to maximizing adaptability and minimizing wear and tear is the efficient operation of the systems that subserve allostasis — efficient turn-on and equally efficient shut-off when no longer needed. Unfortunately, when inefficiently managed or overused by multiple challenges, the systems that normally protect can accelerate the progression towards disease.

What does this all mean for the welfare of animals as well as our own species? One of the most important messages is that our bodies were created to be resilient and possess physiological mechanisms to adapt and show resilience in the face of adversity. Another message is that wear and tear is an almost inevitable aspect of being alive; yet, there are strategies for maximizing the benefits of allostasis and minimizing allostatic load. These include maximizing social support and minimizing social instability.

However, animals and members of our own species seem to like to be challenged and to challenge other conspecifics. Moreover, social hierarchies seem to be inevitable and human societies such as exist in many countries of the Western World share, to varying degrees, the notion that any person can rise to the top of his or her social system. That being the case, conflict and competition are inevitable and constitute the "spice of life" for many. As a result, a certain amount of allostatic load is inevitable.

The recognition that there are gradients of mortality and disease as a function of socioeconomic status and income inequality (Adler et al. 1999) leads to the conclusion that gradients of health favoring the better educated and wealthier members of society are to some degree inevitable consequences of human social organization, at least in industrialized societies. In spite of the wear and tear associated with the competition and conflict of social ordering, a situation that may be worse for social creatures such as humans and many animal species involves social isolation and lack of social support (Kim and Kirkpatrick 1996; Seeman and McEwen 1996). Thus efforts to maximize beneficial social interactions are useful ways of improving welfare, including health, among both humans and members of social animal species and of counteracting the effects of competition and conflict.

REFERENCES

Adler, N., M. Marmot, B.S. McEwen, and J. Stewart, eds. 1999. Socioeconomic status and health in industrial nations: Social, psychological, and biological pathways. *Ann. NY Acad. Sci.* **896**:1–503.

Albeck, D.S., C.R. McKittrick, D.C. Blanchard, R.J. Blanchard, J. Nikulina, B.S. McEwen, and R.R. Sakai. 1997. Chronic social stress alters expression of corticotrophin-releasing factor and arginine vasopressin mRNA expression in rat brain. *J. Neurosci.* **17**:4895–4903.

Blanchard, D.C. 1990. Behavioral correlates of chronic dominance–subordination relationships of male rats in a seminatural situation. *Neurosci. Biobehav. Rev.* **14**:455–462.

Brindley, D., and Y. Rolland. 1989. Possible connections between stress, diabetes, obesity, hypertension and altered lipoprotein metabolism that may result in atherosclerosis. *Clin. Sci.* **77**:453–461.

Bulloch, K. 2000. Regional neural regulation of immunity: Anatomy and function. In: Handbook of Physiology. Vol. IV: Coping with the Environment: Neural and Endocrine Mechanisms, ed. B.S. McEwen pp. 353–379. New York: Oxford Univ. Press.

Cannon, W. 1929. The wisdom of the body. *Physiol. Rev.* **9**:399–431.

Conrad, C.D., L.A.M. Galea, Y. Kuroda, and B.S. McEwen. 1996. Chronic stress impairs rat spatial memory on the Y-Maze and this effect is blocked by tianeptine pre-treatment. *Behav. Neurosci.* **110**:1321–1334.

Conrad, C.D., A.M. Magarinos, J.E. LeDoux, and B.S. McEwen. 1999. Repeated restraint stress facilitates fear conditioning independently of causing hippocampal CA3 dendritic atrophy. *Behav. Neurosci.* **113**:902–913.

DeKloet, E.R., E. Vreugdenhil, M.S. Oitzl, and M. Joels. 1998. Brain corticosteroid receptor balance in health and disease. *Endocr. Rev.* **19**:269–301.

Dhabhar, F.S. 1996. Stress-induced enhancement of antigen-specific cell-mediated immunity: The role of hormones and leukocyte trafficking. New York: Rockefeller Univ.

Dhabhar, F.S., and B.S. McEwen. 1996. Stress-induced enhancement of antigen-specific cell-mediated immunity. *J. Immunol.* **156**:2608–2615.

Dhabhar, F.S., and B.S. McEwen. 1997. Acute stress enhances while chronic stress suppresses cell-mediated immunity *in vivo*: A potential role for leukocyte trafficking. *Brain Behav. Immun.* **11**:286–306.

Dhabhar, F., and B. McEwen. 1999. Enhancing versus suppressive effects of stress hormones on skin immune function. *Proc. Natl. Acad. Sci. USA* **96**:1059–1064.

Dhabhar, F.S., A.H. Miller, B.S. McEwen, and R.L. Spencer. 1995. Effects of stress on immune cell distribution: Dynamics and hormonal mechanisms. *J. Immunol.* **154**:5511–5527.

Dhabhar, F.S., A.R. Satoskar, H. Bluethmann, J.R. David, and B.S. McEwen. 2000. Stress-induced enhancement of skin immune function: A role for ã interferon. *Proc. Natl. Acad. Sci. USA* **97**: 2846–2851.

Eichenbaum, H., and T. Otto. 1992. The hippocampus — What does it do? *Behav. Neur. Biol.* **57**:2–36.

Flugge, G. 1995. Dynamics of central nervous 5-HT$_{1A}$-receptors under psychosocial stress. *J. Neurosci.* **15**:7132–7140.

Galea, L.A.M., P. Tanapat, and E. Gould. 1996. Exposure to predator odor suppresses cell proliferation in the dentate gyrus of adult rats via a cholinergic mechanism. *Soc. Neurosci. Abstr.* **22 474.8**:1196.

Gerin, W., and T.G. Pickering. 1995. Association between delayed recovery of blood pressure after acute mental stress and parental history of hypertension. *J. Hypertension* **13**:603–610.

Gould, E., B.S. McEwen, P. Tanapat, L.A.M. Galea, and E. Fuchs. 1997. Neurogenesis in the dentate gyrus of the adult tree shrew is regulated by psychosocial stress and NMDA receptor activation. *J. Neurosci.* **17**:2492–2498.

Gould, E., P. Tanapat, B.S. McEwen, G. Flugge, and E. Fuchs. 1998. Proliferation of granule cell precursors in the dentate gyrus of adult monkeys is diminished by stress. *Proc. Natl. Acad. Sci. USA* **95**:3168–3171.

Greenough, W.T., and C.H. Bailey. 1988. The anatomy of a memory: Convergence of results across a diversity of tests. *Trends Neurosci.* **11**:142–147.

Jacobson, L., and R. Sapolsky. 1991. The role of the hippocampus in feedback regulation of the hypothalamic-pituitary-adrenocortical axis. *Endocr. Rev.* **12**:118–134.

Kerr, J.E., R.J. Allore, S.G. Beck, and R.J. Handa. 1995. Distribution and hormonal regulation of androgen receptor (AR) and AR messenger ribonucleic acid in the rat hippocampus. *Endocrinology* **136**:3213–3221.

Kim, J.W., and B. Kirkpatrick. 1996. Social isolation in animal models of relevance to neuropsychiatric disorders. *Biol. Psychiatry* **40**:918–922.

Kirschbaum, C., J.C. Prussner, A.A. Stone, I. Federenko, J. Gaab, D. Lintz, N. Schommer, and D.H. Hellhammer. 1995. Persistent high cortisol responses to repeated psychological stress in a subpopulation of healthy men. *Psychosom. Med.* **57**:468–474.

Leibowitz, S.F., and B.G. Hoebel. 1997. Behavioral neuroscience of obesity. In: Handbook of Obesity, ed. G.A. Bray, C. Bouchard, and W.P.T. James, pp. 313–358. New York: Marcel Dekker.

Lopez, J.F., D.T. Chalmers, K.Y. Little, and S.J. Watson. 1998. Regulation of serotonin, glucocorticoid, and mineralocorticoid receptor in rat and human hippocampus: Implications for the neurobiology of depression. *Biol. Psychiatry* **43**:547–573.

Lowy, M.T., L. Wittenberg, and B.K. Yamamoto. 1995. Effect of acute stress on hippocampal glutamate levels and spectrin proteolysis in young and aged rats. *J. Neurochem.* **65**:268–274.

Luine, V., M. Villegas, C. Martinez, and B.S. McEwen. 1994. Repeated stress causes reversible impairments of spatial memory performance. *Brain Res.* **639**:167–170.

Lupien, S.J., and B.S. McEwen. 1997. The acute effects of corticosteroids on cognition: Integration of animal and human model studies. *Brain Res. Rev.* **24**:1–27.

Magarinos, A.M., A. Deslandes, and B.S. McEwen. 1999. Effects of antidepressants and benzodiazepine treatments on the dendritic structure of CA3 pyramidal neurons after chronic stress. *Eur. J. Pharmacol.* **371**:113–122.

Magarinos, A.M., and B.S. McEwen. 1995a. Stress-induced atrophy of apical dendrites of hippocampal CA3c neurons: Comparison of stressors. *Neuroscience* **69**:83–88.

Magarinos, A.M .and B.S. McEwen. 1995b. Stress-induced atrophy of apical dendrites of hippocampal CA3c neurons: Involvement of glucocorticoid secretion and excitatory amino acid receptors. *Neuroscience* **69**:89–98.

Magarinos, A.M., B.S. McEwen, G. Flugge, and E. Fuchs. 1996a. Chronic psychosocial stress causes apical dendritic atrophy of hippocampal CA3 pyramidal neurons in subordinate tree shrews. *J. Neurosci.* **16**:3534–3540.

Magarinos, A.M., M. Orchinik, and B.S. McEwen. 1996b. Oral administration of corticosterone mimics effects of stress on hippocampal CA3c dendritic structure. *Soc. Neurosci.* 764.5.

Manuck, S.B., J.R. Kaplan, M.R. Adams, and T.B. Clarkson. 1995. Studies of psychosocial influences on coronary artery atherosclerosis in cynomolgus monkeys. *Health Psychol.* **7**:113–124.

Manuck, S.B., J.R. Kaplan, M.F. Muldoon, M.R. Adams, and T.B. Clarkson. 1991. The behavioral exacerbation of atherosclerosis and its inhibition by propranolol. In: Stress, Coping and Disease, ed. P.M. McCabe, N. Schneiderman, T.M. Field, and J.S. Skyler, pp. 51–72. Hove and London: Lawrence Erlbaum.

Mason, J. 1959. Psychological influences on the pituitary-adrenal cortical system. In: Recent Progress in Hormone Research, ed. G. Pincus, pp. 345–389. New York: Academic.

Mason, J.W. 1975. A historical view of the stress field. *J. Human Stress* **1**:6–12.

McEwen, B.S. 1998. Protective and damaging effects of stress mediators. *New Engl. J. Med.* **338**: 171–179.

McEwen, B.S. 1999. Stress and hippocampal plasticity. *Ann. Rev. Neurosci.* **22**:105–122.

McEwen, B.S. 2000. Allostasis and allostatic load: Implications for neuropsychopharmacology. *Neuropsychopharmacology* **22**:108–124.

McEwen, B.S., D. Albeck, H. Cameron, H.M. Chao, E. Gould, N. Hastings, Y. Kuroda, V. Luine, A.M. Magarinos, C.R. McKittrick, M. Orchinik, C. Pavlides, P. Vaher, Y. Watanabe, and N. Weiland. 1995. Stress and the brain: A paradoxical role for adrenal steroids. In: Vitamins And Hormones, ed. G.D. Litwack, pp. 371–402. New York: Academic.

McEwen, B.S., C.A. Biron, K.W. Brunson, K. Bulloch, W.H. Chambers, F.S. Dhabhar, R.H. Goldfarb, R.P. Kitson, A.H. Miller, R.L. Spencer, and J.M. Weiss. 1997 Neural-endocrine-immune interactions: The role of adrenocorticoids as modulators of immune function in health and disease. *Brain Res. Rev.* **23**:79–133.

McEwen, B.S., and R.M. Sapolsky. 1995. Stress and cognitive function. *Curr. Op. Neurobiol.* **5**:205–216.

McEwen, B.S., and E. Stellar. 1993. Stress and the individual: Mechanisms leading to disease. *Arch. Intern. Med.* **153**:2093–2101.

McEwen, B.S., J. Weiss, and L. Schwartz. 1968. Selective retention of corticosterone by limbic structures in rat brain. *Nature* **220**:911–912.

McGaugh, J.L. 2000. Memory — A century of consolidation. *Science* **287**:248–251.

McKittrick, C.R., D.C. Blanchard, R.J. Blanchard, B.S. McEwen, and R.R. Sakai. 1995. Serotonin receptor binding in a colony model of chronic social stress. *Biol. Psychiatry* 37:383–393.

McKittrick, C.R., A.M. Magarinos, D.C. Blanchard, R.J. Blanchard, B.S. McEwen, and R.R. Sakai. 1996. Chronic social stress decreases binding to 5HT transporter sites and reduces dendritic arbors in CA3 of hippocampus. *Soc. Neurosci. Abstr.* **22 809.18**:2060.

Michaloudi, H.C., M.E. Majdoubi, D.A. Poulain, G.C. Papadopoulos, and D.T. Theodosis. 1997. The noradrenergic innervation of identified hypothalamic magnocellular somata and its contribution to lactation-induced synaptic plasticity. *J. Neuroendocrinol.* **9**:17–23.

Michelson, D., C. Stratakis, L. Hill, J. Reynolds, E. Galliven, G. Chrousos, and P. Gold. 1996. Bone mineral density in women with depression. *New Engl. J. Med.* **335**:1176–1181.

Phillips, R.G., and J.E. LeDoux. 1992. Differential contribution of amygdala and hippocampus to cued and contextual fear conditioning. *Behav. Neurosci.* **106**:274–285.

Ramboz, S., R. Oosting, D. Ait Amara, H.F. Kung, P. Blier, M. Mendelsohn, J.J. Mann, D. Brunner, and R. Hen. 1998. Serotonin receptor. 1A knockout: An animal model of anxiety-related disorder. *Proc. Natl. Acad. Sci. USA* **95**:14,476–14,481.

Roozendaal, B. 2000. Glucocorticoids and the regulation of memory consolidation. *Psychoneuroendocrinology* **25**:213–238.

Sapolsky, R. 1992. Stress, the Aging Brain and the Mechanisms of Neuron Death. Cambridge, MA: MIT Press.

Seeman, T.E., and B.S. McEwen. 1996. The impact of social environment characteristics on neuroendocrine regulation. *Psychosom. Med.* **58**:459–471.

Sibille, E., C. Pavlides, D. Benke, and M. Toth. 2000. Genetic inactivation of the serotonin$_{1A}$ receptor in mice results in downregulation of major GABA$_A$ receptor subunits, reduction of GABA$_A$ receptor binding, and benzodiazepine-resistant anxiety. *J. Neurosci.* **20**:2758–2765.

Sibille, E., Z. Sarnyai, D. Benjamin, J. Gal, H. Baker, and M. Toth. 1998. Antisense inhibition of 5-hydroxytryptamine$_{2a}$ receptor induces an antidepressant-like effect in mice. *Mol. Pharm.* **52**:1056–1063.

Sousa, N., N.V. Lukoyanov, M.D. Madeira, O.F.X. Almeida, and M.M. Paula-Barbosa. 2000. Reorganization of the morphology of hippocampal neurites and synapses after stress-induced damage correlates with behavioral improvement. *Neuroscience* **97**:253–266.

Spencer, R.L., B.A. Kalman, and F.S. Dhabhar. 2000. Role of endogenous glucocorticoids in immune system function: Regulation and counterregulation. In: Handbook of Physiology. Vol. IV: Coping with the Environment: Neural and Endocrine Mechanisms, ed. B.S. McEwen, pp. 381–423. New York: Oxford Univ. Press.

Spiegel, K., R. Leproult, and E. Van Cauter. 1999. Impact of sleep debt on metabolic and endocrine function. *Lancet* **354**:1435–1439.

Sterling, P., and J. Eyer. 1988. Allostasis: A new paradigm to explain arousal pathology. In: Handbook of Life Stress, Cognition and Health, ed. S. Fisher and J. Reason, pp. 629–649. New York: Wiley.

Stern, J.E., and W.E. Armstrong. 1998. Reorganization of the dendritic trees of oxytocin and vasopressin neurons of the rat supraoptic nucleus during lactation. *J. Neurosci.* **18**:841–853.

Watanabe, Y., E. Gould, H.A. Cameron, D.C. Daniels, and B.S. McEwen. 1992a. Phenytoin prevents stress- and corticosterone-induced atrophy of CA3 pyramidal neurons. *Hippocampus* **2**:431–436.

Watanabe, Y., E. Gould, H. Cameron, D. Daniels, and B.S. McEwen. 1992b. Stress and antidepressant effects on hippocampus. *Eur. J. Pharmacol.* **222**:157–162.

Weiland, N.G., C. Orikasa, S. Hayashi, and B.S. McEwen. 1997. Distribution and hormone regulation of estrogen receptor immunoreactive cells in the hippocampus of male and female rats. *J. Comp. Neurol.* **388**:603–612.

Withers, G.S., and W.T. Greenough. 1989. Reach training selectively alters dendritic branching in subpopulations of layer II-III pyramids in rat motor-somatosensory forelimb cortex. *Neuropsychologia* **27**:61–69.

Standing left to right: Manfred Schedlowski, Jaap Koolhaas, Seymour Levine, Harry Blokhuis, Björn Forkman
Seated left to right: Mike Mendl, Sylvia Kaiser, Don Broom, Paul Plotsky

16

Group Report: Key Sources of Variability in Coping

B. FORKMAN, Rapporteur

H.J. BLOKHUIS, D.M. BROOM, S. KAISER, J.M. KOOLHAAS, S. LEVINE, M. MENDL, P.M. PLOTSKY, and M. SCHEDLOWSKI

INTRODUCTION

Animals respond to environmental challenge by activating a variety of coping systems (see Broom, this volume), which include aspects of the autonomic nervous system, the immune system, behavioral and emotional systems, as well as neuroendocrine systems such as the hypothalamic-pituitary-adrenal (HPA) axis. Indeed, coping systems are an essential part of biological function. There is an understandable tendency to focus on coping systems of laboratory animals and domestic animals, and while most of our discussion centers around these animals we wish to emphasize that the coping systems have evolved in nature, as responses to natural challenges, and their function has to be understood in that context.

There is a variation in coping patterns, both between and within individuals, as well as between species. Part of the variation seen might be due to differences in the early experiences of the individual; however, it is also possible that there are different stable coping strategies that coexist within a population, strategies that are largely innate. When presented with a series of stressors, it is possible that an individual responds by either altering his perception of the stressor or by altering the response to the stressor. Variations in circumstances surrounding the stressing event are also important, not least the presence or absence of social support. An important final aspect on the variation of stress responses is the cost of the different responses, costs that probably vary from one individual to another. This chapter discusses these different sources of variability and what can be learned from them. Both from a basic and an applied point of view, for example, an animal welfare perspective, it is important to understand not only the different coping systems per se, but also the sources of variability in coping.

Both of the two important stress axes, the HPA axis and the sympathetic-adrenal-medullary (SAM) axis, can be activated independently of each other, especially in chronic stress. Discussion in this chapter, however, focuses on the HPA axis, in part because

Coping with Challenge: Welfare in Animals including Humans
Edited by D.M. Broom

of the depth of information about this system currently available but also because it appears to function to promote "self-remodeling" within the other peripheral and central coping systems. By this we mean that glucocorticoids are secreted into the peripheral circulation and, being lipophilic, generally pass easily into cells and the brain (cf. section on WHAT ARE THE CONSEQUENCES AND DYNAMICS OF THE ORGANIZATION OF THE HPA AXIS?). Thus, in addition to their role in regulating the availability of energy sources, glucocorticoids can coordinate the actions and subsequent sensitivity of disparate systems through the induction of plasticity resulting from their regulation of gene expression. Plasticity occurs throughout life and is the basis of development and adaptation. However, the degree of plasticity and permanence of stimulus-induced plasticity varies with a species-typical age dependence. In general, this plasticity is more permanent when it occurs in the perinatal period, becoming less permanent after the age of weaning in mammals, and essentially becoming state- or context-dependent past the pubertal stage. With regard to coping, it is suggested that stimulus-induced activation of the HPA axis leads to an increase in circulating glucocorticoids, which in turn functions to influence the selection of one of a subset of coping strategies. The important general principles here are that the particular outcome of glucocorticoid exposure is highly dependent upon the stage of central nervous system (CNS) development. Glucocorticoid targets shift with the increasing maturity of the CNS and via alterations in gene expression profiles and provide the underlying mechanism for this plasticity in neurocircuit structure and function. Of course, it is also important to acknowledge that the individual's genotype sets the limits of possible plasticity and coping strategies.

The specific questions considered in this chapter are:

- How do early experiences influence brain development and coping abilities of individuals?
- What are the consequences and dynamics of the organization the HPA axis?
- Is it possible to interpret part of the variation seen as the animal reacting with either a perceptive and evaluative system or a response system?
- How well defined are the coping responses or strategies?
- What is social support and what are its consequences?
- What are the costs associated with different coping systems?

HOW DO EARLY EXPERIENCES INFLUENCE BRAIN DEVELOPMENT AND COPING ABILITIES OF INDIVIDUALS?

Organisms are constantly challenged by internal and external stressors. The maintenance of the internal milieu within a normal physiological range (i.e., homeostasis) is very important for the normal pre- and postnatal development. Both pre- and postnatal manipulations have profound and persistent effects on the development of the neural organization, neuroendocrine processes, and behavior, and will consequently affect the coping abilities of the individuals (Ward 1972; Herrenkohl 1979; Sachser and Kaiser 1996; Koehl et al. 1999). Thus, early experiences are one key source of individual variability in coping.

Although evidence on the prenatal influence of substance abuse is extensive, for the purpose of this discussion we focus on the effects of pre- and postnatal stress, in particular postnatal disruptions of mother–infant relationships. Effects of both pre- and postnatal stress have been documented in rodents, primates, and humans, but very little has been done in other species, including farm animals.

The effects of pre- and postnatal stress can be very different. The description of early experiences has therefore been divided into a section on prenatal and one on postnatal stress.

Prenatal Stress

In mammals, the internal environment of the pregnant animal constitutes the embryo's external environment; it is therefore reasonable to suspect that the former can affect the fetal development. The term prenatal stress is taken here to mean both any stress imposed directly on the embryo and any stress imposed on the pregnant female.

Prenatal stress can have long-term effects not only on behavior but also on reproductive functions, immune as well as endocrine system, and on brain development (Herrenkohl 1979; Klein and Rager 1995; Koehl et al. 1999; Anderson et al. 1985; Kay et al. 1998). Increasing evidence indicates that the degree of activation of the HPA axis and the duration of the response can be altered permanently by prenatal stress (cf. Weinstock 1997). Changes in the reactivity of the HPA axis may be associated with an increase in corticotrophin-releasing hormone (CRH) in the amygdala and a decrease in the number of glucocorticoid receptors in the hippocampus. This indicates long-term alterations in the control of glucocorticoids on neuronal functioning. In view of the role of these limbic brain structures in emotional behavior and in learning and memory processes, this implies a long-term change in coping processes as well.

In nonhuman primates, exposure to prenatal stressors is associated with long-term changes in their behavioral regulation, endocrine function, and central neurocircuits. Placing pregnant females in a stressful situation was associated with an increase in miscarriage rates (Sackett 1981) and an increased dermatoglyphic asymmetry, which is indicative of altered fetal development and is associated with increased perinatal mortality (Newell-Morris et al. 1989). Studies by Schneider (1992a, b, c) support the existence of periods of increased vulnerability during early gestation (E45–90) versus mid-to-late gestation (E90–145) with respect to the occurrence of attentional deficits and neuromotor maturation. In addition, both HPA axis function (Clarke et al. 1994) and cerebrospinal fluid concentrations of biogenic amines (Schneider et al. 1998) were affected in these animals. It is interesting to note that the changes in the HPA axis observed after prenatal stress are similar to those seen in humans with endogenous depression.

Prenatal stress in rats also induces enhanced activation of the sympathetic nervous system in response to foot shock stress (Weinstock et al. 1998). It has also been reported that the secretion of different neurotransmitters in specific brain areas (e.g., acetylcholine, dopamine and noradrenaline) is influenced by perinatal exposure (e.g., Moyer et al. 1978).

Ward (1972) showed that prenatal stress (immobilization combined with bright light) had a distinct effect on the behavioral development of males. She showed that prenatal stress in rats leads to a demasculinization and feminization of the behavior of the male offspring. After Ward's fundamental work, numerous studies were conducted which describe different effects of prenatal stress on various aspects of male behavior (e.g., Harvey and Chevins 1985; Kinsley and Svare 1986). In contrast, only a few studies show that prenatal stress also affects the behavior of the female offspring. In one such study, female guinea pigs lived during pregnancy in an unstable social environment, in which the group composition was changed every third day. This had distinct effects on the behavior of the female offspring. Prenatally stressed female guinea pigs showed a behavioral masculinization: they displayed male-typical courtship and play behavior more often than control females (Sachser and Kaiser 1996). Alonso et al. (1991) describe behavioral suppression in stressful environments during adulthood of prenatally stressed female rats. Such animals were less able to cope in aversive situations. Clinical observations on humans suggest that exposure to stress during pregnancy results in changes in the behavior of the children. In particular, hyperactivity, hyperanxiety, and disturbed social behavior are described (Weinstock 1997).

In summary, the animal's prenatal stress experiences induces permanent changes in a variety of physiological systems (neuroendocrine, neurochemical) involved in the coping response indicating a permanent alteration in the later ability to cope with stressors.

There is controversy in the literature as to whether different stressors acting on the same species also have the same outcome or whether the effects are specific for the stressor. As mentioned above, in rats, immobilization combined with bright light leads to a feminization and demasculinization of the behavior of the male offspring (Ward 1972). Using overcrowding as a prenatal stressor, a feminization was also found. However, no demasculinization occurred (Dahlöf et al. 1977). Quite often these different outcomes can be explained by the animals having been stressed at different times. It is important—when comparing the effects of different stressors in the same species—to expose the animals to these stressors at exactly the same time, which results in the animals being at the same developmental stage. On the other hand when comparing the effects of the same stressor in different species, it is necessary to know the different developmental processes of the species. For example, the sexual differentiation of rats starts shortly before birth and continues until ten days after birth (most of it having occurred by three days postpartum), whereas in guinea pigs it occurs around day 36–45 of pregnancy (total duration about 68 days; MacLusky and Naftolin 1981). One method translating neurodevelopmental time across mammalian species involves determining peak times of neurogenesis in developing neuronal systems, as described by Finlay and Darlington (1995) and Clancy and colleagues (2001). Thus, using the same stressor to the same time (e.g., shortly before birth) will lead to a different outcome regarding the behavior, endocrine system, and brain development of the offspring. In conclusion there are many variables that have to be taken into account in the investigation of the effects of perinatal stressors.

Postnatal Stress

The effects of early postnatal manipulations, such as maternal separation, are extensive and range from altering the responsiveness of the adult rodent HPA axis to changing behavioral responses related to fear and anxiety. The effects of maternal separation, however, are dependent upon the number of variables such as the age at which the separation occurs, the duration of the separation period, and the number of repeated separations. The effects of a single 24-hour separation during the initial neonatal week, or shorter repeated maternal separations beginning early in the first neonatal week and continuing through the second neonatal week, have behavioral and HPA axis effects that are opposite to those in response to handling (which also entails a brief maternal separation). The functional significance of early manipulations are extensive and permanent and range from altering the responsiveness of the adult rodent HPA axis to changing behavioral responses related to fear and anxiety. Future research should focus on determining more precisely the extent and nature of the changes occurring in the developing brain that are responsible for the functional changes which have been observed as well as on the particular aspects of maternal behavior and its encoding by the neonate that mediate these changes.

The continuous interplay of genetic and environmental factors "programs" the developing neurocircuitry. These processes influence how an organism perceives and responds to stimuli in its environment, yielding individual differences. Among environmental factors, adverse early experience including abuse, neglect, and severe childhood illness, is a major risk factor during development and may influence the vulnerability to a variety of physio- and psychopathologies later in life. A Long Evans rat model of neonatal maternal separation (180 min daily, during postnatal days 2–14) has been used to explore the consequences of adverse early experience on individual responsiveness (Caldji et al. 2001; Ladd et al. 2000). These animals were compared to animals reared under normal animal facility conditions or animals exposed to brief handling and 15 minutes of maternal separation. As adults, animals in each of these rearing groups displayed a distinct and stable phenotype. The maternally separated rats exhibited enhanced anxiety-like behavior, moderate anhedonia, alcohol preference, resistance to glucocorticoid-mediated negative feedback, and HPA axis hyperresponsiveness to psychological stressors. Underlying these features were regionally specific changes at the level of gene expression, neurochemistry, electrophysiology, and morphology.

Overall, neonatal maternal separation at an early age resulted in rats exhibiting a depression-like syndrome with morbid anxiety that could be reversed by neonatal cross fostering or by chronic antidepressant treatment as adults. Perhaps as important, these animals also exhibited reduced arterial contractility, an increased rate of diabetes with aging, and markers of compromised immune function.

In nonhuman primates, experimental disruption of the maternal–infant dyad by various means, including maternal separation, introduction of social instability to the colony, or presentation of food in an unpredictable manner, has been used as a probe of the effects of postnatal stress. The effects studied include the development of the CNS, endocrine function, and behavioral regulation. Although few studies have been reported on the heritability of individual differences in stress responsiveness in monkeys, a recent report by Lyons et al. (1999) addressed this issue in squirrel monkeys (*Saimiri sciureus*). Combining paternal half-sibling analysis with several postnatal rearing protocols that altered maternal availability, they reported significant postnatal rearing effects with the lowest heritabilities for peak elevations of cortisol 24 hours following removal of the mother in otherwise undisturbed groups. In contrast, individual differences in cortisol 3–7 days later revealed the highest heritabilities with negligible postnatal rearing effects. These studies illustrate the interaction between extrinsic and intrinsic factors in determining individual stress responsiveness.

In the bonnet macaque (*Macaca radiata*), imposition of a variable foraging demand protocol in which the manner of food presentation is altered between periods requiring strenuous foraging with periods of free access to food, has profound consequences on the offspring as a result of altered social and maternal behavior of the mother (Rosenblum and Andrews 1994). Normal physical development of the offspring is observed, but progeny are more fearful and responsive to stressful stimuli and less socially competent. Interestingly, the offspring exhibit persistent alterations in cerebrospinal fluid levels of CRH, somatostatin, and biogenic amines even 4–6 years after this experience (Coplan et al. 1996, 1998). They also exhibit a reduced growth hormone response to the alpha-adrenergic agonist clonidine (Coplan et al. 2000). Elevations in cerebrospinal fluid corticotropin-releasing factor (CRF) levels and a blunted growth hormone response to clonidine challenge are frequently observed in patients with mood and anxiety disorders (Heim et al. 1997).

Rearing rhesus (*Macaca mulatta*) infants under conditions in which animals have auditory, visual, and olfactory but not tactile contact has profound long-term consequences on the animals (e.g., the occurrence of autistic-like behaviors, the inability to develop positive social relationships within the social group, hyperaggressiveness, disoriented sexual behavior, and lack of recognition of social cues). However, if animals are exposed to a social environment at any time prior to 6 months of age, the species-typical repertoire of social and nonsocial behaviors develop, even in the infants reared in total isolation (Harlow et al. 1971).

Adverse early experience such as maternal separation also has long-lasting effects on ingestive behaviors (e.g., polydipsia and polyphagia), immune system function (Lubach et al. 1995), the sympathetic response to stressful stimuli (Martin et al. 1988), central catecholamine levels or innervation patterns (Ginsberg et al. 1993; Higley et al. 1992; Lewis et al. 1990), and cognitive development (Beauchamp et al. 1991; Gluck and Sackett 1976; Sanchez et al. 1998). These cognitive impairments include a higher error rate during the learning phase, which indicates a potential attentional or motivational deficit and strong perseveration, suggesting a failure of systems involved in the cessation and inhibition of ongoing behavior. The neurochemical and neuroanatomical changes underlying such impairments remain unclear.

Structural magnetic resonance imaging (MRI) has been used to evaluate gross neuroanatomical changes in 18-month-old rhesus monkeys separated from the mother at two months of age and nursery reared under conditions of partial isolation (Sanchez et al. 1998). Partial isolation rearing was not associated with differences in the total brain, cerebellar or hippocampal volumes, or in the size of the anterior commissure. MRI scans did reveal a reduction in the midsagittal size of the corpus callosum, in parallel with a decrease in white, but not gray, matter content in the parietal and prefrontal cortices, suggesting the presence of specific reductions in cross-hemispheric projections, possibly resulting from a reduction in the number of callosal axons or in their myelination and growth. Interestingly, these alterations occurred in parallel with cognitive impairments and were not reversed by introduction of these animals into a peer group for six months.

Gender-specific effects of neonatal stress exposure in rodents have been difficult to document as a result of the numerous variations in stressor protocols with respect to age of initiation, duration and number of repetitions of the event, age at testing, and type of testing as well as the tendency for these types of studies to be performed primarily in male progeny. Nonetheless, there are reports of gender-specific effects on neuroendocrine, behavioral, and CNS function in the rodent. Matthews et al. (1996) found that females, but not males, which had experienced repeated neonatal maternal separation, exhibited an attenuation of the acquisition of a conditioned anticipatory locomotor response to the presentation of food. All separated animals displayed enhanced weight gain, a blunted locomotor response to a novel environment, and a blunting of the response to both negative and positive contrast effects. When tested on an elevated plus maze, female Sprague-Dawley rats exposed to neonatal 3 hours of daily maternal separation from postnatal days 1–21 exhibited less anxiety than nonhandled, gender-matched controls (McIntosh et al. 1999). In addition, only the feeding suppressant response to the satiety peptide cholecystokinin (CCK), but not bombesin or amylin, was affected by the early life experience with it being more pronounced in handled and maternally separated males. Wigger and Neumann (1999) found a twofold greater index of anxiety in adult male rats than in females that had experienced three hours of daily neonatal maternal separation from postnatal days 3–10. Neuroendocrine parameters were also altered in these maternally separated rats in a gender-dependent manner. Neither basal adrenocorticotropic hormone (ACTH) or corticosterone levels differed among progeny from the different rearing groups; however, the ACTH response to an emotional stressor was higher in male, but not female, rats of the separation group versus their respective controls. Overall, these studies support the thesis of gender-specific effects of adverse early experience during the neonatal period via mechanisms not yet elucidated.

Possible Adaptiveness of the Perinatal Stress Response

An exciting possibility is that some of the stress responses of the infants might be due to the infants allocating resources based on information gathered from the mother, just as food preferences may be influenced by what the mother eats perinatally (Bilko et al. 1994). A period of starvation during pregnancy might, for example, tell the infant that food is probably going to be scarce in the future. A prerequisite for this adaptive explanation, however, is that the stress response of the infant is specific to what the stressor might predict in the future. If the infant reacts to all stressors in the same way, it does not seem probable that the responses given by the infants are adaptive.

A consideration of the ways in which experimental manipulations can result in a shift in the trajectory of fetal and infant development raises the obvious question of whether these observations are relevant to animals under natural conditions. Here it should be noted that pregnant dams in nature are often exposed to more extreme periods of privation than would ever be employed by researchers. Such challenges may include periods of gestation during extreme climatic conditions (cold or hot), as well as times of limited food availability. In this regard, it is of interest that many seasonally breeding species orient the timing of the birth period towards the optimal time of the year. Often it has been assumed that this temporal orientation has been adapted for the postnatal needs of the infant, but it is equally important to organize the timing around the needs of the pregnant female. For example, as the pregnant female squirrel monkey approaches term, she may drink up to half her body weight of water each day. Thus, the birth of her infant in the rainy season in the Amazon is probably not coincidental.

An example of what might be indicative of the types of patterns we should expect to see if the stress reaction of the offspring were adaptive in the long term can be seen in humans. Population level surveys in several countries have indicated that periods of growth retardation during fetal development, as reflected by birth weight, may be significant predictors for several adult diseases, including cardiovascular illness and diabetes, which are of course not adaptive (Phillips et al. 2000; Godfrey and Barker 2000). Interestingly enough, however, one of the consequences of starvation during pregnancy is obesity in the infant, which is what one would expect from a functional point of view. If food is scarce during the pregnancy of the mother, the risk is that it will be scarce in the future as well. It therefore makes sense for the infant to attempt to build up a fat reserve against further food shortages. The proposed mechanism for these persistent effects into adulthood is an alteration in set points for various aspects of basic metabolism (e.g., glucoregulation, adiposity, blood pressure).

In conclusion, there is little evidence that the majority of the consequences shown, which have been caused by perinatal stress, are adaptive in the long term. It should be noted however, that most of the studies cited here were not done to test this hypothesis explicitly, and that further studies might very well yield different results.

Implications and Conclusions

Overall, the knowledge of existing mechanisms of plasticity and its species-typical age dependence suggests avenues for development of new strategies for management of environmental, including pathological, challenges to promote good welfare of domesticated species, laboratory animals, and humans. Some implications and questions raised by current animal research on glucocorticoid-induced plasticity and the central importance of mother–infant interactions on human and animal welfare include:

1. What genes are involved in conferring susceptibility or resistance to the effects of perinatal experience?
2. How can we translate this knowledge of perinatal plasticity into beneficial practice in animal husbandry as well as in human clinical care? For instance, what does this imply about long-term treatment consequences of current practice in the care of premature infants and neonates, or cow/calf separations in dairy cattle? What might the consequences be of the general failure to acknowledge and prevent the experience of pain in premature infants?
3. When interventions are available, do they promote actual reversal of the stimulus-induced plastic changes in brain circuitry structure or function, or are they associated with a new structure or function on top of the existing adaptation? What might the consequences be of such a "patch" in terms of constraining the range available in coping systems?

WHAT ARE THE CONSEQUENCES AND DYNAMICS OF THE ORGANIZATION OF THE HPA AXIS?

Much research effort is aimed at the HPA axis as the main mediator of the stress-induced changes in behavioral, neuroendocrine, neurochemical, and immunological systems involved in coping. The brain is responsible for the regulation of the HPA axis. Activation of specific neural pathways are required for the neuroendocrine cascade that results in changes in the paraventricular nucleus (PVN), which controls the output of CRH, the major secretagogue for ACTH, and subsequently the secretion and synthesis of glucocorticoids. The neural pathways that are activated and that subsequently activate CRH are, in many cases, stimulus dependent. However, most stimuli that activate the neural pathways result in the activation of multiple neural systems. The brain is a target organ for the glucocorticoids. Although there is controversy concerning which specific glucocorticoid, cortisol, or corticosterone freely enters the brain, the glucocorticoids act on specific structures that result in changes both at the level of the CNS and at the periphery. Three major sites in the CNS appear particularly sensitive to an increased level of the adrenal steroids: the hypothalamus, hippocampus, and amygdala. These areas are rich in steroid receptors. Glucocorticoids act by binding to mineralocorticoid receptors (MR, or "type 1" receptors) or glucocorticoid receptors (GR or "type 2" receptors), which are primarily located in the cytoplasm, evoking activation, dimerization, translocation to the nucleus, and binding to specific sites on the chromatin. The functional action of GR complexes depends upon interaction with concurrently activated intracellular signaling cascades thus imparting a degree of regional specificity of action. It is important to be aware that the distribution patterns of these adrenal steroid receptors may vary among species. In the hypothalamus these steroids serve to reduce the production of CRH effectively, and thus the production of ACTH and glucocorticoids are turned off and return to basal levels. The hippocampus contains a large population of both GR and MR. MR appear to be involved in maintaining basal levels and regulating circadian rhythms. In contrast GR, which are primarily occupied during stress when high levels of circulating glucocorticoids are present, seem to regulate the stress levels of these hormones. In the amygdala, glucocorticoids appear to increase CRH gene expression. CRH in the amygdala has been implicated in the regulation of behavioral expression of emotion and anxiety.

There is a circadian rhythm in the level of cortisol, with a peak at the beginning of the active period of the animal. Cortisol concentrations during the trough of the rhythm are maintained at very low levels because of feedback action of cortisol through MR (high affinity). This trough cortisol level is sufficiently low so that there is not appreciable occupancy of the lower affinity GR at circadian peak. Cortisol feedback regulates CRF secretion through occupancy of both MR and GR.

Chronic stress will elevate through cortisol if mild (by overriding MR-mediated feedback). As the intensity of the stress is increased, the level of trough cortisol will progressively increase until the rhythm flattens finally out at high level. It is the increased GR occupancy (duration and magnitude) that does the damage.

There is a difference between species in the proportions of different steroid hormones produced. For instance, guinea pigs and primates, including humans, have cortisol as their predominant glucocorticoid while rats and most mice have corticosterone. Dogs, on the other hand, produce an approximate 50/50 mixture of cortisol and corticosterone. Among the common farm species, poultry have corticosterone, whereas cattle have cortisol, and finally pigs produce some corticosterone and much cortisol. The functional significance of these species-typical patterns is unknown.

These glucocorticoids act by binding to MR or GR which are primarily located in the cytoplasm, evoking activation, dimerization, translocation to the nucleus, and binding to specific sites on the chromatin. The functional action of these GR complexes depends upon interaction with concurrently activated intracellular signaling cascades, thus imparting a degree of regional specificity of action. It is important to be aware that the distribution patterns of these adrenal steroid receptors may vary among species.

It has long been recognized that the ability of specific glucocorticoids to reach the brain differs; for instance, penetration into the brain of the synthetic glucocorticoid dexamethasone appears to be restricted relative to that of the endogenous rat glucocorticoid, corticosterone (De Kloet et al. 1975). One mechanism regulating glucocorticoid access to the brain appears to be the multidrug-resistant gene product P-glycoprotein (Van Kalken et al. 1993) which is expressed by endothelial cells at the blood-brain-barrier (Cordon-Cardo et al. 1989). The multidrug-resistant P-glycoprotein restricts access of dexamethasone and cortisol to the brain but not access to the pituitary or to peripheral tissues. Genetically engineered mice lacking this protein exhibit penetration of both dexamethasone and corticosterone into the brain (Meijer et al. 1998). This has important implications as it means that treating rodents and primates (including humans) with dexamethasone will saturate peripheral glucocorticoid receptors, shutting down the pituitary-adrenal axis via glucocorticoid-mediated negative feedback, while at the same time possibly creating an apparent glucocorticoid deficit in the CNS akin to an adrenalectomized state. Perhaps the most interesting and controversial findings about the multidrug-resistant P-glycoprotein is that it also appears to restrict access of the primary nonhuman primate and human adrenal glucocorticoid, cortisol, to the brain (Karssen et al. 2000; Pariante et al. 2000). The physiological significance of this finding is unknown but would imply that cortisol may not be the major steroid hormone mediating CNS actions of glucocorticoids in primates, guinea pigs, and other animals in which cortisol is the primary adrenal glucocorticoid product. However, it must be understood that the multidrug-resistant pump does not entirely exclude these substances from the brain but merely impedes their entry. Thus, it is possible for dexamethasone to enter the brain, although its entry is restricted relative to that of corticosterone. Similarly, cortisol does enter the brain and, considering the concentration difference between human adrenal cortisol production versus the minor adrenal steroids, it may still function as the primary glucocorticoid. Further studies are required to assess the importance of minor glucocorticoid products of the adrenal (see Kage et al. 1982) as well as the intricacies of their regulation. Overall, this suggests that it may be important to measure these "minor" steroids in those species primarily producing cortisol if one wishes to use the level of glucocorticoids as a parameter in assessing stress or poor welfare. This is clearly an important area for both basic and clinical research.

IS VARIATION CAUSED BY THE ANIMAL RESPONDING WITH EITHER A PERCEPTIVE OR A RESPONSE SYSTEM?

It is possible to conceptualize variability in coping responses as either being due to differences in the way in which an organism appraises and evaluates the situation or due to differences in the responses that it is able to show, or both (see Mendl, this volume, Figure 14.1). For example, differences between proactive and reactive copers (Koolhaas et al., this volume) may be the result of differences in the range of response patterns that are available to the individuals or due to differences in the way in which they assess and evaluate a challenging situation. This idea of splitting coping responses into (a) appraisal and evaluation and (b) response components is worth exploring more generally, because it has potential relevance for how we interpret data on coping responses in terms of welfare.

In general, when attempting to assess stress or welfare, we measure the response or "readout" component. There are a number of examples in which a lowering of, or decreased reactivity of, a response to a challenge is observed. For instance, one of the actions of oxytocin appears to be to reduce the reactivity of the HPA system to challenges (Carter, this volume).

Several domesticated species appear to show lower HPA and SAM responsiveness to challenges in comparison with their ancestral species. For example, when compared with its wild ancestor, the cavy (*Cavia aperea*), the domestic guinea pig (*Cavia aperea* f. *porcellus*), shows lower reactivities of the HPA and SAM axis (Künzl and Sachser 1999).

When faced with repeated exposure to a predator stimulus (e.g., a model bird of prey flying overhead), a prey animal in a cage may show a lessening of its escape response. What we observe in this example is a change in the activity of the response system. Our interpretation of this change may be that the organism has adjusted or adapted to the challenge or is now less disturbed by the challenge. We could conclude from this that its welfare has been only temporarily poorer. However, this conclusion would probably be strengthened if we found that the change was primarily brought about by an alteration in the appraisal and evaluation system, i.e., the organism no longer evaluates the challenge as a threat. For example, a human may no longer be psychologically disturbed by the experience of flying after becoming used to it and therefore behaves calmly when entering the plane. On the other hand, if the change is primarily in the response system, then our conclusions would be very different. A human may still be terrified of flying but has learned that resisting entry to the plane is a useless strategy and thus behaves calmly when entering the plane. The habituation of the escape behavior of the prey animal in the cage may likewise be explained in these two different ways.

Similarly, the effect of oxytocin on the reactivity of the HPA system may be primarily through local actions in the PVN on CRH synthesis and release, and hence by altering the activity of the response and readout system, or the "final common path" for the HPA system. Alternatively, it may involve some changes to sensory systems or reward systems which could be thought of as part of the appraisal and evaluation component. We might interpret the latter effects as more likely to result in decreased danger or stress being perceived, and thus as having a more beneficial effect on welfare.

It would be useful to develop techniques that could tease apart effects on appraisal and evaluation and on response and readout components. At the neurobiological level, this may be very difficult to do because of the inherent self-adjusting nature of the many neural systems and processes that coordinate the expression of the final (behavioral and neuroendocrine) responses to challenge.

Behavioral techniques, however, may help. For example, an organism faced with repeated exposure to challenge A shows a habituation of its HPA response. If it is then exposed to challenge B, a strong HPA response would indicate that the adjustment to challenge A was occurring at the level of some appraisal system rather than because of a general decrease in activity of the (HPA) response component. However, a continued weak HPA response to challenge B would indicate the opposite.

Another approach would be to use a technique such as aversion learning. For example, if an animal is exposed to a cue predicting an aversive stimulus, it will learn this association and may show an HPA response when the cue is presented. If oxytocin is administered during training, we might predict that the animal shows a smaller HPA response. The animal can then be tested with the cue only in the absence of oxytocin. If its HPA response remains small, this would indicate that the oxytocin minimized the perceived unpleasantness of the stimulus during training and so when the conditioned cue is presented, this is perceived as relatively harmless. However, if the HPA response is increased in this test, this would indicate that the main role of the oxytocin during training was to act directly on the HPA response without affecting perception or appraisal. Once the oxytocin is removed, the perceived danger can then be fully expressed by a large HPA response.

While some of the techniques described above have already been used, there are still very few studies dealing with this specific question.

HOW WELL DEFINED ARE COPING STRATEGIES?

Many coping responses come in packages, which we term as coping strategies, with several responses acting in concert. In some cases this might be explained by the evolutionary inertia of animal's physiology, while in other cases an adaptive explanation seems more likely.

The studies on coping strategies in mice originated in the early 1970s from an intensive collaboration between behavioral geneticists and behavioral ecologists in studies on the role of genetic variation for aggressive behavior in the population dynamics of wild house mice (Koolhaas et al., this volume). Evidence was obtained that the cyclic nature of feral populations of house mice was due to a disruptive selection for high levels of aggressive behavior. The genetic nature of this was confirmed by the successful development of genetic selection lines for high and low levels of aggressive behavior. Subsequently, these lines were studied by behavioral physiologists. A whole range of basically correlative studies resulted in a further characterization of these selection lines by a set of behavioral, neuroendocrine, and neurobiological parameters. Recent behavioral studies suggest that the differences can be reduced to a differential degree of feedforward versus feedback control of behavior. This led to the idea that the two lines in fact represent two different ways of coping with environmental challenges. Over the course of the years, similar individual differences were found in a variety of animal species. Some studies considered the stability of these characteristics over time, while others focused on the consistency across situations. In general, however, apart from rats, mice, and pigs, these aspects need further experimentation in many species. Ideally, the issue of trait characteristics needs a multi-factorial statistical approach to determine the number of dimensions that might explain the individual variation observed. There is evidence in rats and mice that the proactive and reactive coping strategies are also expressed in females.

Recent publications in the field of evolutionary ecology indicate that disruptive selection for behavioral traits such as aggression may occur in other species as well. These findings support the idea that it is possible to generalize the coping strategies shown in the house mice to other taxa (see Koolhaas et al., this volume).

There are a number of studies of farm animals in which the results support some of the predictions of the proactive/reactive strategies. In studies by Ruis et al. (2000), it was shown that pigs differing in behavioral resistance in a so-called "back test" (piglet is put on its back and manually restrained for one minute and the number of escape attempts are measured) at 2–4 days of age also differed in several other situations. Extreme responding gilts (roughly the top and bottom 25% of the distribution) were classified as high resistant (HR) and low resistant (LR). HR gilts showed more aggression in a group-feeding competition test at ten weeks of age. Also, in the competition for the most productive teats at the anterior, a predominant position of HR piglets at this side was observed during the suckling period. In a novel environment test at ten weeks (but not when this was repeated at 24 weeks) LR pigs hesitated longer to leave their home pens and to contact a human. LR pigs had a higher reactivity of the HPA axis, as shown by higher cortisol responses to the first novel environment test, to routine weighing at 25 weeks of age, and to administration of a high dose of ACTH. In another study of Ruis et al. (in preparation) HR and LR pigs also responded differently to social isolation. LR animals showed a higher initial cortisol response, more explorative behavior and less escape behavior. These data give support to consistency of coping strategy over context.

There are, however, also data from a number of other studies on several other species that fail to show either the bimodal distribution or the expected correlations found in the house mice (e.g., pigs: Erhard and Mendl 1999, Forkman et al. 1995; dogs: Scott and Fuller 1965; cattle: Hopster 1998).

There are several probable reasons why the results from these studies are ambiguous and why the results from house mice may be difficult to replicate in domestic species. It is possible that there are domestication effects, with a strong, directed selection for one type of coping strategy and therefore a strong reduction of the individual variation. Another possible explanation involves the ecological situation of the wild ancestors of our domestic species. Although wild populations of many animals show density cycles, the rodent cycles of the temperate zone are extreme in comparison with those of the wild boar or the wild ancestor of cattle. Hence the selection pressure might be disruptive for mice, but stabilizing for other species. This would explain why many other studies that have used factor analyses to look for personality traits have found other patterns of responding (for a review, see Gosling and John 1999). These explanations, however, do not hold for the findings of Hansson (1996). He found that the tendency of bank voles to either freeze or be passive in a stressful situation strongly correlated to age, a result not expected from the data on house mice, in which the coping strategy is relatively inflexible. Additional information, both on the coping patterns of rodents that do not show cyclical fluctuations as well as cyclical nonrodent species, would be helpful to explain this controversial issue.

While data on animals often indicate a strong consistency over time for different personality traits (e.g., Goddard and Beilharz 1986; Visser et al. 2001; Netto and Planta 1997), some studies on humans show less such consistencies. Rather then discussing this in detail, we have considered the possibility that the coping strategies, as observed in various animal species, might be related to the behavioral inhibition theory formulated to characterize stable temperaments in humans. Research on behavioral inhibition with humans has treated social withdrawal and reticence as individual characteristics reflective of a stable temperament (see Kagan et al. 1988). Studies of the stability of these behavioral characteristics, however, have demonstrated mixed results. Although statistically significant correlations have been reported, these correlations tend to be reported only when the groups selected are at the extremes of the distribution during the first measurement stage. When unselected samples are tested at different times, the correlations are usually very small and often approach 0. Similarly, attempts to identify neurophysiological correlates of these behavioral styles have resulted in weak findings. However, although weak, the findings tend to suggest that the behaviorally inhibited individuals express a different autonomic profile with features reflecting sympathetic-adrenal excitation. The basic issue is whether the behavioral typology is useful as a measure of stable, individual differences that might have a genotypic substrate, or is the typology reflecting a more fluid physiological state that can be readily manipulated to potentiate either social withdrawal or social engagement. Research by Coplan et al. (1994) has demonstrated behavioral strategies of children remaining on the periphery of socially interactive children. These reticent children, when measured in the laboratory, appear to have autonomic and EEG response profiles that distinguish them from more spontaneous, engaging children (Schmidt et al. 1999). However, these distinctions in neurophysiological and psychophysiological profiles do not ensure that the differences represent changes in morphology or even neurophysiology. It is possible that research strategies providing the children with methods to regulate neurophysiological state (e.g., methods to increase vagal regulation of heart) might result in a change in behavior, since it has been proposed that manipulation of the behavior state will result in a shift in the profile of emergent behaviors (Porges, this volume). The emphasis here is on the neural regulation of state and the neural mechanisms that can rapidly alter or buffer neurophysiological control of visceral systems (e.g., heart rate, HPA, SAM), which often can be modulated by vagal efferent actions on visceral organs, as well as on the HPA, SAM, and immune systems.

Although the differentiation in coping strategies relates vulnerability to a variety of stress-related diseases, the relevance of the concept of coping strategies for individual fitness and welfare needs further experimentation.

WHAT IS SOCIAL SUPPORT AND WHAT ARE ITS CONSEQUENCES?

Social support can be defined as the degree to which an individual receives "emotional support" (e.g., attention, help) from other individuals (e.g., partners, relatives, friends). Social support appears to have stress-buffering effects, which significantly affect and dampen the effects of stress on behavior, endocrine, and immune functions and disease outcome.

Examples from Work on Humans

Social support given during an acute stress situation reduced the cortisol response to psychological stressors in young healthy subjects (18–20 years old). These stressors consisted of public speaking and mental arithmetic (Kirschbaum et al. 1995; Roy et al. 1998, 2001). Not all persons could function as social supporters, however. In this study, both men and women received a stronger effect of the social support if the supporter was a woman rather than a man.

In an experimental setting, it has been demonstrated that human subjects with a high degree of social support have had the lowest rate of upper respiratory infection and virus load compared to subjects with a low degree of social support (Cohen et al. 1997). Similarly, a prospective, longitudinal study (6 months) demonstrated associations between the degree of social support and disease exacerbation in patients with lupus erythematosus (SLE): the higher the degree of social support, the lower the disease activity (Pawlak et al., submitted).

In addition to experimental work, there is also much literature on social support and psychiatric disorders, particularly depression. Distinctions have been made between wider support from a social network and close support from intimate relationships. There are also distinctions between perceived and actual support. For humans, what is perceived as supportive is in the end likely to be the most important. Social support partly acts by buffering effects of acute stress, and partly its absence may have an independent effect. Mechanisms behind social support still need to be worked out fully. They may include practical help, benefits in cognitive rehearsal with others of solving strategies, and more directly emotional benefits.

Examples from Work on Nonhuman Animals

Several studies in animals show the same pattern. Social animals that are kept alone develop higher frequencies of abnormal behavior (e.g., Broom and Johnson 1993) and infections (e.g., Lewis et al. 2000). In the same way, there are studies showing which rats that have suffered a recent defeat will endure less serious consequences of the defeat, measured as the level of glucocorticoids, if they have access to a companion (Ruis et al. 1999). Just as with humans, not all individuals can give this support. For example, in the guinea pig, young juveniles separated from their mother will accept any other female as a supporting individual. As they age, however, this general acceptance changes so that subadults only accept support from specific known individuals (Sachser et al. 1998).

In animal studies, however, there are several other possible explanations. A rat that receives an electrical shock will show lower levels of ACTH if it has the possibility of attacking a conspecific (Levine et al. 1979). Despite this, it would be wrong to call the role of the attacked individual supportive.

Future research on the effects of social support on behavior, endocrine and immune functions as well as disease outcome should analyze the "acting components" of social support (e.g., strength of relationship, duration, frequency). In addition, the neuroendocrine mechanisms and pathways through which social support affects immune functions and disease outcome need to be analyzed.

There are many clinical and practical applications of this field of research. As soon as the stress-buffering effects of social support are known in more detail, social support strategies can, for example, be precisely "tailored" as intervention strategies to improve welfare in animals and humans.

WHAT ARE THE COSTS ASSOCIATED WITH DIFFERENT COPING SYSTEMS?

Since the various coping systems have evolved, it can reasonably be assumed that they must have a net benefit in fitness terms and, when the individual is selecting among the various possible coping strategies, the cost of the coping attempt as well as the probability that coping will be successful should ideally be taken into account. This would mean that on average in the population, costs which are measured by increased mortality, delayed breeding, or reduced number of offspring produced would be outweighed by benefits. However, any particular attempt to cope may have costs to the individual and, in selecting one coping strategy over another, these costs might be taken into account.

We have little reliable evidence about the costs of many coping methods. The costs of some coping methods can only be estimated in terms of energy utilization. The actual cost of that energy usage will vary according to the availability of energy to the individual. For an animal with an excess of available food and no constraints on its ingestion and processing (e.g., domestic animals under normal husbandry conditions), a little extra energy consumption is a negligible cost. However, to an animal that is short of food, the same amount of energy consumption could lead to death.

Although in assessing the consequences of any particular challenge, the costs of all of the coping systems in operation should be considered together, it is instructive to evaluate examples of the various coping systems in relation to their costs. In each case below, where increased energy requirements represents the cost, actual effects on fitness will depend on the availability of energy from food reserves and from acquiring further food. Any extra energy demand, which a debilitated individual could not replenish, would have much greater effect on chances of surviving and reproducing than the same energy demand in an individual that could replenish any energy easily.

Fever and Sickness Behavior

Fever is an increase in body temperature to some upwardly adjusted set point. This requires extra energy. Other possible costs of fever or sickness behavior are reduced digestive efficiency, reduced efficiency of other enzymes, including those with metabolic functions, costs associated with body shutdown, and reduced gametogenesis. The reduction in feeding associated with sickness is necessary, because pre-feeding during sickness can lead to death but the energy loss associated with the inability to feed is costly.

Immune System Function

It has been suggested that at times the immune system may take up a very high proportion of available non-maintenance energy. However, we have very little clear evidence to substantiate this. The immune system experiences a high turnover of cells when it is not actively combating pathogens, so the energy increase resulting from defense against pathogens is not clear but may be very important. In addition to any energetic cost, where leucocytes have to be synthesized at a high rate, the material needed may reduce or render impossible the synthesis of other cells that are important for fitness. One indication of the costs of the immune system is that specific pathogen-free pigs grow considerably faster than normal pigs of the same breed and with the same management.

Brain Activity

The brain is very active during many types of coping response, and brain activity utilizes a much energy that could have been used in other ways. In the same way, resources might be used up within the brain during an activity, and activity within one part of the brain might prevent or reduce activity in another part. There is related evidence that training of dogs by humans makes them more tired than vigorous exercise, perhaps because of the brain activity.

Adrenal Responses

When glucocorticoids or catecholamines are produced, energy is expended. After production, they affect the modification of food reserves, which involve enzyme activity and which might result in an inefficient use of glucose and other products. Hence, energy may be wasted because of adrenal responses.

A second consequence of adrenal cortex activity is the increased risk of various potential problems, some of them very severe. The effects of glucocorticoids on the immune system are to some extent regulatory, but high and especially prolonged glucocorticoid activity has a substantial suppressive effect on cell-mediated immunity. In addition, high levels of glucocorticoids can disturb water balance, suppress reproductive function, lead to brain remodeling, and perhaps in certain rare circumstances cause brain damage.

Behavioral Responses

All behavior requires some energy, and some behavioral responses to challenge are vigorous or continue for long periods. When a behavioral response to adversity occurs, there is likely to be inhibition of other, potentially useful, behaviors, and there may be harmful consequences of the behavior. It could be directly damaging or could increase risks such as those of predation.

Heart Rate Responses

If an individual responds to challenge by lowering its heart rate (e.g., in the course of a freezing response), there is a risk of dying because of the heart beat control mechanism. If the heart rate increases, to prepare the individual for vigorous activity, myocardial infarction or other heart damage is a possibility. Other cardiovascular responses can also increase risks.

Responses to Poison Ingestion

An immediate consequence of the detection of a noxious substance in the mouth or stomach is retching. This results in loss of ingested food and may also have significant side effects, such as an increased risk of predation. Many ingested toxins can be detoxified, but such processes have a significant energetic cost. Learning about toxins can be somewhat inefficient and lead to the future rejection of nutritious food because of its perception as potentially toxic.

Summary

The individual should select its coping method well because of the costs associated with each method. In some cases there are no alternatives (e.g., when there is too much carbon dioxide in inhaled air, the immediate reaction is straightforward). However, some problems in life can be solved in various ways and the selection, whether or not cognitive processes are involved, has major consequences for fitness. This whole area is not well understood and studies on the costs of coping mechanisms are needed. In practice, overall costs of a variety of methods used at one time would often have to be measured.

A further aspect of this area is the welfare of individuals using the various coping systems to different extents. The welfare (*sensu* Fraser and Broom 1990) of an individual is worse if its fitness is impaired, for example, in the sense that it is more likely to die rapidly. However, welfare is also poor if the individual experiences much difficulty in coping. In many cases that poor welfare is associated with the set of bad feelings that we describe as suffering. Whenever there are alternative coping strategies, we should endeavor to assess the extent of any poor welfare or good welfare using a range of measures such as those of abnormal behavior, emergency physiological responses, injury, etc. One example of this is the work of Duncan and Filshie (1979), which shows that hens disturbed by humans may either show extensive behavioral escape responses or no behavior change but a high increase in heart rate. Where degree of aversion has to be measured in a preference test, such differences in coping methods used should also be taken into account.

CONCLUSIONS AND IMPLICATIONS

Early stressful experiences can influence the development of the brain and consequently also the behavior of the individual as an adult. Stressors affecting the mother while she is pregnant can have a clear effect on the fetus, and later on the adult individual. Precisely how these changes take place and what neurocircuits are modified remains unclear. To what extent these changes are adaptive, i.e., the presence of a stressor is used by an infant to predict its future environment, is unknown. It is clear, however, that there are a large number of studies, the results of which cannot be explained by this hypothesis.

Recent research has shown that the multiple drug-resistant pump may regulate entry of cortisol into the brain. As a consequence, the influence of other adrenal steroids may be of more importance in the primate brain than previously appreciated, at least under certain circumstances. It is not known whether this holds true for other species, for example, the farm animal species. From an applied perspective, it is clearly very important to do further research in this area to clarify the situation, since cortisol levels so often are used as one measure of the level of stress experienced by animals.

An interesting idea is that part of the variation in physiological and behavioral variables seen in stressful situations can be explained by the animal reacting with a different evaluation of the situation (e.g., the noise no longer predicts a predator and is therefore not dangerous) rather than by using the response system (e.g., using oxytocin to lower the stress response). There is very little data either to support or refute this idea, but it is testable and several alternatives have been given in the text.

Stress responses can vary, but certain general principles seem to be visible. In many species, coping strategies, in which the reaction to stress in one context predicts the reaction to stress in another context, can be seen. For stability over time but within context, the evidence is even stronger. There is extensive animal literature suggesting that certain traits (e.g., level of fearfulness) are consistent over time. A very influential theory of coping strategies is the proactive/reactive model. The model has been tested in a number of species with mixed results. On the whole the model seems to be very successful in predicting the behavior of mice and rats over a wide range of different situations but has had less success with domesticated species. Since the model makes strong and important predictions on the ideal composition of a group of animals, it is important to understand what is needed and, if it is at all possible, to implement it on species used by humans, such as farm animals.

One way of coping with stress is through social support. In a number of studies and in a number of species, the effect of social support has been shown to be strong and reliable. Still, we have very little basic knowledge about it and, in fact, know very little about factors that may strengthen or weaken it. Therefore, we can do little to increase the effectiveness of it. Future research in this area is important, both for applications on humans and in farm animals.

Just as a coping system confers a benefit to the individual using it, each coping system also involves a cost. Many times, cost is in the form of energy, something likely to be more important in nature than in well-fed domestic animals, or in western society. Understanding the costs associated with each coping system will help us to understand not only the level of stress an animal suffers but also the cost they must be to withstand it.

REFERENCES

Alonso, S.J., R. Arevalo, D. Afonso, and M. Rodríguez. 1991. Effects of maternal stress during pregnancy on forced swimming test behavior of the offspring. *Physiol. Behav.* **50**:511–517.

Anderson, D.K., R.W. Rhees, and D.E. Fleming. 1985. Effects of prenatal stress on differentiation of the sexually dimorphic nucleus of the preoptic area (SDN-POA) of the rat brain. *Brain Res.* **332**: 113–118.

Beauchamp, A.J., J.P. Gluck, and M.H. Lewis. 1991. Associative processes in differentially reared rhesus monkeys (*Macaca mulatta*): Blocking. *Dev. Psychobiol.* **24**:175–189.

Bilko, A., V. Altbacker, and R. Hudson. 1994. Transmission of food preference in the rabbit: The means of information transfer. *Physiol. Behav.* **56**:907–912.

Broom, D.M., and K.G. Johnson. 1993. Stress and Animal Welfare, pp.135–136. Dordrecht: Kluwer.

Caldji, C., D. Liu, S. Sharma, J. Diorio, D. Francis, M.J. Meaney, and P.M. Plotsky. 2001. Development of individual differences in behavioral and endocrine responses to stress: Role of the postnatal environment. In: Handbook of Physiology: Coping with the Environment, ed. B.S. McEwen, pp. 271–292. New York: Oxford Univ. Press.

Clancy, B., R.B. Darlington, and B.L. Finlay. 2001. Translating developmental time across mammalian species. *Neuroscience*, in press.

Clarke, A.S., D.J. Wittwer, D.H. Abbott, and M.L. Schneider. 1994. Long-term effects of prenatal stress on HPA axis activity in juvenile rhesus monkeys. *Dev. Psychobiol.* **27**:257–269.

Cohen, S., W.J. Doyle, D.P. Skoner, B.S. Rabin, and J.M. Gwaltney, Jr. 1997. Social ties and susceptibility to the common cold. *J. Am. Med. Assn.* **277**:1940–1944.

Coplan, J.D., M.W. Andrews, L.A. Rosenblum, M.J. Owens, S. Friedman, J.M. Gorman, and C.B. Nemeroff. 1996. Persistent elevations of cerebrospinal fluid concentrations of corticotropin-releasing factor in adult nonhuman primates exposed to early-life stressors: Implications for the pathophysiology of mood and anxiety disorders. *Proc. Natl. Acad. Sci. USA* **93**:1619–1623.

Coplan, R.J., K.H. Rubin, N.A. Fox, S.D. Calkins, and S.L. Stewart. 1994. Being alone, playing alone, and acting alone: Distinguishing among reticence and passive and active solitude in young children. *Child Dev.* **65(1)**:129–137.

Coplan, J.D., E.L.P. Smith, R.C. Trost, B.A. Scharf, M. Altemus, L. Bjornson, M.J. Owens, J.M. Gorman, C.B. Nemeroff, and L.A. Rosenblum. 2000. Growth hormone response to clonidine in adversely reared young adult primates: Relationship to serial cerebrospinal fluid corticotropin-releasing factor concentrations. *Psychiatry Res.* 95:93–102.

Coplan, J.D., R.C. Trost, M.J. Owens, T.B. Cooper, J.M. Gorman, C.B. Nemeroff, and L.A. Rosenblum. 1998. Cerebrospinal fluid concentrations of somatostatin and biogenic amines in grown primates reared by mothers exposed to manipulated foraging conditions. *Arch. Gen. Psychiatry* **55**:473–477.

Cordon-Cardo, C., J.P. O'Brian, D. Casals, G.L. Rittman, J.L. Biedler, M.R. Melmed, and J.R. Bertino. 1989. Multidrug-resistance gene (P-glycoprotein) is expressed by endothelial cells at blood-brain barrier siyes. *Proc. Natl. Acad. Sci. USA* **86**:695–698.

Dahlöf, L.G., E. Hård, and K. Larsson. 1977. Influence of maternal stress on offspring sexual behavior. *Animal Behav.* **25**:958–963.

De Kloet, E.R., G. Wallach, and B.S. McEwen. 1975. Differences in corticosterone and dexamethasone binding to rat brain and pituitary. *Endocrinology* **96**:598–609.

Duncan, I.J.H., and J.H. Filshie. 1979. The use of radiotelemetry devices to measure temperature and heart rate in domestic fowl. In: A handbook on biotelemetry and radio tracking, ed. C.J. Amlaner and D.W. Macdonald, pp. 579–588. Oxford: Pergamon.

Erhard, H.W., and M. Mendl. 1999. Tonic immobility and emergence time in pigs: More evidence for behavioral strategies. *Appl. Animal Behav. Sci.* **61**:227–237.

Finlay, B.L., and R.B. Darlington. 1995. Linked regularities in the development and evolution of mammalian brains. *Science* **268**:1578–1584.

Forkman, B., I.L. Furuhaug, and P. Jensen. 1995. Personality, coping patterns, and aggression in piglets. *Appl. Animal Behav. Sci.* **45(1–2)**:31–42.

Fraser, A.M., and D.M. Broom. 1990. Farm Animal Behavior and Welfare. London: Bailliere Tindall.

Ginsberg, S.D., P.R. Hof, W.T. McKinney, and J.H. Morrison. 1993. The noradrenergic innervation density of the monkey paraventricular nucleus is not altered by early social deprivation. *Neurosci. Lett.* **158**:130–134.

Gluck, J.P., and G.P. Sackett. 1976. Extinction deficits in socially isolated rhesus monkeys (*Macacca mulatta*). *Dev. Psychol.* **12**:173–174.

Goddard, M.E., and R.G. Beilharz. 1986. Early prediction of adult behavior in potential guide dogs. *Appl. Animal Behav. Sci.* **15**:247–260.

Godfrey, K.M., and D.J.P. Barker. 2000. Fetal nutrition and adult disease. *Am. J. Clin. Nutr.* **71**: 1344–1352.

Gosling, S.D., and O.P. John. 1999. Personality dimensions in non-human animals. *Curr. Dir. Psychol. Sci.* **8**:69–75.

Hansson, L. 1996. Regional differences in behavior in bank voles (*Clethrionomys glareolus*): Dyadic encounters. *Behav. Ecol. Sociobiol.* **39**:331–334.

Harlow, H.F., M.K. Harlow, and S.J. Suomi. 1971. From thought to therapy: Lessons from a primate laboratory. *Am. Sci.* **59**:538–549.

Harvey, P.W., and P.F.D. Chevins. 1985. Crowding pregnant mice affects attack and threat behavior of male offspring. *Horm. Behav.* **19**:86–97.

Heim, C., M.J. Owens, P.M. Plotsky, C.B. Nemeroff. 1997. The role of early adverse life events in the etiology of depression and posttraumatic stress disorder: Focus on corticotropin-releasing factor. *Ann. NY Acad. Sci.* **821**:194–207.

Herrenkohl, L.R. 1979. Prenatal stress reduces fertility and fecundity in female offspring. *Science* **206**:1097–1099.

Higley, J.D., S.J. Suomi, and M. Linnoila. 1992. A longitudinal assessment of CSF monoamine metabolite and plasma cortisol concentrations in young rhesus monkeys. *Biol. Psychiatry* **32**:127–145.

Hopster, H., J.T.N. Van der Werf, and H.J. Blokhuis. 1998. Short- and long-term consistency of behavioral and physiological stress responses in dairy cows during a novel environment test. Coping strategies in dairy cows. Ph.D. diss., pp 67–84. Wageningen University, Wageningen, The Netherlands

Kagan, J., J.S. Reznick, and N. Snidman. 1988. Temperamental influences on reactions to unfamiliarity and challenge. *Adv. Exp. Med. Biol.* **245**:319–339.

Kage, A., A. Fenner, B. Weber, and M. Schoneshofer. 1982. Diurnal and ultradian variations of plasma concentrations of eleven adrenal steroid hormones in human males. *Klin. Wochenschr.* **60**:659–666.

Karssen, A.M., O.C. Meijer, A.G. De Boer, I.C.J. Van der Sandt, E.C.M. De Lange, and E.R. De Kloet. 2000. The multidrug resistance 1a p-glycoprotein affects penetration of glucocorticoids, except corticosterone, in the brain. Abstract 152.17. 30th annual meeting of the Soc. for Neuroscience, New Orleans.

Kay, G., N. Tarcic, T. Polytrev, and M. Weinstock. 1998. Prenatal stress depresses immune function in rats. *Physiol. Behav.* **63**:397–402.

Kinsley, C.H., and B. Svare. 1986. Prenatal stress reduces intermale aggression in mice. *Physiol. Behav.* **36**:783–786.

Kirschbaum, C., T. Klauer, S.H. Filipp, and D.H. Hellhammer. 1995. Sex-specific effects of social support on cortisol and subjective responses to acute psychological stress. *Psychosom. Med.* **57(1)**:23–31.

Klein, S.L., and D.R. Rager. 1995. Prenatal stress alters immune function in the offspring of rats. *Dev. Psychobiol.* **28**:321–336.

Koehl, M., M. Darnaudery, J. Dulluc, O. Van Reeth, M. Le Moal, and S. Maccari. 1999. Prenatal stress alters circadian activity of hypothalamo-pituitary-adrenal-axis and hippocampal corticosteroid receptors in adult rats of both gender. *J. Neurobiol.* **40**:302–315.

Künzl, C., and N. Sachser. 1999. The behavioral endocrinology of domestication: a comparison between the domestic guinea pig (*Cavia aperea* f. *porcellus*) and its wild ancestor, the cavy (*Cavia aperea*). *Horm. Behav.* **35**:28–37.

Ladd, C.O., R.L. Huot, K.V. Thrivikraman, C.B. Nemeroff, M.J. Meaney, and P.M. Plotsky. 2000. Long-term behavioral and neuroendocrine adaptations to adverse early experience. In: Progress in Brain Research: The Biological Basis for Mind Body Interactions, vol. 122, ed. E.A. Mayer and C.B. Saper, pp. 81–103. Amsterdam: Elsevier.

Levine, S., J. Weinberg, and L.P. Brett. 1979. Inhibition of pituitary-adrenal activity as a consequence of consummatory behavior. *Psychoneuroendocrinology* **4**:275–286.

Lewis, M.H., J.P. Gluck, A.J. Beauchamp, M.F. Keresztury, and R.B. Mailman. 1990. Long-term effects of early social isolation in *Macaca mulatta*: Changes in dopamine receptor function following apomorphine challenge. *Brain Res.* **513**:67–73.

Lewis, M.H., J.P. Gluck, J.M. Pettito, L.L. Hensley, and H. Ozer. 2000. Early social deprivation in nonhuman primates: Long-term effects on survival and cell-mediated immunity. *Biol. Psychiatry* **47(2)**:119–126.

Lubach, G.R., C.L.Coe, and W.B. Ershler, 1995. Effects of early rearing environment on immune responses of infant rhesus monkeys. *Brain Behav. Immun.* **9(1)**:31–46.

Lyons, D.M., F.L. Martel, S. Levine, N.J. Risch, and A.F. Schatzberg. 1999. Postnatal experiences and genetic effects on squirrel monkey social affinities and emotional distress. *Horm. Behav.* **36(3)**:266–275.

MacLusky, N.J., and F. Naftolin. 1981. Sexual differentiation of the central nervous system. *Science* **211**:1294–1303.

Martin, R.E., G.P. Sackett, V.M. Gunderson, and B.M. Goodlin-Jones. 1988. Auditory evoked heart rate responses in pigtailed macaques (*Macaca nemestrina*) raised in isolation. *Dev. Psychobiol.* **22**:251–260.

Matthews, K., F.S. Hall, L.S. Wilkinson, and T.W. Robbins. 1996. Retarded acquisition and reduced expression of conditioned locomotor activity in adult rats following repeated early maternal separation: Effects of prefeeding, d-amphetamine, dopamine antagonists, and clonidine. *Psychopharmacology* **126(1)**:75–84.

McIntosh, J., H. Anisman, and Z. Merali. 1999. Short and long periods of neonatal maternal separation differentially affect anxiety and feeding in adult rats: Gender-dependent effects. *Brain Res. Dev. Brain Res.* **113(1–2)**:97–106.

Meijer, O.C., E.C.M. De Lange, D.D. Breimer, A.G. De Boer, J.O. Workel, and E.R. De Kloet. 1998. Penetration of dexamethasone into brain glucocorticoid targets is enhanced in mdr1A P-glycoprotein knockout mice. *Endocrinology* **139**:1789–1793.

Moyer, J.A., L.R. Herrenkohl, and D.M. Jacobowitz. 1978. Stress during pregnancy: Effect on catecholamines in discrete brain regions of offspring as adults. *Brain Res.* **144**:173–178.

Netto, W.J., and D.J.U. Planta. 1997. Behavioral testing for aggression in the domestic dog. *Appl. Animal Behav. Sci.* **52**:243–263.

Newell-Morris, L.L., C.E. Fahrenbruch, and G.P. Sackett. 1989. Prenatal psychological stress, dermatoglyphic asymmetry, and pregnancy outcome in the pigtailed macaque (*Macaca nemestrina*). *Biol. Neonate* **56**:61–75.

Pariante, C.M., R. Dayanandan, A. Heyden, S. Ackerley, A.J. Grierson, G. Nebbia, S. Lovestone, A.M. Makoff, M. Danielsen, B.D. Pearce, A.H. Miller, and R.W. Kerwin. 2000. A new green fluorescent protein-glucocorticoid receptor (GR) chimera to study steroid-independent regulation of the GR in cell lines and in rat cortical and hippocampal neurons. Abstract 152.1. 30th annual meeting of the Soc. for Neuroscience, New Orleans.

Phillips, D.I.W., B.R. Walker, R.M. Reynolds, D.E.H. Flanagan, P.J. Wood, C. Osmond, D.J.P. Barker, and C.B. Whorwood. 2000. Low birth weight predicts elevated plasma cortisol concentrations in adults from 3 populations. *Hypertension* **35**:1301–1306.

Rosenblum, L.A., and M.W. Andrews. 1994. Influences of environmental demand on maternal behavior and infant development. *Acta Paediatr.* **Suppl. 397**:57–63.

Roy, M.P., C. Kirschbaum, and A. Steptoe. 2001. Psychological, cardiovascular, and metabolic correlates of individual differences in cortisol stress recovery in young men. *Psychoneuro-endocrinology* **26**:375–391.
Roy, M.P., A. Steptoe, and C. Kirschbaum. 1998. Life events and social support as moderators of individual differences in cardiovascular and cortisol reactivity. *J. Personality & Soc. Psychol.* **75**:1273–1281.
Ruis, M.A., J.H.A. te Brake, B. Buwalda, S.F. De Boer, P. Meerlo, S.M. Korte, H.J. Blokhuis, and J.M. Koolhaas. 1999. Housing familiar male wildtype rats together reduces the long-term adverse behavioral and physiological effects of social defeat. *Psychoneuroendocrinology* **24(3)**:285–300.
Ruis, M.A.W., J.H.A. te Brake, J.A. van der Burgwal, I.C. de Jong, H.J. Blokhuis, and J.M. Koolhaas. 2000. Personalities in female domesticated pigs: Behavioral and physiological indications. *Appl. Animal Behav. Sci.* **66**:31–49.
Sachser, N., D. Hierzel, and M. Dürschlag. 1998. Social relationships and the management of stress. *Psychoneuroendocrinology* **23**:891–904.
Sachser, N., and S. Kaiser. 1996. Prenatal social stress masculinizes the females' behavior in guinea pigs. *Physiol. Behav.* **60**:589–594.
Sackett, G.P. 1981. A nonhuman primate model for studying causes and effects of poor pregnancy outcomes. In: Preterm Birth and Psychological Development, ed. S. Friedman and M. Sigman, pp. 41–63. New York: Academic.
Sanchez, M.M., E.F. Hearn, D. Do, J.K. Rilling, and J.G. Herndon. 1998. Differential rearing affects corpus callosum size and cognitive function of rhesus monkeys. *Brain Res.* **812**:38–49.
Schmidt, L.A., N.A. Fox, J. Schulkin, and P.W. Gold. 1999. Behavioral and psychophysiological correlates of self-presentation in temperamentally shy children. *Dev. Psychobiol.* **35(2)**:119–135.
Schneider, M.L. 1992a. Delayed object permanence development in prenatally stressed rhesus monkeys infants (*Macaca mulatta*). *Occ. Therapy J. Res.* **12**:96–110.
Schneider, M.L. 1992b. Prenatal stress exposure alters postnatal behavioral expression under conditions of novelty challenge in rhesus monkey infants. *Dev. Psychobiol.* **25**:529–540.
Schneider, M.L. 1992c. The effect of mild stress during pregnancy on birth weight and neuromotor maturation in rhesus monkey infants (*Macaca mulatta*). *Infant Behav. Dev.* **15**:389–403.
Schneider, M.L., A.S. Clarke, G.W. Kraemer, E.C. Roughton, G.R. Lubach, S. Rimm-Kaufman, D. Schmidt, and M. Ebert. 1998. Prenatal stress alters brain biogenic amine levels in primates. *Dev. & Psychopathol.* **10**:427–440
Scott, J.P., and J.L. Fuller. 1965. Genetics and the Social Behavior of the Dog. Chicago: Univ. of Chicago Press.
Van Kalken, C.K., H.J. Broxterman, H.M. Pinedo, N. Feller, H. Dekker, J. Lankelma, and G. Giaccone. 1993. Cortisol is transported by the multidrug resistance gene product P-glycoprotein. *Br. J. Cancer* **67**:284–289.
Visser, E.K., C.G. van Reenen, H. Hopster, M.B.H. Schilder, J.H. Knaap, A. Barneveld, and H.J. Blokhuis. 2001. Quantifying aspects of young horses' temperament: Consistency of behavioral variables. *Appl. Animal Behav. Sci.* **74**:241–258.
Ward, I. 1972. Prenatal stress feminizes and demasculinizes the behavior of males. *Science* **175**:82–84.
Weinstock, M. 1997. Does prenatal stress impair coping and regulation of hypothalamic-pituitary-adrenal axis? *Neurosci. Biobehav. Rev.* **21(1)**:1–10.
Weinstock, M., T. Polytrev, D. Schorer-Apfelbaum, D. Men, and R. McCarty. 1998. Effect of prenatal stress on plasma corticosterone and catecholamines in response to footshock in rats. *Physiol. Behav.* **64(4)**:439–444.
Wigger, A., and I.D. Neumann. 1999. Periodic maternal deprivation induces gender-dependent alterations in behavioral and neuroendocrine responses to emotional stress in adult rats. *Physiol. Behav.* **66(2)**:293–302.

17

How Are Stress and Depression Interrelated?

M. IRWIN

Cousins Center for Psychoneuroimmunology, Department of Psychiatry and Biobehavioral Sciences, University of California, Los Angeles, CA 90037, U.S.A.

Keywords: major depression, life stress, mortality, psychoneuroimmunology, aging

ABSTRACT

Stress and depression are highly positively correlated in clinical populations. Numerous studies implicate stress in the onset of major depression, and there is little question that depression is profoundly stressful and also often precipitates external life events. In this chapter, the interrelationships between life stress and depression in humans and the pathological consequences of these conditions are considered. There is a relationship between life stress and depression onset while individual differences in coping, social support, and cognitive appraisal of stress can have an impact on emotional health. The effects of psychological stress and depression are not limited to decrements in emotional functioning. Rather, stress and depression are generally associated with poor welfare, in particular that resulting from progression of chronic diseases (e.g., rheumatoid arthritis, chronic obstructive lung disease, and cardiovascular disease), onset of infectious illness (e.g., herpes zoster), and increased mortality. The physiological mechanisms that account for individual differences in chronic morbidity in relation to stress and depression are not fully appreciated, although considerable evidence suggests that psychological stress and depression influence multisystem physiological processes. Moreover, sleep and sleep depth are hypothesized to have a critical role in mediating changes in autonomic and immunological function in stressed and depressed patients. Finally, the notion that the effects of stress accumulate is evaluated using the concept of allostasis, and the important contribution that stress reactivity studies make in identifying dynamic variability in sympathetic, neuroendocrine, and immunologic responsivity in depressed and stressed persons in relation to disease outcomes, is addressed.

Coping with Challenge: Welfare in Animals including Humans
Edited by D.M. Broom

INTRODUCTION

The relationship between stress and depression has been suggested since the early work of Kraepelin, who postulated that individuals with a genetic predisposition to a mood disorder are likely to develop depression when they experience acute life stress as compared to those who do not undergo stress. Furthermore, once triggered, depression may then progress independently of psychosocial stressors.

Acute stress generally refers to any physical or psychological change that harmfully disrupts the organism's balance or homeostasis. In other words, acute stress is defined as a threatening of homeostasis to which the organism responds with a large number of adaptive responses and which has some harmful effect.

Psychological stressors are typically emphasized as key elements in psychiatric illness such as depression, although physical stressors (e.g., viral infections, autoimmune disorders) are becoming increasingly recognized as potential triggers.

RELATIONSHIP BETWEEN STRESS AND DEPRESSION

The correlation between depression and psychological and psychosocial stress has face validity, not only because being profoundly depressed is unquestionably stressful but because acute external stressful events often precipitate depressive episodes (Post 1992). Indeed, epidemiological studies strongly suggest a role for both genetic factors and psychosocial stressors in depression. Adverse life events are likely to occur around the onset of affective illness (Brown and Harris 1982), and the presence of psychosocial stressors is associated with poor outcome in depressed patients (Sargeant et al. 1990). Moreover, even in the absence of precipitating events, the occurrence of a depressive episode can lead to substantial psychosocial stress, secondary to the disruption of family and work relations. Interestingly, in families who have a genetic predisposition to affective disorder, Kendler et al. (1995) have shown further that there is increased likelihood of life events, even those events that are independent and can not be attributed to a depressive episode.

Undesirable events show uniformly adverse effects on mental health and depressive symptom onset, although the relationships are not linear, as discussed below. In older adults experiencing the major stressors of conjugal bereavement and physical disability, everyday events and self-reports of mental health evaluated across a three-month time span revealed a causal association between undesirable events and increases in depressive symptoms (Zautra et al. 1990). Indeed, marked life adversities are more likely to occur in depressed patients and might precipitate depression in the first place (McNaughton et al. 1992).

However, the relationship of life stress to depression is complex and determined by multiple biological and social factors. Kendler et al. (2000) have shown, for example, that the strength of the association between stressful life event and onset of depressive episodes declines with increasing number of previous depressive episodes, consistent with the kindling hypothesis. Furthermore, the number of previous depressive episodes affects the association between stressful event and depression in a biphasic pattern. When the number of previous depressive episodes was nine or fewer, each additional previous depressive episode conveyed a large increased risk for a depressive onset. After nine previous episodes, each additional depression conveyed a smaller risk. Importantly, none of the existing animal models of depression fully takes into account the nonlinear causal consequences of stress and the interplay proposed here between biological and social factors in the onset and recurrence of depression.

It is not simply the occurrence of acute life events that increases risk for depression. Among those who experience life events, social factors moderate the association between stress and depression. Reduced number of social supports and greater use of emotion-focused coping (self-blame, wishful thinking, and avoidance) identifies those who are likely to have an exacerbation of their depression (McNaughton et al. 1992, 1995) and predicts those elderly who are more symptomatic and psychologically distressed (Smith et al. 1990). These findings are particularly evident in older adults who experience a major health problem; efficacy in coping is longitudinally associated with less psychological distress (Zautra and Wrabetz 1991), whereas negative social ties predict future depression (Finch and Zautra 1992). In addition, person variables, such as beliefs that external factors control one's destiny and endorsement of irrational beliefs in the context of chronic life stress, are related to declines in mental health (McNaughton et al. 1995).

PREVALENCE OF DEPRESSIVE SYMPTOMS, CHRONIC STRESS, AND DEPRESSIVE DISORDERS

Depressive syndromes and major depression are exceedingly common. Lifetime prevalence rates of major depression and dysthymia are 13% and 5%, respectively (Kessler et al. 1994). In primary care outpatients who suffer the stress of chronic disease burden, point prevalence rates range from 9% to 20% for all depressive disorders (Blacker and Clare 1987; Barrett et al. 1988; Barry et al. 1998). In medical inpatients who likely experience even greater stress related to acute or chronic disease morbidity, rates are even higher, with 8% for major depression and 15% to 36% for all depressive disorders (Magni et al. 1986; Feldman et al. 1987).

In older adults who are at increased risk of life events, such as chronic disease and personal losses, the clinical course of depression is one that is chronic, marked by recurrent and remitting episodes (Berger et al. 1998; Jorm 1995) The development of major depression can take many years in the older adult with prodromal and progressive elevation of depressive symptoms, in contrast to the relatively short clinical onset in younger patients (Berger et al. 1998). Moreover, despite up-to-date clinical management, there is often incomplete resolution of depressive symptoms and an increased risk of relapse in old adults who continue to experience substantial increased rates of life adversity (Lyness et al. 1993; Blazer 1989). The factors that contribute to the more chronic nature of depression in old age are not fully known, although the presence of chronic disease, increase in risk of exposure to stressful life events and personal losses, diminished social supports, and declines in self-concepts of efficacy and mastery are suggested (Blazer 1989). It is also well documented that severity of physical illness is one of the most important variables associated with depression in patients with chronic medical illness (Blazer 1989).

Thus, to understand the relationships between depression and stress, assessment of multiple domains is needed, as shown in Figure 17.1, modified from McEwen (1998). Together, such a model takes into account the reciprocal and cascading relationships between severity of chronic disease, other life stresses, social support, self-concept, and coping at the onset of affective disturbance and depressive disorders.

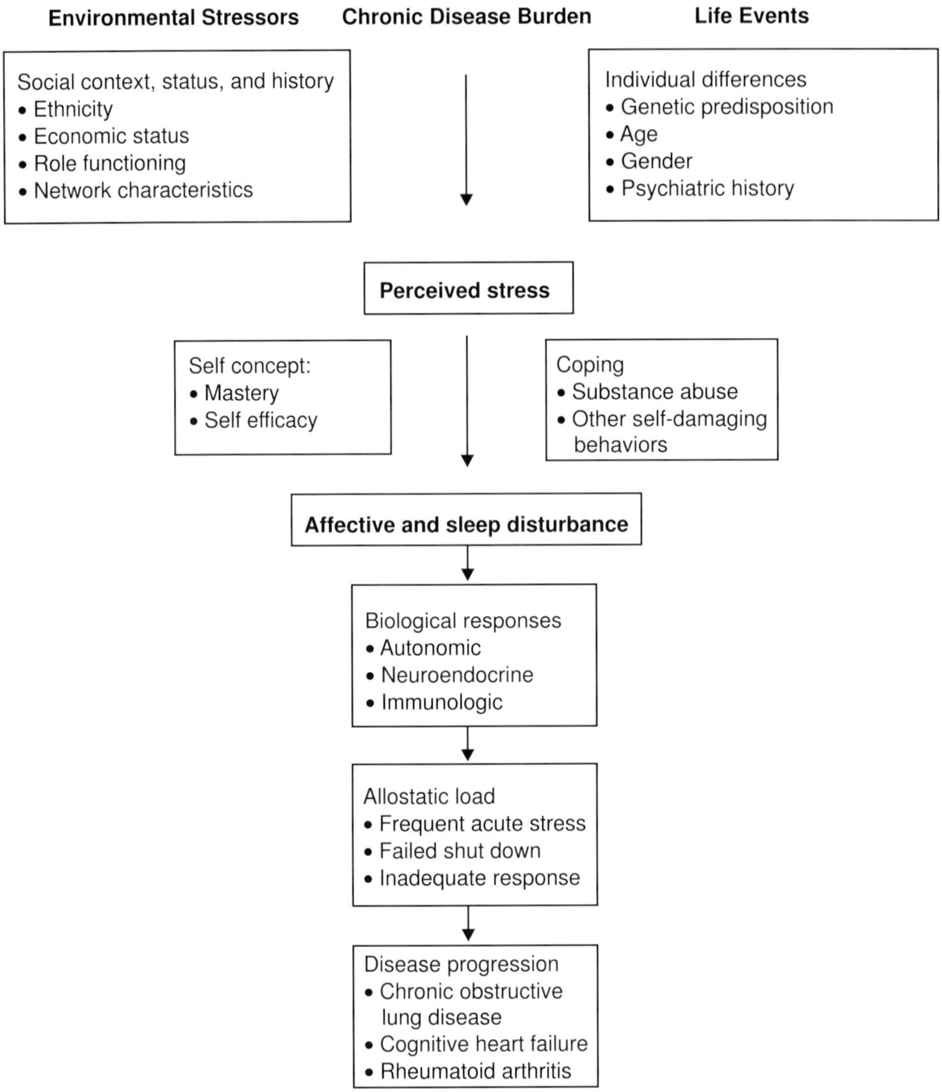

Figure 17.1 Assessment of the relationships between chronic disease, other life stresses, social support, self-concept, and coping at the onset of affective disturbance and depressive disorders (after McEwen 1998).

STRESS AND DEPRESSION: IMPACTS ON HEALTH FUNCTIONING

Both psychological distress and depression are associated with declines in various domains of living, such as transportation, family relationships, relationships with friends, financial, safety, and living arrangements (McAvay et al. 1996). Depression also leads to an increased risk of onset of disability in activities of daily living and physical health status (Bruce,

Seeman et al. 1994) and an overall decline in quality of life assessment (Pyne et al. 1997). Again, in support of the complexity of the interacting relationships between psychosocial stress and depression, emotional support from a social network and self-efficacy has been found to buffer functional declines, particularly in the elderly who face diminished physical capacity (Seeman, Berkman et al. 1995a).

The impact of depression and subthreshold depressive symptoms associated with life stress is not trivial or limited to effects on social functioning. First, impairments in measures of physical functioning, pain, and general health occur along with decrements in emotional health (Bruce, Seeman et al. 1994; Broadhead et al. 1990; Wells et al. 1989). Second, limitations in functioning of depressed persons across physical domains are equal to or greater than those found in many major chronic medical conditions (Wells et al. 1989). Third, while impairments can improve over time, as might be expected for an episodic disorder, persistent limitations are generally found consistent with the view of depressive disorders as chronic and recurrent with residual disability (Hays et al. 1995). Bruce, Leaf et al. (1994) have reported that both depressive symptoms and stress of physical disability can initiate a spiraling decline in physical and psychological health.

DEPRESSION, STRESS, AND DISEASE OUTCOMES

Psychological stress and depressive symptoms impact a variety of health outcomes and result in increased rates of cardiovascular and noncardiovascular diseases (McEwen and Stellar 1993; Bruce, Leaf et al. 1994). Again, the older adult may be particularly vulnerable to the health risks associated with stress and depression, as the elderly show substantial alterations in the regulation of autonomic, neuroendocrine, and immune systems (Kiecolt-Glaser and Glaser 1995; Sapolsky et al. 1986; Doherty 1990). For example, among older adults who are undergoing the chronic stress of caregiving for a spouse with Alzheimer's disease, Shaw et al. (1997) found that greater caregiving burden was associated with an increased risk for serious illness in a six-year longitudinal evaluation of 150 spousal caregivers and 46 married control participants.

The notion that stress or depressive disorders can increase one's risk for developing disease or influence the severity or rate of progression of disease is controversial and is often "explained" by the hypothesis that patients with a life stress or a psychiatric disorder generally have other risk factors for the development of chronic disease (Hayward 1995). Alternatively, depression may adversely affect compliance with medical therapy and rehabilitation and increase medical comorbidity via this route (Blumenthal et al. 1982; Carney et al. 1995). However, in the example of cardiovascular disease, analyses of other risk factors (such as hypertension, hypercholesterolemia, nicotine and other substance abuse, and physical inactivity) along with control of age, gender, and socioeconomic status implicate depression as an independent risk factor (Glassman and Shapiro 1998; Simonsick et al. 1995; Everson et al. 1996; Pratt et al. 1996; Wassertheil-Smoller et al. 1996). The presence of depression preceding the onset of ischemic heart disease, the greater risk of sudden death among postmyocardial infarction patients with depression and arrhythmia, and the predisposition to increased platelet aggregation among depressed patients all point to a causal relationship.

Cardiovascular disease is also impacted by psychosocial stress, in addition to the independent effects of depression. In elderly people who already show considerable cardiac morbidity and are undergoing hospitalization for heart failure, lack of emotional support is associated with a significant one-year risk of fatal and nonfatal cardiovascular outcomes, after adjustment for clinical severity, social ties, and instrumental support (Krumholz et al. 1998). Conversely, presence of social integration has a highly beneficial effect on postmyocardial infarction prognosis as reflected by functioning and longevity (Seeman 1996).

Preclinical data further support the link between psychosocial stress and cardiovascular disease risk. Psychological stress is reported to be an independent risk factor in the development of atherosclerosis (Manuck et al. 1986; Marmot 1988), which may result from repeated physiological arousal in response to psychological challenges (Manuck et al. 1986). Extensive work with monkeys has further shown that psychosocial stress increases cardiovascular disease risk factors (Skantze et al. 1998). For example in group-living cynomolgus monkeys, chronic social separation was associated with increases of heart rates that could exacerbate atherosclerosis (Watson et al. 1998). Moreover, socially unstable male cynomolgus monkeys showed a significant increase in the number of injured endothelial cells in the circumostial coronary arteries (Skantze et al. 1998).

The effects of stress and depression are not limited to cardiovascular disease risk. Although not as firmly supported by the existing literature, the progression and severity of two other chronic diseases, namely chronic obstructive lung disease (COPD) and rheumatoid arthritis, also appear to be altered by psychosocial stress and depressive symptoms. For example, in patients with COPD, Kaplan et al. (1994) found that assessment of psychological adaptation by self-report of self-efficacy was a significant predictor of five-year survival, even when the physiological measures were controlled. In another investigation, the relationship between social support and survival for patients with COPD was considered. Among the patients with COPD followed over the course of six years, there were 68 survivors and 42 nonsurvivors. Survival analysis demonstrated that quality of social support was related to longer survival; those who were satisfied with their social support experienced a 21% increase in the chance of survival over six years (Grodner et al.1996).

In rheumatoid arthritis patients, Zautra et al. (1997) prospectively examined the effects of everyday interpersonal stress on disease activity in women ($n = 41$) with rheumatoid arthritis. Using a series of time series analyses, increases in number of interpersonal stressors in the same week and one week prior were associated with increases in disease activity as reflected by a clinician's rating of disease and self-report of joint tenderness. Person–environment interactions are also clearly important in determining the effects of stress on rheumatoid arthritis disease activity. Self-beliefs such as helplessness predicted greater flare activity in a group of patients ($n = 50$) undergoing a three-month clinical drug trial (Nicassio et al. 1973). Furthermore, depressed rheumatoid arthritis patients exhibit negative self-description, negative biasing of illness-related information, and greater functional disability (Clemmey and Nicassio 1997). Finally, negative marital interactions are associated with poor psychological adjustment in women ($n = 103$) with rheumatoid arthritis (Manne and Zautra 1990), and self-beliefs are especially vulnerable to such interpersonal stress in those women with external locus of control (Reich and Zautra 1995).

ELEVATED MORTALITY RISK IN ASSOCIATION WITH DEPRESSIVE SYMPTOMS

Beyond the effects on functioning and morbidity, stress and depression predict mortality whether depression is considered as a disorder or depressive symptoms are treated as a continuous risk factor (Bruce, Leaf et al. 1994; Herrmann et al. 1998; Rovner 1993). Social factors are also implicated in mortality risk as older caregivers show a mortality risk that is 63% higher than noncaregivers (Schulz and Beach 1999).

With respect to subsyndromal depression found in association with chronic disease stress, increased risk of mortality has also been reported. For example, in mixed patient populations typical of a hospitalized setting (Herrmann et al. 1998) and in elderly persons residing in a nursing home (Rovner 1993), rates of mortality are nearly twofold of those who are not depressed. While conflicting results are also reported, with no higher risk of mortality in depressed populations with cancer or end-stage renal disease (Richardson et al. 1990; Devins et al. 1990; Christensen et al. 1994), substantial data in the instance of patients recovering from a myocardial infarction demonstrate that depression significantly predicts mortality even after the effects of previous cardiovascular disease severity and extent of cardiac compromise are considered (Frasure-Smith et al. 1993, 1995).

DEPRESSION, STRESS, AND BIOLOGICAL MARKERS

Autonomic, neuroendocrine, and immune abnormalities are found in stress and depressive disorders and are thought to contribute to the increased risk of cardiovascular, infectious, and inflammatory disorders (Sternberg et al. 1992; Gold et al. 1988; Kiecolt-Glaser and Glaser 1995). However, there is little agreement beyond this point. Is the dysregulation of these multiple physiological systems primary to depression, or is it secondary to the stress of being profoundly disturbed emotionally? Is this simply a state-dependent dysregulation, or is it an underlying imbalance exacerbated when the patient is in the episode?

In the instance of autonomic activity, depressed or psychologically stressed persons show elevations of sympathoadrenal and pituitary adrenal activity at rest and in response to acute challenge (Mills, Ziegler et al. 1997; Pike et al. 1997; Irwin et al. 1991; Lake et al. 1982). While the clinical implications of these abnormalities are not yet fully understood, such cardiovascular and sympathetic abnormalities correlate with hypertensive cardiovascular disease (Ziegler et al. 1997). In addition, elevations in sympathetic tone and dysregulation of β-adrenergic receptor expression are associated with risk of hypertension in African Americans (Mills et al. 1995). Finally, increases in 12-hour urinary-free cortisol excretion in older adults identify those at increased risk for poorer memory performance at baseline and greater declines over a 2.5-year follow-up period (Seeman et al. 1997).

Elevations in sympathetic tone also correlate with reductions in cellular and natural immunity, possibly via changes in the expression of β-receptors on lymphocytes (Mills et al. 1997; Dimsdale et al. 1994; Irwin et al. 1991) Indeed, alterations of enumerative and functional measures of immunity are now well recognized to occur in depression and in stress, are exacerbated by comorbid alcohol or tobacco abuse, and normalize once depressive symptoms resolve (Irwin, Caldwell et al. 1990; Irwin, Patterson et al. 1990; Irwin, Lacher et al. 1992; Jung and Irwin 1999). Persons who experience significant depressive symptoms in the context of life stress are at greatest risk to show changes in immune function (Zisook et al. 1994). Extension of these immunological studies to disease-relevant measures, such as varicella zoster virus specific immunity, indicates that depressed patients, especially older adults with depression, may be at substantial risk for reactivation of latent viral infections such as herpes zoster (Irwin et al. 1998). Nevertheless despite the promise of this work, interpretation of immune changes in depression is not yet clear, and it is not known whether such immune changes translate into clinically relevant outcomes. Cohen et al (1997) have found that psychological stress increases vulnerability to upper respiratory infections; however, whether depressive symptoms or immune changes mediate this relationship is not known.

Beyond the immunosuppressive changes related to stress and depression, depression has been demonstrated to be associated with activation of certain aspects of the immune system, leading to increases in circulating concentrations of soluble interleukin-2 (IL-2) receptor and the pro-inflammatory cytokine IL-6 (Zorrilla et al. 2001). As reviewed earlier, there is also considerable evidence linking depression to cardiovascular disease. Maisel and colleagues have recently generated data that have implications for understanding the association between depression and the clinical course of myocardial infarction as well as the possible involvement of immunological mechanisms (Maisel et al. 1998).

Neurohumoral activation early in the postinfarction period triggers a series of specific inflammatory and immunological events accompanied by myocardial necrosis. The greater the inflammatory responses, as might be the case in postmyocardial infarction patients with depression or stress, the greater the propensity for cardiac-specific cellular infiltration and the development or progression of congestive heart failure (Maisel et al. 1998). Importantly, many of these immunological changes can be attenuated by antagonism of β-adrenergic receptor activation and excessive sympathetic output (Maisel 1994).

Immune activation may also play a role in the hypothesized relationship between stress, depression, and disease activity in rheumatoid arthritis patients. Harrington et al. (1993) found that increases in mood disturbance were linked with decreases in sIL-2R levels and increases in joint pain.

Vulnerability to the effects of stress increases with age. Aged organisms show an exaggerated autonomic and neuroendocrine responsivity to acute stress in which responses are delayed in shutting off, with attendant suppressive effects on natural and cellular immunity (Palmer et al. 1978; Hauger et al. 1994; Irwin, Hauger et al. 1992).

DOES DISORDERED SLEEP MEDIATE SYMPATHETIC AND IMMUNE ALTERATIONS?

The impact of sleep on physiological systems and as mediator of sympathetic and immune changes in depression deserves attention, although in general this notion has not been widely discussed in the literature. Sleep disturbance is prominent in older adults (Ancoli-Israel et al. 1996; Blazer et al. 1995), stressed persons (Hall et al. 1998), and depressed patients (Benca et al. 1992). For example, subjective insomnia for two weeks or more identifies those individuals who are at increased risk for the onset of a major depression. Moreover, in those persons who have had a major depression, increasing subjective sleep problems often signals relapse (Gillin 1998).

Numerous studies have now suggested sleep disturbance uniquely contributes to the multisystem physiological effects of depression and stress. Nocturnal elevations in plasma and urinary catecholamines occur in patients with sleep-disordered breathing which contribute to day-time hypertension (Ziegler et al. 1997; Dimsdale et al. 1995, 1997). Furthermore, even a modest loss of sleep following a night of partial sleep deprivation or the occurrence of disordered sleep architecture with loss of slow wave sleep, as commonly occurs in depression (Benca et al. 1992), serves to elevate nocturnal catecholamines and contribute to cardiovascular disease (Irwin et al. 1999).

Measures of immune function (e.g., NK activity, IL-2 production) are also related to sleep disturbance. Indeed, disordered sleep is hypothesized to mediate the relationship between depression and immune alterations. Within depressed patients, a reduction of NK activity correlates with severity of insomnia as measured by subjective report and by objective EEG (Cover and Irwin 1994; Irwin, Smith et al. 1992). Within controls, a similar relationship is found between immune function and objective EEG measures (Irwin, Smith et al. 1992), and insomniac patients who have no affective disturbance show a pattern of immune alterations similar to that found in depressed patients (Irwin 1999). Moreover, Hall et al. (1998) have shown that disordered sleep mediates the decline of immune function found in bereaved persons. Finally, in an experimental model, partial night sleep deprivation induces a reduction of NK activity, lymphokine-activated killer cell activity, as well as T-cell and monocyte-dependent cytokine production. (Irwin et al. 1994, 1996).

Feedback between disease severity, pain, and behavioral symptoms also occurs. Nicassio and Wallston (1992) followed 242 patients with rheumatoid arthritis over a two-year interval and found that pain exacerbates sleep difficulties and that both factors contributed to depression.

In community dwelling elderly (n = 426), sleep-disordered breathing, another aspect of sleep disturbance, identified those men and women with a significantly shorter survival. This increase in mortality is especially high in those with preexisting cardiovascular or pulmonary disease who also have sleep-disordered breathing (Ancoli-Israel et al. 1996). Not only is sleep-disordered breathing highly prevalent in the elderly, but subgroups of healthy elderly are at elevated risk. Elderly African American men and women show a relative risk twofold as great as white older adults (Ancoli-Israel et al. 1995).

ALLOSTATIC LOAD AS A PREDICTOR OF HEALTH OUTCOMES

Much research has focused on the effects of psychological stress and depressive disorders on individual biological parameters. However, a more cumulative, multisystem view of the physiological toll exacted on the body through adaptation is also needed. McEwen introduced the term allostatic load to capture the notion that stress effects are accumulated over time and impact the regulation of autonomic, neuroendocrine, and metabolic systems (McEwen and Stellar 1993). Although further work is needed to validate this notion and there are no studies that have examined allostasis in depression, allostatic load is believed to reflect the wear and tear on the body resulting from chronic overactivity or inactivity of physiological systems that are normally involved in adaptation to environmental challenge (McEwen and Stellar 1993; McEwen 1998). Seeman et al. (1997) have extended this conceptualization, generated an operational definition of allostatic load, and presented evidence of the predictive validity of allostatic load scores in salient outcomes in high-functioning older men and women (n = 1189). Higher allostatic load was associated with poorer physical function and predicted larger decrements in physical function over time, independent of sociodemographic and health status risk factors. In addition, older adults with higher allostatic load scores showed an increased risk for the incidence of cardiovascular disease. This notion of allostatic load is accurate as far as it goes, but is probably incomplete. It is not known, for example, whether allostatic load is a measure of physiological risk that can be extended to older adult populations who already show substantial disease. Furthermore, a more complete account of allostatic load should include aspects of immune function.

STRESS REACTIVITY AS AN INDEX OF DYNAMIC REGULATION OF VARIOUS PHYSIOLOGICAL SYSTEMS

An integral part of adaptation to a potential stressor is the protection of the organism against an overreaction and curtailment of the response following termination of stress exposure. Thus, a core hypothesis is that counterregulatory mechanisms are essential if overreaction is to be prevented,and that termination of the stress response is defective in the depression. In other words, the imbalance in physiological systems that promote adaptation is thought to be the result of too much repeated stress and this leads to physiological systems either failing to shut off or, alternatively, failing to turn on adequately (McEwen and Stellar 1993).

Understanding the dynamic variability in levels of activity in the various regulatory systems is needed to evaluate the psychobiological predictors of adaptation. First, psychobiological reactivity to stress uncovers physiological abnormalities that are not always captured at rest or at a single point in time. Second, individual differences in response to acute challenges (e.g., sympathoadrenal or cardiovascular reactivity) provide a stable disposition that is replicable over time and across different stressful events; Cohen and Manuck (1995) and others suggest that this has implications for successful characterization of vulnerability risk in the stress–disease relationship. Finally, one would further expect that inclusion of data concerning reactivity of the various physiological systems that respond to allostatic load would result in a more precise estimation of the association between psychological, pathological, and other biological outcomes of interest.

By examining the responsivity and recovery of autonomic, neuroendocrine, and immune systems in response to acute challenge, stress reactivity studies provide valuable insight into levels of activity and regulation of various physiological systems. Chronic stress, depression, and aging are associated with elevated resting levels of sympathetic and HPA activity, and also result in an imbalance in these systems; either responses are attenuated and fail to turn on or they are exaggerated and fail to shut off adequately (Pike et al. 1997; Irwin et al. 1991, Irwin, Hauger et al. 1992). Dysregulation of these systems leads, in turn, to abnormalities in lymphocyte migration (Rehman et al. 1997; Mills, Karnik et al. 1997; Mills and Dimsdale 1996) and greater impairments in functional measures of immune function (Pike et al. 1997; Irwin, Hauger et al. 1992).

Individual differences in responsivity are thought to have implications for disease risk as reviewed above, and further studies have evaluated additional factors that moderate reactivity. Older women, as a group, are more likely to show larger HPA responses than older men (Seeman, Singer et al. 1995), a response that could be an important factor in explaining the genesis of memory deficits (Lupien et al. 1997). Furthermore, low self-esteem (Seeman, Berkman et al. 1995b) and the presence of nonsupportive social interactions in older adults are important factors associated with enhanced reactivity (Seeman and McEwen 1996). Other studies have revealed exaggerated responses in African American men (Gillin et al. 1996) and women (Mills et al. 1996) suggesting that reactivity and risk for hypertensive disease may be socioculturally mediated (Dellehanty et al. 1991).

FUTURE DIRECTIONS

The interaction between stress and depression may be due to the fact that depression is a chronic stressor. Alternatively, it is also possible to conceive of a disruption in stress control as a critical feature in the development or maintenance of depression (i.e., the hypersecretion of corticotrophin-releasing hormone). Nevertheless, the occurrence of stress or life events does not necessarily lead to depression. Rather, depression and associated physiological alterations reflect an interrelationship between stressor, cognitive appraisal of stress, affective integration, and translation of the cognitive and affective processes into physiological responses. Feedback systems are also operating in which one's appraisal of stressors elicits affective and physiological arousal. Conversely, physiology can influence one's thinking.

If we are to advance the welfare of humans by examining these relationships, a critical question concerns the impact of stress management on mental and physical health outcomes. For example, if cognitive appraisal is modified, can we ameliorate emotional distress and impact disease outcome? Alternatively, if psychophysiological arousal is reduced, can this impair effective response to stress and associated negative affective expression? In a model that is transactional and synergistic, the causal priority of any intervention is less of a concern than the view that effective stress management is related to its impact on the interaction of thoughts, feelings, physiological activity, and behavior (Parker 1995). Clinical studies are needed that will provide an integrated assessment of psychological and physiological adaptation, identify individual differences in response to stress, and evaluate the impact of stress-management programs on emotional and physical health functioning.

ACKNOWLEDGMENTS

This work was supported in part by N.I.H. grants AA10215, MH5T32-18399, MH55253, 2P30-MH30914, AG18367-01, AT00255-01, AR/AG41687, and M01 RR00827.

REFERENCES

Ancoli-Israel, S., M.R. Klauber, C. Stepnowsky, E. Estline, A. Chinn, and R. Fell. 1995. Sleep-disordered breathing in African-American elderly. *Am. J. Respir. Crit. Care Med.* **152**:1946–1949.

Ancoli-Israel, S., D.F. Kripke, M.R. Klauber, R. Fell, C. Stepnowsky, E. Estline, N. Khazeni, and A. Chinn. 1996. Morbidity, mortality, and sleep-disordered breathing in community-dwelling elderly. *Sleep* **19**:277–282.

Barrett, J.E., J.A. Barrett, T.E. Oxman, and P.D. Gerber. 1988. The prevalence of psychiatric disorders in a primary care practice. *Arch. Gen. Psychiatry* **45**:1100–1106.

Barry, K.L., M.F. Fleming, L.B. Manwell, L.A. Copeland, and S. Appel. 1998. Prevalence of and factors associated with current and lifetime depression in older adult primary care patients. *Family Med.* **30**:366–371.

Benca, R.M., W.H. Obermeyer, R.A. Thisted, and J.C. Gillin. 1992. Sleep and psychiatric disorders: A meta analysis. *Arch. Gen. Psychiatry* **49**:651–668.

Berger, A.K., B.J. Small, Y. Forsell, and L. Backman. 1998. Preclinical symptoms of major depression in very old age: A prospective longitudinal study. *Am. J. Psychiatry* **155**:1039–1043.

Blacker, C.V., and A.W. Clare. 1987. Depressive disorder in primary care. *Br. J. Psychiatry* **150**:737–751.

Blazer, D. 1989. Depression in the elderly. *New Engl. J. Med.* **320**:164–166.

Blazer, D.G., J.C. Hays, and D.J. Foley. 1995. Sleep complaints in older adults: A racial comparison. *J. Gerontol.* **50A**:M280–M284.

Blumenthal, J.A., R.S. Williams, A.G. Wallace, R.B. Williams, Jr., and T.L. Needles. 1982. Physiological and psychological variables predict compliance to prescribed exercise therapy in patients recovering from myocardial infarction. *Psychosom. Med.* **44**:519–527.

Broadhead, W.E., D.G. Blazer, L.K. George, and C.K. Tse. 1990. Depression, disability days, and days lost from work in a prospective epidemiologic survey [see comments]. *J. Am. Med. Assoc.* **264**:2524–2528.

Brown, G.W., and T.O. Harris. 1982. Life events and psychiatric disorders II. *Psychosoc. Med.* **3**:159–176.

Bruce, M.L., P.J. Leaf, G.P. Rozal, L. Florio, and R.A. Hoff. 1994. Psychiatric status and 9-year mortality data in the New Haven Epidemiologic Catchment Area Study [see comments]. *Am. J. Psychiatry* **151**:716–721.

Bruce, M.L., T.E. Seeman, S.S. Merrill and D.G. Blazer. 1994. The impact of depressive symptomatology on physical disability: MacArthur studies of successful aging. *Am. J. Pub. Health* **84**:1796–1799.

Carney, R.M., K.E. Freedland, S.A. Eisen, M.W. Rich, and A.S. Jaffe. 1995. Major depression and medication adherence in elderly patients with coronary artery disease. *Health Psychol.* **14**:88–90.

Christensen, A.J., J.S. Wiebe, T.W. Smith, and C.W. Turner. 1994. Predictors of survival among hemodialysis patients: Effect of perceived family support. *Health Psychol.* **13**:521–525.

Clemmey, P.A., and P.M. Nicassio. 1997. Illness self-schemas in depressed and nondepressed rheumatoid arthritis patients. *J. Behav. Med.* **20**:273–290.

Cohen, S., W.J. Doyle, D.P. Skoner, B.S. Rabin, and J.M. Gwaltney. 1997. Social ties and susceptibility to the common cold. *J. Am. Med. Assoc.* **277**:1940.

Cohen, S., and S.B. Manuck. 1995. Stress, reactivity, and disease [editorial] *Psychosom. Med.* **57**:423–426.

Cover, H., and M. Irwin. 1994. Immunity and depression: Insomnia, retardation, and reduction of natural killer cell activity. *J. Behav. Med.* **17**:217–223.

Dellehanty, S.G., J.E. Dimsdale, and P. Mills. 1991. Psychosocial correlates of reactivity in black and white men. *J. Psychosom. Res.* **35**:451–460.

Devins, G.M., J. Mann, H. Mandin, L.C. Paul, R.B. Hons, E.D. Burgess, K. Taub, S. Schorr, P.K. Letourneau, and S. Buckle. 1990. Psychosocial predictors of survival in end-stage renal disease. *J. Nerv. & Mental Dis.* **178**:127–133.

Dimsdale, J.E., T. Coy, S. Ancoli-Israel, P. Mills, J. Clausen, and M.G. Ziegler. 1997. Sympathetic nervous system alterations in sleep apnea. The relative importance of respiratory disturbance, hypoxia, and sleep quality. *Chest* **111**:639–42.

Dimsdale, J.E., T. Coy, M.G. Ziegler, S. Ancoli-Israel, and J. Clausen. 1995. The effect of sleep apnea on plasma and urinary catecholamines. *Sleep* **18**:377–381.

Dimsdale, J.E., P. Mills, T. Patterson, M. Ziegler, and E. Dillon. 1994. Effects of chronic stress on beta-adrenergic receptors in the homeless. *Psychosom. Med.* **56**:290–295.

Doherty, J.R. 1990. Cardiovascular responses in ageing: A review. *Pharmacol. Rev.* **42(2)**:103–125.

Everson, S.A., D.E. Goldberg, G.A. Kaplan, R.D. Cohen, E. Pukkala, J. Tuomilehto, and J.T. Salonen. 1996. Hopelessness and risk of mortality and incidence of myocardial infarction and cancer. *Psychosom. Med.* **59**:113–121.

Feldman, E., R. Mayou, K. Hawton, M. Ardern, and E.B. Smith. 1987. Psychiatric disorder in medical in-patients. *Q. J. Med.* **63**:405–412.

Finch, F.J., and A.J. Zautra. 1992. Testing latent longitudinal models of social ties and depression among the elderly: A comparison of distribution-free and maximum likelihood estimates with nonnormal data. *Psychol. & Aging* **7**:107–118.

Frasure-Smith, N., F. Lespérance, and M. Talajic. 1993. Depression following myocardial infarction. Impact on 6-month survival [see comments] *J. Am. Med. Assoc.* **270**:1819–1825. [Published erratum appears in *J. Am. Med. Assoc.* 1994 **271**:1082.]

Frasure-Smith, N., F. Lespérance, and M. Talajic. 1995. The impact of negative emotions on prognosis following myocardial infarction: Is it more than depression? *Health Psychol.* **14**:388–398.

Gillin, J.C. 1998. Are sleep disturbances risk factors for anxiety, depressive, and addictive disorders? *Acta Psychiatr. Scand. Suppl.* **393**:39–43.

Gillin, J.L., P.J. Mills, R.A. Nelesen, E. Dillon, M.G. Ziegler, and J. Dimsdale. 1996. Race and sex differences in cardiovascular recovery from acute stress. *Intl. J. Psychophysiol.* **23**:83–90.

Glassman, A.H., and P.A. Shapiro. 1998. Depression and the course of coronary artery disease. *Am. J. Psychiatry* **155**:4–11.

Grodner, S., R.M. Kaplan, and A.L. Ries. 1996. The impact of social support in pulmonary rehabilitation of patients with chronic obstructive pulmonary disease. *Ann. Behav. Med.* **18**:139–145.

Gold, P.W., F.K. Goodwin, and G.P. Chrousos. 1988. Clinical and biochemical manifestations of depression. *New Engl. J. Med.* **319**:413–420.

Hall, M., A. Baum, D.J. Buysse, H.G. Prigerson, D.J. Kupfer, and C.F. Reynolds, III. 1998. Sleep as a mediator of the stress-immune relationship. *Psychosom. Med.* **60**:48–51.

Harrington, L., G. Affleck, S. Urrows, H. Tennen, P. Higgins, A. Zautra, and S. Hoffman. 1993. Temporal covariation of soluble interleukin-2 receptor levels, daily stress, and disease activity in rheumatoid arthritis. *Arthr. & Rheum.* **36**:199–203.

Hauger, R.L., K.V. Thrivikraman, and P.M. Plotsky. 1994. Age-related alterations of hypothalamic-pituitary-adrenal axis function in male Fischer 344 rats. *Endocrinology* **134**:1528–1536.

Hays, R.D., K.B. Wells, C.D. Sherbourne, W. Rogers, and K. Spritzer. 1995. Functioning and well-being outcomes of patients with depression compared with chronic general medical illnesses. *Arch. Gen. Psychiatry* **52**:11–19.

Hayward, C. 1995. Psychiatric illness and cardiovascular disease risk. *Epidemiol. Rev.* **17**:129–138.

Herrmann, C., S. Brand-Driehorst, B. Kaminsky, E. Leibing, H. Staats, and U. Rüger. 1998. Diagnostic groups and depressed mood as predictors of 22-month mortality in medical inpatients. *Psychosom. Med.* **60**:570–577.

Irwin, M. 1999. The immune correlates of depression. *Adv. Exp. Med. Bio.* **461**:1–24.

Irwin, M., M. Brown, T. Patterson, R. Hauger, A. Mascovich, and I. Grant. 1991. Neuropeptide Y and natural killer cell activity: Findings in depression and Alzheimer caregiver stress. *FASEB J.* **5**:3100–3107.

Irwin, M., C. Caldwell, T.L. Smith, S. Brown, M.A. Schuckit, and J.C. Gillin. 1990. Major depressive disorder, alcoholism, and reduced natural killer cell cytotoxicity: Role of severity of depressive symptoms and alcohol consumption. *Arch. Gen. Psychiatry* **47**:713–719.

Irwin, M., C. Costlow, H. Williams, K.H. Artin, M.J. Levin, A.R. Hayward, and M.N. Oxman. 1998. Cellular immunity to varicella-zoster virus in depression. *J. Infect. Dis.* **178**:S104–S108.

Irwin, M., R. Hauger, and M. Brown. 1992. Central corticotropin releasing hormone activates the sympathetic nervous system and reduces immune function: Increased responsivity of the aged rat. *Endocrinology* **131**:1047–1053.

Irwin, M., U. Lacher, and C. Caldwell. 1992. Depression and reduced natural killer cytotoxicity: A longitudinal study of depressed patients and control subjects. *Psychol. Med.* **22**:1045–1050.

Irwin, M., A. Mascovich, J.C. Gillin, R. Willoughby, J. Pike, and T.L. Smith. 1994. Partial sleep deprivation reduces natural killer cell activity in humans. *Psychosom. Med.* **56**:493–498.

Irwin, M., J. McClintick, C. Costlow, M. Fortner, J. White, and J.C. Gillin. 1996. Partial night sleep deprivation reduces natural killer and cellular immune responses in humans. *FASEB J.* **10**:643–653.

Irwin, M., T. Patterson, T.L. Smith, C. Caldwell, S.A. Brown, J.C. Gillin, and I. Grant. 1990. Reduction of immune function in life stress and depression. *Biol. Psychiatry* **27**:22–30.

Irwin, M., T. L. Smith, and J.C. Gillin. 1992. Electroencephalographic sleep and natural killer activity in depressed patients and control subjects. *Psychosom. Med.* **54**:10–21.

Irwin, M., J. Thompson, C. Miller, J.C. Gillin, and M. Ziegler. 1999. Effects of sleep and sleep deprivation on catecholamine and interleukin-2 levels in humans: Clinical implications. *J. Clin. Endocrinol. Metab.* **84**:1979–1985.

Jorm, A.F. 1995. The epidemiology of depressive states in the elderly: Implications for recognition, intervention and prevention. *Soc. Psychiatr. Epidemiol.* **30**:53–59.

Jung, W., and M. Irwin. 1999. Reduction of natural killer cytotoxic activity in major depression: Interaction between depression and cigarette smoking [see comments]. *Psychosom. Med.* **61**:263–270.

Kaplan, R.M., A.L. Ries, L.M. Prewitt, and E. Eakin. 1994. Self-efficacy expectations predict survival for patients with chronic obstructive pulmonary disease. *Health Psychol.* **13**:366–368.

Kendler, K.S., R.C. Kessler, E.E. Walters, C. MacLean, M.C. Neale, A.C. Heath, D. Phil, and L.J. Eaves. 1995. Stressful life events, genetic liability, and onset of an episode of major depression in women. *Am. J. Psychiatry* **152**:833–842.

Kendler, K.S., L.M. Thornton, and C.O. Gardner. 2000. Stressful life events and previous epsidoes in the etiology of major depression in women: An evaluation of the "kindling" hypothesis. *Am. J. Psychiatry* **157**:1243–1251.

Kessler, R.C., K.A. McGonagle, S. Zhao, and C.B. Nelson. 1994. Lifetime and 12-month prevalence of DSM-III—R psychiatric disorders in the United States: Results from the National Comorbidity Study. *Arch. Gen. Psychiatry* **51**:8–19.

Kiecolt-Glaser, J. K., and R. Glaser. 1995. Psychoneuroimmunology and health consequences: Data and shared mechanisms. *Psychosom. Med.* **57**:269–274.

Krumholz, H.M., J. Butler, J. Miller, V. Vaccarino, C.S. Williams, C.F. Mendes de Leon, T.E. Seeman, S.V. Kasl, and L.F. Berkman. 1998. Prognostic importance of emotional support for elderly patients hospitalized with heart failure. *Circulation* **97**:958–964.

Lake, C.R., D. Pickar, M.D. Ziegler, S. Lipper, S. Slater, and D.L. Murphy. 1982. High plasma norepinephrine levels in patients with major affective disorders. *Am. J. Psychiatry* **139**:1315–1318.

Lupien, S.J., S. Gaudreau, B.M. Tchiteya, F. Maheu, S. Sharma, N.P. Nair, R.L. Hauger, B.S. McEwen, and M.J. Meaney. 1997. Stress-induced declarative memory impairment in healthy elderly subjects: Relationship to cortisol reactivity. *J. Clin. Endocrinol. Metab.* **82**:2070–2075.

Lyness, J.M., E.D. Caine, Y. Conwell, and D. King. 1993. Depressive symptoms, medical illness, and functional status in depressed psychiatric inpatients. *Am. J. Psychiatry* **150**:910–915.

Magni, G., F. Schifano, and D. de Leo. 1986. Assessment of depression in an elderly medical population. *J. Affect. Disord.* **11**:121–124.

Maisel, A., D. Cesario, S. Baird, J. Rehman, P. Haghighi, and S. Carter. 1998. Experimental autoimmune myocarditis produced by adoptive transfer of splenocytes after myocardial infarction. *Circul. Res.* **82**:458–463.

Maisel, A.S. 1994. Beneficial effects of metroprolol treatment in congestive heart failure. Reversal of sympathetic-induced alterations of immunologic function. *Circulation* **90**:1774–1780.

Manne, S.L., and A.J. Zautra.1990. Couples coping with chronic illness: Women with rheumatoid arthritis and their healthy husbands. *J. Behav. Med.* **13**:327–342.

Manuck, S., R. Morrison, and A. Bellack. 1986. Psychological factors in hypertension. *Kardiologiia* **26**:92–100.

Marmot, M. 1988. Psychosocial factors and cardiovascular disease: Epidemiological approaches. *Eur. Heart J.* **9**:690–697.

McAvay, G.J., T.E. Seeman, and J. Rodin. 1996. A longitudinal study of change in domain-specific self-efficacy among older adults. *Js. Gerontol. Series B* **51**:243–253.

McEwen, B.S. 1998. Stress, adaptation, and disease. Allostasis and allostatic load. *Ann. NY Acad. Sci.* **840**:33–44.

McEwen, B.S., and E. Stellar 1993. Stress and the individual. *Arch. Internal Med.* **153**:2093–2101.

McNaughton, M.E., T.L. Patterson, M.R. Irwin, and I. Grant. 1992. The relationship of life adversity, social support, and coping to hospitalization with major depression. *J. Nerv. & Mental Dis.* **180**:491–497.

McNaughton, M. E., T.L. Patterson, T.L. Smith, and I. Grant. 1995. The relationship among stress, depression, locus of control, irrational beliefs, social support, and health in Alzheimer's disease caregivers. *J. Nerv. & Mental Dis.* **183**:78–85.

Mills, P.J., and J. Dimsdale. 1996. The effects of acute psychologic stress on cellular adhesion molecules. *J. Psychosom. Res.* **41**:49–53.

Mills, P.J., J.E. Dimsdale, M.G. Ziegler, and R.A. Nelesen. 1995. Racial differences in epinephrine and beta 2-adrenergic receptors. *Hypertension* **25**:88–91.

Mills, P.J., R.S. Karnik, and E. Dillon. 1997. L-selectin expression effects T-cell circulation following isoproterenol infusion in humans. *Brain Behav. Immun.* **11**:333–342.

Mills, P.J., R. A. Nelesen, M.G. Ziegler, B.L. Parry, C.C. Berry, E. Dillon, and J. Dimsdale. 1996. Menstrual cycle effects on catecholamine and cardiovascular responses to acute stress in black but not white normotensive women. *Hypertension* **27**:962–967.

Mills, P.J., M.G. Ziegler, T. Patterson, J.E. Dimsdale, R. Hauger, M. Irwin, and I. Grant. 1997. Plasma catecholamine and lymphocyte beta 2-adrenergic receptor alterations in elderly Alzheimer caregivers under stress. *Psychosom. Med.* **59**:251–256.

Nicassio, P.M., V. Radojevic, M.H. Weisman, A.L. Culbertson, C. Lewis, and P. Clemmey. 1973. The role of helplessness in the response to disease modifying drugs in rheumatoid arthritis. *J. Rheumatol.* **20**:1114–1120.

Nicassio, P.M., and K.A. Wallston. 1992. Longitudinal relationships among pain, sleep problems, and depression in rheumatoid arthritis. *J. Abn. Psychol.* **101**:514–520.

Palmer, G.J., M.G. Ziegler, and C.R. Lake. 1978. Response of norepinephrine and blood pressure to stress increases with age. *J. Gerontol.* **33**:482–487.

Parker, J.C. 1995. Stress management. In: Managing Chronic Illness, a Biopsychosocial Perspective, ed. P.M. Nicassio and T.W. Smith, pp. 285–312. Washington, D.C.: Am. Psychol. Assoc.

Pike, J.L., T.L. Smith, R. Hauger, P.M. Nicassio, T.L. Patterson, J. McClinitick, C. Costlow, and M.R. Irwin. 1997. Chronic life stress alters sympathetic, neuroendocrine, and immune responsivity to an acute psychological stressor in humans. *Psychosom. Med.* **59**:447–457.

Post, R.M. 1992. Transduction of psychosocial stress into the neurobiology of recurrent affective disorder. *Am. J. Psychiatry* **149**:999–1010.

Pratt, L.A., D.E. Ford, R.M. Crum, H.K. Armenian, J.J. Gallo, and W.W. Eaton 1996. Depression, psychotropic medication, and risk of myocardial infarction. *Circulation* **94**:3123–3129.

Pyne, J. M., T. L. Patterson, R. M. Kaplan, J. C. Gillin, W. L. Koch, and I. Grant. 1997. Assessment of the quality of life of patients with major depression. *Psychiatric Serv.* **48**:224–230.

Rehman, J., P. J. Mills, S. M. Carter, J. Chou, J. Thomas, and A. Maisel. 1997. Dynamic exercise leads to an increase in circulating ICAM-1: Further evidence for adrenergic modulation of cell adhesion. *Brain Behav. Immun.* **11**:343–351.

Reich, J.W., and A.J. Zautra. 1995. Spouse encouragement of self-reliance and other-reliance in rheumatoid arthritis couples. *J. Behav. Med.* **18**:249–260.
Richardson, J.L., Z. Zarnegar, B. Bisno, and A. Levine. 1990. Psychosocial status at initiation of cancer treatment and survival. *J. Psychosom. Res.* **34**:189–201.
Rovner, B.W. 1993. Depression and increased risk of mortality in the nursing home patient. *Am. J. Med.* **94**:19S–22S.
Sapolsky, R.M., L.C. Krey, and B.S. McEwen. 1986. The neuroendocrinology of stress and aging: The glucocorticoid cascade hypothesis. *Endocr. Rev.* **7**:284–300.
Sargeant, J.K., M.L. Bruce, L.P. Florio, and M.M. Weissman. 1990. Factors associated with 1-year outcome of major depression in the community. *Arch. Gen. Psychiatry* **47**:519–26.
Schulz, R., and S.R. Beach. 1999. Caregiving as a risk factor for mortality: The Caregiver Health Effects Study [see comments]. *J. Am. Med. Assoc.* **282**:2215–2219.
Seeman, T.E. 1996. Social ties and health: The benefits of social integration. *Ann. Epidemiol.* **6**:442–451.
Seeman, T.E., L.F. Berkman, P.A. Charpentier, D.G. Blazer, M.S. Albert, and M.E. Tinetti. 1995a. Behavioral and psychosocial predictors of physical performance: MacArthur studies of successful aging. *Js. Gerontol. Series A* **50**:M177–183.
Seeman, T.E., L.F. Berkman, B.I. Gulanski, R.J. Robbins, S.L. Greenspan, P.A. Charpentier, and J.W. Rowe. 1995b. Self-esteem and neuroendocrine response to challenge: MacArthur studies of successful aging. *J. Psychosom. Res.* **39**:69–84.
Seeman, T.L., and B.S. McEwen. 1996. Impact of social environment characteristics on neuroendocrine regulation. *Psychosom. Med.* **58**:459–471.
Seeman, T.E., B.S. McEwen, B.H. Singer, M.S. Albert, and J.W. Rowe. 1997. Increase in urinary cortisol excretion and memory declines: MacArthur studies of successful aging. *J. Clin. Endocrinol. Metab.* **82**:2458–2465.
Seeman, T.E., B. Singer, and P. Charpentier. 1995. Gender differences in patterns of HPA axis response to challenge: MacArthur studies of successful aging. *Psychoneuroendocrinology* **20**:711–725.
Shaw, W.S., T.L. Patterson, S.J. Semple, S. Ho, M.R. Irwin, R.L. Hauger, and I. Grant. 1997. Longitudinal analysis of multiple indicators of health decline among spousal caregivers. *Ann. Behav. Med.* **19**:101–109.
Simonsick, E.M., R.B. Wallace, D.G. Blazer, and L.F. Berkman. 1995. Depressive symptomatology and hypertension-associated morbidity and mortality in older adults. *Psychosom. Med.* **57**:427–435.
Skantze, H.B., J. Kaplan, G. Bondjers, S. Manuck, and K. Pettersson. 1998. Endothelial injury and leukocyte adherence in Häutchen preparations from coronary arteries and aorta of cynomolgus monkeys. *Atherosclerosis* **136**:33–42.
Smith, L.W., T.L. Patterson, and I. Grant.1990. Avoidant coping predicts psychological disturbance in the elderly. *J. Nerv. & Mental Dis.* **178**:525–530.
Sternberg, E.M., G.P. Chrousos, R.L. Wilder, and P.W. Gold. 1992. The stress response and the regulation of inflammatory disease. *Ann. Internal Med.* **117**:854–866.
Wassertheil-Smoller, S., W.B. Applegate, K. Berge, C.J. Chang, B.R. Davis, R. Grimm, Jr., J. Kostis, S. Pressel, and E. Schron. 1996. Change in depression as a precursor of cardiovascular events. SHEP Cooperative Research Group (systoloc hypertension in the elderly). *Arch. Internal Med.* **156**:553–561.
Watson, S.L., C.A. Shively, J.R. Kaplan, and S.W. Line. 1998. Effects of chronic social separation on cardiovascular disease risk factors in female cynomolgus monkeys. *Atherosclerosis* **137**:259–266.
Wells, K.B., A. Stewart, R.D. Hays, M.A. Burnam, W. Rogers, M. Daniels, S. Berry, S. Greenfield, and J. Ware. 1989. The functioning and well-being of depressed patients. *J. Am. Med. Assoc.* **262**:914–919.
Zautra, A.J., J. Hoffman, P. Potter, K.S. Matt, D. Yocum, and L. Castro. 1997. Examination of changes in interpersonal stress as a factor in disease exacerbations among women with rheumatoid arthritis. *Ann. Behav. Med.* **19**:279–286.

Zautra, A.J., W. Reich, and C.A. Guarnaccia. 1990. Some everyday life consequences of disability and bereavement for older adults. *J. Personality & Soc. Psychol.* **59**:550–561.

Zautra, A.J., and A.B. Wrabetz. 1991. Coping success and its relationship to psychological distress for older adults. *J. Personality & Soc. Psychol.* **61**:801–810.

Ziegler, M.G., R. Nelesen, P. Mills, S. Ancoli-Israel, B. Kennedy, and J.E. Dimsdale. 1997. Sleep apnea, norepinephrine-release rate, and daytime hypertension. *Sleep* **20**:224–231.

Zisook, S., S.R. Schuchter, M. Irwin, D.F. Darko, P. Sledge, and K. Resovsky. 1994. Bereavement, depression, and immune function. *Psychiatry Res.* **52**:1–10.

Zorrilla, E.P., L. Luborsky, J.R. McKay, R. Rosenthal, A. Houldin, A. Tax, R. McCorkle, D.A. Seligman, and K. Schmidt. 2001. The relationship of depression and stressors to immunological assays: A meta-analytic review. *Brain Behav. Immun.* **15**:199–226.

18

What Are the Neurobiological Consequences of Stress?

E. FUCHS, M. KRAMER†, and G. FLÜGGE

Division of Neurobiology, Deutsches Primatenzentrum, Kellnerweg 4, 37077 Göttingen, Germany

Keywords: adult neurogenesis, aging, blood–brain barrier, gender differences, glia–neuron interaction, hippocampal formation, lateralization, processive and systemic stressors, structural plasticity

ABSTRACT

The brain responds to stress in a complex, orchestrated manner that is related to the activation or inhibition of structures involved in sensory, motor, autonomic, cognitive, and emotional processes. Neuronal circuits that respond to stress are strongly dependent on the type of stressor. Even in the mature differentiated central nervous system, neurons and glial cells undergo permanent dynamic changes. The resulting structural plasticity of the brain, its ability to change and to respond reversibly to environmental challenges, is clearly demonstrated in the hippocampal formation, a brain structure that is important for learning, memory, and emotional processing, and which is involved in the control of vegetative and autonomic circuits. The effects of stress on the hippocampal formation seem to range from initial deficits in cognitive and memory function, possibly mediated by a significant and reversible retraction of apical dendrites of pyramidal neurons in the CA3 region, to an increased vulnerability to metabolic insults, reduced neurogenesis, and reversible reduction of the hippocampal volume as a whole. Since not only adrenocortical hormones, such as glucocorticoids, but also gonadal hormones have an impact on various central nervous circuits, this chapter describes gender- and age-dependent modulations of the neurobiological mechanisms of responsivity to stress. Furthermore, the negative impact of stressful manipulations on the protective function of the blood–brain barrier is discussed. The body's stress response not only involves the modification of brain circuits but also other systems, such as autonomic, endocrine, and immune. These links between the brain and peripheral systems provide the basis for the adaptive as well as the maladaptive responses of an organism to stressful challenges.

Coping with Challenge: Welfare in Animals including Humans
Edited by D.M. Broom

INTRODUCTION

In animals, including humans, the nervous system is the sensing, thinking, and control system of the body. To perform these functions, it collects sensory information from the body and the environment. Via the nerve endings from eyes, ears, olfactory and gustatory organs, the vestibular system, the skin, the deep tissues, and other sensory organs, this information is transmitted into the central nervous system (spinal cord and brain). Spinal cord and brain may immediately react to sensory information and transmit signals to muscles and internal organs, or they may not, transmitting instead the sensory information into the brain's "memory banks." Here, the input is compared and combined with other available information. Thus, within the nervous system, the brain is the principal integrative center. It is the site where functions related to complex control of the body are performed, memories are stored, and emotions and thoughts are generated.

Of course, the brain does not work in isolation. Together with the rest of the body, it depends on the circulation of blood that provides oxygen and nutrients for survival and hormones from the endocrine glands to shape neural cells and guide them to their specific functions. On the other hand, the brain is not only a target of peripherally derived hormones but produces a large number of bioactive compounds whose effects within the brain depend strongly on the brain area and, at the cellular level, on the co-localization with other neuromodulators and the interaction with different classes of receptors.

When Hans Selye formulated his stress theory some 70 years ago, stress was thought to have a merely endocrine character, and noxious stimuli of physical or chemical nature were the primary stressors discussed (Selye 1936). Subsequent research, however, demonstrated that psychological stimuli are also strong activators of the endocrine system (Mason 1968), and there was increasing evidence that hypothalamic-pituitary-adrenal (HPA) hormones, such as adrenocorticotrophin (ACTH) and adrenocortical steroids, profoundly influence brain excitability (Torda and Wolff 1952). Since then, great progress has been made toward understanding the biology of reactions to challenge and the physiological and behavioral alterations that occur as a result of exposure to events in the environment. In this large body of research, the HPA response is usually depicted as adaptive physiological and psychological processes activated in humans and animals. It is regarded as an alarm system that is initiated whenever there is a discrepancy between what an organism is expecting and what really exists (Levine and Ursin 1991). Loss of control, lack of information, or uncertainty induces alarm reactions whereas the presence of information, control, and social support reduces the alarm reactions. Therefore, the HPA and sympathetic-adrenal-medullary (SAM) responses per se are not harmful or pathological in themselves. Only when demanding, prolonged, and sustained can body and brain homeostasis be threatened and health may be endangered. Since this is a consistent finding across species, including humans, it is important to understand the relationship between stressors and diseases.

There are two major mechanisms mediating most components of the stress response. The first mediating link is the HPA system, which stimulates the adrenal cortex to release glucocorticoids, such as cortisol or corticosterone, into the blood. The second major mechanism mediating the stress response is the SAM system, which influences the stress response through two different pathways working in parallel. One pathway is built up by the nerve endings that trigger the release of adrenaline from the adrenal medulla into the blood. The second pathway comprises the sympathetic nerve endings that supply essentially every organ in the body where they release noradrenaline.

NEURONAL CIRCUITS AND LATERALIZATION

Different brain regions are involved in the perception of challenge. If the neuronal response immediately follows the stress exposure, it may represent the sum of various steps within the neuronal network. Thus, using the presently available histological, physiological, molecular, pharmacological, and behavioral tools, investigators can only identify the regions that are activated by stress.

Within the brain, reactions to environmental and psychological stressors are mediated by specific neurotransmitter and neuropeptide pathways. Rapid neuronal responses to stress are elicited by activation of excitatory (e.g., glutamate) or inhibitory (e.g., gamma-amino butyric acid, GABA) receptor and ion channel complexes that directly control neuronal firing. However, this fast transmission of information is modulated by slow-acting neurotransmitters, such as serotonin and noradrenaline, and neuropeptides that exert their actions via specific receptors located in the membranes of neurons and, for some of them, in glia cells. Many of these neurotransmitter and neurohormone-binding sites belong to the family of G-protein-coupled receptors, which are linked to intracellular second messenger systems via guanosine 5′-triphosphate (GTP)-binding proteins. Such signal transduction pathways mediate many of the short- and long-term effects on neuronal functioning, and second messenger-coupled intracellular pathways regulate most aspects of neuronal function, including neuronal activity, metabolism, and gene expression. Thus, these signal transduction systems play a key role, by which systems detecting challenge regulate neuronal functioning and structure, and represent the first step in a cascade of events in the brain. The adverse effects of stress could result from (a) sustained activation or inhibition of certain neurotransmitter and intracellular pathways or (b) from chronic exposure to neuroactive substances (e.g., cytokines) or hormones (e.g., glucocorticoids) released during stress exposure. Like all steroid hormones, glucocorticoids activate intracellular receptor complexes which, after translocation into the cell nucleus, enhance or repress expression of a variety of glucocorticoid-responsive genes.

Dependent on the type of stressor, different neuronal circuits are activated. Limbic circuits connecting, for example, hippocampus, amygdala, and prefrontal cortex are sensitive to stressors such as restraint, fear, or exposure to a novel environment. Common to these stressors is that they require — prior to the initiation or inhibition of stress responses — a processing of information from different sensory organs, and they are strongly dependent on previous experience. In contrast, physiological threats, such as exposure to ether, result in an activation of efferent visceral pathways that are directly relayed to the paraventricular nucleus of the hypothalamus. In this case, the rapid activation of brainstem and hypothalamus circumvents the cognitive processing in limbic and cortical areas. The marked difference between limbic-sensitive and limbic-insensitive stressors led Herman and Cullinan (1997) to the postulation of two generalized stress pathways: the "processive" and the "systemic" pathway. According to their hypothesis, respiratory, cardiovascular, or immune stimuli represent "systemic" stressors that require immediate reactions for survival but no further interpretation from higher-order brain structures. In contrast, multimodal stimuli, such as those resulting from psychological challenges, are regarded as "processive" stressors. They need cortical processing and, depending on previous experience or ongoing activation, the information is assembled within higher-order brain structures to induce neuroendocrine and behavioral responses.

The cerebral cortex exerts strong influences on many aspects of stress responses. Several findings favor the idea that stress responses are mainly controlled by the right cerebral hemisphere. In addition, at the neurochemical level, the right hemisphere clearly differs from the left. For example, the neurotransmitter noradrenaline, being involved in the cerebral regulation of arousal, attention-related functions, and adaptive responses to stress, shows an asymmetric distribution favoring the right brain side and, in particular, the right hemisphere. This coincides with findings in humans showing that the somatic stress response, including both the HPA and sympathetic-adrenomedullary system, are primarily under control of the right hemisphere (for further discussion, see Wittling 1997). This idea of a lateralization in the brain's responses to stress is supported by recent findings in rats investigating the regulation of stress responses to acute or repeated restraint stress on the level of the medial prefrontal cortex (mPFC). This region is known to modulate neuroendocrine function during stress by regulating the HPA axis and by acting as a site for glucocorticoids to exert negative feedback modulation of HPA activity. The mPFC also regulates a variety of autonomic functions associated with stress in both rats and humans. From their studies in rats, Sullivan and Gratton (1999) suggested a preferential role for the right mPFC in activating physiological stress responses. These results have also clinical relevance to major depressive disorders, where either left frontal hypofunction or right frontal hyperfunction are associated with depressed states. Therefore, understanding the neurobiology of stress by investigating the brain circuits that are associated with or altered by stress responses will provide clues to the mechanisms by which stress affects behavior, cognition, mood, and health.

STRUCTURAL PLASTICITY OF THE MATURE BRAIN

Approximately one century ago, it was finally established that the brain consists of neural cells (neurons and glia) and thus is a cellular assembly rather than a continuous syncytium. Since this discovery, a rather static view of the brain has prevailed in which electrical and chemical information was thought to be processed through a fixed system of neuronal circuits. Ramon ý Cajal stated, "In adult centers the nerve paths are something fixed, ended, immutable. Everything may die, nothing may be regenerated" (Cajal and May 1959, p. 750). In recent years, this view has been gradually revised on the basis of studies showing that neuronal circuits and connections are subject to lifelong modifications and reorganizations. This ability of the brain to perform functionally relevant adaptations following various challenges is called plasticity. Neuronal plasticity is absolutely necessary for adequate functioning of an individual in the continuously changing environment. These dynamic processes are based on the capability of neuronal systems, brain nuclei, single neurons, synapses, and receptors to adapt to alterations in the internal and external environment by modifying specific structures and functions (Zilles 1992). Adaptive or experience-dependent plasticity reveals itself in many forms, ranging from changes in gene expression to changes in neurotransmitter release and in behavior. Even under normal and undisturbed conditions, contacts between neurons are continuously replaced and renewed in the adult and differentiated brain. Enhanced axonal outgrowth and collateral sprouting on the presynaptic site may lead to the formation of new synapses, and existing ones may be eliminated by terminal retrograde degeneration. The number of postsynaptic sites can be increased or decreased by alterations in the size of the dendritic tree, its spine density, or through changes in the glial organization. Hormones from the gonads, adrenals, and thyroid gland seem to play a particular role as triggers of such structural reorganization. Through their transcriptional activity they regulate gene expression in the neural cells (for further discussion, see McEwen et al. 1991). An impressive example of steroid-mediated reorganization of neuronal connectivity in the adult brain is the fluctuation of synaptic contacts in different brain areas of female rats during the menstrual cycle (Woolley 1999). Furthermore, as a consequence of changes in sensory inputs and in circulating hormones, synaptic contacts of mossy fibers with CA3 hippocampal pyramidal neurons in the brain of ground squirrels are altered in many aspects of their structure during different stages of the torpor-activity cycle (Popov and Bocharova 1992). These two examples demonstrate the capacity of the adult brain to reorganize and adjust its neuronal circuits in response to altered hormonal and/or sensory inputs.

Based on such findings, it is reasonable to postulate that similar structural modifications may also occur during stressful life events when hormonal levels are altered (e.g., elevated glucocorticoids, reduced gonadal steroids) and sensory inputs are changed. In this context, one of the most intensively investigated brain circuits is the limbic-HPA system. This system combines both brain and endocrine components and can be regarded as a classic neuroendocrine circuit. Limbic and hypothalamic brain structures integrate emotional, cognitive, neuroendocrine, and autonomic inputs, and determine the magnitude and duration of behavioral, hormonal, and neural responses to stressful experiences. In concert with other neurohormonal components of the stress systems, such as the sympathetic-adrenomedullary system, the adrenocortical system is activated and glucocorticoid hormones (cortisol in primates, including humans, corticosterone in rodents) are released into the blood. By influencing gene transcription and altering electrical activity of excitable cells, these corticosteroids are potent modulators of cell physiology and behavior. They serve both to alert the organism to an environmental or physiological challenge and to defend its homeostasis (for references, see de Kloet et al. 1998; Sapolsky et al. 2000).

Within the limbic-HPA system, the hippocampus plays a central role. This area, which is part of the processive stress circuit, has a high concentration of receptors for glucocorticoids and modulates the release of this steroid hormone via a negative feedback loop exerting inhibitory effects on the HPA axis. Besides its intimate involvement in neuroendocrine regulation, the hippocampus plays an important role in emotional processing and in key aspects of learning and memory.

While short-term activation of the limbic-HPA system is essential for vital functions, the deleterious effects of a constant hyperactivity of the circuit resulting in chronically elevated glucocorticoids have long been recognized and can be considered as maladaptive and pathological. As demonstrated in nonhuman studies, chronic stress leads to impaired feedback mechanisms, which are associated with a decreased gene expression of hippocampal corticoid receptors. The feedback abnormalities in chronically stressed animals resemble those seen in patients with a depressive symptomatology, where a constant hyperactivity in the limbic-HPA system has been most consistently demonstrated on the basis of elevated circulating glucocorticoid hormones. These findings indicate that the hippocampus (a) is an important center for the integration of neurochemical, neurohormonal, and cognitive responses to stress, and (b) probably plays a role in stress-related psychiatric disorders.

A number of studies have investigated the effects of stress and glucocorticoids on hippocampal function in laboratory animals and in humans (for references, see McEwen 1997 and 1999). Increasing evidence suggests a connection between structural alterations and volume reduction of the hippocampal formation, cognitive impairments, as well as disturbances of the activity of the limbic-HPA axis. Although it was initially proposed that glucocorticoids cause hippocampal neuronal loss (Sapolsky et al. 1986), later studies have reformulated the glucocorticoid cascade concept of stress and hippocampal damage. Loss of CA3 neurons under extreme or prolonged stress or extremely high concentrations of glucocorticoids, which may further extend to regions CA1 and CA4, may possibly be mediated through apoptosis. The glucocorticoid cascade concept of stress and hippocampal damage was challenged, however, by recent stereological studies, which failed to detect significant neuronal loss following prolonged hypercortisolism resulting from either aging, stress exposure, or corticosteroid administration (Rapp and Gallagher 1996; Vollmann-Honsdorf et al. 1997; Sousa et al. 1998; Leverenz et al. 1999).

The deleterious effects of glucocorticoids within the cornu amonis (CA) of the hippocampal formation are presumably mediated in part via neuronal glucocorticoid receptors, the mineralocorticoid receptors (MR), and glucocorticoid receptors (GR), which functionally complement each other to maintain homeostasis. The MR has a tenfold higher affinity to cortisol or corticosterone than the GR, resulting in an extensive occupation of this receptor under basal or glucocorticosteroid levels. Since GR is extensively occupied only during times of high concentrations of circulating glucocorticoids (e.g., at the circadian peak or under stress), it has been suggested to be the primary mediator of feedback inhibition, i.e., it underlies autoregulation, whereas the MR exerts tonic influences on the brain, mainly driven by the circadian rhythmicity of glucocorticoid secretion. In this context, it is interesting to note that in the rhesus monkey hippocampal formation, in contrast to rats, the expression of GR mRNA and the amount of the receptor protein is very low, whereas high levels of MR expression were detected (Mar Sanchez et al. 2000). Thus the hypothesis that the hippocampal GR:MR ratio reflects susceptibility to stress, and MR protects the neuron against excitotoxic challenges, such as those induced by high glutamate concentrations, reinforces investigations on the role of the high density of MRs in the primate hippocampus. The differences in levels of GR and MR expression in the hippocampus of primates versus rats also points to the fact that caution should be exerted when extrapolating neurochemical findings from rodents to humans, in particular with respect to higher brain functions.

Within the hippocampal formation, the dentate gyrus recently received much attention since it is one of the few areas of the mature brain, including the adult human brain, where neurogenesis has been demonstrated. The most prominent neuronal cell type of the dentate gyrus is the granule cell, which is formed in a germinal zone along the border between the granule cell layer and the hilus. In this germinal zone, progenitor cells persist well into adulthood and continuously produce granule cells that become incorporated into the granule cell layer. Several neuroendocrine and experimental factors have been identified that regulate neurogenesis in the adult dentate gyrus (for references, see Fuchs and Gould 2000). Adrenal hormones have been shown to inhibit the production of new granule neurons by suppressing the proliferation of granule cell precursors. On the other hand, removal of circulating glucocorticoids by adrenalectomy results in a clear increase in neurogenesis in the dentate gyrus. Interestingly, the ovarian steroid estrogen has been shown to stimulate the proliferation of granule cell precursors in the dentate gyrus of adult female rats. This increase in the rate of cell proliferation occurs naturally across the rat estrous cycle, with maximal levels of cell production during proestrus, a time when estrogen levels are highest. Besides the hormonal influences, neural mechanisms such as glutamatergic and serotonergic inputs have also been demonstrated to alter the production of new granule cells in the dentate gyrus.

The biological relevance of neurogenesis in the adult brains is suggested by the observation that cell proliferation in the dentate gyrus can be modulated by experience. Several different types of stressful experiences, such as exposure to the odor of a predator, subordination stress, and resident-intruder stress, have been shown to inhibit granule cell production in the dentate gyrus of rats, tree shrews, and monkeys.

Because of their presumed ability to form new synapses rapidly, newly generated neurons might be responsible for forming more new connections with other neurons than mature cells. Factors and conditions that inhibit production of new neurons, such as adrenal steroids, stress and aging, are associated with diminished performance on hippocampal-dependent tasks. However, although these findings suggest a positive association between neurogenesis and hippocampal-dependent learning, yet no study has demonstrated that new neurons are necessary for learning.

Cross-sectional studies in humans demonstrated that both a *hyperactivity* of the limbic-HPA system with elevated levels of circulating adrenal glucocorticoid hormones (e.g., depressed patients, patients with Cushing's disease), as well as a *hypoactivity* of the system with low glucocorticoid plasma levels (e.g., posttraumatic stress disorder patients) are accompanied by shrinkage of the hippocampus (for references, see McEwen 1997). The hippocampal atrophy has also been shown to be reversible in patients with Cushing's disease upon normalization of glucocorticoid levels. The fact that glucocorticoid-induced volumetric changes are reversible makes it unlikely that the volume reduction results from neuronal loss. A more plausible explanation for the reversibility is that alterations occur in the dendritic, axonal, synaptic, and glia components of the hippocampal neural network. In this context, Ongur et al. (1998) report on the reduction in the number of glia, but not neurons, in a part of the prefrontal cortex of patients with major depressive disorder. Although these clinical findings can be explained, in part, by the basic research experiments, postmortem studies will be necessary to characterize the cellular mechanisms that underlie the reduction in hippocampal volume and function in humans. Thus, based on today's knowledge, the deleterious effects of stress on the hippocampal formation seem to range from initial deficits in cognitive and memory function, via significant and reversible retraction of apical dendrites of CA3 pyramidal neurons towards an increased vulnerability to metabolic insults, reduced neurogenesis, and reversible volume reduction of the hippocampus as a whole.

Both the duration and frequency of the stress period are important determinants for the induction of the cascade of stress-triggered neurobiological processes (for references, see Koolhaas et al. 1997). However, most available techniques in preclinical neurobiology are limited by their invasiveness, which precludes repeated investigations of the same individual. Magnetic resonance imaging (MRI) and magnetic resonance spectroscopy (MRS) can overcome these disadvantages and offer new perspectives for neuroscientific research. In humans, neuroimaging is an important tool to evaluate degenerative processes which lead, for example, to volumetric and metabolic changes within defined brain structures, such as the hippocampus. Although the cross-sectional approaches of these studies could provide important information about particular states of neuropathologies, there is still a lack of information about the etiology and how frequently and with what time course these diseases are accompanied by hippocampal atrophy. Thus, MRI and MRS should be applied in preclinical experiments as noninvasive tools allowing repeated measurements of changes in brain anatomy and metabolism within the same animal throughout experiments over extensive time courses.

NEURON–GLIA INTERACTION

The most abundant cells within the nervous system are the glia cells. In the adult mammalian brain, there are 10 to 50 times more glia cells than neurons. Within the group of glia cells, different types can be distinguished: oligodendrocytes, microglia, and astrocytes. Oligodendrocytes are small cells whose processes wrap themselves like a lamella around axons, and they have the same function as the Schwann cells in the peripheral nervous system. Microglia cells represent a type of central nervous glia characterized by small cells of distinct shape. In the adult brain, microglia cells exist in at least three states. The resting state, represented by ramified glia, is present in the healthy adult central nervous system. In pathological states, there is a reactive microglia and phagocytotic microglia with the latter representing brain macrophages. When activated, these microglial cells secrete cytokines and produce substances such as nitric oxide, which may have toxic effects. Glucocorticoids and gonadal steroids can regulate the functional state of microglia. Since activated microglia can influence neuronal sprouting, the effects of the steroid hormones on neuronal morphology may be mediated by glia.

Astrocytes are the most numerous glia cells and account for about one-third of the mass of the brain. They are characterized by many ramifications and are found associated with axons as well as cell bodies. Essential roles are the supply of energy and nutrients to the neurons, the metabolism of neurotransmitters, maintenance of ion homeostasis in the extracellular space, regulation of neuronal migration, secretion of growth factors, and participation in immune and inflammatory responses. Further, they form close connections with blood capillaries and are thus part of the blood–brain barrier. A recent and unexpected discovery is that astrocytes also possess receptors for neurotransmitters and steroid hormones, which like the receptors in neurons, can trigger electrical and biochemical events inside the astrocyte. Moreover, astrocytes may coordinate their function through Ca^{2+}-dependent signaling mechanisms.

These findings shed light on the physiological and structural importance to investigate neuron–glia interactions. Depending on the local microenvironment, stress-related changes in glial and neuronal physiology could be a sign of dysfunction contributing to maladaptation or could represent compensatory responses in adaptation that are essential to neuroplasticity. The complexity of the neural tissue — which includes neurons and glia — and the availability of adequate tools seem to be major obstacles for detailed studies of neuron–glia interaction in response to stressful life events.

GENDER- AND AGE-DEPENDENT MODULATIONS OF THE STRESS RESPONSE

Gender-specific modulation of the mechanisms that lead to the stress response have been described for various central nervous circuits. Gender differences or effects of gonadal hormones have been reported for anatomical and functional characteristics of several neurotransmitter and neuromodulatory systems, such as the gamma-aminobutyric acid system as well as the dopaminergic, noradrenergic, and serotonergic systems. On the structural level, in the hippocampi of chronically stressed female rats — unlike those of stressed male rats — no significant reduction of apical dendrites of CA3 pyramidal neurons was observed. Female rats, however, manifest a decrease in basal dendritic branching, indicating another structural and functional stress-induced sex difference in the rat brain (Galea et al. 1997). Until now, the influence of gonadal hormones on the characteristics of the response to stress has only been partially elucidated. Nevertheless, they seem to modify the efficacy of glucocorticoids in the brain, either by acting directly as transcription factors or by acting transsynaptically (for references, see Patchev and Almeida 1998).

The clear gender differences in neurobiological components of the stress response and the pronounced differences in the reaction of the HPA axis to various challenges suggests the need for more research on stress effects in females. Only a very small number of studies have utilized female animals in preclinical research on the psychopharmacology of depression (for references, see Blanchard et al. 1995). However, important differences in the bioavailibility of drugs between women and men have been described. Factors such as drug absorption and metabolism, percentage of body fat, renal clearance, and a variety of other physiological mechanisms have been reported to be gender specific (for references, see Weissmann and Olfson 1995).

Depression in women of childbearing age can be a serious medical problem because of the antidepressants secreted into breast milk. Thus, antidepressant treatment in the postpartum period poses a problem for nursing women, since the long-term effects of the treatment for the infant are unknown. Therefore, preclinical studies on stress and the actions of antidepressants ought to include pregnant and lactating female in addition to male animals. However, the urgent need to use female individuals in preclinical psychopharmacological research programs is not met. The rationale for the predominant use of males in preclinical research on stress and stress-related psychopathologies is simple. First, it is easier and cheaper to use only males. Second, there is a large discrepancy in the ease of inducing social stress in the two genders. While there are several well-documented models of social stress in males, there are virtually no similar models in females. The most widely used social stressors, such as defeat in aggressive encounters, social isolation, and crowding, are obviously ineffective in females.

Aging is characterized by a general decline in organ function, decreased adaptability to environmental changes, and increasing deficits in the ability to maintain homeostasis. The vulnerability of the brain to stress-induced impairments will thus increase with age. The rate of decline and vulnerability varies between individuals. Part of this individual variability may be due to differences in the stress levels experienced during life.

On the neuronal level, physiological aging processes are characterized by an increasing imbalance among the various neurotransmitters and neuromodulators resulting in significant biochemical and structural changes. The predominant change that occurs in glia during aging is glial activation, which can progress to reactive gliosis in response to neurodegeneration. Astrocytes show hypertrophy and exhibit signs of metabolic activation; microglial cells become activated and subsets of them may enter the phagocytotic or reactive state. A marker for astrocyte activation is glial fibrillary acidic protein (GFAP). Steroids such as glucocorticoid hormones regulate the interactions between glia and neurons, including GFAP expression. Recent data indicate that corticosterone treatment can decrease the expression of GFAP in the aged brain, yet GFAP is resistant to downregulation by endogenous glucocorticoids (for references, see Nichols 1999). Thus, it was proposed that the effect on glial cell products is adaptive, mediating a return to basal expression after physiological disturbances and contributing to functional recovery that may reverse synaptic loss and cognitive impairment during aging.

Dysregulation of the limbic-HPA system appears to be a widespread phenomenon in aged mammals, including humans. As a consequence, research on the effects of high concentrations of the adrenocortical steroids cortisol and corticosterone has received much attention, and the above mentioned correlations of the hormone concentrations with structural brain changes and with decline of cognitive functions have been suggested.

IMPACT OF STRESS ON BLOOD–BRAIN BARRIER PERMEABILITY

The blood–brain barrier protects the brain from rapid and large fluctuations in the circulating levels of ions, neuromodulators, growth factors, bacteria, toxins, and other substances that can affect the normal function of the central nervous system. The blood–brain barrier consists of a continuous layer of endothelial cells that are joined by tight junctions with very high electrical resistance. This hinders the flow of ions and large molecules across the membrane of the endothelial cells. Small molecules can cross the blood–brain barrier, and the more lipophylic the molecules are, the better they are able to penetrate into and out of the brain. In addition, specific transporters for molecules such as glucose and amino acids are located within the blood–brain barrier. Recent findings suggest that stress may alter the permeability of the blood–brain barrier. Acute immobilization in rats, low temperature exposure, or isolation in mice resulted in increased penetration of the blood–brain barrier by drugs, neurotransmitters, and viruses that are normally excluded. In soldiers exposed to extreme emotional stress during the Gulf War, it was observed that the blood–brain barrier became unexpectedly more permeable to the cholinergic drug pyridostigmine (Friedman et al. 1996). This compound is an acetylcholinesterase inhibitor routinely employed in the treatment of the autoimmune disease myasthenia gravis. Because of its protective effect against organophosphate poisoning, pyridostigmine has been recommended by Western armies as a prophylactic drug under threat of chemical weapons. The finding that pyridostigmine acts centrally in severely stressed individuals suggests that (a) peripherally acting drugs administered under stress may reach the brain and (b) we must approach the blood–brain barrier from a different perspective other than the classical view, which states that it is a very effective barrier and protects the brain from undesirable molecules. Increased awareness of this important phenomenon is essential, not only for drug design but also for the understanding of the neurobiological consequences of stress.

ACKNOWLEDGMENTS

The authors' work is supported by the German Science Foundation (DFG), the German Academic Exchange Service (DAAD), the Studienstiftung des Deutschen Volkes, and grants from the European Commission.

REFERENCES

Blanchard, D.C., G. Griebel, and R.J. Blanchard. 1995. Gender bias in the preclinical psychopharmacology of anxiety. Male models for (predominantly) female disorders. *J. Psychopharmacol.* **9**:79–82.

Cajal, R.S., and R.T. May. 1959. Degeneration and Regeneration of the Nervous System, vol. 2, p. 750. New York: Hafner.

de Kloet, E.R., E. Vreugdenhil, M.S. Oitzl, and M. Joëls. 1998. Brain corticosteroid receptor balance in health and disease. *Endocr. Rev.* **19**:269–301.

Friedman, A., D. Kaufer, J. Shemer, I. Hendler, H. Soreq, and I. Tur-Kaspa. 1996. Pyridostigmine brain penetration under stress enhances neuronal excitability and induces early immediate transcriptional response. *Nature Med.* **2**:1382–1385.

Fuchs, E., and E. Gould. 2000. *In vivo* neurogenesis in the adult brain: Regulation and functional implications. *Eur. J. Neurosci.* **12**:2211–2214.

Galea, L.A.M., B.S. McEwen, P. Tanapat, T. Deak, R.L. Spencer, and F.S. Dhabhar. 1997. Sex differences in dendritic atrophy of CA3 pyramidal neurons in response to chronic restraint stress. *Neuroscience* **81**:689–697.

Herman, J.P., and W.E. Cullinan. 1997. Neurocircuitry of stress: Central control of the hypothalamo-pituitary-adrenocortical axis. *Trends Neurosci.* **20**:78–84.

Koolhaas, J.M., P. Meerlo, S.F. De Boer, J.H. Strubbe, and B. Bohus. 1997. The temporal dynamics of the stress response. *Neurosci. Biobehav. Rev.* **21**:775–782.

Leverenz, J.B., C.W. Wilkinson, M. Wamble., S. Corbin, J.E. Grabber, M.A. Raskind, and E. Peskind. 1999. Effect of chronic high-dose exogenous cortisol on hippocampal neuronal number in aged nonhuman primates. *J. Neurosci.* **19**:2356–2361.

Levine, S., and H. Ursin. 1991. What is stress? In: Stress: Neurobiology and Neuroendocrinology, ed. M.R. Brown, G. Koob, and C. Rivier, pp. 3–21. New York: Marcel Dekker.

Mar Sanchez, M., L.Y. Young, P.M. Plotsky, and T.R. Insel. 2000. Distribution of corticosteroid receptors in the rhesus brain: Relative absence of glucocorticoid receptors in the hippocampal formation. *J. Neurosci.* **20**:4657–4668.

Mason, J.W. 1968. A review of psychoneuroendocrine research on the pituitary-adrenal cortical system. *Psychosom. Med.* **30**:576–609.

McEwen, B.S. 1997. Possible mechanisms for atrophy of the human hippocampus. *Mol. Psych.* **2**:255–262.

McEwen, B.S. 1999. Stress and hippocampal plasticity. *Ann. Rev. Neurosci.* **22**:105–122.

McEwen, B.S., H. Coirini, A. Danielson, M. Franfurt, E. Gould, S. Mendelson, M. Schumacher, A. Segarra, and C. Woolley. 1991. Steroid and thyroid hormones modulate a changing brain. *J. Steroid Biochem. Mol. Biol.* **40**:1–14.

Nichols, N.R. 1999. Glial responses to steroids as markers of brain aging. *J. Neurobiol.* **40**:585–601.

Ongur, D., W.C. Drevets, and J.L. Price. 1998. Glia reduction in the subgenual prefrontal cortex in mood disorders. *Proc. Natl. Acad. Sci. USA* **95**:13,290–13,295.

Patchev, V.K., and O.F.X. Almeida. 1998. Gender specificity in the neural regulation of the stress response: New leads from classical paradigms. *Mol. Neurobiol.* **16**:63–77.

Popov, V.I., and L.S. Bocharova. 1992. Hibernation-induced structural changes in synaptic contacts between mossy fibres and hippocampal pyramidal neurons. *Neuroscience* **48**:53–62.

Rapp, P.R., and M. Gallagher. 1996. Preserved neuron number in the hippocampus of aged rats with spatial learning deficits. *Proc. Natl. Acad. Sci. USA* **93**:9926–9930.

Sapolsky, R., L. Krey, and B.S. McEwen. 1986. The neuroendocrinology of stress and aging: The glucocorticoid cascade hypothesis. *Endocr. Rev.* **7**:284–301.

Sapolsky, R.M., L.M. Romero, and A.U. Munck. 2000. How do glucocorticoids influence stress response? Integrating permissive, suppressive, stimulatory, and preparative actions. *Endocr. Rev.* **21**:55–89.

Selye, H. 1936. A syndrome produced by diverse nocuous agents. *Nature* **138**:32.

Sousa, N., O.F.X. Almeida, F. Holsboer, M.M. Paula-Barbosa, and M.D. Madeira. 1998. Maintenance of hippocampal cell numbers in young and aged rats submitted to chronic unpredictable stress. Comparison with the effects of corticosterone treatment. *Stress* **2**:237–250.

Sullivan, R.M., and A. Gratton. 1999. Lateralized effects of medial frontal cortex lesions on neuroendocrine and autonomic stress response in rats. *J. Neurosci.* **19**:2834–2840.

Torda, C., and H.G. Wolff. 1952. Effects of various concentrations of adrenocorticotrophic hormone on electrical activity of brain and on sensitivity to convulsion-induced agents. *Am. J. Physiol.* **168**: 406–415.

Vollmann-Honsdorf, G.K., G. Flügge, and E. Fuchs. 1997. Chronic psychosocial stress does not affect the number of pyramidal neurons in tree shrew hippocampus. *Neurosci. Lett.* **233**:121–124.

Weissman, M.M., and M. Olfson. 1995. Depression in women: Implications for health care research. *Science* **269**:799–801.

Wittling, W. 1997. The right hemisphere and the human stress response. *Acta Physiol. Scand.* **161**: 55–59.

Woolley, C.S. 1999. Effects of estrogens in the CNS. *Curr. Op. Neurobiol.* **9**:349–354.

Zilles, K. 1992. Neuronal plasticity as an adaptive property of the central nervous system. *Ann. Anat.* **174**:383–391.

19

Stress and Sickness Decrease Food Intake and Body Weight

How Does This Happen? When Does This Adaptive Response Progress to Pain and Suffering?

M.F. DALLMAN

Department of Physiology, University of California at San Francisco, San Francisco, CA 94143–0444, U.S.A.

Keywords: distress, corticosterone, catecholamines, insulin, CRF, stressors

ABSTRACT

Challenges are a normal part of the life of all individuals, and organisms are equipped to cope well with environmental and psychological insults. There is a central neural system (corticotropin-releasing factor) activated by stimuli that, in turn, programs behavioral, autonomic, and neuroendocrine mechanisms that modulate the effects of the stimuli on the organism. Stress that overwhelms the large repertoire of physiological, biochemical, and behavioral coping mechanisms usually progresses to distress, pain, and suffering. Because stress response systems are centrally integrated, there are many markers of stress that usually all change in predictable directions. Changes in body weight are very sensitive markers of stress that are easily applied and provide an objective measure of the degree and duration of the response. In laboratory studies, this endpoint should be followed to determine when studies progress to an unacceptable degree of stress.

INTRODUCTION

Sick, depressed, and stressed animals, including humans, all have in common the responses of negative energy balance, usually decreased food intake and impaired growth. Food

---

Coping with Challenge: Welfare in Animals including Humans
Edited by D.M. Broom

consumption and change in body weight are commonly measured to determine the presence of pain (FELASA 1994), distress (Morton and Griffiths 1985), and suffering (Fraser 1990) as well as to study regulation of behavior (Koolhaas et al. 1999; Arnsten 1998), food intake (Weingarten 1996), metabolism (Laugero and Moberg 2000), and function in the hypothalamo-pituitary-adrenal (HPA) axis (Dallman and Bhatnagar 2000). These are highly specific, predictable, and sensitive measures of welfare that are simple and easy to apply to many species.

Most forms of stress induce at least a transient decrease in food intake and body weight, increased metabolic rate, and negative energy balance, apparently through a common mechanism involving the centrally mediated action(s) of corticotropin-releasing factor (CRF) and cytokines. Decreased food intake, by itself, can cause many other effects that are also taken as measures of pain, distress, and suffering; however, these responses appear to be physiologically adaptive and are geared to allow the organism to cope with the threat of starvation.

My conclusion from this overview is that the adequate assessment of chronic pain, distress, and suffering in each species of animal and in humans depends on knowledge of the initial conditions and the degree and duration of negative energy balance. Reduction in food intake is typical of "sickness behavior." This was initially described as resulting from the action of elevated cytokines in animals with bacterial and viral infections or after being given endotoxin (Hart 1988; Aubert 1999). Other persistent stresses as well as chronic infusion of CRF (Linthorst et al. 1997) have subsequently been shown to stimulate sickness behavior and similar endocrine responses.

Examples of responses to stimuli that elicit components of sickness behavior will be shown below. These reveal the similarities in response and also clarify differences in degree and duration of responses to increasingly intense provocative stimuli, or altered initial conditions. Results cited are primarily from rats, because a large literature is available for this species including data from our laboratory. Moreover, because of their small size and high metabolic rates, rats and other small rodents show rapid and substantial responses to stimuli that may seem surprisingly trivial to those who impose them. However, the phenomena demonstrated in rodents appear to be general to all species studied. The examples of chronic stress used here range in severity. They include the normal physiological stimulus of lactation, as well as repeated physical restraint, chronic exposure to cold, and tumor-induced illness.

In controlled laboratory studies, ethical constraints limit the duration of any experimental period and the degree to which negative energy balance, decreased food intake, and loss in body weight are allowed to progress. I submit that three of the four examples are permissible stressors, and that the last is not, if it is allowed to proceed to death. Measurement of changes in body weight throughout the course of the experiment, and the increased slope in weight loss, has been shown to predict death in the case of tumor-induced illness, allowing an earlier alternative endpoint that shortens the duration of distress (Redgate et al. 1991).

The framework developed by the Federation of European Laboratory Animal Science Associations (FELASA 1994 and expanded to species-specific frameworks by Jones et al. 2000) is a useful guide, and it serves well to describe both various categories of responses to laboratory experiments and their permissible limits. It takes into account both the intensity and duration of insults.

HOW ENERGY BALANCE IS ALTERED BY STRESSORS

Table 19.1 represents a composite list generated from many references examined by the Committee on Pain and Distress in Laboratory Animals (1992). There is general agreement that poor welfare can be recognized using the variables shown in Table 19.1.

Congruence of alterations in more than one of the categories is strong evidence of a problem (Morton and Griffiths 1985). However, the three categorical columns in Table 19.1 are interdependent and, consequently, there *must* be alterations from normal in each category. Changes in the behavioral and physiological columns result from changes in the biochemical column, based on results from studies performed on animals exhibiting sickness behavior or exposed to chronic stressors (for a review, see Koolhaas et al. 1999; Arnsten 1998; Dallman and Bhatnagar 2000; Hart 1988; Aubert 1999). Therefore, the expected temporal sequence of effects is first biochemical, and then physiological and behavioral, and the order of presentation has been changed from the original.

Initial Responses to Stressors

The first, or initial, response to a stress in rats (e.g., surgery, handling, transport, social stress, cold, restraint, immobilization, induction of diabetes, immune system challenges, shock paradigms) results in rapid neural and hormonal responses. The relative magnitudes and exact nature of the responses depend on individual differences and past history (Koolhaas et al. 1999; Arnsten 1998). Nonetheless, responses will occur in all three systems: biochemical, physiological, and behavioral. Within seconds to minutes there is elevated activity in the sympathetic nervous system (norepinephrine and epinephrine), HPA axis (CRF,

Table 19.1 Departures from normal behaviors and characteristics are suggestive of changes in well-being. A knowledge of species-typical and individual-specific behaviors and clinical values is essential. The table has been modified by reversing columns 1 and 3 (see Committee on Pain and Distress 1992, p. 47).

Biochemical	Physiological	Behavioral
Corticosteroids	Temperature	Grooming
Catecholamines	Pulse	Appetite
Thyroxin	Respiration	Activity
Prolactin	Weight loss	Aggression
β-endorphin	Blood-cell count	Facial expression
Adrenocorticotropin	Cardiac output	Vocalization
Glucagon	Blood flow	Appearance
Insulin		Posture
Vasopressin		Response to handling
Substance P		

adrenocorticotropin (ACTH), and corticosteroids (cortisol, corticosterone)), and immune system (cytokines).

This set of acute biochemical responses is sufficient to cause the values of variables shown in columns 2 and 3 of Table 19.1 to change in a consistent direction. The initial effects of increases in catecholamines and glucocorticoids are physiological. They increase metabolic rate, substrate mobilization, heart rate and blood pressure, plasma protein, and hematocrit (Gärtner et al. 1980), preparing the organism for active motor responses to the challenge. Elevated proinflammatory cytokines, acting either peripherally or centrally, may cause increased core temperature as well as alter hypothalamic CRF and other neuropeptide release, expression, and synthesis. The increased activity of CRF, central catecholaminergic systems, and cytokines alters activity in other neural and neuroendocrine systems, insulin and leptin secretion, gut motility, food intake, body weight gain, and behavior during the 24-h period after a stressor is applied. Behaviorally, there is a programmed retreat from exploration, more likelihood of "freezing" and immobility, as well as more sleep and less of the usual daily "housekeeping" behavior (Koolhaas et al. 1999; Arnsten 1998; Hart 1988; Aubert 1999; Linthorst et al. 1997).

These stress response systems have existed throughout evolution and are clearly of vital importance. Lack of a stress-induced response in the sympathetic, HPA axis, or cytokine systems is likely to be fatal. The transient, but marked increases in autonomic and HPA activity initiated by a single, even mild stimulus are sufficient to change the physiology and behavior of the organism for at least a few hours in a manner that prepares it for altered response patterns, should the stressor recur or persist. The set of behavioral responses is not rigidly adopted or imposed on the individual, since acutely or chronically stressed organisms will emerge from being "sick" and choose alternative behaviors when it is in their interests (Aubert 1999).

The stress-induced stimulation of brain CRF secretion appears to promote many of the physiological and behavioral effects of stressors (Linthorst et al. 1997). Administration of a CRF antagonist at the time of a single restraint stress significantly diminishes the post-restraint reduction in food intake (Krahn et al. 1986). CRF receptor down regulation and the decrease in CRF with concomitant increase in vasopressin in the paraventricular nucleus (PVN), which is characteristic of chronic stress (e.g., Dallman and Bhatnagar 2000), may modify the initial anorexogenic effect of stress. Caloric intake increases again, although a persistent increase in sympathetic outflow may maintain increased energy expenditure and decreased body weight.

Chronic Responses to Stressors

The overall magnitude of chronic responses to stressors is proportional to the intensity and duration of stimuli (Dallman and Bhatnagar 2000). Chronic stressors recruit a set of brain pathways that acts to modulate feeding and other behavior, temperature rhythms, and autonomic and endocrine responses (Bhatnagar and Dallman 1998, 1999). The afferent signals of everyday input and new stressors appear to be processed to a greater degree than usual through pathways that involve the limbic brain. In particular, activity of the "limbic thalamus" (the PVN of the thalamus), the amygdala, and the powerful aminergic systems involved with attention, the locus coeruleus (norepinephrine) and dorsal raphe (serotonin) nuclei is increased (Bhatnagar and Dallman 1998). Persistent glucocorticoid secretion induced by chronic stress would be expected to inhibit further HPA activity through the potent negative feedback effects of the glucocorticoids. There are diminished responses in the HPA axis to repetition of the same stressor, but the chronically stressed organism responds at a normal or supranormal level to a new type of stimulus, showing state-dependent changes in regulation of the HPA axis as a function of chronic stress (Dallman and Bhatnagar 2000; Bhatnagar and Dallman 1998).

Chronically stressed animals and people display alterations in the biochemical, physiological, and behavioral variables listed in Table 19.1. In the chronically stressed individual, anabolic functions (growth and reproductive activity) are damped, and catabolic functions (sympathetic and HPA activity) are slightly to markedly elevated. Immune function, probably because of persistent overall elevation in glucocorticoids, may be damped. Behavior may become withdrawn or aggressive (Koolhaas et al. 1999; Arnsten 1998).

General Metabolic Effects of Chronic Stress

On all of the occasions where it causes overall elevation in sympathetic and HPA activity and a diminished level of parasympathetic and anabolic hormonal activity, chronic stress results in an increased rate of catabolism. Food intake, although initially decreased, may return to normal, but is frequently not sufficient in adults to restore body weight. Body weight is either decreased or drops further with respect to control. Elevated sympathetic activity increases metabolic rate and thus caloric usage. Elevated catecholaminergic activity and glucocorticoids result in loss of muscle mass, with some protection of fat mass. However, as the specific examples below show, chronically stressed individuals remain in net negative energy balance even though their food intake may be increased.

The Measure of Body Weight

Figure 19.1 shows normal growth rates in male rats that are intact or adrenalectomized. Many maneuvers (alterations in diet, endocrinectomies) applied to animals can shift them to a new, but reliable weight gain curve. The figure shows that body weight gain in male rats is highly predictable even after bilateral adrenalectomy (left), and that deviations from normal are easily detected (right). This is a simple and not unduly onerous measurement to make for either the investigator, the animal technician, or the compliance officer. Knowledge of normal growth curves for the species and strain being studied is essential for this evaluation. Records of body weight are exceedingly useful, both as an index of unintended stress and for compliance purposes.

The Measure of Caloric Efficiency and Its Usefulness

Even if food intake is increased, net energy balance may become negative because food intake is decreased compared with the increase in metabolic rate. If both food intake and body weight are routinely measured, this can be determined by calculation of the caloric efficiency (weight gained/Cal ingested). We have found changes in this calculated value to be the single most sensitive measure of welfare or its lack in our experiments (see below).

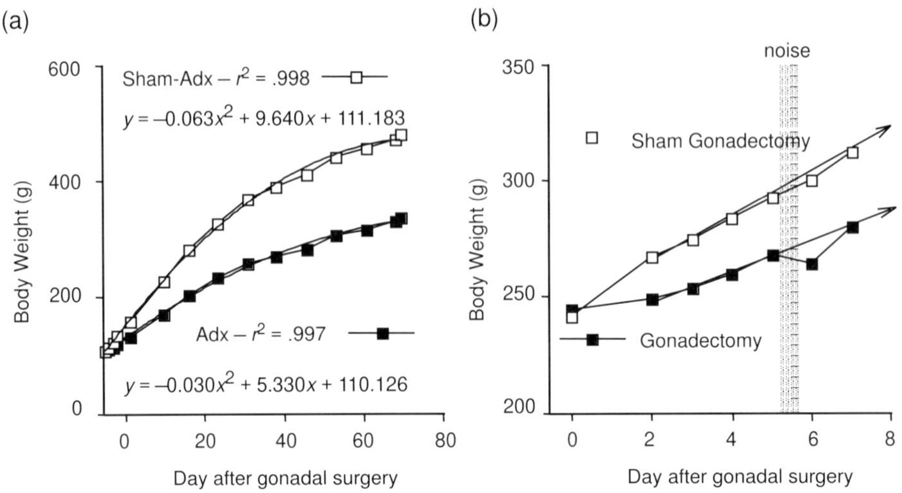

Figure 19.1 Body weight is predictable and is a measure that is highly sensitive to stress. (a) Normal growth rate in male Sprague-Dawley rats with or without bilateral adrenalectomy and supplementation with 0.5% saline (unpublished data from A.M. Strack). (b) Unanticipated environmental noise transiently decreases the normal growth rate of male rats after sham gonadectomy or gonadectomy. Body weight decreased in gonadectomized group between day 5 and 6 after noise occurred, but all rats had recovered their normal growth trajectory (arrows) by day 7 (unpublished data from V.G. Viau).

Metabolic Effects of Specific Stressors

Lactation as a Chronic Stressor

Most of the variables listed in Table 19.1 are altered by lactation, i.e., lactation induces the biochemical, physiological, and behavioral measures that indicate distress in response to other stimuli. However, there is a significant difference between lactation and most other stressors. Provided that there is good nutrition and *ad lib* access to food, positive energy balance and increased fat stores accrue during pregnancy in both rats and women (see Dallman and Bhatnagar 2000; Naismith et al. 1982). Thus, at the end of pregnancy there are high levels of stored energy. Postpartum, most of the daily energy is expended on lactation, and lactating rats use significantly less energy in housekeeping activities (Roberts and Coward 1984). Their core temperatures are higher (Aubert 1999), similar to the sickness behavior exhibited by animals with infections (Hart 1988).

Suckling by newborns stimulates ACTH and corticosteroid secretion in both rats and women (see Walker et al. 1992) and both enter a typical negative energy balance during lactation (Naismith et al. 1982; Roberts and Coward 1984). The elevated prolactin that occurs during suckling stimulates food intake and lactating mothers do not reduce their food intake in response to suckling. However, they do not eat adequately to maintain the fat stores accumulated during pregnancy, and caloric efficiency is decreased below that of mothers without pups or virgin rats. Perhaps unique to lactating rats, the usual glucocorticoid-induced protein loss is inhibited, and rats lose fat but not protein over the course of lactation. However, typical of other chronic stressors, activity in the hypothalamic-pituitary-gonadal axis is decreased, and in rats this has been shown to be a result of suckling-induced CRF secretion (Walker et al. 1993). By the end of lactation, female rats have a nearly normal body composition; they have lost most of the pregnancy-accrued fat stores during the period of lactation (Naismith et al. 1982; Roberts and Coward 1984). If suckling is not allowed after delivery, rats that increased food intake and fat stores during pregnancy remain obese (Rolls et al. 1984).

If lactating rats are challenged with lipopolysaccharide (LPS), they exhibit the classic cytokine-induced sickness behavior. However, this is modifiable by other behavioral demands. Although a lactating rat made sick with LPS will not build nests when housed with her litter at room temperature, she will build a nest if the environmental temperature is lowered. Thus, competing demands can cause the mother to emerge from sickness behavior to engage in other appropriate behaviors when the need arises. This result suggests strongly that sickness behavior can be sought as a coping strategy in response to stress (Aubert 1999).

This example of "chronic stress" in response to "physiological" circumstances is one example of a beneficial outcome of chronic stress. A similar sequence of events appears to characterize nesting or migrating birds. This combination of responses resembling chronic stress has clear advantages, both for survival of the mothers and perpetuation of the subsequent generation. In mammals the placental and pituitary hormones secreted during pregnancy stimulate markedly increased food intake and positive energy balance, despite the metabolic burden of the growing fetuses (Naismith et al. 1982). Thus, the mother at the onset of lactation is ideally poised to undergo the chronic stress of lactation and the accompanying net negative energy balance.

Chronic Stress during Repeated Bouts of Restraint

Repeated restraint in a tube from which the rat cannot escape (from 30 min to 3 h/day) serves as a mild chronic stressor. Nonetheless, rats restrained for 3h/day for only three days exhibit reduced feeding for up to 6 days and reduced body weight for more than 40 days after the end of the restraint sessions (Harris et al. 1998; Zhou et al. 1999). The reduction in food intake and in caloric efficiency is due to the central neural actions of CRF. Injection of a CRF antagonist into the third brain ventricle prior to each bout of restraint blocks the restraint-induced reduction in all three variables (Smagin et al. 1999). Although increasingly damped with repetition, the corticosterone response to restraint occurred daily (Smagin et al. 1999).

Handling of small rodents elicits repeated stress responses that affect energy balance (Laugero and Moberg 2000). It is interesting that in the studies of restraint cited above, the increased handling of the rats for intraventricular injections also caused a transient decrease in food intake and a plateau in body weight that persisted for a week. This effect was additive to the effects of repeated restraint; injection of the CRF antagonist, but not vehicle, blocked the effect of restraint on food intake and body weight, but not the effects of handling for injection of the CRF antagonist (Smagin et al. 1999).

The amplitude of the circadian rhythm in core body temperature is reduced in repeatedly restrained rats, just as it is with many stresses and CRF treatment (see Bhatnagar and Dallman 1999; Linthorst et al. 1997). In the cases in which it is measured, mean body temperature is maintained (see Figure 19.2). Repeated restraint results in a reduction in lean body mass but not fat stores on day 4, one day after the end of the repeated restraint period. This redistribution of energy stores probably represents the distinct catabolic actions of increased

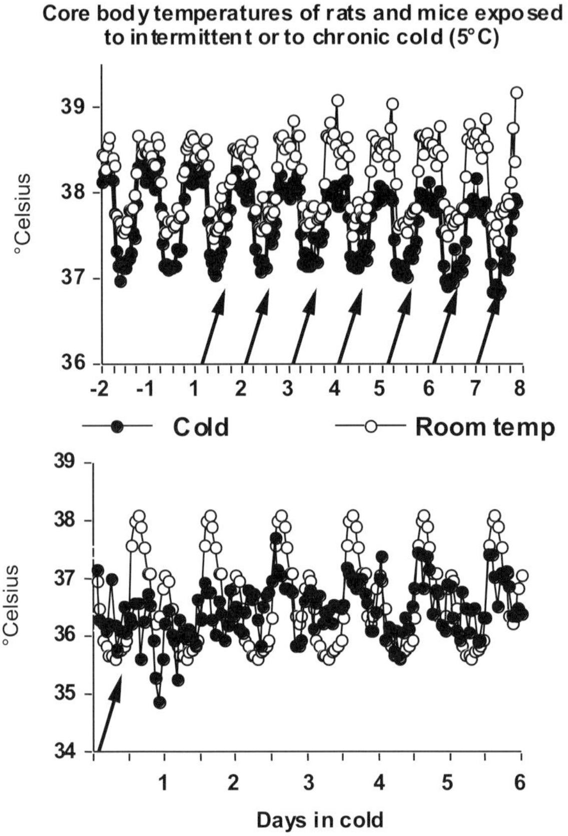

Figure 19.2 Intermittent or continuous exposure to cold decreases the amplitude of the circadian rhythm in core body temperature. The top panel shows core temperature rhythms of rats exposed to four hours of cold daily (arrows) for seven days (data from Bhatnagar and Dallman 1999). The bottom panel shows a similar phenomenon in mice living in the cold for five days (mice entered the cold box at the arrow; unpublished data from A. Strack). Mean core temperature was maintained, but the amplitude of the temperature rhythm in cold was decreased as part of the integrated response of the central nervous system to the state of chronic stress (Bhatnagar and Dallman 1999).

glucocorticoids in the presence of reduced insulin (Zhou et al. 1999). Once the daily restraint bouts are over, body composition is restored to normal within 8 days, so that fat is decreased and protein increased (Harris et al. 1998; Zhou et al. 1999; see below). It is likely that this restorative effect on body composition occurs at the end of the stressful period because the stress-induced changes in corticosterone, insulin, growth hormone, and gonadal hormones revert to normal, unstressed values. Thus, rather like the period of lactation, during the period immediately following repeated restraint, the animal is prepared to live on easily mobilizable, relatively compact fat stores, perhaps in preparation for further stress. With time, the previously stressed animal normalizes almost completely, although it still has a reduced body weight compared with its nonstressed counterparts.

The Chronic Stress of Cold

Rats or mice placed into the cold (5° C) either intermittently (4 h/day) or chronically (continually for 5 days), also exhibit changes in all variables shown in Table 19.1. Figure 19.2 shows responses in core temperature in rats challenged intermittently (top) or in mice continuously housed in the cold (bottom). Both groups of animals maintain core temperatures well, and the results show the characteristic (Harper et al. 1996) damping effect of chronic stress on the amplitude of the circadian rhythm in temperature. This damping effect also results from other chronically applied stressors, such as restraint and challenging social interactions. The damping effect of stress on the amplitude of the temperature rhythm persists beyond the end of stress application. The damped temperature rhythm in rats exposed to intermittent cold is blocked by lesions of the "limbic" thalamus (Bhatnagar and Dallman 1999). This appears to be an expression of the reorganization of brain pathways that accompanies the state of chronic stress (Bhatnagar and Dallman 1998).

It is not necessary to demonstrate a damped temperature rhythm to conclude that animals are undergoing responses to a chronic stress. The simple measurement of body weight change is sufficient. Figure 19.3 shows ten variables measured in young male rats at room temperature and after five days in cold (5° C). All but PM (peak circadian) corticosterone levels are significantly different from control in the cold-stressed rats. Metabolic responses are shown on the top row, and indices of increased adrenocortical (and autonomic nervous system) activity are shown on the bottom row. In the cold, body weight does not increase as it does in controls at room temperature, even though food intake is increased. Although food intake is elevated, it is not maximal and the animals are clearly eating less than they could in cold. Caloric efficiency shows a marked reduction. Cold activates the HPA axis to a slight, but

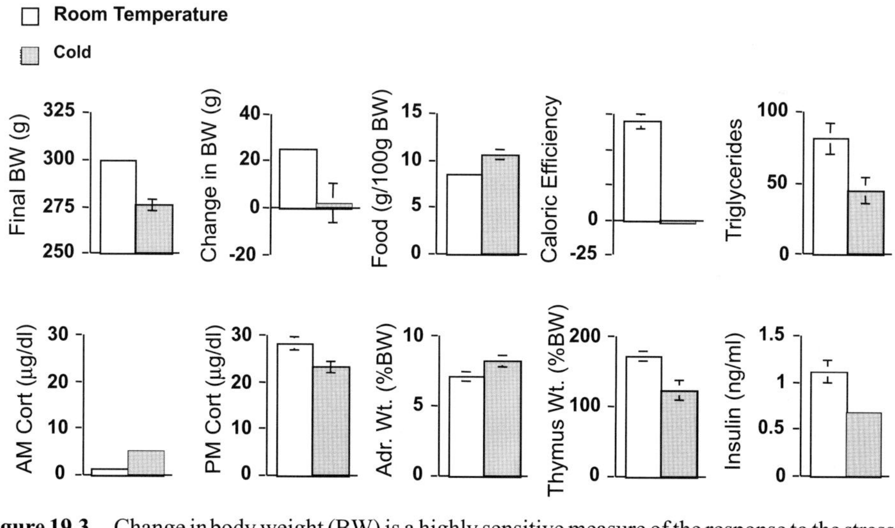

Figure 19.3 Change in body weight (BW) is a highly sensitive measure of the response to the stress of chronic cold in growing male Sprague-Dawley rats. The top row shows cold-induced changes in variables associated with metabolism, while the bottom row shows the increase in AM corticosterone and decrease in AM insulin that occurs in cold-exposed animals. All of the variables shown are significantly changed by cold except for PM corticosterone concentrations. (Unpublished data from A. Strack.)

persistent degree. This is seen in the elevated trough levels of plasma corticosterone (AM Cort), the increased adrenal weight and the decrease in thymus weight. Plasma insulin concentration is decreased in cold.

The decrease in caloric efficiency occurs, at least in part, because of the persistent elevation in sympathetic output. Increased sympathetic outflow generates heat through activation of thermogenesis in brown adipose tissue and resists the loss of core temperature in cold through peripheral vascular constriction. The decreased caloric efficiency explains the decrease in body weight gain because of inadequately elevated food intake in the cold. This is possibly a consequence of the elevated activity of central CRF exerting its anorexogenic effect (Linthorst et al. 1997; Krahn et al. 1986; Smagin et al. 1999).

Adrenal weight is elevated as a consequence of persistent stimulation by elevated ACTH. The reduction in thymus weight provides an integrated measure of the persistent hypersecretion of corticosterone. Insulin concentrations are decreased, probably as a consequence of the inhibitory effect of sympathetic activity or reduction of the stimulatory effect of the parasympathetic activity on the B-cells of the pancreas. Not measured in these experiments, but measured in others, increased glucagon and decreased growth hormone and testosterone concentrations in plasma are also results of the cold.

The relative negative energy balance in cold-stressed rats represents loss of muscle and fat. This is a consequence of elevated HPA and sympathetic activity, and decreased activity in the anabolic hormonal systems. Unlike rats at room temperature, which grow each day, rats in cold tend toward a catabolic state. Nonetheless, they are able to maintain core body temperature and tend to maintain a stable body weight.

The combined measures of body weight and food intake, together with calculation of caloric efficiency demonstrate that in each of the above examples the animals were stressed. These measures are at least as sensitive as other, more expensive and difficult to measure biochemical and other behavioral endpoints.

In all three examples, the animals coped well during the period of persistent challenge. Although they exhibited all of the signs and measures attributed to pain, distress, and suffering shown in Table 19.1, they used their physiological and behavioral repertoires to counteract the potentially harmful effects of the stress or challenge. They were able to weather the stress, using phylogenetically old and engrammed mechanisms to cope. It is clear that mammalian and "lower" species are well equipped to handle many challenges without undue distress, pain, and suffering. Although purpose-bred animals live under extremely stable conditions, the normal case for animals and humans outside of laboratory conditions is exemplified by far less environmental stability and, thus, the continual use of their innate capacity to cope with change and threat.

Overwhelming Chronic Stresses

When challenge becomes sufficiently severe or intense, the capacity to respond is exceeded and distress, pain, and suffering may ensue, prior to death. However, even under these extreme conditions, there are built-in protective mechanisms (such as coma) that appear to diminish pain and suffering. Efforts must be made by all concerned (physicians, investigators, oversight committees, and veterinary and animal care staff) to reduce the intensity and, if possible, the duration of the pre-moribund state. All of the variables in Table 19.1 will be markedly abnormal under conditions of overwhelming challenge in rats and in humans (Shaw and Wolfe 1988).

Not surprisingly, changes in body weight again are also a simple, sensitive, and accurate index of overwhelming stress. In a retrospective study, Redgate et al. (1991) analyzed data from 149 rats implanted with tumor cells into the cerebral ventricles or into brain parenchyma. There were a control and three treatment groups, and the studies had used death as an endpoint. Three phases of change in body weight were characterized with time after tumor implantation. Overall weight loss (<15%) occurred during the first phase, and the duration depended on the treatment group. Rats grew during the second phase, and its duration also depended on the treatment. In the third phase, weight loss was persistent, treatment independent, and death occurred 9.8 ± 0.3 days after the onset of phase three, when the rats had lost 30% of their body weight (Redgate et al. 1991). This analysis of changes in body weight provided a very robust model that allowed the authors to substitute phase three weight loss for death as an endpoint without losing the statistical power to determine treatment efficacy. Most of the animals could be humanely killed one or two days before predicted death would have occurred (Redgate et al. 1991).

There are occasions when changes in body weight do not provide an accurate assessment of welfare, for instance, in animals bearing growing tumor masses. In this case, a simple index of metabolism is useful to assess the animal's welfare (Ullman-Cullere and Foltz 1999). The animal is examined from the rear (tail to lower back and hind legs), while it is sitting; the degree of spinal and hip coverage by muscle and fat is visually evaluated. This approach circumvents the problem of the increasing weight of the tumor burden by assessing the degree of calorie storage in the host animal (Ullman-Cullere and Foltz 1999).

If there is a compelling reason to produce overwhelming stress in laboratory animals, limiting the duration of the experiment by continued assessment of body weight, or a corollary of this, serves to limit the distress, pain, and suffering experienced by the individual. However, maintaining the stress responses even under conditions of intense stress may result in net good to the organism. In a recent report (Kress et al. 2000), it was shown that heavily sedated patients in an intensive care unit were weaned more rapidly from ventilatory support and left the unit earlier when they were awakened from sedation for two hours each day. Moreover, the usual sedatives employed for people in intensive care units include narcotics that block, at least HPA, responses to stress. In another recent study, it was shown that addition of high levels of cortisol to the infusions given to patients with sepsis in the intensive care unit also increased recovery and hastened the time to discharge (Briegel et al. 1999). Both studies of severely stressed patients suggest that there is physiological utility to even major and more chronic stress responses.

ROLE OF GLUCOCORTICOIDS IN METABOLISM IN THE ABSENCE OF CHRONIC STRESSORS

Administration of glucocorticoids has different metabolic effects from those of naturally elevated glucocorticoids in the presence of stress. This is probably because treatment with

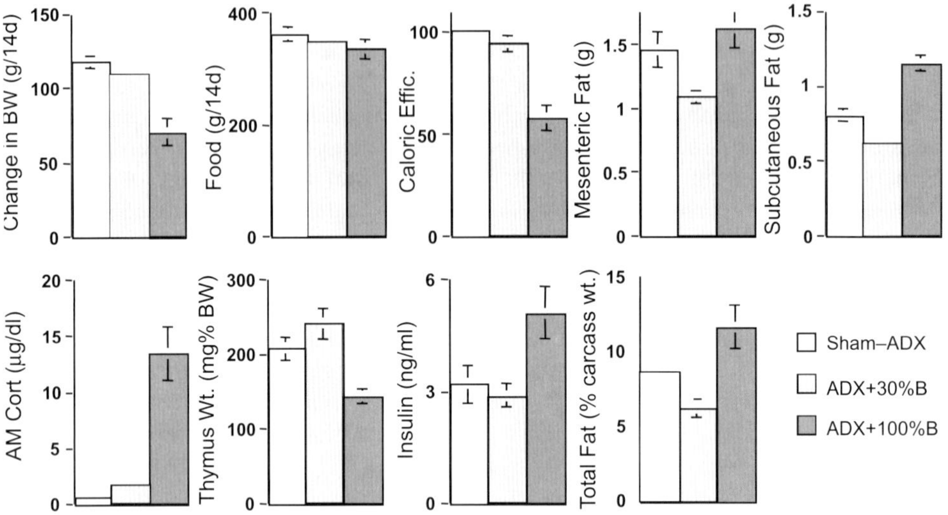

Figure 19.4 Elevated levels of both corticosterone and insulin result in redistribution of energy stores from muscle to fat. Groups of male Sprague-Dawley rats were adrenalectomized (Adx) and replaced with 2 levels (30%B and 100%B) of corticosterone (Cort). Sham-adrenalectomized rats were used as control. The experiment lasted for 14 days, during which all rats were allowed to drink a choice of 0.5% saline or a solution of saccharin. Although the rats provided with a high level of corticosterone (100%B) became fat, there was a high rate of metabolism as shown by the reduced gain in weight (BW) and diminished caloric efficiency. (Data from Bhatnagar et al. 2000.)

exogenous glucocorticoids inhibits CRF and inflammatory cytokine production, whereas stress stimulates both of these.

Figure 19.4 shows the effects of providing near-replacement levels of corticosterone (30%B) or stress levels of corticosterone (100%B) to adrenalectomized rats. For comparison, sham-adrenalectomized rats were used. Plasma corticosterone concentrations (bottom left) are elevated and constant throughout the day in the 100%B group, and thymus weight is reduced. Body weight gain is reduced in the 100%B group compared with the others, although total food intake over the 14-day experiment did not differ. Caloric efficiency was significantly reduced in the 100%B group compared with the others, suggesting a higher metabolic rate in this group. Elevated glucocorticoids in the absence of stress inhibit activity in the sympathetic nervous system, therefore there must be another explanation for the decreased caloric efficiency. This is probably a consequence of the well-known effects of glucocorticoids on both inhibition of protein synthesis and mobilization of amino acids from skeletal muscle and increased gluconeogenesis by liver. Unlike stress-induced elevations of glucocorticoids, which inhibit insulin secretion (see Figure 19.2), high glucocorticoid concentrations in the absence of stress promote insulin secretion, probably as a result of hepatic insulin resistance. Together, the elevated glucocorticoids and insulin cause marked lipogenesis resulting in obesity, both in individual fat depots (mesenteric and subcutaneous) and in total body fat.

The circulating concentrations of both insulin and corticosterone are very important for body weight gain and composition. The hormones interact in a complex way to affect metabolism. Figure 19.5 shows the relationships between corticosterone, insulin, and increase in

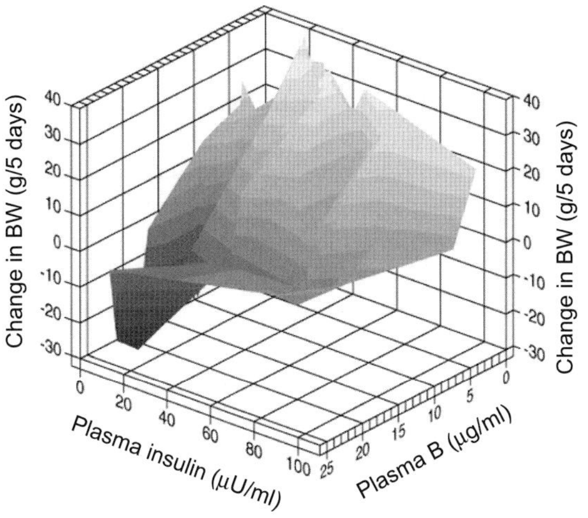

Figure 19.5 Interrelationships between corticosterone and insulin on body weight (BW) gain in male Sprague-Dawley rats. High corticosterone (B) and low insulin (conditions that occur in stressed individuals) are associated with decreased body weight gain, whereas no B or insulin results in modest weight gain. High B and high insulin result in maintenance of body weight, but shift stores from protein to fat (see Figure 19.4). The B and insulin concentrations measured in unstressed rats (about 5 μg/dl and 20 μU/ml, respectively) represent the best growth rate. The 82 rats that provided data for this figure were all adrenalectomized and B-replaced; many of them were also made diabetic and replaced with insulin infusions. (Data from Strack et al. 1995.)

body weight over a 5-day period in male rats. All rats were adrenalectomized and provided with a constant source of corticosterone to provide different concentrations in the circulation. About half of the rats were also made diabetic with injection of streptozotocin; they then received constant infusions of different amounts of insulin. The area of the figure where corticosteroids are high and insulin is low represents the typical effect of exposure to chronic stress, and weight loss generally ensues. The greatest weight gain occurs when mean corticosterone and insulin concentrations are in the range observed in unstressed rats (a daily average of ~5 μg/dl corticosterone and ~20–30 μU/ml insulin) and appears to represent good rat welfare. When corticosteroids and insulin are both high, body weight is relatively stable but calories are stored primarily as fat (see Figure 19.4).

The combination of high glucocorticoids and high insulin results in obesity in rats. Similar results have been obtained in humans after glucocorticoid infusions (Gelfand et al. 1984), and are also characteristic of people with Cushing's syndrome resulting from excessive glucocorticoids. The overall metabolic effect of glucocorticoids in the presence of normal or high insulin is to shift the site of calorie storage from muscle to fat.

In our affluent society, people may differ in their responses to chronic stressors. Whereas most challenged with major physical stressors (infection, cardiac infarction, stroke) exhibit all of the characteristics outlined above (Shaw and Wolfe 1988), purely psychological stressors may provoke different responses. People diagnosed with typical unipolar depression (DSM-IV) also respond as do most animals to the chronic stressor. They eat less and lose weight from muscle and, less so, from fat. Although absolute fat stores are decreased, abdominal fat stores are relatively increased, even when overt obesity is absent (see Dallman and Bhatnagar 2000).

However, some individuals with atypical depression and self-assessed stress increase their food intake and gain weight, primarily as fat. Moreover, they may change their eating patterns and shift food intake from primarily during the active, light period to ingesting the bulk of their caloric intake in the dark (Birketvedt et al. 1999). Apparently, the continual availability of highly palatable food, and the use of eating for "comfort," allows this alternative response which is not generally available to animals or people living under less affluent conditions. Under these conditions, stressors may result in marked metabolic pathology (Rosmond et al. 1998).

When stress-induced overeating occurs, so does relatively increased insulin secretion and this, together with the stress-induced cortisol secretion directs the extra calories to storage in fat. Moreover, the excess calories appear to be stored to a greater extent in intra-abdominal than subcutaneous depots (Rosmond et al. 1998). Increased abdominal fat stores are strongly correlated with the increased morbidity and mortality resulting from hypertension, coronary artery disease, stroke, and type II diabetes (Must et al. 1999).

CONCLUSION

Commonly encountered challenges that lead to states of stress activate basic central mechanisms built into the brains of vertebrate species that act to alter activities in the behavioral, autonomic nervous and hormonal systems allowing organisms to cope. It is likely that many of these responses are a result of activation of central CRF pathways. The outputs of these pathways serve to modify the biochemistry, physiology, and behavior of the organism under stress. The state of stress is often characterized by the loss, or lack of gain in body weight and this is an excellent and convenient measure to use in following the degree and duration of stress.

The constellation of responses (Table 19.1) that is widely agreed to represent distress in animals is common to lactation and other relatively minor or even major challenges that are part of living. These changes can indicate stress, but do not necessarily imply distress. However, very high intensity single stimuli, or persistent moderate intensity stimuli may progress to distress, pain, and suffering. Pain and suffering are even more likely as stress starts to become overwhelming and the animal can no longer muster successful behavioral or physiological coping mechanisms. In all conditions, but particularly in experimental situations, all efforts should be made to prevent the progression to pain and suffering.

REFERENCES

Arnsten, A.F.T. 1998. The biology of being frazzled. *Science* **280**:1711–1712.

Aubert, A. 1999. Sickness and behavior in animals: A motivational perspective. *Neurosci. Biobehav. Rev.* **23**:1029–1036.

Bhatnagar, S., M.E. Bell, J. Liang, L. Soriano, T.R. Nagy, and M.F. Dallman. 2000. Corticosterone facilitates saccharin intake in adrenalectomized rats: Does corticosterone increase stimulus salience? *J. Neuroendocrinol.* **12**:453–460.

Bhatnagar, S., and M.F. Dallman. 1998. Neuroanatomical basis for facilitation of hypothalamic-pituitary-adrenal responses to a novel stressor after chronic stress. *Neuroscience* **84**:1025–1039.

Bhatnagar, S., and M.F. Dallman. 1999. The paraventricular nucleus of the thalamus alters rhythms in temperature and energy balance in a state-dependent manner. *Brain Res.* **851**:66–75.

Birketvedt, G.S., J. Florholmen, J. Sundsfjord, B. Osterud, D. Dinges, W. Bilker, and A. Stunkard. 1999. Behavioral and neuroendocrine characterization of the night-eating syndrome. *J. Am. Med. Assn.* **282**:657–673.

Briegel, J., F. Forst, M. Haller, G. Schelling, E. Kilger, G. Kuprat, B. Hemmer, T. Hummel, A. Lenhart, M. Heyduck, C. Stoll, and K. Peter. 1999. Stress doses of hydrocortisone reverse hyperdynamic septic shock: A prospective, randomized, double-blind, single center study. *Crit. Care Med.* **27**:723–732.

Committee on Pain and Distress in Laboratory Animals. 1992. Recognition and Alleviation of Pain and Distress in Laboratory Animals, p. 47. Institute of Laboratory Animal Resources. National Research Council, National Academy Press, Washington, D.C.

Dallman, M.F., and S. Bhatnagar. 2000. Chronic stress and energy balance: Role of the hypothalamo-pituitary-adrenal axis. In: Handbook of Physiology, Section 7: The Endocrine System, vol. 4, Coping with the Environment: Neural and Endocrine Mechanisms, ed. B.S. McEwen, pp. 179–210. Oxford: Oxford Univ. Press.

FELASA Working Group on Pain and Distress. 1994. Pain and distress in laboratory rodents and lagomorphs. *Lab. Animals* **28**:97–112.

Fraser, A.F. 1990. An analysis of suffering. In: The Experimental Animal in Biomedical Research, ed. B.E. Rollin, pp. 217–229. Boca Raton: CRC Press.

Gärtner, K., D. Büttner, K. Döhler, R. Friedel, J. Lindena, and I. Trautschold. 1980. Stress response of rats to handling and experimental procedures. *Lab. Animals* **14**:267–274.

Gelfand, R.A., D.E. Matthews, D.M. Bier, and R.S. Sherwin. 1984. Role of counterregulatory hormones in the catabolic response to stress. *J. Clin. Invest.* **74**:2238–2248.

Harper, D.G., W. Tornatzky, and K.A. Miczek. 1996. Stress induced disorganization of circadian and ultradian rhythms: Comparisons of effects of surgery and social stress. *Physiol. & Behav.* **59**: 409–419.

Harris, R.B.S., J. Zhou, L. Youngblood, B.D. Rybkin II, G.N. Smagin, and D. Ryan. 1998. Effect of repeated stress on body weight and body composition of rats fed low- and high-fat diets. *Am. J. Physiol.* **275**:R1928–R1938.

Hart, B.L. 1988. Biological basis of the behavior of sick animals. *Neurosci. Biobehav. Rev.* **12**:123–137.

Jones, H.R.P., J. Oates, and B.A. Trussell. 2000. An applied approach to the assessment of severity. In: Proc. of Intl. Conf. on Humane Endpoints in Animal Experiments for Biomedical Research, ed. C.F.M. Hendriksen and D.B. Morton, pp. 40–47. London: Royal Society of Medicine. http:/www.lal.org.uk/cont.htm (2001).

Koolhaas, J.M., S.M. Korte, S.F. De Boer, B.J. Van Der Vegt, C.G. Van Reenen, H. Hopster, I.C. De Jong, M.A.W. Ruis, and H.J. Blokhuis. 1999. Coping styles in animals: Current status in behavior and stress-physiology. *Neurosci. Behav. Rev.* **23**:925–935

Krahn, D.D., B.A. Gosnell, A. Grace, and S. Levine. 1986. CRF antagonist partially reverses CRF- and stress-induced effects on feeding. *Brain. Res. Bull.* **17**:285–289.

Kress, J.P., A.S. Pohlman, M.F. O'Connor, and J.B. Hall. 2000. Daily interruption of sedative infusions in critically ill patients undergoing mechanical ventilation. *New Engl. J. Med.* **342**:1471–1477.

Laugero, K.D., and G.P. Moberg. 2000. Energetic response to repeated restraint stress in rapidly growing mice. *Am. J. Physiol.* **279**:E33–E43.

Linthorst, A.C., C. Falachskamm, S.J. Hopkins, M.E. Hoadley, M.S. Labeur, F. Holsboer, and J.M. Reul. 1997. Long-term intracerebroventricular infusion of corticotropin-releasing hormone alters neuroendocrine, neurochemical, autonomic, behavioral, and cytokine responses to a systemic cytokine challenge. *J. Neurosci.* **17**:4448–4460.

Morton, D.B., and H.M. Griffiths. 1985. Guidelines to the recognition of pain, distress, and discomfort in experimental animals and a hypothesis for assessment. *Vet. Record* **116(16)**:431–438.

Must, A., J. Spadano, E.H. Coakley, A.E. Field, G. Colditz, and W.H. Dietz. 1999. The disease burden associated with overweight and obesity. *J. Am. Med. Assn.* **282**:1523–1529.

Naismith, D.J., D.P. Richardson, and A.E. Pritchard 1982. The utilization of protein and energy during lactation in the rat, with particular regard to the use of fat accumulated during pregnancy. *Br. J. Nutr.* **48**:433–441.

Redgate, E.S., M. Deutsch, and S.S. Boggs, 1991. Time of death of CNS tumor-bearing rats can be reliably predicted by body weight-loss patterns. *Lab. Animal Sci.* **41**:269–273.

Roberts, S.B., and W.A. Coward. 1984. Lactation increases the efficiency of energy utilization in rats. *J. Nutr.* **114**:2193–2200.

Rolls, B.J., P.M. van Duijvenvoorde, and E.A. Rowe. 1984. Effects of diet and obestity on body weight regulation during pregnancy and lactation in the rat. *Physiol. & Behav.* **32**:161–168.

Rosmond, R., M.F. Dallman, and P. Björntorp. 1998. Stress-related cortisol secretion in men: Relationships with abdominal obesity, endocrine, metabolic, and hemodynamic abnormalities. *J. Clin. Endocrinol. Metab.* **83**:1853–1859.

Shaw, J.H.F., and R.R. Wolfe. 1988. An integrated analysis of glucose, fat, and protein metabolism in severely traumatized patients. *Ann. Surg.* **209**:63–72.

Smagin, G.N., L.A. Howell, S. Redmann, Jr., D.H. Ryan, and R.B.H. Harris. 1999. Prevention of stress-induced weightloss by third ventricle CRF receptor antagonist. *Am. J. Physiol.* **276**: R1461–R1469.

Strack, A.M., R.J. Sebastian, M.W. Schwartz, and M.F. Dallman. 1995. Glucocorticoids and insulin: Reciprocal signals for energy balance. *Am. J. Physiol.* **268**:R142–R149.

Ullman-Cullere, M.H., and C.J. Foltz. 1999. Body condition scoring: A rapid and accurate method for assessing health status in mice. *Lab. Animal Sci.* **49**:319–323.

Walker, C.-D., M.F. Dallman, A.A. Palmer, and M.K. Steele. 1993. Involvement of central corticotropin-releasing factor (CRF) in suckling-induced inhibition of luteinizing hormone secretion in lactating rats. *J. Neuroendocrinol.* **5**:451–459.

Walker, C.-D., S.L. Lightman, M.K. Steele, and M.F. Dallman. 1992. Suckling is a persistent stimulus to the adrenocortical system of the rat. *Endocrinology* **130**:115–125.

Weingarten, H.P. 1996. Cytokines and food intake: The relevance of the immune system to the student of ingestive behavior. *Neurosci. Biobehav. Rev.* **20**:163–170.

Zhou, J., X. Yan, D. Ryan, and R.B.H. Harris. 1999. Sustained effects of repeated restraint stress on muscle and adipocyte metabolism in high-fat-fed rats. *Am. J. Physiol.* **277**:R757–R766.

20

Social Stress in Wild Mammals in Their Natural Habitat

D. VON HOLST

Department of Animal Physiology, Universität Bayreuth,
95440 Bayreuth, Germany

Keywords: social stress, mammals, season, social stability, social rank, adrenocortical activity, immune measures, parasitic diseases, life expectancy

ABSTRACT

Mammals usually live in social groups that are established and maintained by interactions between the individuals. These interactions can negatively affect the welfare of individuals and lead to stress of varying intensity, depending on the situation and the multitude of behavioral and physiological coping mechanisms employed by the individuals.

First, the modern stress concept and its relevance to understand the welfare, including health, of animals is outlined, focusing on the two main stress systems — the pituitary-adrenocortical and the sympathetic-adrenomedullary systems. The activities of these systems are frequently triggered by emotions, important examples of which occur in situations characterized by uncertainty or unpredictability. Depending on the behavioral strategy used by an animal to cope with the situation, different physiological stress reactions are induced. Active attempts at obtaining control (e.g., fight or flight) are predominantly characterized by an activation of the sympathetic-adrenomedullary system and may, in the long term, lead to cardiovascular diseases (active stress). Passive perception of defeat or loss of control is characterized by pituitary-adrenocortical activation and, in the long term, by negative effects on almost all bodily functions (passive stress).

Those social situations thought to be especially relevant for the welfare of animals in their natural habitat are outlined in more detail and are based primarily on studies of European wild rabbits: elevated aggression and social instability during the mating season, differences in social rank and rank shifts, and the social integration of subadults into groups of adult conspecifics. The results of these studies clearly demonstrate that the lives of mammals in their natural habitat can be stressful, and that stability and predictability of social relationships are most important for the welfare of individuals.

Coping with Challenge: Welfare in Animals including Humans
Edited by D.M. Broom

INTRODUCTION

In 1859, *The Origin of Species by Means of Natural Selection* was published, in which Charles Darwin showed that evolution had taken place on earth as a result of natural selection caused by heritable variation and differential reproductive success. Darwin's view was based on two facts, already well known during his time, namely, that (a) as a result of their tremendous reproductive potential, populations of all creatures should increase exponentially and (b) over long periods the number of animals of each species remains relatively constant. This stability of populations must result in an intense competition to survive between the individuals, a competition that results in a higher survival rate of individuals with heritable characteristics better suited for survival. That means, for whatever reason, high degrees of stress, including suffering and premature death, are an inevitable part of the life of all animals living in their natural environment.

The regulation of populations and especially the marked periodic fluctuations of the populations of many rodents in arctic regions have interested ecologists since the beginning of the nineteenth century. As already suggested by Darwin, most authors proposed extrinsic factors, such as weather, food, predation, and disease, as the predominant or sole factor influencing animal numbers in natural populations. Without doubt these factors all act on populations, either alone or in combination, and can cause a population to decline or even become extinct, in particular instances. However, it is unclear whether such factors limit population increases under natural conditions (for a review, see, e.g., Watson and Moss 1970; Krebs and Myers 1974; Lochmiller 1996). Therefore, alternative hypotheses were proposed, claiming population regulation by intrinsic factors.

For the present discussion, the hypothesis of John J. Christian, that regulation of population densities of small mammals might be a result of social stress, is of particular relevance. Increasing population density, according to this hypothesis, results in qualitative and quantitative changes in the behavior of the animals, which in turn stimulate pituitary–adrenocortical activity and decrease pituitary–gonadal activity. As a consequence, the mortality rate of the animals increases while natality decreases, thus counteracting the increase in population density. In his first paper, Christian (1950) suggested adrenocortical exhaustion as a direct cause for the mortality. In subsequent years, however, it became evident that increased susceptibility to infectious and parasitic diseases, probably brought about by increased adrenocortical activity, is much more important (Christian 1980).

Much support for this concept is based on the results of experimental crowding of small mammal populations in the laboratory, which demonstrated all the changes in reproduction, growth, and mortality typically found in natural populations of high densities. However, the relationships between behavior, stress, and density under natural conditions remain unclear. This is for many different and often methodological reasons, such as lack of detailed behavioral data, indirect measurements of population densities, and insufficient measurement of physiological indicators of stress (Watson and Moss 1970; Myers et al. 1971; Christian 1980; Nowell 1980; Krebs 1996).

It is only in recent years that a few ecologically oriented field research projects have included physiological stress markers in their analyses, such as on dasyurid marsupials in Australia, European wild rabbits in Europe and elsewhere, as well as mongooses, cheetahs, and baboons in Africa. These studies demonstrated that social interactions may result in strong stress responses in nonhuman mammals, which can reduce the vitality and fertility of individuals and even lead to their death.

Before discussing those social situations, which are obviously most important for the welfare of wild animals, the present stress concept and its relevance in understanding the health impact of social interactions is briefly outlined.

STRESS CONCEPT

Research in the last few decades has revealed that many, if not all, neuroendocrine systems respond to challenge. According to our present knowledge, two systems are especially relevant in understanding the relationship between a challenge and the behavioral and physiological response pattern of mammals: the hypophyseo-adrenocortical system and the sympathetic-adrenomedullary system (Henry and Stephens 1977; Henry 1986).

Active and Passive Stress

The activities of these two stress systems are frequently triggered by emotions occurring in situations characterized by uncertainty or unpredictability, i.e., when the individual's control over the situation is threatened or impossible. If the stressful situation persists, different physiological stress reactions are induced, depending on the behavioral strategy used by an animal to gain control and cope with the situation. Active attempts at obtaining control over a situation (e.g., fight or flight) are characterized by a predominant activation of the sympathetic-adrenomedullary system and may, in the long term, lead to cardiovascular diseases (active stress). Passive perception of defeat or loss of control is characterized by pituitary-adrenocortical activation and, in the long term, by negative effects on almost all body functions (passive stress). These stress responses have a bipolar aspect: the sympathetic-adrenomedullary anger-fear response is opposed by the serenity-relaxation state, the latter of which is characterized by enhanced grooming and resting. The opposite pole to the pituitary-adrenocortical depression, and loss-of-attachment response, is probably a subjective feeling of elation, such as in dominant animals in the presence of clear subordinate individuals — and probably also in animals with strong bonds to partners. Low sympathetic-adrenomedullary and adrenocortical activities are, therefore, indications of good welfare.

These two coping strategies resulting in active or passive stress responses depend on the appraisal of the situation by the animals and on the quality and duration of the stressor. For example, to avoid the attacks of a dominant rival, active escape may, in the short term, have the same result as quiet concealment in a hiding place, especially if escape is impossible. Furthermore, the two coping patterns may play different roles depending on the social structures and dynamics of the population. For example, aggressive wild house mice are more successful in settled stable demes, whereas nonaggressive mice do better in growing colonies (van Oortmerssen et al. 1985).

Passive stress responses are found in subordinate individuals, while active stress responses can occur in subordinate as well as in dominant animals. It appears that gonadal activity is usually reduced by passive stress whereas, dependent on the context, active stress can have inhibiting effects in subordinate individuals or activating effects in dominant individuals striving for their position. However, immune functions appear, at least in the long term, to be more or less suppressed by all stress reactions, although divergent hormonal patterns probably have differential effects on the immune subsystems.

Relevance of Individual Expectations

Appraisal of a stimulus or situation, as well as the resulting coping behavior, are psychological processes that are based on expectations. Deviations from expectations result in poorer welfare and, if the animals are not able to remove this discrepancy, will result in stress. These expectations are influenced by various factors such as sex, age, and individual characteristics depending on the genetics, prenatal influences, and just about all learning processes of the animals.

Some expectations are very stable, as can be deduced from the rigid responses of animals to dangerous predators. However, particularly in the social context, expectations of individuals must be flexible and are, indeed, usually changed again and again during their life. Probably the first serious disappointment in the expectations of all mammals is the weaning period, when the mother denies suckling of her offspring. At puberty, even the behavior of the former friendly parents toward their offspring can become offensive and can often result in expulsion from the family unit. In mammals living in groups, the relationships to other individuals may be organized in a hierarchical system necessitating different responses to the individuals depending on their social rank positions. After losing a major fight in its resident group, an animal may transfer into another group. This is usually a rather harrowing period for the individual, with high rates of aggression.

All these periods are certainly stressful for the animals; nevertheless, they usually adapt by building up new expectations, which allow them to cope successfully with the new situations, although sometimes at the cost of poorer welfare. However, social expectations that had been fulfilled for a long time may become so fixed that they cannot be changed any more. Thus, in our colony of European wild rabbits, adult individuals that had lived for years with a bonded partner usually died within a few weeks after the death of their partners. Furthermore, while rank shifts occurred frequently in adult rabbits without serious health consequences, adult rabbits that had maintained a stable high-ranking position for a long time usually became apathetic after a rank loss and succumbed.

Nonsocial and Social Stressors

Without doubt, many nonsocial influences, such as heavy predation, extremely harsh winter conditions, or summer droughts with food shortage, can be stressful to individuals and even lead to death. However, in evolution, species adapt to these factors in their morphology, physiology, and behavior. For example, during winter periods, energetically expensive behaviors such as courtship and aggression are reduced, the animals hardly move at all, and concentrate instead on feeding. Furthermore, nonsocial stressors are usually limited to shorter spans of time. In contrast, social interactions take place all the time and, as far as stress responses are concerned, can hardly be avoided. This means that there is intense competition between the individuals of each population to survive and maximize their individual reproductive success, even at the cost of reduced health or premature death, as shown in a particularly impressive way in the Australian insectivorous marsupials of the genus *Antechinus*.

Postmating Mortality in *Antechinus*

All species of *Antechinus* examined so far exhibit an extremely synchronous life cycle. At the end of September, the females give birth to their young, which are weaned in January but continue to live with their mothers for a few more months. At the end of May, the males disperse, establish territories, and become sexually mature. The short reproductive season commences in August. During the search for females, the males invade the territories of others and are frequently involved in vigorous fights with other males and long-lasting copulations with the females. Following the 2–3 week reproductive season and before the end of the first year of their life, virtually all the males "die off." The females survive and after a one-month gestation period give birth to their young — a new cycle ensues.

The death of the males is due to typical stress reactions characterized by a tenfold increase in the plasma levels of free glucocorticoids, based on a failure of the negative feedback system controlling glucocorticoid release from the adrenals and a plasma androgen-induced suppression of corticosteroid binding proteins, and a simultaneous breakdown of the immune and inflammatory responses. As a consequence, gastrointestinal hemorrhaging associated with gastroduodenal ulcers, bacterially induced hepatic necrosis, heavy parasitic, and other diseases lead to the death of all males within a short period of time (McDonald et al. 1981; Lee and Cockburn 1985).

The physiological changes causing death of these marsupials are caused mainly by the increased levels of aggression between the males. Accordingly, if males are captured before the breeding season and housed singly, they may survive up to two years longer, as do females under natural conditions.

The same life-history pattern with mortality after the first reproductive period has also been found among males of the Australian dasyurid marsupial *Phascogale calura* and two South American marsupials, *Monodelphis dimidiata* and *Marmosa incana* (Bradley 1997). However, this life-history pattern is not restricted to marsupials, but has been found also in the Australian bush rat, *Rattus fuscipes*, although in this rodent some adult males may also survive beyond their first breeding season (McDonald et al. 1988).

METHODOLOGICAL PROBLEMS IN DETERMINING STRESS RESPONSES

The effects of stressors can basically be measured on two levels, which of course are not exclusive: the epidemiological approach is usually taken in sociologically oriented medicine; its actual aim is to clarify the role of certain social (or other) situations in the development of diseases, from cardiovascular, gastrointestinal, or renal diseases to tumor growth and infertility. The physiological approach mainly involves investigation of those structures of the central nervous system involved in stress reactions and their effects on peripheral neuroendocrine and immune parameters which may affect the fertility and health of animals. In this chapter, I focus on changes in peripheral measures, which not only indicate a stressful situation but also allow limited statements on possible pathophysiological consequences of the situation on the individuals.

A multitude of very different physiological measures can and should be assessed as individuals can react to the same stimuli in very different ways. However, in most field studies, only the activity of the adrenocortical system is assessed. Since glucocorticoid concentrations increase within minutes after each disturbance (e.g., trapping), "true" baseline levels cannot be obtained under field conditions. Therefore, standardized challenge tests are usually used to gain information on the adaptive state of the adrenocortical system (e.g., determination of the serum glucocorticoid levels at definite times after trapping, darting, or ACTH treatment). These challenge tests give a suitable measurement of an animal's adrenocortical activity or secretory capacity.

It must be emphasized, however, that an alteration of a challenge value must not be taken as an indication of an equivalent alteration of serum baseline levels of the glucocorticoid (or other) hormones. The secretory activity of the adrenal glands is dependent on the nominal value of the hormone levels as well as on the metabolism and clearance of the hormones, all of which may be influenced by stressors. Therefore, increased secretory capacities of the adrenal glands (as well as higher excretion rates with urine or feces) without a simultaneous increase of baseline adrenocortical hormone levels can be found as the consequence of increased sympathetic-adrenomedullary activity and higher blood pressure during active stress, which makes the differentiation between active and passive stress responses impossible (von Holst 1998). Nevertheless, when a social situation of low predictability and control is associated with increased adrenocortical activity, poorer welfare can usually be assumed.

One largely neglected aspect that is particularly relevant to stress research is the relationship between free and protein-bound hormone levels. In nearly all studies only the overall glucocorticoid concentration (bound and free) is determined. However, the small biologically active fraction are the free hormones, not bound to glucocorticoid-binding proteins (CBG). Therefore, a decrease in CBG concentrations, as a consequence of a glucocorticoid-induced protein mobilization or as the result of increased testosterone levels, can increase the free hormone levels, although the total concentration of hormones remains more or less the same (as in *Antechinus spp.*).

SOCIAL CHALLENGES THAT MAY AFFECT THE WELFARE OF MAMMALS

Dominance relationships are usually of overwhelming importance in all aspects of the life of group-living mammals, as they influence behavior, reproductive success, and health. However, dominance hierarchies based on agonistic behavior are only one aspect of social systems. Although much less conspicuous, social bonds, usually based on attachment between individuals, are at least as necessary for the establishment and stability of social systems as are dominance relationships. Attachment between mothers and their infants is usually a prerequisite for the survival of infants. In later life, social bonds to peers and mates, as well as to adults of the same sex, may develop which profoundly influence the behavior of the animals. Such bonds, in terms of social support, can also play a positive role in welfare, including health, presumably by altering the way in which a potentially stressful situation is perceived.

However, research into the relevance of social bonds and the stress-buffering effects of sociopositive interactions are usually neglected in laboratory studies, and there are no field studies on nonhuman mammals.

Since field studies on the behavior and stress of wild animals over the life span of individuals are almost completely lacking, I shall focus on our own studies of European wild rabbits (*Oryctolagus cuniculus*). In contrast to the marsupial species mentioned above, wild rabbits live in small territorial groups of several males and females. During the reproductive period, the territories are intensively defended against external rivals by the males. Within the groups both sexes exhibit separate linear rank orders (Lockley 1961; Myers and Poole 1961; Mykytowycz and Hesterman; 1975; Cowan 1987; Thompson and King 1994).

Season and Stress

Fights over limited resources such as food, shelter, territories, and/or rank are among the most conspicuous behaviors in animal societies. All mammals react to acute stress, such as that which is induced by a fight, by immediate activation of the sympathetic-adrenomedullary and pituitary-adrenocortical systems. If such fights only occur infrequently, they have no negative health consequences; however, if they occur more frequently, they can have detrimental effects on health and may even result in rapid death, as has already been shown for the Australian marsupials of the genus *Antechinus*.

In other species with seasonally restricted reproduction, the reproductive season is also characterized by increased aggression and correspondingly increased adrenocortical and sympathetic-adrenomedullary activities in individuals, as our investigations on European wild rabbits demonstrate.

In a first step, wild rabbits in a large natural population on a North Sea island (Sylt) were investigated at the beginning of the reproductive period, when aggression was high, and again after the end of the reproductive period with low levels of aggression. At these two times of the year, a total of 500 animals were shot between 18:00 and 19:00 hours; within less than 3 minutes of their death, blood samples were taken and subsequently different organs were dissected for endocrinological and other investigations.

As expected, we found greatly increased adrenocortical and sympathetic-adrenomedullary activities as well as many other physiological changes in adults of both sexes in spring, indicating high levels of stress. That is, under natural conditions wild rabbits of both sexes show endocrine responses of both stress systems of the same magnitude as those demonstrated mainly in rodents under laboratory conditions (Figure 20.1).

Since detailed behavioral observations were not possible in this study, we have subsequently investigated the relationships between social behavior and physiological stress responses more closely. This research was conducted over a period of 12 years in a population of wild rabbits living under natural conditions in a 22,000 m^2 area, which allowed natal and breeding dispersal. Emigration and immigration over long distances was prevented by a fence; mammalian and avian predators could, however, enter the area. The area contained a pasture grass mixture, groups of trees, and a pond. In addition to several rabbit-dug burrows, there were 14 artificial concrete burrow systems with 6–10 chambers each located in the enclosure, which the rabbits used as main warrens of their territories and also for breeding. Over the whole area more than 100 wooden traps were evenly distributed. The study area was

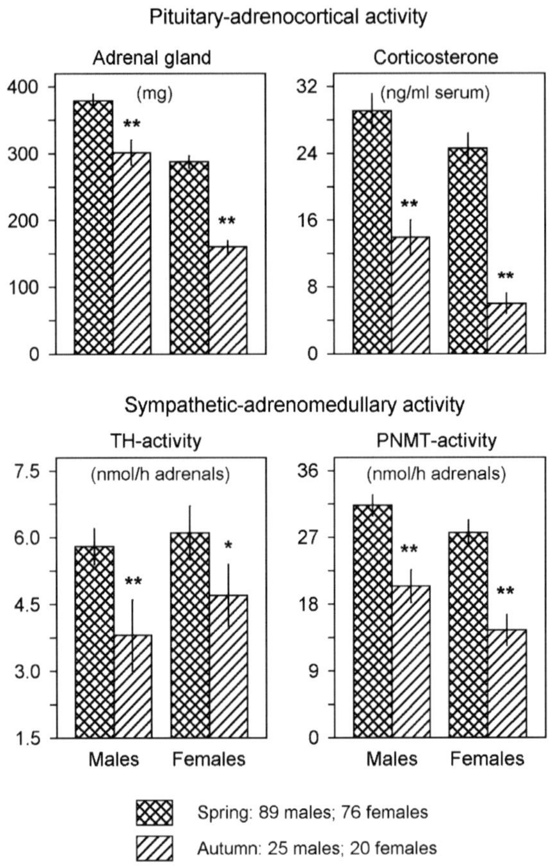

Figure 20.1 Adrenocortical and sympathetic-adrenomedullary activities of adult male and female wild rabbits (means + SEM). Tyrosine hydroxylase (TH) and phenyethanolamine-N-methyltransferase (PNMT) are pace-limiting enzymes of the catecholamine biosynthesis in the adrenal medulla. Their activities are measures of the adaptive state of the sympathetic-adrenomedullary system; corticosterone data show baseline levels. Significant differences between spring and autumn are indicated — t-test: $^{*}p < 0.05$; $^{**}p < 0.01$.

encompassed by a 10 × 10 m grid system, delineated by numbered poles; the location of any rabbit could be accurately recorded to a 5 × 5 m grid reference. Every morning the entire enclosure was patrolled and checked for deceased animals. Human interference, except for monthly capture, did not occur. All animals were individually tagged and could be observed from two separate outlook towers.

The behavior of all adult (and sometimes also of subadult) animals was observed during their peak activities in the last 4 h before dawn (about 6 h observation per animal and month). Adult animals were numbered from the outset of their first reproductive season at March 1st of the year following their birth. Since 1987, an average of 65 ± 6 adult individuals (range 33–93) have lived with their young in 8–14 territorial groups within the area (for details, see von Holst et al. 1999).

In this population, both sexes engaged in heavy fighting in the spring and many injuries resulted. In the males, these fights were over territories and rank positions and decreased only slightly over the course of the reproductive period. Females exhibited a bimodal pattern of aggression, with maxima at the beginning and the end of the reproductive period. At the beginning, fights usually occurred over rank positions within their groups, while in autumn most aggression was directed against subadult animals attempting to enter the groups. In concert with these behavioral changes, there were marked changes in adrenocortical and sympathetic-adrenomedullary activities in both sexes (Figure 20.2).

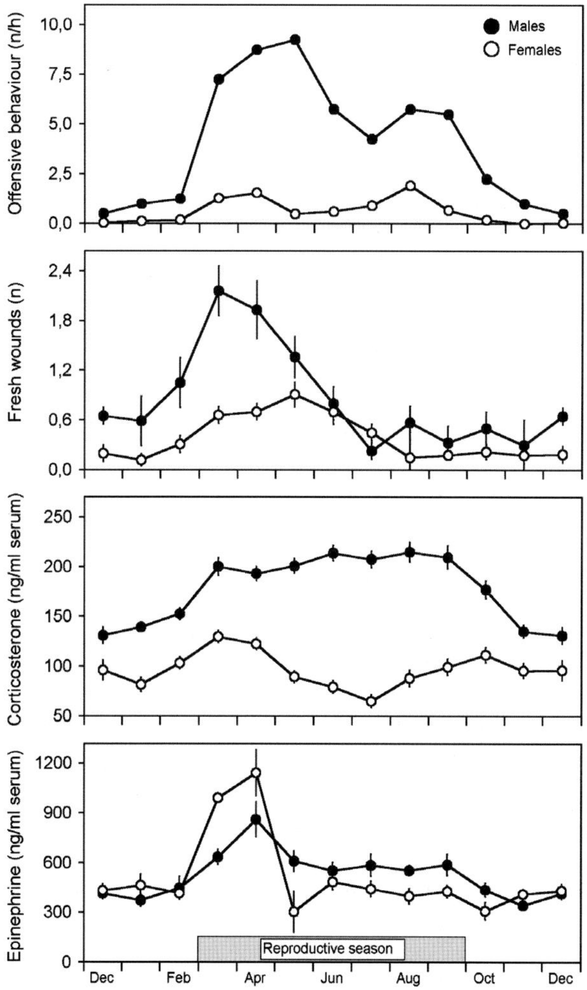

Figure 20.2 Annual variations in offensive behavior (medians) and some physiological parameters (means + SEMs) of adult wild rabbits (behavioral data from 25 males and 48 females; other parameters from about 50 males and 50 females (age >1 year). Blood samples were taken monthly for the determination of endocrine and other parameters. Corticosterone and epinephrine are challenge values, determined from blood samples taken after handling (epinephrine) and 1 h after an ACTH injection (corticosterone).

Social Rank and Stress

In most species, hierarchical systems with dominant and subordinate animals develop in such a way that subordinates can usually live in the presence of dominants with only slight or no signs of stress. In situations of continuing conflict, which result from social instability within the group, life may be stressful for subordinates as well as for dominants.

European Wild Rabbits

An overall analysis of our wild rabbit population revealed only small differences in endocrine and immune parameters of males and females with different rank positions during the reproductive season. However, at the group level, marked differences existed: Generally, males and females living without rivals within their groups had the lowest corticosterone challenge test values. Yet depending on the number of subordinate rivals, the adrenocortical activities of dominant males and females raised up to the values of subordinate individuals, thus rank-related differences disappeared. However, the heart rates in dominant individuals of both sexes, living in stable groups of 2–3 males and as many females, were always significantly lower than in subordinates, and every change of rank resulted in a corresponding change of heart rate within a few weeks (von Holst 1998; von Holst et al. 1999).

Furthermore, all individuals of both sexes from unstable social groups, with intense fighting and rank shifts, always had significantly higher serum corticosterone levels (after ACTH challenge) as well as lower *in vitro* lymphocyte proliferation rates than individuals from socially stable groups without rank shifts during the reproductive season (Figure 20.3). In

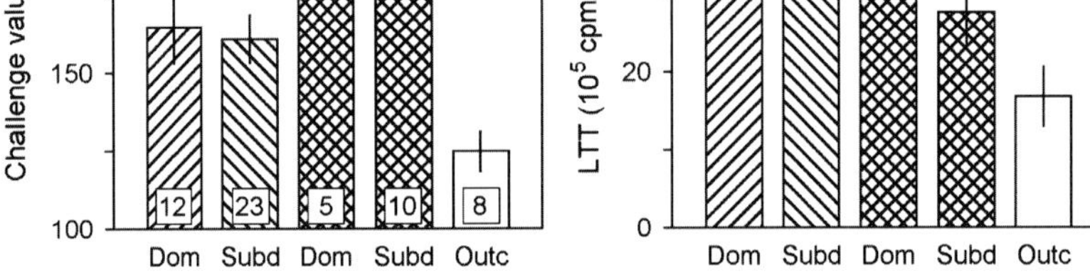

Figure 20.3 The effects of social stability or instability during the reproductive season on adrenocortical activity and lymphocyte proliferation (after concanavalin A (Con A)-stimulation) of adult male rabbits. Only individuals that had been trapped several times during the reproductive season were included in the analysis. The means of the physiological data from 3–6 months of each individual were used to calculate the group means. Dom: dominant males; Subd: subdominant males; Outc: outcasts; the number of individuals per group is given at the bottom of the bars of the figure on the left. ANOVA: Corticosterone: $F(4,53) = 11.83$; $p < 0.001$; lymphocyte proliferation: $F(4,53) = 20.12$; $p < 0.001$.

addition, in unstable social groups, serum immunoglobulin G and catecholamine challenge levels (after trapping and handling) were significantly increased in all individuals (data not shown).

A shocking phenomenon are the outcasts (Figure 20.3). Outcasts were always males with a long history of a stable high social rank, which had been defeated by a rival. In all cases, these individuals did not actively try to cope with this situation: they became completely apathetic, did not interact with any conspecific, did not groom themselves, and restricted their living area to a few square meters, even though the new dominant rival did not inhibit their free movement (Figure 20.4).

All outcasts died within a few weeks after the loss of their position. As expected, they showed typical indications of severe persistent stress such as a dramatic loss in body mass and a marked decrease in immune functions; however, contrary to expectations, their corticosterone challenge values also decreased after the defeat, reaching the lowest average

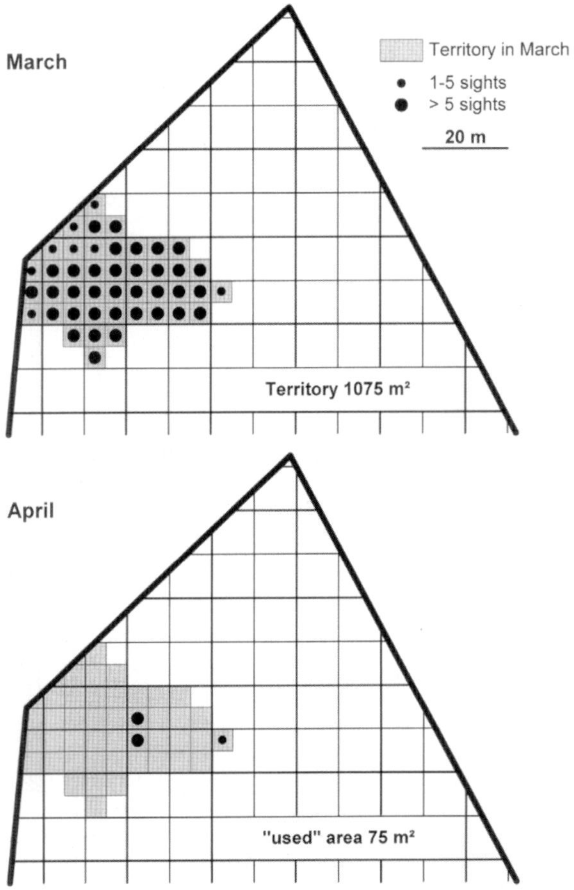

Figure 20.4 Spatial orientation and use of the territory by a dominant male before (March) and after loss of his dominant position (April).

levels ever found in our colony. This is a striking contrast to the normal response of individuals to severe stressors (as seen in the individuals of unstable social groups). The same effect on the adrenocortical system has so far only been found in humans with posttraumatic stress disorder (Yehuda 2000). This psychiatric condition is found in people after a traumatic exposure to life threat and is accompanied by extreme fear, helplessness, and horror, continuing even decades after the event occurred. The same dreadful psychological deterioration may also take place in these outcasts. Furthermore, we suppose that due to their long history of social stability and high control, these rabbits were unable to cope with this completely unexpected situation of a defeat.

Differences in adrenocortical activities between dominant and subordinate individuals were usually very small or even nonexistent. In contrast, the strong relationship between the social ranks of the individuals and their mortality rates is a most important indication for the poorer welfare of subordinate individuals. In both sexes, mortality increased with the degree of social subordination (Figure 20.5). The causes and physiological mechanisms underlying these rank-dependent differences in mortality are not known, because moribund individuals were usually killed and partially or completely eaten by predators before they were found on the daily control walks. Nevertheless, we suppose that the outbreak of parasitic diseases, especially intestinal coccidiosis, was the major cause of death, because the animals usually showed indications of severe diarrhea weeks before their death.

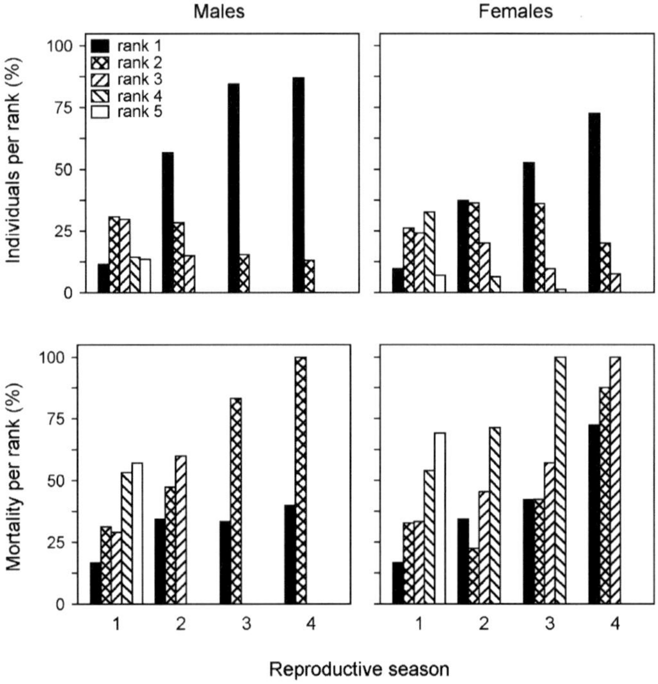

Figure 20.5 Percentage of individuals per social rank and reproductive season (top) and their annual mortality rates (bottom); data cover 12 years. Animal numbers: year 1: 104 males, 186 females; year 2: 67 males, 110 females; year 3: 39 males, 72 females; year 4: 23 males, 40 females.

In wild rabbits, the first reproductive season and the spring of the second reproductive year are apparently of paramount importance for the health and life expectancy of both sexes. This period is characterized by especially high levels of aggression during the attempts of the yearlings to gain high rank positions; unsuccessful individuals regularly change their groups. During the second reproductive season, the social positions of the adults stabilize and subordinate individuals usually upgrade their ranks only after death of higher-ranking conspecifics. This period of extreme social instability correlates with significantly higher corticosterone values, especially outside the breeding season, and extremely high mortality rates compared to later times (Figure 20.6).

Furthermore, the social ranks achieved by the individuals during their first reproductive season strongly influenced their life expectancies. The life expectancies of both sexes decreased linearly with the degree of social subordination: males with rank 4 and 5 survived only about 40% as long as males with rank 1 while in females the difference was approximately 60% (von Holst et al.1999).

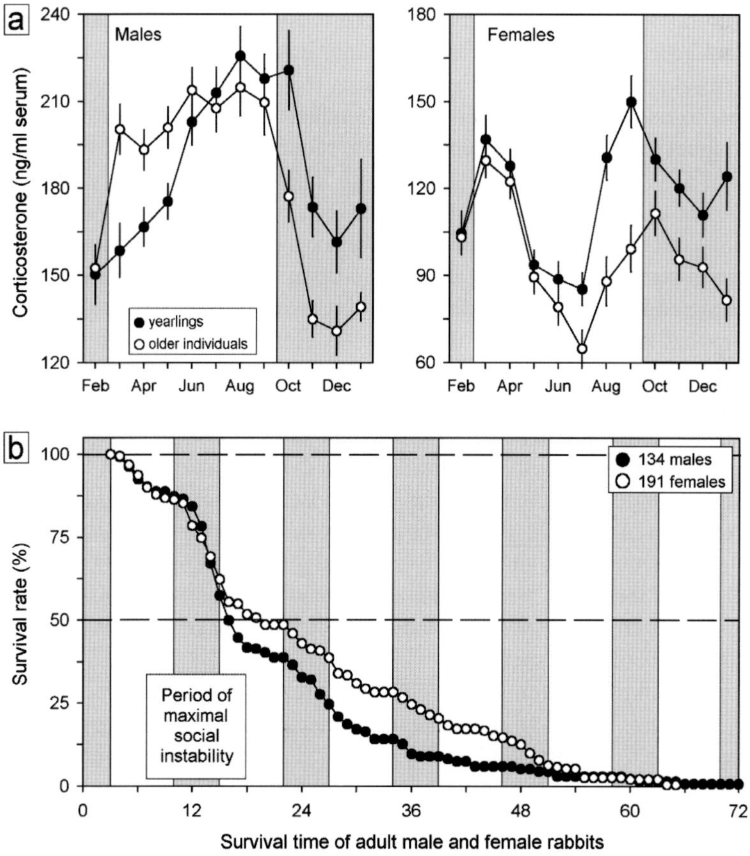

Figure 20.6 (a) Corticosterone levels (after ACTH challenge) of yearlings (about 50 males and 55 females each month) and older individuals (about 50 males and 50 females). (b) Survival rates of adult males and females. The winter periods are indicated by dark bars.

Overall, these data indicate that the welfare of wild rabbits depends predominantly on the composition and stability of their groups. In a stable social system, life is predictable, with the dominants having the highest freedom to move. In unstable social systems, where there are fights for high social ranks and many rank shifts, adrenocortical activities are increased and immune measures decreased in all individuals, indicating poor welfare.

Other Species

Similar results have also been found in field studies on olive baboons (*Papio anubis*) living in groups in the East African savanna. Dominant males living in stable groups exhibited lower "initial" levels of cortisol (10 minutes after darting) compared with subordinate individuals. Additionally, subordinate olive baboons were less responsive to dexamethasone-induced cortisol suppression than dominant males, due to a selective decrease of glucocorticoid receptors in the hippocampus. Since dexamethasone resistance is a typical indicator of reactive depression in humans, these results may indicate a similar state in animals with a long history of social instability and lack of control. Furthermore, compared with dominant individuals, subordinate animals exhibited significant reductions in high-density cholesterol, which can promote arteriosclerosis and coronary heart disease. However, during competition for dominance, plasma cortisol increased in all males. In addition, significantly lower lymphocyte counts and a higher basal cortisol concentration were found in an adult male that had entered a stable group of olive baboons, as well as in those individuals that were victims of the intruder's aggression, than in noninvolved individuals (review Sapolsky 1993).

Furthermore, according to Hausfater and Watson (1976), high-ranking individuals in groups of free-living yellow baboons (*Papio cynocephalus*) of both sexes exhibited a higher fecal parasite ova emission than subordinate individuals. Subadult individuals generally held lower ranks and had lower egg counts than older ones. However, examination of the mean egg counts, in adult individuals only, continued to show a correlation between egg output and dominance rank in males, but not in adult females.

In contrast to these studies, Creel and associates (1996) described higher fecal glucocorticoid levels in dominant African wild dogs (*Lycaon pictus*) of both sexes, as well as in samples of urine of dominant female dwarf mongooses (*Helogale parvula*), both living under natural conditions in the wild. It has to be emphasized again that increased glucocorticoid excretion rates must not be a result of increased basal glucocorticoid levels, they can also be the consequence of increased sympathetic-adrenomedullary activities, as found in active stress responses. Nevertheless, these data indicate (active and passive) stress responses in the dominant individuals of these species.

Data on relationships between social status and stress in females are largely lacking. As outlined by Christian (1980) in an extensive review, subordinate females in most rodent species are characterized by larger adrenal glands, indicating an increased adrenocortical activity.

Social Integration and Stress

Social integration into an existing group and social bonds to group members are prerequisites for good welfare for all group-living mammals, as is for example evident in our work on wild rabbits. Throughout the reproductive phase, each female usually produced 5–6 litters at monthly intervals, resulting in up to 30 progeny per year. Depending on the number of females in our population, up to 1100 animals were born each year. However, about 70% of the young were taken by predators (e.g., cats, martens, weasels, hawks) and diseases (mostly liver coccidiosis) before the onset of the winter season; on average, only 5% of the original number actually survived the winter and became adults. The number of surviving juveniles varied from one year to the next (from 0 to more than 40 individuals) and was approximately equivalent to the number of adults that had died.

Predation played only a minor role in mortality during the winter months (November to February). Death usually occurred after an extreme loss of body mass (mean >30%) within 3–4 weeks, caused by the breakdown of all fat reserves and large quantities of muscle tissue culminating in hypoglycemic shock.

Although these findings point to starvation, a general lack of food cannot be the reason for death, because all the adults as well as those juveniles that survived the winter showed only a very small loss in body mass, if at all. Rather, the moribund juveniles were incapacitated, in spite of increased food intake, by extensive parasitic damage to their intestinal epithelium, which prevented the digestion or reabsorption of food. In addition, toxins produced by the changed intestinal flora probably also contributed to the death of the animals. Compared with the surviving individuals, this parasitic infestation was probably due to a reduced immune resistance against the parasites, as indicated by a reduced *in vitro* lymphocyte proliferation, a decreased number of T-lymphocytes in the blood (Figure 20.7), and a 50% reduction in the *in vitro* phagocytic capacity of the granulocytes and in serum immunoglobulin G levels (data not shown). As can be seen in Figure 20.7, these changes in the excretion rates of oocysts and immune measures developed several weeks before the death of the individuals.

Based on our current findings, mortality during the winter months appears to be a result of socially induced immune suppression. Young animals usually left their native groups in autumn and attempted to join other groups. In this process, all immigrants were initially attacked and chased away by members of the groups. However, after a while, some juveniles were tolerated and integrated into the group, although most did not achieve this social integration. The successful integration of a juvenile was indicated by its spatial position within the group and its behavior towards the adults. Integrated animals restricted their whereabouts more or less exclusively to the existing territory of a group of adults, while nonintegrated animals tended to roam over a wide area and move from one group to the next. In addition, integrated animals were usually observed in either close proximity to, or in direct contact with, individuals of a group, while this was seldom the case in nonintegrated individuals. Although initially aggressive reactions by adults were directed against all juveniles with almost equal frequency, integrated animals were more often involved in friendly interactions with adults. While in integrated animals two out of three interactions with adults were of a friendly nature, nonintegrated animals were only involved in one sociopositive interaction for every three aggressive ones. Finally, both groups also differed significantly from each other in immunological measures: nonintegrated individuals clearly exhibited lower immune values (lymphocyte numbers, IgG levels, *in vitro* lymphocyte proliferation rates) than integrated animals. These immunological differences between integrated and nonintegrated individuals became evident during the period of social integration and thus weeks before the winter mortality (Figure

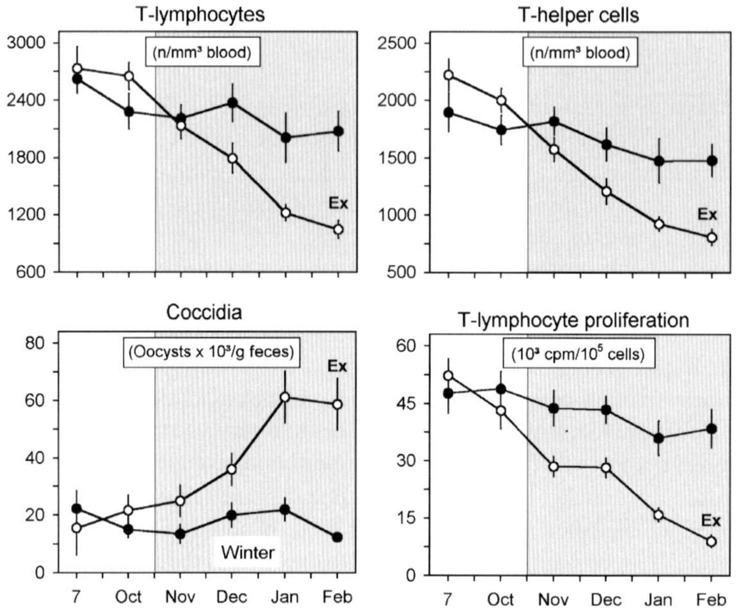

Figure 20.7 Some immune measures and the oocyst excretion rates of subadults rabbits that died during the winter (open circles) and other individuals of the same age group that survived (filled circles). Between 15 and 25 individuals per each data point. T-lymphocyte proliferation capacity was determined *in vitro* after stimulation with the mitogen Con A. Ex: last values taken about 8 days before the death of the individuals.

20.7). Correspondingly, if a change in integration state occurred during the winter season, it was always followed by changes in physiological measures: if integration status deteriorated, then immune parameters were reduced, but if integration status improved, that is, in the case of successful integration into a group even during winter, immune parameters improved (von Holst 1998).

Based on these findings, an improved immune status and reduced parasitic infestation in juveniles surviving the winter would appear to be the result of successful integration into existing social groups. Accordingly, out of more than 100 juveniles observed in detail over the last years, only those animals capable of successful social integration into groups during the autumn and winter months actually survived the winter.

Unfortunately, nothing is known about the physiological effects of the successful integration of subadult individuals into existing groups in other mammalian species in natural habitats, although this is certainly an important aspect of the life history of most mammals, especially of males. However, Caro et al. (1989) found strong differences between territory owner and those without a territory in male cheetahs (*Acinonyx jubatus*): the latter had very poor body condition and high serum cortisol challenge levels (10 min after darting) and showed changes in hematological measures indicative of stress.

CONCLUSIONS

Mammals usually live in social organizations, which are established and maintained by interactions between the individuals. Contact with conspecifics not only influences the behavior of individuals but is also associated with marked physiological changes, which can influence the vitality and fertility of the individuals in positive or negative ways, depending on the type of interaction. The term used to describe these negative effects is social stress. In species that are solitary for most of their life, social stress can occur predominantly during their reproductive season, although encounters with rivals outside the breeding season also occur. However, in group-living species conspecifics constantly influence the behavior and physiology of the group members. This can result in stress responses of various degrees depending on age, sex, genetics, and experiences of the individuals as well as on the season and group composition.

Overall, the data indicate that the stability of social systems and the relationships among the individuals are decisive for the welfare, and in particular the health, of individuals. In a stable social system, life is predictable, with highest freedom to move with the dominants. Therefore, life in well-established social systems may be possible without any apparent physiological stress effects on any group members. However, differences in life expectancies, as shown in wild rabbits, indicate rank-dependent differences in welfare, even if they cannot be discerned at the physiological level. In contrast, social instability, such as occurs during establishment of a territory or a dominance hierarchy, represents a potent psychosocial stressor in mammals, and may more or less equally affect the welfare of all animals.

Furthermore, subadults of many species leave their natal group around puberty and must integrate into other social groups. As shown in wild rabbits, this process is very stressful for the individuals and most animals die. Later in life, a breakdown of group composition or attempts of adults to associate with members of other groups are also often associated with high social instability and strong stress responses.

Since social relationships can influence the physiological state of individuals in so many positive or negative ways, it is not surprising that social status alone cannot always predict stress-related measures in individuals, particularly in natural conditions where there are many uncontrollable social and nonsocial influences. This means that to understand the physiological consequences of social interactions, an integrated approach is required to assess what factors, including rank and social bonds, interact to affect an individual's fertility and health. As mentioned before, such studies, which include detailed and life-long measures of behavior and physiology, are scarce.

Finally, it has to be emphasized that in field studies, the adaptive state of the adrenocortical system is usually taken as a measure of stress. As mentioned above, increased serum glucocorticoid levels after standardized challenge tests or increased excretion rates of these hormones with urine or feces do not allow conclusions about the basal levels; a differentiation between active and passive stress is therefore not possible. However, this differentiation is of great relevance for the assessment of the coping mechanism of the individuals and its consequences for their health and other aspects of welfare. For example, an active stress response, such that probably present in a dominant individual striving successfully for control of its social rank, may be associated with strong feelings of elation and good welfare, although it may lead to high blood pressure and the risk of developing cardiovascular diseases. This situation is certainly completely different from those passive coping patterns, associated with the feelings of frustration, anxiety, and depression. One of the most important goals concerning further research is therefore to find markers that allow the differentiation between different coping styles.

ACKNOWLEDGMENTS

I thank my undergraduate and graduate students who helped to collect the data on wild rabbits over the past years, especially Hans Hutzelmeyer, Paul Kaetzke, and Martin Kaschei. This study was supported by the Bayerische Staatsministerium für Ernährung, Landwirtschaft und Forsten and by grants from the Deutsche Forschungsgemeinschaft and the Volkswagenstiftung.

REFERENCES

Bradley, A.J. 1997. Reproduction and life history in the red-tailed phascogale, *Phascogale calura* (*Marsupialia: Dasyuridae*): The adaptive stress-senescence hypothesis. *J. Zool. Lond.* **241**: 6739–755.

Caro, T.M., C.D. Fitzgibbon, and M.E. Holt. 1989. Physiological costs of behavioural strategies for male cheetahs. *Animal Behav.* **38**:309–317.

Christian, J.J. 1950. The adreno-pituitary system and population cycles in mammals. *J. Mammal.* **31**:247–259.

Christian, J.J. 1980. Endocrine factors in population regulation. In: Biosocial Mechanisms of Population Regulation, ed. N.N. Cohen, R.S. Malpass, and H.G. Klein, pp. 55–116. New Haven: Yale Univ. Press.

Cowan, D.P. 1987. Aspects of the social organization of the European wild rabbit, *Oryctolagus cuniculus*. *Ethology* **75**:197–210.

Creel, S., N.M. Creel, and S.L. Monfort. 1996. Social stress and dominance. *Nature* **379**:212.

Darwin, C. 1859. The Origin of Species. London: Murray.

Hausfater, G., and D.F. Watson. 1976. Social and reproductive correlates of parasite ova emissions by baboons. *Nature* **262**:688–689.

Henry, J.P. 1986. Neuroendocrine patterns of emotional response. In: Emotion: Theory, Research and Experience, ed. R. Plutchik, vol. 3, pp. 37–60. New York: Academic.

Henry, J.P., and P.M. Stephens. 1977. Stress, Health, and the Social Environment. A Sociobiologic Approach to Medicine. New York: Springer.

Krebs, C.J. 1996. Population cycles revisited. *J. Mammal.* **77**:8–24.

Krebs, C.J., and J.H. Myers. 1974. Population cycles in small mammals. *Adv. Ecol. Res.* **8**:267–399.

Lee, A.K., and A. Cockburn. 1985. *Antechinus* as a paradigm in evolutionary ecology. In: Evolutionary Ecology of Marsupials, ed. A.K. Leeand and A. Cockburn, pp. 162–183. Cambridge: Cambridge Univ. Press.

Lochmiller, R.L. 1996. Immunocompetence and animal population regulation. *Oikos* **76**:594–602.

Lockley, R.M. 1961. Social structure and stress in the rabbit warren. *J. Animal Ecol.* **32**:385–423.

McDonald, I.R., A.K. Lee, A.J. Bradley, and K.A. Than. 1981. Endocrine changes in dasyurid marsupials with differing mortality patterns. *Gen. Comp. Endocrinol.* **44**:292–301.

McDonald, I.R., A.K. Lee, K.A. Than, and R.W. Martin. 1988. Concentration of free glucocorticoids in plasma and mortality in the Australian bush rat (*Rattus fuscipes* Waterhouse). *J. Mammal.* **69**:740–748.

Myers, K., C.S. Hale, R. Mykytowycz, and R.L. Hughes. 1971. The effects of varying density and space on sociality and health in animals. In: Behavior and Environment. The Use of Space by Animals and Men, ed. A.H. Esser, pp. 148–187. New York: Plenum.

Myers, K., and W.E. Poole. 1961. A study of the biology of the wild rabbit, *Oryctolagus cuniculus (L.)* in confined populations. II. The effect of season and population increase on behaviour. *CSIRO Wildl. Res.* **6**:1–41.

Mykytowycz, R., and E.R. Hesterman. 1975. An experimental study of aggression in captive European rabbits, *Oryctolagus cuniculus L. Behaviour* **52**:104–123.

Nowell, N.W. 1980. Adrenocortical function in relation to mammalian population densities and hierarchies. In: General, Comparative and Clinical Endocrinology of the Adrenal Cortex, vol. 3, ed. I. Chester Jones and I.W. Henderson, pp. 349–393. London: Academic.

Sapolsky, R.M. 1993. Endocrinology alfresco: Psychoendocrine studies of wild baboons. *Rec. Prog. Horm. Res.* **48**:437–468.

van Oortmerssen, G.A., I. Benus, and D.J. Dijk. 1985. Studies in the wild house mice: Genotype-environment interactions for attack latency. *Neth. J. Zool.* **35**:155–169.

von Holst, D. 1998. The concept of stress and its relevance for animal behavior. *Adv. Study Behav.* **27**:1–131.

von Holst, D., H. Hutzelmeyer, P. Kaetzke, M. Kaschei, and R. Schönheiter. 1999. Social rank, stress, fitness, and life expectancy in wild rabbits. *Naturwissenschaften* **86**:388–393.

Thompson, H.V., and C.M. King, eds. 1994. The European Rabbit: The History and Biology of a Successful Colonizer. Oxford: Oxford Univ. Press.

Watson, A., and R. Moss. 1970. Dominance, spacing behaviour and aggression in relation to population limitation in vertebrates. In: Animal Populations in Relation to Their Food Resources, ed. A. Watson, pp. 167–218. Oxford: Blackwell Scientific.

Yehuda, R. 2000. Biology of posttraumatic stress disorder. *J. Clin. Psychiatry* **61(7)**:14–21.

Standing left to right: Mary Dallman, Dirk Hellhammer, Eberhard Fuchs, Stephan von Hörsten, Pierre Mormède, Dietrich von Holst
Seated left to right: Eugene Paykel, Michael Irwin, Cobi Heijnen, Christine Heim

21

Group Report: Pathological Consequences of Stress

P. MORMÈDE, Rapporteur

M.F. DALLMAN, E. FUCHS, J.J. HEIJNEN, C. HEIM, D.H. HELLHAMMER, M. IRWIN, E.S. PAYKEL, D. VON HOLST, and S. VON HÖRSTEN

Keywords: stress markers, body–brain interactions, pathology

INTRODUCTION: WELFARE, STRESS, AND PATHOLOGY

Stress generally refers to the response to any physical or psychological stimulus that harmfully disrupts ongoing homeostasis. The threat to homeostatic systems mobilizes behavioral, neuroendocrine, autonomic, and immune systems that respond to the challenge. Such challenges are integral to life, and challenge *per se* does not equate with poor welfare. With respect to the integral of challenge, welfare is conceived as an inverted U-shaped curve (Figure 21.1). The absence of challenging stimulation may impede good welfare as does overwhelming stress. In humans, it is common for perceived boredom (i.e., absence of challenge) to be counteracted behaviorally by thrill-seeking or even addictive behaviors (e.g., alcohol or psychostimulant use). Thrill-seeking behaviors activate the neuroendocrine and autonomic systems and may, in some instances, move the individual upward on the welfare curve, whereas addictive behaviors almost universally lead to poorer welfare and health functioning. Similarly, in understimulated farm animals, various behavioral pathologies may appear, possibly as an attempt to self-stimulate. On the other hand, very high-intensity single stimuli, such as in precipitation of posttraumatic stress disorder, or persistent moderate-intensity stressors, including addictive behaviors, can overwhelm the response systems and result in pathology.

Markers for stress are behavioral, physiological, and biochemical. They include sickness behavior, changes in autogrooming, eating and drinking (and thus in body weight), changes in blood pressure and respiration, as well as changes in many blood-borne substances, such as hormones, glucose, and white blood cell distribution and function. Generally, if one of these

Coping with Challenge: Welfare in Animals including Humans
Edited by D.M. Broom

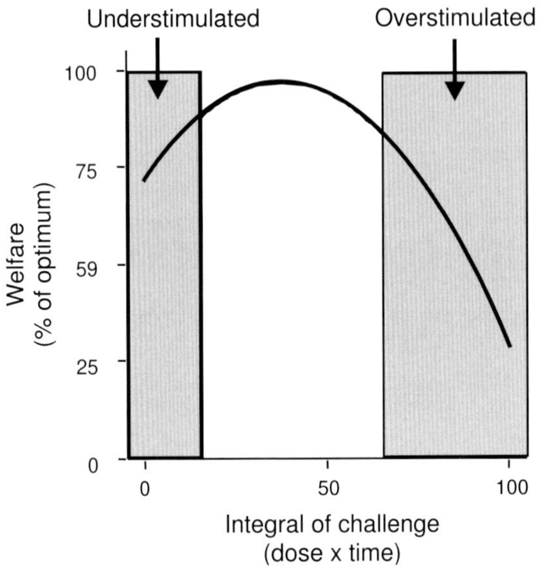

Figure 21.1 Theoretical relationship between challenge, which may lead to stress, and welfare. The shaded areas show the levels of under- and overstimulation where pathology can be encountered.

measures changes as a result of stress, several of the others will as well. Consequently, several markers should be used to assess stress; a single marker is not sufficient.

Interpretation of the changes in these markers should take into account the degree and duration of the deviation from normal. Many of these markers of stress change after transient, mild challenges, which are a normal part of life and which may lead to better, not worse, overall welfare. However, if changes in markers that indicate stress persist, with no trend toward normal within a reasonable time (days to weeks), poor welfare of the individual may be assumed and action should be taken, if appropriate. Whenever possible, the normal range of markers in the individual, or at least the strain under observation, should be determined so that deviations can be sensitively detected. However, persistent alterations in markers do not need, by definition, to be detrimental for the welfare of an animal, at least in the short term (McEwen, this volume).

The brain is the key integrator of the stress response and diseases may also result from modifications of the stress response systems rather than from the stress itself. Brain structures are the sites at which environmental and psychosocial demands are recognized and from which coordinated stress responses are initiated. However, the brain is also the site that is primarily affected in the course of the stress response. The stress response triggers adaptive modifications of structure and function in the brain and thus serves to adjust the behaviors of an individual to the requirements of its external world. The characteristics of adaptive responses towards a challenge depend on genetic (including gender) and nongenetic (environment and life history, including age) factors, thereby defining individuality. Most likely, therefore, stress-related diseases are highly dependent on the individual and are often driven by acquired characteristics of the stress response system itself.

STRESS-RELATED PATHOLOGY

Stress and Pathology in Animals

Much evidence concerning environmental stress and pathology comes from the study of animals. Stress is a fundamental part of the life of all animals in nature. Although factors such as predation, climatic conditions, or food availability can affect the welfare of animals, the influence of intraspecific social factors on vitality and health is much more important in many mammals (von Holst, this volume). Stressful situations that may lead to diseases and premature death in nonhuman mammals are mostly characterized by social instability, for example, overcrowded populations, establishment of social hierarchies, and integration of new individuals into existing groups. Although the physiological mechanisms leading to these diseases are usually not well understood, for example, due to many methodological difficulties in field work, two groups of stress-related diseases seem to be most relevant (for a more recent review, see von Holst 1998):

1. Diseases that are associated with reduced immune functions leading to the outbreak of parasitic, bacterial, viral, and other diseases: Under extreme conditions, as in the males of several species of Australian marsupials, these diseases can lead to the death of all individuals after a period of less than four weeks of fighting for females and territories (von Holst, this volume). In other species, death due to diseases is also common, such as in subadult wild rabbits, which usually die of coccidiosis, a common parasitic disease of vertebrates, after social isolation due to the lack of acceptance of adult individuals within existing groups.
2. Diseases of the cardiovascular system or kidneys, which are mostly demonstrated on the basis of postmortem examinations, such as in crowded populations of some species of voles, mice, hare, and deer: Long-term laboratory studies and some evidence from the field indicate that the development of these diseases is associated with long periods (weeks to months) of heightened sympathetic-adrenomedullary activity, for example, due to social instability. Thus dominant males in large artificial colonies of laboratory mice are characterized by vigilant patroling behavior associated with heightened adrenomedullary activities and a moderate hypertension, which lead within 5–8 months to a fixed hypertension, increased heart weights, aortic and intramural coronary arteriosclerosis, and increased renin activity. Similar results are known from several monkey species that had been kept artificially or lived under natural conditions for months under socially unstable conditions. The animals usually show increased heart rates; develop essential hypertension, coronary insufficiency, and myocardial infarction; they also die prematurely.

In farm animals, a number of disorders have been related to bad environmental conditions and are considered as markers for poor welfare. These include usually nonharmful stereotypic behaviors, such as bar biting in sows, but also behavioral disturbances that may have major consequences for the welfare and especially health of animals, for example, navel sucking in calves, feather pecking in poultry, or tail biting in pigs. On the other hand, several infectious diseases, such as diarrhea in piglets, respiratory diseases in calves, and reproduction problems in sows, have in common different characteristics, including the involvement of germs with low pathogenicity, chronic course of the disease in a given production unit, and high incidence of the symptoms with a low growth rate but few deaths. Furthermore, epidemiological studies have shown that these pathological entities are usually associated with poor environmental conditions and document the multifactorial nature of these "stress-related" syndromes. Indeed, antibiotics and other anti-infectious treatments are usually inefficient in the long run, and actions correcting the environment must be taken to get rid of the disease. The current movement towards the production of clean animals with no antibiotic supplementation of food and no growth factors will undoubtedly push towards a more careful consideration of the animal as a living organism sensitive to the interactions with its physical and social environment, and with humans.

Stress and Human Pathology

There is evidence that stress may play a role in the initiation and progression of a variety of human disorders: psychiatric, addictive, and physical. The strength of these relationships varies among different disorders and is dependent on the methodological adequacy and difficulties of the studies.

In psychiatric disorders, the most specific relationship is with posttraumatic stress disorder (PTSD), which can be initiated by a single overwhelming life-threatening experience, although in some cases chronic repeated traumatic experiences may be implicated, for example, torture. Even here, other factors may be important; not everyone exposed to these experiences develops the disorder (Yehuda and McFarlane 1995). There is also strong evidence suggesting that stressful experiences represent one of the major risk factors for the development of depression, in addition to genetic disposition and female gender among other risk factors. Episodes of depression are often preceded by or mutually related to acute or chronic stress in adulthood (Irwin, this volume). A role for early adverse experience, for example, child abuse and neglect or parental loss, has also been demonstrated for the development of depression in adulthood. Stress early in life also appears to predispose an individual for several anxiety disorders which often coincide with major depression (e.g., Kendler et al. 2000). Stress can similarly impact risk for substance dependence and is thought to be key in triggering relapse in alcohol dependence in men and women. Increasing evidence also suggests that stress, including stress early in life, plays a role in several physical disorders, for example, cardiovascular disorders, immune-related disorders, and cancer, among others (McEwen 1998). For future research it will be important to investigate individual susceptibility to these stressors and whether we can define markers for susceptibility (or poor adaption) to stress experienced either early in life or during adulthood.

Stress may contribute in varying degrees to a variety of psychiatric and physical disorders. To what degree are these disorders the result of the same morbid process? To answer this question properly, we need a clear understanding of underlying mechanisms and processes, which is not yet fully available (cf. below). Within psychiatric disorders, whether apparent comorbidities such as anxiety and depression represent two disorders or one mixed disorder is to some extent a definitional issue depending on where the boundaries are set. Within physical disorders or between physical and psychiatric disorders there are often clearly different pathologies. Especially in the instance of PTSD and depression, stress generally appears to increase the susceptibility to mental and physical disorders which have additional specific causative factors and pathologies.

It is important to emphasize that the simple coexistence of pathology and stress symptomatology is not sufficient to prove any causal relationships. Rather, it is necessary to obtain knowledge on specific mechanisms by which stress can possibly affect the respective pathology in order to develop successful intervention strategies. An example for such an approach is research on male infertility. Here, some patients report depressed mood and suggest to the andrologist that the patient's fertility may profit from antidepressants or psychotherapy. Studies suggest that social dominance and active coping but not depression are associated with low sperm count and low testosterone levels in both men and experimental animals.

ARE THERE DISTINCTIVE RESPONSE SYSTEMS FOR PATHOLOGY?

Characteristics of Stress Responses

Interactions and Balances among and within Response Systems

The outputs for the central stress response system include behavioral, endocrine, autonomic, and immune responses. Generally, because of the nature of the brain stress circuitry, more than one output system is activated. The output of each system affects and modulates responses in the others, and the range of responses varies from lower to higher than normal. For instance, in PTSD, basal activity in the hypothalamic-pituitary-adrenocortical (HPA) system is low, and anxiety and sympathetic outflow are high. By contrast, in early starvation, HPA activity is high and sympathetic outflow is low, although motor activity is increased (Dallman 1999). Compared with normal situations, a high level of integrated stress may exist with low or high levels of behavioral signs, endocrine and autonomic output, and immune activity.

The outputs of the various stress response systems interact, allowing mutual communication between the different organs of the body. For instance, elevated glucocorticoid secretion affects learning and memory, inhibits central sympathetic outflow, and has strong immunomodulatory effects. In principle, the secretion of glucocorticoids as a response to a stressor may function as an adaptive response, often reflected by increases in immune function or recruitment of immune cells from the peripheral circulation to other organs, such as the skin, where these cells can defend the organism. Persistent secretion of glucocorticoids as a result of chronic stress may be nonadaptive and can, in some instances, be detrimental for immune functioning (McEwen, this volume), although in humans the immunosuppressive effects of cortisol have not been reliably demonstrated in depressed or stressed populations. Increased sympathetic activity and epinephrine secretion affect immune function and increase central arousal and can alter adrenocortical responsivity to ACTH (Ulrich-Lai 2000). Cytokines, produced and released by the immune response, alter outputs of the behavioral, endocrine, and autonomic systems.

Assessment of multiple stress response systems provides a more accurate evaluation of the integrated response of the organism to stress than measurement of the changes in a single system (Dallman, this volume), especially since there are different response patterns depending on the situation and the coping behavior of the individuals. In regard to cardiovascular disease risk, it is well recognized that acute psychological stress is often associated with increases in the release of catecholamines with attendant effects on blood pressure and possibly atherogenic processes. However, to understand the effects of psychological stressors on cardiovascular disease risk fully, it is essential to evaluate changes in other systems of organization including, for example, behavior and immunological responses. One behavior that is critical in understanding cardiovascular disease risk is quantity and depth of sleep. Loss of sleep or disordered sleep architecture with loss of delta sleep in humans is associated with profound elevations of catecholamines and day-time levels of blood pressure and heart rates. Such increases of catecholamines have cascading effects on parameters of immune function that are increasingly implicated in initiation and progression of atherosclerosis. Increases in pro-inflammatory cytokines and increases in cellular adhesion molecule expressions likewise occur in concert with sympathetic activation, which together are hypothesized to drive inflammatory changes at the level of the endothelium with the development of atherogenic lesions.

Responsivity to Challenges

Understanding the normal range of responsivity of various regulatory systems is necessary for evaluating potential pathological consequences associated with stress. Rather than simply assessing behavioral, physiological, and biochemical measures under static conditions or at a single point in time, measurement of reactivity to stressors in standardized challenge tests characterizes the adaptive state of stress responsive systems, yielding insight into the magnitude of the response and the extent to which the responses are exaggerated or attenuated or shortened or prolonged in duration. Furthermore, reactivity to stressors provides information about whether such systems return to a baseline that may differ from that prior to the stressful stimuli. Deviations from an established normal baseline may indicate an ongoing dysregulation with implications for pathology. Such reactivity studies have been extremely useful in identifying defects in the integrated regulation of autonomic, neuroendocrine, and immune systems. For example, in aged animals, the central administration of corticotrophin-releasing hormone (CRH) leads to an exaggerated release of circulating catecholamines which is prolonged in duration; simultaneous assessment of parameters of immune function, such as natural killer cell activity suggests, that the administration of CRH "resets" resting levels of immune function (Irwin, this volume). Likewise, in some human studies, chronically stressed individuals show an exaggerated release of catecholamines and cortisol in response to acute challenge, and these differences are only revealed following standardized challenge tests of the individual; baseline, resting levels of these parameters are similar.

Body–Brain Interactions

The linear study of the links between environmental inputs and behavioral, autonomic, and endocrine outputs is the most frequent approach to exploring the pathophysiology of stress-related pathology. However, the integration of these responses in functional loops may be necessary to the understanding of the critical mechanisms where action can be taken. There are classic examples of communication pathways from the periphery to the brain, such as the effect of epinephrine on brain arousal, memory, and learning by its action on the central nucleus of the amygdala and other brain sites; the action of steroid hormones on various brain functions, such as the effect of glucocorticoids on learning and memory; response to psychostimulants; and the effect of sex steroids on a number of behaviors. Examples of integrated functional loops include:

- The vagus nerve as a bidirectional communication pathway (Porges, this volume): From the brain to the periphery, the vagus nerve has developed throughout evolution as a double system, a primitive inhibitory system of unmyelinated fibers originating from the dorsal motor nucleus of the vagus (dorsal vagal complex), involved in the withdrawal/conservation responses (stress-induced defecation). It is complemented in mammals with a system of myelinated fibers originating from the nucleus ambiguus (ventral vagal complex), inhibiting the sympathetic nervous system at the level of the heart, linked to the expression of emotions receiving inputs from the frontal area of the cortex and from the facial and trigeminal nerves, involved in the facial expression of emotions and oral motor stimulations.

  However, 70% of the fibers of the vagus nerve originate in the peripheral organs. They project massively to the nucleus tractus solitarius, and from there they provide the primary afferent input to the visceromotor component of the nucleus ambiguus and the dorsal motor nucleus of the vagus and radiate as well to other portions of brain and even influence nocioception via solitary spinal pathways. This system conveying peripheral information to the brain is involved in the regulation of the sympathetic-adrenomedullary system (SAM) activity but has more recently been shown to be operative in other conditions (e.g., visceral pain, immune signals).
- Central state and metabolic balance: The two main stress systems, the HPA axis and SAM, are primarily involved in the regulation of energy fluxes. They have been hijacked as stress responsive systems due to their ability to mobilize the energy necessary for behavioral responses. As a result, a decrease in body mass, which reflects this mobilization of energy stores, is often a very sensitive index of stress in nonhuman mammals, and ruptures of adaptation, most probably at the source of bad welfare, are accompanied by an increased rate of weight loss (Dallman, this volume).

  Inversely, energy balance is a major regulator of HPA axis and SAM activity. Perturbations of ingestive behaviors will have neuroendocrine consequences, such as increased HPA axis activity (gluconeogenesis) and reduced activity of the peripheral noradrenergic system (energy-saving mechanism), in case of low food intake (Landsberg and Young 1978; Dallman 1993; Garcia-Belenguer et al. 1993). Consequently, some neuroendocrine changes may result from these behavior-related metabolic changes rather than being the direct consequence of psychological states (e.g., depression and HPA axis hyperactivity, early weaning in pigs and reduced urinary excretion of norepinephrine (NE); Hay et al. 2001).

  Other metabolic inputs to the brain have pleiotropic effects (e.g., the effect of leptin, a hormone originating from the adipose tissue, on food intake, reproductive physiology, ANS activity). Chronic stress in general inhibits production of leptin through sympathetic inhibition; the consequences include decreased gonadal hormones, increased food intake, and increased SAM activity (Ahima 2000).
- Impact of stress on blood–brain barrier permeability: Recent findings suggest that stress may alter the permeability of the blood–brain barrier. Acute immobilization in rats, low-temperature exposure, or isolation in mice resulted in increased penetration of

the blood–brain barrier by drugs, neurotransmitters, and viruses that are normally excluded. In soldiers exposed to extreme emotional stress during the Gulf War, the blood–brain barrier has become unexpectedly more permeable to the cholinergic drug pyridostigmine (Feldman et al. 1996). This compound is an acetylcholinesterase inhibitor routinely employed in the treatment of the autoimmune disease myasthenia gravis. Because of its protective effect against organophosphate poisoning, pyridostigmine has been recommended as a prophylactic drug by Western armies under threat of chemical weapons. The finding that pyridostigmine acts centrally in severely stressed individuals suggests that (a) peripherally acting drugs administered under stress may reach the brain and (b) we must analyze the blood–brain barrier from a perspective different to the classic view, which states that it is a very effective barrier and protects the brain from undesired molecules. Increased awareness of this important phenomenon is essential, not only for drug design but also to gain an understanding of the neurobiological consequences of stress.

RESPONSE SYSTEMS IN THE BRAIN

The CRF Response System as an Example

Increased susceptibility to various disorders may partially be based on alterations of the corticotrophin-releasing factor (CRF) system within the central nervous system (CNS). This may particularly be true for depression and anxiety disorders, which occur in relation to stress.

Immunohistochemical and radioimmunoassay mapping studies revealed a heterogeneous distribution of CRF neurons throughout the CNS (Owens and Nemeroff 1991). High densities of CRF-containing cell bodies were found in the parvocellular region of the hypothalamic paraventricular nucleus (PVN), and these neurons mainly form the central component of the neuroendocrine stress response system. Outside the hypothalamus, CRF-containing neurons were found in the cortex and the limbic system. CRF neurons in the central nucleus of the amygdala send direct and indirect projections — via the bed nucleus of the stria terminalis (where there are also CRF neurons) — to the parvocellular division of the PVN and to brain stem regions associated with autonomic regulation (Swanson 1983). A similarly widespread distribution of two types of CRF receptors has been found (Behan 1996). These findings suggest that CRF acts not only as a neuroendocrine-releasing hormone but also as a neurotransmitter in integrating the endocrine, autonomic, immune, and behavioral responses to stress, and that persistent hyperactivity of one or more CNS CRF systems, as a result of stress, may be implied in depression or anxiety (Koob 1999).

In preclinical studies, it has been shown that administration of CRF directly into the CNS of laboratory animals produces many physiological and behavioral changes reminiscent of stress, depression, and anxiety. Behavioral, autonomic, endocrine, and immune responses to stress have been shown to be blocked by central administration of CRF antagonists (Koob 1999). Evidence from studies using antisense oligonucleotides against CRF_1 and CRF_2 receptors suggest that the anxiogenic effects of CRF are mediated via binding to the CRF_1 receptor, whereas CRF_2 receptors, which bind with higher affinity to urocortin than to CRF, may mediate stress coping (Liebsch et al. 1999). Accordingly, transgenic mice lacking the

CRF_1 receptor are less fearful and show decreased pituitary-adrenal stress responses (Timpl et al. 1998), whereas transgenic mice lacking the CRF_2 receptor display increased anxiety-like behavior and stress sensitization (e.g., Bale et al. 2000). Evidence from preclinical studies also suggests that acute and chronic stress increases CRF concentrations in various brain regions. Moreover, early developmental stress, for example, maternal separation, induces persistent hyper(re)activity of CNS CRF systems in rodents and nonhuman primates (see Plotsky, this volume, and Kaufman et al. 2000, for reviews).

Evidence from clinical studies suggests increased CNS CRF activity in patients with major depression and patients with anxiety disorders, including PTSD, as reflected for example by increased cerebrospinal fluid (CSF) CRF concentrations (Nemeroff 1984). It should be noted that not all studies in depression found increased CSF CRF concentrations, although the negative studies most likely included cases of atypical or nonmelancholic depression (Wong 2000). It should also be noted that while these disorders are similar with respect to central alterations, there are marked differences with respect to adrenal activity in depression and PTSD. As opposed to depression, PTSD has been associated with relative hypocortisolemia in most studies, and the mechanisms may, at least in part, involve alterations of HPA feedback sensitivity or changes at the adrenal level (Buijs 1999).

There is also evidence for an association between alterations of the CNS CRF system and increased vulnerability to arthritis. Sternberg et al. (1989a, b) demonstrated that the difference between Lewis rats, who are highly susceptible to develop arthritis in response to a streptococcal cell wall stimulus, and Fischer rats, who are not susceptible to develop the disease, is a lack of CRF mRNA expression and corticosterone response to stress. Pharmacological doses of dexamethasone completely abolish the development of arthritis in Lewis rats and lower doses reduce its severity, suggesting that the lack of corticosterone is indeed causally involved in the development of arthritis in this rat strain.

Taken together, the relationships between stress and depression or anxiety, but some physical disorders as well, may be mediated, at least in part, by changes in CRF neural circuits. Importantly, other neuropeptide/neurotransmitter systems known to modulate CRF activity may be involved, and the role of interactions between these systems needs to be elucidated in the pathophysiology of clinical disorders. Similarities of changes in several neurobiological response systems may represent one possible pathway in the expression of comorbidities among several stress-related disorders.

Therefore, CRF appears to be a key integrator of stress responses in the brain. However, other neurotransmitters may modulate the course and outcome of CRF-induced behavioral, neuroendocrine, and immunological alterations. Such a concept of stress-induced alterations of the peptide balance — not exclusively focusing on CRF alone — would have the advantage of explaining specificity in the stress response and in stress-related disease (Henry 1986; Henry and Stephens 1977). It has been demonstrated that endogenous opioids (EOP), neuropeptide Y (NPY), and oxytocin (OT) specifically exert anti-CRF-like effects, such as reduced sympathetic outflow, metabolism, feeding, and anxiolysis. It is therefore evident that a decrease of endogenous anti-CRF-like acting peptides results in similar alterations as seen after CRF infusion or as found in, e.g., anxiety-related disorders. The infusion of, for example, NPY Y1 receptor antagonists is anxiogenic. It should be kept in mind that classical neurotransmitter systems, such as serotonergic, GABAergic, and noradrenergic systems, all contribute to an internal balance within the brain. However, neuropeptides have a high specificity of action. Thus, a detailed knowledge of specific alterations of the internal neuropeptide balance in response to stress would have the advantage that it could result in a specific explanation of what stress does in the brain, which specific disease susceptibility may be increased, and which treatment strategies are most effective for the individual. Thus, several other central mechanisms may account for a certain phenotype and a specific disease susceptibility.

Drug-induced Stress Reduction

Male tree shrews exposed to social conflict and treated with an antidepressant over a clinically relevant period of three weeks showed normal behavior (locomotor activity, marking behavior) as well as HPA axis activity (urinary cortisol). At the end of the experiment, brains were analyzed for various types of binding sites (α_2-adrenergic, β_1- and β_2-adrenergic, serotonergic ($5HT_{1A}$), CRF). Despite the fact that the readout (behavior, endocrine system) had returned to the prestress range, the receptor characteristics were very different as compared to control animals. This finding indicates that, within the brain, drug-induced (plastic) alterations occur that result in a new equilibrium within the neuronal network, although behavior and endocrine systems returned to normal. A more complete account of these mechanisms can be found in McEwen (this volume). With regard to the pathological consequences, a question remains: How many times can the brain undergo such reorganizations without significant impact on the readout (behavior, autonomic endocrine system)?

FACTORS OF VARIABILITY

Schematic Appraisal

Variability reveals itself in many forms ranging, for example, from changes in gene expression, to changes in neurotransmitter release, in immune system, and in behavior. It became clear from our discussion that neuronal plasticity is both a prerequisite for adequate functioning of an individual in the continuously changing environment and one of the bases of variability. This is demonstrated by findings that the body's stress response involves multidimensional modifications of brain circuits and the activation/inhibition of autonomic, endocrine, and immune components. Furthermore, the links between brain and periphery (mind–brain–body) provide the basis for the adaptive as well as maladaptive responses of an organism to challenge and the basis for variability. Based on present knowledge and the discussion in the different groups, we offer the following model (Figure 21.2).

The three different levels/structures of the CNS (cortical, limbic, midbrain, and brainstem areas), which represents main evolutionary levels in the interaction with the environment, are interconnected. Within the three areas, behavioral capacities ranging from reflex to appraisal are located and are interacting with the periphery. This model may also help us understand the pathological consequences of the stress responses. Violation of the central interpretation (discrepancy between reality and goal) on the different levels may result in a pathological outcome.

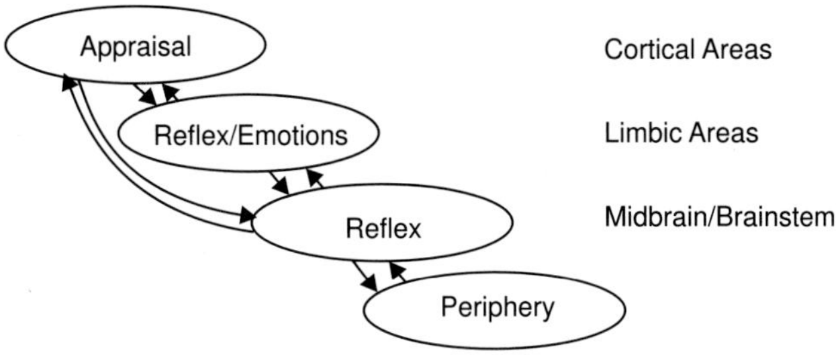

Figure 21.2 Schematic relationship between different levels of the nervous system, central processing and the periphery.

Some Factors of Variability

From the previous discussion about the complex relationships between environmental inputs and pathology, it is clear that there is no direct link between the stimulus and the appearance of the disease. At every level — perception and integration of the signals from the environment, activation of the neuronal circuits leading to behavioral and biological responses, and eventually to pathological outcomes — a whole range of individual differences can be found. Numerous genetic differences have been demonstrated, including sex factors; early experiences have a profound and durable effect on stress reactivity and can sensitize the individual to further insults; age must also be considered as a major intervening factor; more general factors, such as the socioeconomic status in humans as well as dietary habits, are of primary importance in the pathological consequences of life events. Furthermore, the "natural" history of the disease, including individual factors of vulnerability, must also be taken into account when studying the influence of stressors.

The group discussed general strategies for future research on pathology. One strategy proposed was to investigate why different individuals respond quite differently to the same challenges with respect to pathology. The other approach discussed referred to the complexity and heterogeneity of pathogenetic mechanisms in individuals with similar pathology. As an example, rather characteristic behavioral patterns (type A behavior of humans) have been considered relevant in the development of essential hypertension. However, such behavioral coping strategies become clinically relevant when they are associated with specific biological mechanisms (e.g., catecholaminergic and autonomic systems), especially if they meet the organism in a vulnerable status (obesity, vascular premorbidity) (Schmidt et al. 1986). Such complex patterns may strongly differ between individuals with essential hypertension, thus constituting a broad heterogeneity of pathological mechanisms and eventually resulting in similar pathology. Whatever the research strategy adopted, the identification of these factors of variability/vulnerability to disease in clinical studies and their integration in animal models is probably a major challenge for future research in the field.

Gender as a modulating factor of the stress response is taken as an example to illustrate this point. Gender-specific modulations of the mechanisms that lead to the stress response have been described for various central nervous circuits (see Fuchs et al., this volume). Gender differences or effects of gonadal hormones have been reported for anatomical and functional characteristics of several neurotransmitter and neuromodulatory systems (Patchev and Almeida 1998; Viau et al. 1999). On the structural level, in the hippocampi of chronically stressed female rats — unlike those of stressed male rats — no significant reduction of apical dendrites of CA3 pyramidal neurons was observed. Female rats, however, manifest a decrease in basal dendritic branching, indicating another structural and functional stress-induced sex difference in the rat brain (Galea et al. 1997). Until now, the influence of gonadal hormones on the characteristics of the response to stress has only been partially elucidated. Nevertheless, they seem to modify the efficacy of stressful inputs and glucocorticoids in the brain, either by acting directly as transcription factors or by acting transsynaptically.

The clear gender differences in neurobiological substrates of the stress response and the pronounced differences in the reaction of the HPA axis to various challenges suggests the need for more research on stress effects in females. Only a very small number of studies has utilized female animals in preclinical research of the psychopharmacology of depression. This is important since differences in the bioavailability of drugs between women and men have been described. Factors such as drug absorption and metabolism, percentage of body fat, renal clearance, and a variety of other physiological mechanisms are gender specific (Weissman and Olfson 1995). Furthermore, it is worth noting that sex factors interact with the rest of the genome, so that the magnitude of their influence may be very variable from one strain to the other (e.g., the sensitizing effect of cocaine in the Roman rat strains; Haney et al. 1994).

Decline and deterioration in bodily functions associated with a decreased ability to maintain homeostasis are hallmarks of aging (Everitt and Meites 1989). Thus, the vulnerability of the organism to stress-induced impairments will increase with age. The rate of decline and vulnerability varies between individuals. Part of this individual variability may be due to differences in the stress levels experienced during life. Aging is commonly associated with dysregulation of the HPA axis and, consequently, increased glucocorticoid levels are thought to play a crucial role in brain aging and the decline in cognitive functions (for references, see McEwen et al. 1999; Lupien et al., 1998). From studies mainly in rats and humans, there is evidence that plasma glucocorticoid concentration increases with age in rats and humans, but this does not appear to occur invariably (van Eekelen et al. 1995; van Kampen and Fuchs 1998).

FUTURE RESEARCH

Interventions

Alternative Pharmacological Approaches for Intervention in Stress-related Diseases

The possibility of intervention by pharmacological treatments modifying stress responsiveness in several different pathologies other than anxiety-related diseases is certainly an outcome from the field of stress research.

An example is the susceptibility to periodontal disease (periodontitis, gum disease). In humans, periodontal disease is associated with genetics, aging, diabetes, heavy smoking, and major depression. On the one hand, these associations have a variety of underlying and causal mechanisms. On the other hand, all of these predisposing factors have in common that the reactivity of the HPA axis and possibly other stress responsive systems (e.g., the sympatho-adrenomedullary axis activity) is increased. Rats with a genetically determined high neuroendocrine responsiveness to stress, i.e., the Fischer 344 (F344) strain, develop significantly more periodontal disease when compared to the low responding but histocompatible Lewis strain (Breivik et al. 2000). Surprisingly, postnatal experiences such as stimulation (handling) or maternal deprivation were found to modulate the disease susceptibility in adulthood of Lewis but not in F344 rats, although both strains showed specific alterations in the response to novelty (open field) and in anxiety on the elevated plus maze. Handled Lewis rats develop less periodontal disease, while maternal deprivation of Lewis pups leads to a more severe course of periodontal disease in comparison with undisturbed controls. Thus, genetic factors, postnatal experiences, and anxiety in adulthood specifically interact in the modulation of the susceptibility to an inflammatory disease. Prophylactic treatment strategies, such as additional tactile stimulation of the dorsum (backside) of pups or treatment using antidepressants in adulthood (imipramine via drinking water), were both successful in protecting against the maternal deprivation-induced aggravation of the disease. Underlying the behavioral and clinical outcome are probably changes in the characteristics and functions of multiple brain areas and neurocircuits, including the CRF, NE, and GABA systems (see Plotsky et al., this volume).

Future research should prove whether the approach of using "stress responsiveness modulating" drugs in the treatment of inflammatory disease is also promising in other stress-related diseases, for example, autoimmune and cardiovascular diseases. The exact mechanisms activated by the protective-like acting pharmacological treatments should be elucidated and should not focus exclusively on the CRF system.

Behavioral Interventions

Behavioral interventions are readily designed to target at least two levels of central organization of stress responses: cognition and emotion. For example, cognitive behavioral interventions are being developed for a variety of medical disorders which have the possibility to alter appraisal of stress and behavior in response to illness. Thus far, controlled trials suggest that cognitive behavioral approaches might impact immunological responses to the human immunodeficiency virus or following vaccination with influenza. Other behavioral interventions, which have relied on group psychotherapy, have provided some of the first evidence that outcomes in breast cancer and malignant melanoma patients can be affected. Finally, interventions that target the level of emotional arousal are showing promise for future research. Writing about a stress event is thought to decrease levels of emotional arousal, and such interventions are reported to result in decreases in the severity of rheumatoid arthritis and in increases in immunological responses to hepatitis B vaccination. Other interventions that decrease levels of psychological and physiological arousal, such as mindfulness mediation, have been reported to initiate improvements in hypertension and in dermatological conditions such as psoriasis.

Prophylaxis versus Interventions

Animals including humans can suffer from a multitude of conditions which may be due to inadequate physical as well as social surroundings. If suffering is evident (indicated by behavioral, physiological, or pathological measures), the situation must be changed to reduce these stress responses and reach a state of better welfare.

A clear distinction must be made as to whether these indications of poorer welfare or even pathophysiological and pathopsychological indications are only evident in certain individuals (as is usually the case in humans) or in (nearly) all individuals (as is often the case in animals kept in farms, zoos, or laboratories):

- In the first case, the only way is often the attempt to improve welfare by behavioral or pharmacological intervention (see above). Although pharmacological treatment, in particular, may improve welfare of nonhuman mammals under otherwise stressful conditions, this is not an appropriate way to handle welfare problems in animals: it would require animals to exist under highly stressful conditions, which is questionable for ethical reasons.
- A further perspective is offered by the study of genetic factors responsible for extremes of individual reactivity to environmental factors and their link to pathological outcomes. The consideration of such parameters in selection programs should improve the welfare of farm animals and reduce adverse economic consequences of stress.
- The only adequate intervention in the latter case is to change the housing conditions in such a way that these symptoms of inadequate housing diminish. This may be difficult since it may, and often does, necessitate changes in the social surroundings, for example, housing of highly social animals in adequate groups.

Sometimes behavioral indications of poor welfare (such as many stereotypies observed in nonhuman animals) are not supported by physiological indications of poor welfare; physiological markers may even indicate good welfare (e.g., increased opioid levels). In these situations, deviant behavioral patterns may indicate stressful situations during the past, which were (possibly) compensated by these stereotypies. In these cases, one has to discover which situations were responsible for the development of these behavioral stereotypies. More generally, although clinical and experimental studies have pointed out the strong and long-lasting influences of environmental factors during pregnancy and early postnatal life in the handling and consequences of stress, little is known about these factors in farm animals. Such knowledge would have an immediate application for shaping the emotional reactivity of the animals through controlled environmental influences during development.

In conclusion, to improve the welfare of animals kept in farms, laboratories, and zoos, the surroundings must be planned in such a way as to fulfil the physical (e.g., temperature, food), behavioral (e.g., locomotion, exploration), and social (e.g., mutual grooming and resting) needs of the individuals. This requires detailed knowledge of the behavioral and physiological adaptations of the species during evolution or human selection, and of the ontogeny of the individuals.

Markers: New Directions

Although sequential measurement of markers with time after stress may be informative, this may not be a sufficiently sensitive method to use in the delineation of the mechanisms underlying stress-induced pathologies. Far more specific information can be gained by testing the reactivity of the various output systems in standardized challenge tests (von Holst, this volume and 1998). Chronically stressed animals may behave normally when undisturbed but are likely to reveal abnormal freezing behavior when examined in the open field. Similarly, autonomic indices, such as heart rate or blood pressure, which may be in the normal range when the animal is undisturbed, may show distinctly abnormal responses to a controlled provocative challenge. The HPA axis is reliably hyperresponsive to a novel stimulus in mildly chronically stressed animals, although basal values may be in the normal range. Under some conditions, the HPA axis response to novel stress may be markedly blunted, an unexpected result whose mechanism needs delineation. The same under- or over-responsiveness can be revealed in the apparently normal immune system of the chronically stressed.

The altered responses in all systems to acute challenge during and after chronic stress provide a route that will expose the central mechanisms underlying adaptations or maladaptations to chronic stress. It may also eventually reveal the pathological mechanisms.

In addition to the usual markers for stress, new tools are available and should be pursued. In the brain and immune system, particularly, there are alterations in the expression of receptors. Their numbers may increase or decrease, and these changes may, together with the measurement of ligands, explain more fully the changes that occur in the organism headed toward stress-induced pathology. In this respect, it will be even more important to measure the reactivity of a given organ system to stress hormones, such as receptor sensitivity and receptor signaling cascades, than the actual level of the corresponding hormone. It has been shown that hormone or neuropeptide receptors can become supersensitized, especially in models of chronic pain, chronic inflammation, or chronic stress, where G protein coupled receptors can acquire this state (Lombardi 1999). On the other hand, as a result of chronic stress, neuroendocrine receptors can become less sensitive to stimulation by hormone or neurotransmitters as a result of desensitization or downregulation of the receptors by various intracellular mechanisms (Kavelaars 2000).

Finally, changes that occur with chronic stress are dynamic, and the duration of investigation is important. To date, most animal studies of stress explore the phases of initiation, whereas most clinical and ethological studies examine established pathologies. In experimental models of stresses leading to pathology, the longitudinal progression, together with underlying mechanisms, should be determined to demonstrate causality.

Animal Models for Understanding and Treating Stress-related Pathology

Stress leading to poorer welfare and eventually to pathology will often first manifest itself by a change in the balance of various physiological markers. Stress will often lead or contribute to pathology when the intensity and duration increase. However, another important factor is the timing of the stressor, with a special emphasis on the period of early development.

Animal models represent experimental preparations developed in one species for the purpose of studying phenomena occurring in another species. In the case of animal models of human psychopathology, one seeks to develop syndromes in animals that resemble those in humans in certain ways in order to study selected aspects of human psychopathology (e.g., Willner 1991).

The complexity of the stress system is also mirrored by the fact that in other animals, as in humans, gender- and age-dependent variations are described among individuals in response to challenges. Rearing and housing conditions play a crucial role in the establishment of individual variability in response to stimuli.

In *perinatal stress models,* the maternal influences on the fetus (e.g., induced by antigens (infections), nutrients, neuroendocrine mediation via prenatal stress, or medication such as high doses of glucocorticoids) are investigated.

Postnatal stress models investigating the stress input via the brain to the body can be subdivided into two categories:

1. Conditions of *understimulation*: unstructured living conditions, social isolation in group living or socially bonded animals.
2. Conditions of *overstimulation*: light, noise, conditions of long-lasting social disturbances (fighting, overcrowding).

A second group of *postnatal stress models* investigates the stress input via the body to the brain. This group can be subdivided in three categories:

1. *Antigenic stimulation* of the immune system due to viral, microbial, or allergic responses to external components, such as food, leading to a cytokine or chemokine production that can trigger the brain via afferent inputs (e.g., vagal stimulation).
2. *Mitogenic stimulation* via bacterial polysaccharides.
3. *Autoantigen stimulation* via activation of cross-reactive antigens in the body leading to autoimmune (anti-self) diseases, such as arthritis and experimental allergic encephalomyelitis (EAE) which is a model for multiple sclerosis in humans.

Depending on the specific question one wishes to answer with respect to stress-induced pathology, the genetic background of the animals can be of particular interest. In rodents like rats and most certainly mice, one can choose among several genotypes, allowing the study of the interaction between individual and environmental variables:

- wild type animals,
- genetically selected lines of animals,
- genetically modified animals (transgenic or knockout animals),
- genetically modified animals (transgenic or knockout animals restricted to genetic manipulation of the gene in a specific organ/tissue or cell type by the use of organ/tissue or cell type specific promotor sequences).

Models are also available to study the influence of stress on the chronicity of the disease. By now, immunologists have a reasonable insight into the processes involved in the induction of, for example, autoimmune diseases in animals. The autoimmune disease is usually induced by immunization of the animal with the auto- or cross-reactive antigen in the presence of a highly immunogenic adjuvant. The autoimmune disease in animal models is usually of a transient nature, dissimilar to what we see in humans. In humans, we still do not know why a given autoimmune disease process remains chronic, which can also be explained as a lack of feedback control of the immune system.

It would be of great importance to investigate whether stress can promote chronicity of the disease and, conversely, whether psychotherapy can help promote the remission of these diseases. For this purpose, the chronic relapsing/remitting form of EAE induced by the autoantigen myelo-oligodendrocyte glycoprotein can be used in rats and mice.

CONCLUSIONS

Pathological consequences of stress occur in many species and under a variety of environmental conditions. If and what kind of pathology develops depends very much upon environmental and respective psychological and biological response systems. Furthermore, these response systems interact in a complex way with each other, including a permanent multidirectional communication between the brain and the body. Mental and physical pathology may develop as a consequence of the duration or intensity of such stress responses. Future research should focus primarily on a better assessment of response systems and factors constituting response variability, to improve diagnostic, preventive, and therapeutic interventions of stress pathology.

REFERENCES

Ahima, R.S., and C.B. Saper, et al. 2000. Leptin regulation of neuroendocrine systems. *Frontiers Neuroendocrin.* **21**:263–307.

Bale, T.L., A. Contarino, G.W. Smith, R. Chan, L.H. Gold, P.E. Sawchenko, G.F. Koob, W.W. Vale, and K.F. Lee. 2000. Mice deficient for corticotropin-releasing hormone receptor-2 display anxiety-like behaviour and are hypersensitive to stress. *Nat. Genet.* **24**:410–414.

Behan, D.P., D.E. Grigoriadis, et al. 1996. Neurobiology of corticotropin releasing factor (CRF) receptors and CRF-binding protein: Implications for the treatment of CNS disorders. *Mol. Psychiatry* **1**:265–277.

Breivik, T., P.S. Thrane, P. Gjermo, P.K. Opstad, R. Pabst, and S. von Hörsten. 2001. Hypothalamic-pituitary-adrenal (HPA) axis activation by experimental periodontal disease in rats. *J. Periodont. Res.* **5**:301–308.

Buijs, R.M., J. Wortel, et al. 1999. Anatomical and functional demonstration of a multisynaptic suprachiasmatic nucleus adrenal (cortex) pathway. *Eur. J. Neurosci.* **11**:1535–1544.

Dallman, M.F., A.M. Strack, S.A. Akana, M.J. Bradbury, E.S. Hanson, K.A. Scribner, and M. Smith. 1993. Feast and famine: Critical role of glucocorticoids with insulin in daily energy flow. *Frontiers Neuroendocrin.* **14**:303–347.

Everitt, A., and J. Meites. 1989. Aging and anti-aging of hormones. *J. Gerontol.* **44**:B139–B147.

Feldman, A., D. Kaufer, J. Shemer, I. Hendler, H. Soreq, and I. Tur-Kaspa. 1996. Pyridostigmine brain penetration under stress enhances neuronal excitability and induces early immediate transcriptional response. *Nature Med.* **2**:1382–1385.

Galea, L.A.M., B.S. McEwen, P. Tanapat, T. Deak, R.L. Spencer, and F.S. Dhabhar. 1997. Sex differences in dendritic atrophy of CA3 pyramidal neurons in response to chronic restraint stress. *Neuroscience* **81**:689–697.

Garcia-Belenguer, S., C. Oliver, and P. Mormède. 1993. Facilitation and feedback in the hypothalamo-pituitary-adrenal axis during food restriction in rats. *J. Neuroendocrinol.* **55**:663–668.

Haney, M., N. Castanon, M. Cador, M. Le Moal, and P. Mormède. 1994. Cocaine sensitivity in Roman high and low avoidance rats is modulated by sex and gonadal hormone status. *Brain Res.* **645**:179–185.

Hay, M., P. Orgeur, F. Lévy, J. Le Dividich, D. Concordet, R. Nowak, B. Schaal, and P. Mormède. 2001. Neuroendocrine consequences of very early weaning in swine. *Physiol. Behav.* 72:263–269.

Henry, J.P. 1986. Neuroendocrine patterns of emotional response. In: Emotion: Theory, Research, and Experience, ed. R. Plutchik, vol. 3, pp. 37–60. New York: Academic.

Henry, J.P., and P.M. Stephens. 1977. Stress, Health, and the Social Environment. A Sociobiologic Approach to Medicine. New York: Springer.

Kavelaars, A., W. Kuis, L. Knook, G. Sinnema, and C.J. Heijnen. 2000. Disturbed neuroendocrine–immune interactions in chronic fatigue syndrome. *J. Clin. Endocrinol. Metabol.* **85**:692–696.

Kendler, K.S., C.M. Bulik, J. Silberg, J.M. Hettema, J. Myers, and C.A. Prescott. 2000. Childhood sexual abuse and adult psychiatric and substance use disorders in women: An epidemiological and cotwin control analysis. *Arch. Gen. Psychiatry* **57(10)**:953–959.

Koob, G.F. 1999. Corticotropin-releasing factor, norepinephrine and stress. *Biol. Psychiatry* **46**:1167–1180.

Landsberg, L., and J.B. Young 1978. Fasting, feeding and regulation of the sympathetic nervous system. *N. Engl. J. Med.* **298**:1295–1301.

Liebsch, G., R. Landgraf, M. Engelmann, P. Lorscher, and F. Holsboer. 1999. Differential behavioural effects of chronic infusion of CRH-1 and CRH-2 receptor antisense oligonucleotides into the rat brain. *J. Psychiatry Res.* **33**:153–163.

Lombardi, M.S., A. Kavelaars, M. Schedlowski, J.W.J. Bijlsma, K.L. Okihara, M. van de Pol, S. Ochsmann, C. Pawlak, R.E. Schmidt, and C.J. Heijnen. 1999. Decreased expression and activity of G-protein-coupled receptor kinases in peripheral blood mononuclear cells of patients with rheumatoid arthritis. *FASEB J.* **13**:715–725.

Lupien, S.J., M. de Leon, S. de Santi, A. Convit, C. Tarshish, N.P.V. Nair, M. Thakur, B. McEwen, R.L. Hauger, and M.J. Meaney. 1998. Cortisol levels during human aging predict hippocampal atrophy and memory deficits. *Nature Neurosci.* **1**:69–73.

McEwen, B.S. 1998. Stress, adaptation, and disease. Allostasis and allostatic load. *NY Acad. Sci.* **40**:33–44.

McEwen, B.S. M.J. de Leon, S.J. Lupien, and M.J. Meaney. 1999. Corticosteroids, the aging brain and cognition. *Trends Endocrinol. Metab.* **10**:92–96.

Nemeroff, C.B., E. Widerlov, and et al. 1984. Elevated concentrations of CSF corticotropin-releasing factor-like immunoreactivity in depressed patients. *Science* **226**:1342–1344.

Owens, M.J., and C.B. Nemeroff. 1991. Physiology and pharmacology of corticotropin-releasing factor. *Pharmacol. Rev.* **43**:425–473.

Patchev, V.K., and O.F.X. Almeida. 1998. Gender specificity in the neural regulation of the stress response: New leads from classical paradigms. *Mol. Neurobiol.* **16**:63–77.

Schmidt, T.H., T.M. Dembroski, and G. Blümchen, eds. 1986. Biological and Psychological Factors in Cardiovascular Disease. Berlin: Springer.

Sternberg, E.M., J.M. Hill, G.P. Chrousos, T. Kamilaris, S.J. Listwak, P.W. Gold, and R.L. Wilder. 1989a. Inflammatory mediator-induced hypothalamic-pituitary-adrenal axis activation is defective in streptococcal cell wall arthritis-susceptible Lewis rats. *Proc. Natl. Acad. Sci. USA* **86**:2374–2378.

Sternberg, E.M., W.S. Young, R. Bernardini, A.E. Calogero, G.P. Chrousos, P.W. Gold, and R.L. Wilder. 1989b. A central nervous system defect in biosynthesis of corticotropin releasing hormone is associated with susceptibility to streptococcal cell wall induced arthritis in Lewis rats. *Proc. Natl. Acad. Sci. USA* **86**:4771–4775.

Swanson, L.W., P.E. Sawchenko, et al. 1983. Organization of ovine corticotropin-releasing factor immunoreactive cells and fibers in the rat brain: An immunohistochemical study. *Neuroendocrinology* **36**:165–186.

Timpl, P., R. Spanagel, I. Sillaber, A. Kresse, J.M. Reul, G.K. Stalla., V. Blanquet, T. Steckler, F. Holsboer, and W. Wurst. 1998. Impaired stress response and reduced anxiety in mice lacking a functional corticotropin-releasing hormone receptor. *Nat. Genet.* **19**:162–166.

van Eekelen, J.A.M., M.S. Oitzl, and E.R. de Kloet. 1995. Adrenocortical hyporesponsiveness and glucocorticoid feedback resistance in old male brown Norway rats. *J. Gerontol.* **50A**:B83–B89.

van Kampen, M., and E. Fuchs. 1998. Age-related levels of urinary free cortisol in the tree shrew. *Neurobiol. Aging* **19**:363–366.

von Holst, D. 1998. The concept of stress and its relevance for animal behavior. *Adv. Study Behav.* **27**:1–131

Viau, V., A. Chu, L. Soriano, and M.F. Dallman. 1999. Independent and overlapping effects of corticosterone and testosterone on CRF and AVP mRNA expression in the paraventricular nucleus of the hypothalamus and stress-induced ACTH release. *J. Neurosci.* **19**:6684–6693.

Yehuda, R., and A.C. McFarlane. 1995. Conflict between current knowledge about posttraumatic stress disorder and its original conceptual basis. *Am. J. Psychiatry* **152**:1705–1713.

Weissman, M.M., and M. Olfson. 1995. Depression in women: Implications for health care research. *Science* **269**:799–801.

Willner, P. 1991. Behavioural models in psychopharmacology: Theoretical, Industrial and Clinical Perspectives. Cambridge: Cambridge Univ. Press.

Wong, M.L., M.A. Kling, et al. 2000. Pronounced and sustained central hypernoradrenergic function in major depression with melancholic features: Relation to hypercortisolism and corticotropin-releasing hormone. *Proc. Natl. Acad. Sci. USA* **97**:325–330.

Author Index

Blokhuis, H.J. 249–270
Broom, D.M. xv–xvii, 1–9, 249–270
Buwalda, B. 197–209

Carere, C. 197–209
Carter, C.S. 11–30, 79–100
Coe, C.L. 151–168

Dallman, M.F. 301–316, 337–355
Dantzer, R. 101–110, 151–168
Dawkins, M.S. 63–76
de Boer, S.F. 197–209

Flügge, G. 289–300
Forkman, B. 249–270
Fraser, D. 79–100
Fuchs, E. 289–300, 337–355

Gärtner, K. 79–100
Groothuis, A.B.B. 197–209

Heijnen, J.J. 337–355
Heim, C. 337–355
Hellhammer, D.H. 337–355

Irwin, M. 271–287, 337–355

Jensen, P. 123–134, 151–168

Kaiser, S. 249–270
Knierim, U. 79–100
Koolhaas, J.M. 197–209, 249–270
Kramer, M. 289–300

Levine, S. 169–196, 249–270
Lightman, S.L. 151–168
Lutgendorf, S.K. 49–62, 79–100

McEwen, B.S. 229–246
Mendl, M. 211–228, 249–270
Mineka, S. 79–100
Mormède, P. 337–355

Nelson, R.J. 111–122

Panksepp, J. 79–100
Paykel, E.S. 337–355
Plotsky, P.M. 169–196, 249–270
Porges, S.W. 135–149, 151–168

Reul, J.M.H.M. 151–168
Rushen, J. 151–168

Sachser, N. 31–48, 79–100
Sanchez, M.M. 169–196
Schedlowski, M. 249–270
Stefanski, V. 151–168

van der Vegt, B.J. 197–209
von Holst, D. 317–335, 337–355
von Hörsten, S. 337–355

Zanella, A.J. 151–168

Subject Index

abuse
- child 144, 253, 340
- infant 183, 184
- substance 25, 187, 231, 250, 275, 277

acetylcholine 16, 118
acetylcholinesterase inhibitor 298, 344
active coping 52, 91, 197, 216, 341
active stress 317–322, 330, 333
adaptation 2, 36, 84, 102, 152, 169–172, 280, 338, 351
addictive behavior 337, 340
adrenaline 116, 117, 202
adrenals 2, 27, 119, 135, 264, 293, 310
- adrenocortical cells 143, 144
- cortex 19, 116, 137, 175
- glands 89, 111, 175
- medulla 137–140, 154, 155, 163, 231, 324

adrenocorticotrophin (ACTH) 12, 15–19, 27, 39, 115–118, 163–165, 175, 203, 290, 322–326, 329, 341
Affect Balance Scale 95
affective disorders 186, 272, 274
affective states 33, 47, 79, 81–85, 96, 105
age 31–33, 42–45, 53, 85, 151, 338, 347
aggression 18, 40, 43, 44, 88, 199, 201, 205, 213
aging 236, 271, 280, 289, 294–298, 348
- Alzheimer's 106, 275
- sleep disturbance 278, 279

MacArthur studies on 59
alcohol use 185, 231, 277, 337, 340
allostasis 229–233, 241, 271
- in depression 279

allostatic load 59, 171, 232, 279, 280
α-adrenergic receptor 118, 179
Alzheimer's 58, 106, 275
amphibians 40, 139, 143
amygdala 18, 115, 119, 186, 204, 213, 342
androgens 15, 17, 21, 90
anhedonia 82, 178, 253
anterior olfactory nucleus (AON) 18
anticipation 64–68, 74, 95, 115, 231
antidepressant treatment 164, 180, 253, 297, 346
anti-stress 12, 87, 145, 146
Antonovsky, A. 51, 95
anxiety 24, 82, 91, 113, 136, 186, 200, 340, 344
- elevated plus maze 349
- posttraumatic syndrome 295, 328, 337, 340–345

apnea 142, 155, 163
area postrema 146
arginine vasopressin (AVP) 11–18, 21–27, 175, 176, 205
arousal 45, 86, 117
- theory 136, 137

arteriosclerosis 221, 233, 276, 330, 339, 342
arthritis 271, 345, 352
- rheumatoid 276–279, 349

artificial selection 37, 39, 89, 197
astrocytes 296, 298
atherosclerosis 221, 233, 276, 342
attachment 13, 17, 22, 24, 92, 174, 180, 181
- loss of 85, 319
- role in development 50

autoimmune disease 241, 272, 298, 349, 352
autonomic nervous system (ANS) 3, 25, 71, 87, 116, 135–138, 162, 171, 172, 179
- evolution of 137, 141
- neural regulation of 154

aversion learning 73, 74, 259
avoidance behavior 68, 69, 112
awareness 3, 7, 136

baboons 214, 219, 223, 318, 330
- hamadryas 36

beavers 36
bed nucleus of the stria terminalis 17, 116, 156, 175, 179, 204, 344

behavior 26, 36, 42, 44, 91, 95, 124, 125, 136, 155, 160, 161, 170, 201, 347
addictive 337, 340
avoidance 68, 69, 112
courtship 38, 41
fight–flight 155, 199
individual differentiation in 207
maternal 177, 178, 180, 186
mineralocorticoid receptors in 157
phylogenetic origin of 137
reconciliation 222
social 12, 19, 33–45, 87, 216, 221, 225, 341
stereotypic 44, 162, 339
submissive 41, 215, 216
behavioral plasticity 197, 201
benzodiazepines 119
bereavement 57, 65, 105, 113
β-adrenergic receptor 139, 277, 278
β-endorphin 74, 106
biorhythms 31, 43
birds 11, 20, 39, 40, 66, 69, 92, 140, 204, 307
chaffinch 33, 34
migrating 43
birth 20, 26, 176, 255
defects 174, 180
blood–brain barrier 26, 106, 289, 296, 298, 343, 344
body–brain interactions 342, 343
body mass, loss of 331
impact from repeated restraint 307
stress-induced 327, 343
body temperature 308
body weight 184, 301–314, 337
induced change in 309
loss of 67, 238, 240
obesity 255, 312–314, 347
bonding 15, 16, 21, 39, 83
maternal–infant 26
pair 23, 25
partners 39, 44–46
social bonds 12, 15, 21, 24, 39, 104, 319–323, 330, 333
boredom 1, 72, 73, 160, 337
Bowlby, J. 50, 174, 181
brain 72, 86, 95, 106, 111, 112, 338, 346
effect of steroid hormones on 342
neuronal plasticity 171–173, 185, 186, 292
relationship to immune system 101, 108
response to inflammation 103
brain stem 18, 108, 117, 137, 139, 140, 143, 155
breast cancer 52, 57, 58, 105, 349
breast-feeding 26, 27
breeding 20, 38, 94, 120, 181, 205, 263, 323
inbreeding 23, 197
selective 79, 92, 205

cancer 46, 56, 94, 340
breast 52, 57, 58, 105, 349
canids 20, 21
dogs 20, 33, 39, 43
captivity 36, 37, 40, 89, 123, 212, 220–224
cardiovascular disease 6, 85, 214, 255, 275, 277, 280, 317, 333, 339, 340
risk of 276, 341
catecholamines 11, 12, 27, 117, 171, 303, 304, 341, 342
epinephrine (EPI) 19, 39, 111, 116–118, 140, 144, 166, 325, 341, 342
norepinephrine (NE) 19, 39, 40, 86, 117, 118, 137, 144, 165, 177, 179, 304
cats 39, 91, 341
cellular immunity 103–105, 265, 277, 278
cheetahs 318, 332
child abuse 144, 253, 340
infant 183, 184
cholecystokinin 87, 116, 255
chromaffin tissue 138–140, 143, 144
chronic disease 271, 274, 277
chronic obstructive lung disease 271, 276
coherence 51, 52, 57–59
Sense of Coherence Scale 95
Committee on Pain and Distress in Laboratory Animals 303
communication 137–142, 146, 155
conceptual motivational model 128, 129
conservation–withdrawal response 199
contentment 4, 73, 81,82, 95
control 1–4, 7, 32, 35, 53, 57, 63, 85, 114, 159
coping
active 52, 91, 197, 216, 341
defined 63, 64, 131, 198
passive 215, 216
problems defining 126, 127
strategies 91, 126, 133, 199, 200, 207, 211
neuroendocrinology of 202–204
systems 1–3, 6, 37, 40
genetic variability in 8
physiological 169, 170
coronary atherosclerosis 221, 233, 276, 342

corticosterone 15–19, 23, 118, 119, 143, 156, 163, 175, 202, 238, 257, 290, 298, 307–313, 324–329, 345
corticotrophin-releasing factor (CRF) 87, 154, 165, 175, 179, 302–307, 310–314
  receptors 164, 344
  system 156, 345
corticotrophin-releasing hormone (CRH) 12, 19, 27, 80, 87, 107, 111, 115–120, 251, 281, 342
  receptors 117, 156, 164, 304, 344
cortisol 18, 36, 39–41, 45, 46, 136, 137, 144, 147, 156, 163, 165, 215, 341, 342
  response to separation 181–183
courtship behavior 38, 41
cow 2, 33, 256
cytokines 103, 107, 111, 119, 152, 153, 278, 341

death 35, 40, 64–67, 263, 302, 331, 339
  feigning 138, 163
dehydration 65–67, 85, 116
dementia
  Alzheimer's 106, 275
  HIV-induced 106
dendrites 236–241, 295, 297, 348
depression 6, 49, 82, 85, 91, 105, 106, 113, 271, 274–277, 340, 344, 345
  animal models of 253, 272
  genetic predisposition to 272
  HPA activity in 163, 280
  impact of sleep on 278
  major 273
  in male infertility 341
  relationship to stress 271–274, 340, 345
dexamethasone 164, 345
diabetes 178, 253, 255, 303, 314
distress 82, 126, 301–306, 310, 311, 314
dogs 20, 33, 39, 43
domestication 14, 15, 20, 31, 32, 37–39, 79, 89, 123–125, 132
dominance 38–42, 211, 339
dopamine 19, 25, 119, 162, 182, 205
dorsal horn of the spinal cord 18, 115
dust-bathing 127, 161
dysthymia 273

early experiences 40, 79, 249, 253, 340, 347
  adverse 172, 177–179
  deprivation 90
ecological niche 31, 35, 39, 88, 173
elephants 32
elevated plus maze 349
emotions 33, 47, 69, 71–75, 86, 103–105, 112, 113, 289, 319, 344, 349
  defined 3
  effect on immune system 102, 104
  evolution of 68
  facial expression 343
  physiological measures of 72
  regulation of 158
encephalomyelitis 352
endocrine stress responses 41, 42, 44
endogenous opioids 345
energy 2, 51, 65, 71, 111–117, 124, 132, 156, 174, 232, 263–266
  balance 160, 301, 303–311
  negative balance 85, 302, 343
environmental challenges 1, 35, 46, 90, 123, 141, 165, 170, 199, 231, 249, 260, 289
environmental demands 51, 111, 113, 120, 198, 338
environmental enrichment 44, 45, 75, 186
environmental niche 88, 173
environmental stimuli 31, 32, 44, 171–173, 177, 201
epinephrine 19, 39, 111, 116–118, 140, 144, 166, 325, 341, 342
escape 7, 33, 66, 87, 112, 157, 202, 222–226, 259, 319
ethics 47, 68, 350
European wild rabbits 318, 326–329, 333
eustress 50, 126
evolution 11, 14, 21, 32–35, 64–69, 123, 124, 132–136, 141, 318
  of autnomic nervous system 137, 141
  of emotions 68
  impact on coping strategies 198
  regressive 89
evolutionary adaptations 112, 124, 125, 129, 132
exercise 27, 58, 72, 114, 186, 264
expectations 31–42, 54, 56, 170, 320
exploration 84
external causal factors 53, 123–131, 133
exteroceptive sensory system 156, 175
extreme traits 92

FACES scale 94
farm animals 65, 75, 89, 123, 133, 161, 197, 202, 257, 266, 337, 339, 350

fear 16, 52, 71, 73, 82, 87, 113, 136, 158, 177, 183, 200, 238, 252, 328
role of amygdala in 186
Federation of European Laboratory Animal Science Associations 302
feedforward control 2, 35, 210, 217
feelings 4, 12, 36, 47, 68, 73, 85, 281, 345
defined 3
fight–flight behavior 155, 199
Fischer rats 345, 348
fish 40, 43
fitness 36, 64, 65, 67, 84, 124, 125, 173, 205, 212
defined 80
male 20
flight 85, 89, 130, 132, 136, 138, 144, 155, 158, 161, 231, 317
food
availability 39, 65, 67, 89, 120, 174, 197, 207, 255, 263, 331, 339
competition for 212, 213
deprivation 33, 95, 113
intake 117, 301–314, 343
search for 65–67, 90, 112, 125, 132, 151, 184
Freud 50, 51
frustration 1, 81, 105, 114, 334
future research 80, 96, 166, 167, 281, 334, 340, 347–349

gamma aminobutyric acid (GABA) 16, 27, 118, 119, 172, 179, 237, 238, 349
GABAergic system 345
gender 28, 90, 338, 340
differences 14, 289, 297, 347
genetically modified animals 22, 344, 352
genetic disorders 92
genetic factors 347
genetic predisposition 31, 32, 35, 36, 43, 125, 338, 340
genetic selection 199, 205
gestation 20
gibbons 36
glial cells 103
glia–neuron interaction 289, 292–296
glucocorticoid receptor 156, 157, 175, 179, 294
glucocorticoids 12, 18, 19, 71, 90, 106, 111, 112, 119, 120, 153, 156, 163–165, 223, 289–297, 341, 342
cascade concept 294
immunosuppressive effects of 102
gonads 119
good life 32, 44, 46
good welfare 31–49, 133
characteristics of 55
defined 80, 82
indicators of 94
role of affective states in 81
grooming 84
growth 112
guinea pigs 32, 36–46
Gulf War 344
gut defense system 152

hamster 36
handling 39
happiness 51, 53, 54
hardiness 51, 159
health 33, 42, 64, 274, 275
predictors of 279, 280
heart
coronary atherosclerosis 221, 233, 276, 342
myocardial infarction 339
rate 71, 72, 171
regulation of 138–141
heritability 91
hierarchical model of human needs 49, 50
hippocampus 115, 119, 172, 289, 293
HIV-induced dementia 106
homeostasis 19, 112, 136, 147, 337
homeostatic model 32, 34, 128
hormones 11, 15–19, 36, 70, 111, 119
corticotrophin-releasing (CRH) 12, 19, 27, 80, 87, 107, 111, 115–120, 251, 342
luteinizing-releasing (LH–RH) 87
peptide 90
response to separation 182
steroid 119, 342
human motivation 59
human pathology 340
husbandry 214
hyperpolarization 119
hypertension 279
in African Americans 277
hypothalamic-hypophyseal blood portal system 116
hypothalamic paraventricular nucleus (PVN) 17, 115, 344
hypothalamic-pituitary-adrenal (HPA) system 85, 111, 112, 125, 135, 211, 290, 341

axis 49, 56, 59, 117–120, 137, 153, 156, 163, 164, 174–177, 202, 203, 303
reactivity 24
hypothalamus 87, 115
hypothalamus-pituitary-thyroid (HPT) axis 117

immobilization 138, 139, 142–146, 163, 303
immune system 33, 50, 56, 101, 102, 106, 135–138, 143, 145, 152, 153, 165, 215, 339
effect of positive states on 56, 57
immunosuppression 108, 120, 320, 327, 341
relation to coping 102–105
response to social factors 331, 332, 325, 326
response to stress 305, 321, 330, 331, 340, 341, 344
immunity 49, 56, 57, 101–108, 278
cellular immunity 103–105, 265, 277, 278
natural immunity 277
imprinting-like processes 40
inbreeding 23, 197
individual differences 43, 90, 104, 105, 107, 211, 214, 215
evolutionary ecology 197
infant abuse 183, 184
infertility 115, 139, 341
instrumental conditioning 95
insulin 303, 304, 308–314
intensity 31, 32
interferon 102
interleukin-2 278
internal causal factors 123–133
internal determinants of welfare 53
interoceptive sensory system 152, 153, 156, 175
intrapopulation variation 132
introversion 91
ionophore 119
isolation 19, 24, 36, 39, 53, 86, 89, 158, 181–185, 212, 222, 243, 254, 261, 290, 297, 339, 352

Jung 50

Karnofsky scale 55, 94
kidney disease 339
Kobasa 51

lactation 20, 26, 27, 120, 302, 306–308, 314
lateralization 289, 291, 292
learned helplessness syndrome 90
learning 136, 341
ability 89
effect of glucocorticoids on 342
leptin 343
Lewis rats 345, 349
life events 105, 338
life expectancy 329, 333
life orientation test 95
life satisfaction 53, 54
index 95
limbic structures 89, 291
limbic system 293, 344
lion 33
lipopolysaccharide (LPS) 307
locus coeruleus 18, 115, 120, 177, 179, 186, 304
Long Evans hooded rats 178
losers 40
lust 86
luxuries 33
lymphocyte proliferation 102–105
lymphocytes
Th1-Th2 106

magnetic resonance imaging (MRI) 295, 296
magnetic resonance spectroscopy (MRS) 295, 296
maladaptations 172, 296, 351
markers for susceptibility 340
marsupials 318, 321, 339
masculinization 44, 90
Maslow's model of human needs 49, 50
Mason, J.W. 170
maternal behavior 177, 178, 180, 186
effects on offspring 187
in macques 184
maternal–infant bonding 26
maternal separation 177–182, 186, 187, 345
mating 20–22, 35, 36, 40
courtship behavior 38, 41
partner preference 22
measures of variability 95
median eminence 116, 118
memory 3, 51, 111, 157, 169, 185, 225, 251, 272, 289, 341
effect of epinephrine on 342
mental health 28, 43, 272, 273, 340
mesotocin 17, 87
metabolic balance 343

mice 42–45, 339
 transgenic 22, 344
microglia 103, 106, 296, 298
mineralocorticoid receptor 156, 157, 175, 179
mobilization 138–146, 155, 163
mongooses 21, 318
 Helogale parvula 330
monogamy 20, 21, 36
mortality 277
mother–infant relationship 174–176, 181
motivation 17, 89, 123–132, 160, 161
 defined 127, 128
 in domesticated animals 89
 in relation to coping 129
 theories 50
motor activity 341
muscimol 119
myelinated vagus 138, 145, 155, 162
myocardial infarction 339

natural habitats 37–43, 46
natural killer cell activity 342
natural killer cytotoxicity 103
natural life 31, 35, 36, 46, 85
natural selection 31, 36–39, 43, 88, 111
natural strategy 133
needs 33, 161
 hierarchical model of 49, 50
negative affectivity 91
negative energy balance 301, 302
negative feedback control 2
nest-building 35, 123, 127
 in sows 130, 131
neural circuitry 111–113
neural networks 86
neuroendocrine stress response system 170, 344
neurogenesis 186, 289
 adult 294, 295
neuroimmunomodulation 103
neuronal plasticity 171–173, 185, 186, 292, 346
neuron–glia interaction 296
neuropeptides 86, 102, 137, 345
 pathways 291
 pro-opiomelanocortin 103
neurotransmitters 111, 291
nitric oxide 153
N-methyl-D-aspartate (NMDA) 172, 186, 236–238
noninvasive monitoring 119
noradrenaline 116, 117, 156, 202, 291
noradrenergic system 117, 177, 343, 345
norepinephrine 19, 39, 40, 86, 117, 118, 137, 144, 165, 177, 179, 304
nucleus ambiguus 138–140, 146, 155, 343
nucleus of the solitary tract 115, 146, 163

obesity 255, 312–314, 347
olfactory-limbic-hypothalamic system 18
ontogeny 11, 31, 32, 33, 39, 42, 43
opioids 11, 12, 16, 18, 23–25, 87
optimism 51, 56, 58
 heritability of 52
overstimulation 44, 337, 338, 352
oxytocin 11–19, 21–27, 87, 116, 135, 137, 145, 146, 155, 345
 development of receptor system 187

pain 3, 35, 68, 74, 82, 127, 275, 301, 310, 351
 Committee on Pain and Distress in Laboratory Animals 303
pair bonding 20–25
panic 27, 91
parabrachial nucleus 115
Paradox of Animal Welfare 67
parasitic disease 318, 321, 328–332, 339
paraventricular nucleus of the hypothalamus (PVN) 16–18, 25, 115–118, 344
parental investment theory 174
parrot 33, 34
passive stress 317, 319, 322, 330, 333
pathology 3, 126, 211, 338–340
peptides 24–26, 90
perinatal stress models 88, 352
periodontal disease 348
personality 90, 126, 217
Philadelphia Geriatric Center Morale Scale 95
physiological stress responses 36, 43
pigs 32, 39, 339
pituitary-adrenal hormonal secretion 171
pituitary-adrenocortical 38, 45
plants 68, 69, 72
plasticity
 behavioral 197, 201
 neuronal 292, 346
 structural 289, 292
pleasure 44, 82, 84
polyvagal theory 137, 138, 141–147, 154, 155, 162

Positive and Negative Affect Scale 95
positive psychology 49, 50, 52
effects on physiology 56
postnatal ontogenetic development 92
postnatal stress models 352
posttraumatic stress disorder 295, 328, 337, 340, 341
poultry 35, 339
prairie voles 14, 15, 21–23
oxytocin in 18, 19
predation 37, 65, 67, 71, 339
predictability 31, 32, 44, 131, 160
lack of 114
pregnancy 44, 120
prenatal maternal stress 180
primates 36, 39, 104, 142, 146, 158, 176, 212, 214, 258
New World 15, 21
nonhuman 176, 180–185, 251, 253, 345
proactive coping strategy 43, 199
prolactin 87, 130, 131
pro-opiomelanocortin 103
prostaglandins 107, 108, 153
psychoneuroimmunology 101–103, 271
psychosocial demands 338
psychostimulants 337, 343
puberty 36, 40–43, 46, 92
pyridostigmine 344

quality of life 55, 56, 80, 83, 152
humans 81
laboratory animals 151
medical patients 55

rage 86
raphe 18
rats 14, 20, 23, 27, 36, 46, 345
oxytocin in 18, 19
reactive coping 43, 154, 164, 165, 199
rearing conditions 41
reconciliation behavior 222
red deer 43
regressive evolution 89
repair mechanisms 67, 69
reproduction 33, 35, 36, 111, 112, 120, 212, 339
reptiles 40
restraint 303
repeated 302, 307, 308
reticular activating system 115
rhesus monkeys 39, 183, 185
rheumatoid arthritis 276–279, 349
Ryff's model of well-being 49, 55, 83

sadness 86
Saimiri sciureus 37
Second Law of Thermodynamics 111
security 40, 42, 45, 92
selective breeding 79, 92, 205
self-soothing behavior 155
Selye, Hans 50, 102, 125, 290
sensitive periods 40, 92
separation 36, 82, 182, 184
cortisol levels in 181
hormonal response to 182
maternal 177–182, 186, 187, 345
septum 115
serotonergic system 345
serotonin 86, 118, 172
set values 32
sex 18, 31–33, 38, 40, 42, 347
differences 90
drive 117
steroids 343
sexual dimorphism in nervous system 25
sheep 73, 74
sickness behavior 101, 107, 302–306, 337
cytokine-induced 307
skin defense system 152, 153
sleep 158, 225, 271, 342
disordered 278, 279, 342
rhythm 84
social attachment 13, 24
social behavior 12, 19, 33, 36–42, 45, 87, 216, 221–225, 341
instability in 339
neurobiological mechanisms for 11, 16
neurological mechanisms for 15
sources of variation 13, 14
social bonds 11, 12, 15, 17, 21, 24, 39, 104, 319–323, 330, 333
social deprivation 184, 185
social experience 11, 15, 19, 42, 88, 143, 217, 218
social integration 317, 330–332, 339
social isolation 181, 222, 339
social rank 320, 325–330, 333, 339
social status 42, 44, 211–213, 330, 333
social stress 44, 303, 317–321, 333
social support 36, 42–46, 53, 56–59, 93, 211
squirrel monkeys 32, 36, 37

starvation 7, 65, 67, 127, 302, 331, 341
  during pregnancy 255
stereotypic behavior 44, 162, 339
steroids 24, 25, 87, 119, 342
stress 5, 7, 25, 31–38, 41–46, 51, 70, 101, 123–127, 136, 152, 186, 211, 213, 271, 343, 346
  chronic 158, 163, 302, 309, 310
  defined 6, 7, 131, 136, 170, 337
  gender variation in 23
  hypo-responsive period (SHRP) 176, 178, 186
  induced immunosuppression 108, 120, 320, 327, 341
  management 57, 281, 337, 338, 350
  passive 317–322, 330, 333
  pathological consequences 274, 275, 278, 337–340
  phylogenetic approach to 146, 147, 154
  predictor of mortality 277
  reactivity 271, 280
  relationship to anxiety 345
  relationship to depression 271–274, 340, 345
  response system 35, 39, 44, 49, 56, 112, 341
  theory of 102
  transactional model of 93
stressors 33, 34, 38, 46, 103, 111, 115, 170, 302
  alteration in energy balance 303–311
  behavioral control over 125
  nonspecificity of 113
  type of 291
structural plasticity 289, 292
submissive behavior 41, 215, 216
substance abuse 25, 187, 231, 250, 275, 277, 340
suckling 306, 307
suffering 35, 68, 73
suprachiasmatic nuclei (SCN) 116, 156, 205
supraoptic nucleus 16, 17
survival 35, 111, 112, 120
sympathetic-adrenal system 33, 38, 45, 49, 85, 136, 211, 290, 339
sympathetic nervous system 125, 137, 303

telencephalon 115
temperament 90, 105
territorial 36
testosterone 213, 341
thalamus 115
Th1-Th2 106
thyroid gland 117
titi monkey 32, 36, 37
transgenic mice 22, 344
tree shrews 346
tumor 102, 103, 302
twin studies 52, 54

uncertainty 44, 85
uncontrollability 103, 104
understimulation 352
unmyelinated vagus 138, 155, 162
unpredictability 44, 85
urocortin 156, 176, 344

vagal brake 141, 144, 155, 163
vagal nerve 107, 108, 153
vaginal cervical stimulation 22
vagus 18, 139, 140, 145
vagus nerve 115, 137, 138, 343
  dorsal motor nucleus of 138
vasopressin 16, 87, 107, 116, 135, 137, 145, 146, 155
vasotocin 87
ventral medial hypothalamus (VMH) 18
ventral subiculum (SUBv) 115
vertebrates 31
vigilance 24, 117
visceral processes 17
vocalizations 95, 158, 184
  distress 181, 182
voles 21, 23, 25
vomeronasal 25
V3 receptors 116

Weiss 125, 126, 130
welfare
  assessment of 3, 4, 66, 68, 72, 74, 75, 84
  of captive animals 214
  defined 3, 4, 12, 65, 81
  indicators of 65, 66, 70
  neurobiology of 11, 12, 14, 15
  of plants 68
  poor 31, 133
well-being 79, 83
  Ryff's model of 49, 55
wild boar 32, 39
wild mammals 317, 319
wild rabbits 339
wolves 20, 39, 42

zebras 33,36
zoo animals 44, 67, 75, 123, 350

Forthcoming Dahlem Workshop Reports

DWR 89 Energy and Mass Transfer in Marine Hydrothermal Systems

P. Halbach, V. Tunnicliffe, and J. Hein, eds.

What happens to the heat released when magma rises from within the Earth to generate new mid-ocean ridge crust — a fundamental component of plate tectonics? How does seawater transport this heat to the oceans? How deep does water penetrate beneath the seafloor; under what range of pressure and temperature conditions does it react; and with which rock types? How do the resulting fluids rise to the surface and what range of mineral deposits do they produce? How do the resultant fluxes of energy — both thermal and chemical — serve to support unique deep-sea hydrothermal ecosystems? Is there a valid link to the origin of life? What impact do hydrothermal fluxes have on the chemical balance of the oceans and what is the fate of the materials they produce? Does a predicted ~100km-wide high-productivity corridor — a larval superhighway — exist along ridge-crests? Can we identify hydrothermal signatures in the geological (including paleontological) record; where do 100 million tonne ore deposits form in the modern day; and have hydrothermal sediments played an important role, throughout Earth history, in the transfer of material from the mantle to the continental crust?

These are just some of the questions highlighted during the 89th Dahlem Workshop. Full discourse of the various subjects will be published in June, 2002, including new multi-branched and interlinking process-flow model — the Dahlem Hydrothermal Reference Model (DaHRM) —which will be taken forward as the new international state-of-the-art for hydrothermal fluxes at mid-ocean ridges, within the broader whole-Earth Geochemical Earth Reference Model (GERM).

Publication date: June, 2002 *in press* dahlem university press

DWR 88 Global Desertification: Do Humans Cause Deserts?

J.F. Reynolds and D.M. Stafford Smith, eds.

Do humans cause deserts? Surprisingly, the answer to this question is contentious. Climate (particularly drought) is obviously a controlling influence, and it is equally certain that humans and their activities have caused desertification in some places. However, a great deal of disagreement exists as to the causes and extent of this land degradation, and consequently about how much of its impact on human well-being is manageable. The resulting arguments create confusion in policies and management programs intended to help many of the world's poorest people, and have had a direct effect on the implementation of the United Nation's Convention to Combat Desertification.

There is a complex of socioeconomic and biophysical causal factors involved in land degradation that have differing levels of influence in different regions of the world at different times, and it links with other issues, such as vulnerability and poverty alleviation, in various ways. Failure of the scientific community to develop a consensus about this complexity has resulted in simplistic interpretations being passed on to practitioners and policy makers. Consequently, attempts to apply the same "solutions" to diverse problems has heightened the sense of confusion and led to a situation where there is not even agreement on the extent of desertification. A first step toward creating such a consensus was recently taken by a group of international researchers from a wide range of disciplines — ecology, atmospheric sciences, economics, social sciences, policy, and integrated assessment. Recognizing the need for novel interdisciplinary approaches to address the pressing global problem of desertification, they explored a new paradigm for a synthetic assessment framework beyond regional and disciplinary concerns, which is presented in this volume. The paradigm explicitly accommodates the various linkages between socio-economic and biophysical factors, as well as the fact that these linkages evolve over time in disparate ways and at different scales. It thus provides the basis for a new approach to assess the extent of desertification and to tailor appropriate solutions to the myriad of problems encompassed by that term.

Publication date: June, 2002 *in press* dahlem university press

Recently published Dahlem Workshop Reports

DWR 86 Rational Decision-making in the Preservation of Cultural Property
N.S. Baer and F. Snickars, eds.

The term *value* is most often used in the context of relative monetary worth. In this context, economists and other social scientists have sought to define the value of such environmental goals as clean air and clean water, leading to concepts of "green values." Yet the term implies more than simple financial valuation, especially when applied to often intangible public goods, as it also considers the relative place such goals hold in the choices made by individuals and society as a whole make. Those charged with the stewardship of cultural property, be it an individual document, an historic landscape or an entire art city on the World Heritage List, are faced daily with decisions based on often conflicting value systems.

It has become obvious that public policy — as determined by such socioeconomic concepts as environmental accounting, cost-benefit analysis, sustainable development, valuation of loss of cultural property, and intergenerational equity — plays a major role in determining what cultural properties will be saved: which books, which documents, which artifacts, and which monuments. So also, public policy determines which portions of our cultural history will be lost to decay and development.

In a synthetic approach, an international interdisciplinary group of experts drawn from the social and physical sciences together with specialists from the fields of architectural preservation, museums, and library and archive preservation confronted these issues. The usefulness of "World Heritage" as a planning concept, the mechanisms by which society sets priorities, the balance between short- and long-term risk factors, the definition of what constitutes the artifact in the information age, and the validity of the application of the value systems of industrialized nations to the cultural property of developing nations are considered in the examination of the applicability of economic and other models to rational decision-making in the preservation of cultural property at several levels of society and institutions.

ISBN 3-934504-07-8 304 pp. Published: Sept., 2001 DM 79.00 dahlem university press

DWR 85 Science and Integrated Coastal Management
B. von Bodungen and R.K. Turner, eds.

Today more than 60% of the world's population lives within a narrow strip of about 100 km along the world's seashores and probably much more in the entire drainage basins of the coastal seas. Most of the world's megacities are located near the sea, and urbanization within the coastal zone is expected to increase in the future. Coastal development is growing out of control, and destruction of the coastal environment is ever increasing. The immediate issue is how to regain control of the coastal environment and mitigate the degradation of valuable resources. Integrated coastal management (ICM) has been accepted globally as an appropriate approach for conserving and sustaining the uses of marine and coastal resources. It aims to integrate government and stakeholders, science and management, as well as sectoral and public interests.

This volume reports on the findings of a multidisciplinary meeting between international experts from environmental, cultural, social, and economic sciences. Success and failure in transboundary issues, shoreline development, ICM in developing countries, and unifying concepts are addressed. Lack of communication, neglect of scientific information and the precautionary and subsidiarity principles in the policy cycle of ICM were identified as the main causes for failures in the ICM process. It is emphasized that management should take a form of an adaptive approach to cope with uncertainties in prediction and outcome of management and to respond to changes in the coastal environments. The state of science relevant to ICM issues at the different stages of the policy cycle should be periodically evaluated and communicated widely. This volume offers interesting reading for scientists, managers, and decision makers.

ISBN 3-934504-02-7 378 pp. Published: April, 2001 DM 79.00 dahlem university press

DWR 83 Simplicity and Complexity in Proteins and Nucleic Acids

H. Frauenfelder, H. Deisenhofer, and P.G. Wolynes, ed.

All living systems are built from the same macromolecules. Nucleic acids (DNA and RNA) store and transfer the information for the construction; proteins do essentially all the work. One goal of the biosciences is to formulate a predictive theory of how these biomolecules work. Studies dedicated to find the universal characteristics of biological macromolecules are essential, but they are rare.

The goal of this Dahlem Workshop was to strive towards a unified picture by examining the structural, dynamic, and functional features of proteins and nucleic acids that lead to an apparent simplicity of many functions, but which show a surprising complexity when studied in detail. Such an attempt requires a close collaboration of scientists from different disciplines. Thirty-five biologists, chemists, biophysicists, and physicists from Europe and the U.S.A. spent a week in Berlin trying to find answers, often in heated exchanges, to the following questions: What do we need to know about proteins and nucleic acids?·What are the common structural and dynamical properties of biological macromolecules?·How does complexity lead to an apparently simple function?·What determines the diversity of proteins and nucleic acids?

The papers and summary reports resulting from the exchanges provide excellent surveys on many unsolved problems. This book will be of interest to a broad range of scientists, and it will stimulate more collaboration among the disciplines that are needed to solve the riddles of life.

ISBN 393 4504 019 396 pp. Published: March, 2000 DM 79.00 dahlem university press

---

To order a Dahlem Workshop Report from the dahlem university press, please contact:

dahlem university press
der Freien Universität Berlin
S. Haske, Customer Service
Thielallee 66
14195 Berlin
Germany

Tel: +49 (30) 838 55053
Fax: +49 (30) 841 09103
Email: dahlemup@zedat.fu-berlin.de

Available from MIT Press

DWR 84 Bounded Rationality: The Adaptive Toolbox

G. Gigerenzer and R. Selten, eds.

In a complex and uncertain world, humans and animals make decisions under the constraints of limited knowledge, resources, and time. Yet models of rational decision making in economics, cognitive science, biology, and other fields largely ignore these real constraints and instead assume agents with perfect information and unlimited time. About forty years ago, Herbert Simon challenged this view with his notion of "bounded rationality." Today, bounded rationality has become a fashionable term used for disparate views of reasoning.

This book promotes bounded rationality as the key to understanding how real people make decisions. Using the concept of an "adaptive toolbox," a repertoire of fast and frugal rules for decision making under uncertainty, it attempts to impose more order and coherence on the idea of bounded rationality. The contributors view bounded rationality neither as optimization under constraints nor as the study of people's reasoning fallacies. The strategies in the adaptive toolbox dispense with optimization and, for the most part, with calculations of probabilities and utilities. This book extends the concept of bounded rationality from cognitive tools to emotions; it analyzes social norms, imitation, and other cultural tools as rational strategies; and it shows how smart heuristics can exploit the structure of environments. It brings together experts from cognitive science, economics, evolutionary biology, and anthropology to create an interdisciplinary basis for understanding the rationality of the adaptive toolbox.

ISBN 0-262072-14-9 377 pp. Published: April, 2001 MIT Press.

To order this Dahlem Workshop Report, please contact:

The MIT Press
Orders and Book Information
Five Cambridge Center
Cambridge, MA 02142-1493, U.S.A.

Tel: +1 (617) 625-8569
Fax: +1 (617) 625-6660
Email: mitpress-order@mit.edu

Additional Dahlem Workshop Reports available from John Wiley & Sons:

Integrating Hydrology, Ecosystem Dynamics, and Biogeochemistry in Complex Landscapes
J.D. Tenhunen and P. Kabat, eds. (0471 984744)

The Eradication of Infectious Diseases
W.R. Dowdle and D.R. Hopkins, eds.
(0471 980897)

Mechanistic Relationships between Development and Learning
T.J. Carew, R. Menzel, and C.J. Shatz, eds.
(0471 977020)

Saving Our Architectural Heritage: The Conservation of Historic Stone Structures
N.S. Baer and R. Snethlage, eds. (0471 96526X)

The Evolution of the Universe
S. Gottlöber and G. Börner, eds. (0471 965243)

Regulation of Body Weight: Biological and Behavioral Mechanisms
C. Bouchard and G.A. Bray, eds. (0471 963739)

Upwelling in the Ocean: Modern Processes and Ancient Records
C.P. Summerhayes, K.-C. Emeis, M.V. Angel, R.L. Smith, and B. Zeitzschel, eds. (0471 960411)

Aerosol Forcing of Climate
R.J. Charlson and J. Heintzenberg, eds.
(0471 956937)

Molecular Aspects of Aging
K. Esser and G.M. Martin, eds. (0471 956899)

The Role of Nonliving Organic Matter in the Earth's Carbon Cycle
R.G. Zepp and Ch. Sonntag, eds. (0471 954632)

Flexibility and Constraint in Behavioral Systems
R.J. Greenspan and C.P. Kyriacou, eds.
(0471 952524)

Cellular and Molecular Mechanisms Underlying Higher Neural Functions
A.I. Selverston and P. Ascher, eds. (0471 943045)

Acidification of Freshwater Ecosystems: Implications for the Future
C.E.W. Steinberg and R.F. Wright, eds.
(0471 942065)

Twins as a Tool of Behavioral Genetics
T.J. Bouchard, Jr., and P. Propping, eds.
(0471 941743)

Motor Control: Concepts and Issues
D.R. Humphrey and H.-J. Freund, eds.
(0471 929190)

Fire in the Environment: The Ecological, Atmospheric, and Climatic Importance of Vegetation Fires
P.J. Crutzen and J.G. Goldammer, eds. (0471 936049)

Exploring Brain Functions: Models in Neuroscience
T.A. Poggio and D.A. Glaser, eds. (0471 936022)

Neurodegenerative Disorders: Mechanisms and Prospects for Therapy
D.L. Price, H. Thoenen, and A.J. Aguayo, eds.
(0471 929794)

Towards a New Pharmacotherapy of Pain
A.I. Basbaum and J.-M. Besson, eds. (0471 928542)

Organic Acids in Aquatic Ecosystems
E.M. Perdue and E.T. Gjessing, eds. (0471 926310)

Structure and Function of Biofilms
W.G. Characklis and P.A. Wilderer, eds.
(0471 924806)

Complex Organismal Functions: Integration and Evolution in Vertebrates
D.B. Wake and G. Roth, eds. (0471 923753)

The Changing Atmosphere
F.S. Rowland and I.S.A. Isaksen, eds. (0471 920479)

Etiology of Dementia of Alzheimer's Type
A.S. Henderson and J.H. Henderson, eds.
(0471 920754)

Durability and Change: The Science, Responsibility, and Cost of Sustaining Cultural Heritage
W.E. Krumbein, P. Brimblecombe, D.E. Cosgrove, and S. Staniforth, eds. (0471 952214)

Global Changes in the Perspective of the Past
J.A. Eddy and H. Oeschger, eds. (0471 936030)